Toxicology of Nanoparticles and Nanomaterials in Human, Terrestrial and Aquatic Systems

Toxicology of Nanoparticles and Nanomaterials in Human, Terrestrial and Aquatic Systems

Edited by

Marc A. Williams B.Sc. (Hons.), PhD, FAAAAI
Directorate of Toxicology – Health Effects Division
U.S. Army Public Health Center
Aberdeen Proving Ground
USA

WILEY

This edition first published 2022

Registered Offices
John Wiley & Sons, Inc., 111 River Street, Hoboken, NJ 07030, USA
John Wiley & Sons Ltd, The Atrium, Southern Gate, Chichester, West Sussex, PO19 8SQ, UK

Editorial Office
John Wiley & Sons, Inc., 111 River Street, Hoboken, NJ 07030, USA

For details of our global editorial offices, customer services, and more information about Wiley products visit us at www.wiley.com.

Wiley also publishes its books in a variety of electronic formats and by print-on-demand. Some content that appears in standard print versions of this book may not be available in other formats.

For general information on our other products and services or for technical support, please contact our Customer Care Department within the United States at (800) 762-2974, outside the United States at (317) 572-3993 or fax (317) 572-4002.
Wiley also publishes its books in a variety of electronic formats. Some content that appears in print may not be available in electronic formats. For more information about Wiley products, visit our web site at www.wiley.com.

Library of Congress Cataloging-in-Publication Data
Names: Williams, Marc A. (Biologist), editor.
Title: Toxicology of nanoparticles and nanomaterials in human, terrestrial and aquatic systems / Marc A. Williams.
Description: Hoboken, NJ : Wiley, 2022. | Includes bibliographical references and index.
Identifiers: LCCN 2022008008 (print) | LCCN 2022008009 (ebook) | ISBN 9781119316336 (cloth) | ISBN 9781119316350 (adobe pdf) | ISBN 9781119316374 (epub)
Subjects: LCSH: Nanoparticles–Toxicology. | Nanoparticles–Environmental aspects. | Nanoparticles–Health aspects.
Classification: LCC RA1270.N36 T695 2022 (print) | LCC RA1270.N36 (ebook) | DDC 615.7/040724–dc23/eng/20220404
LC record available at https://lccn.loc.gov/2022008008
LC ebook record available at https://lccn.loc.gov/2022008009

Cover Design: Wiley
Cover Image: © Olemedia/Getty, FG Trade/Getty, Hans Henning Wenk/Getty

Set in 9.5/12.5pt STIXTwoText by Straive, Chennai, India

Contents

List of Contributors *xv*
Foreword *xxi*
Editor Biography *xxiii*

Section I Introduction *1*

1 A One Health Perspective and Introduction *3*
Marc A. Williams and Gunda Reddy
1.1 Background *3*
1.2 Structural and Logical Organization of the Book *9*
Acknowledgments *15*
References *15*

Section II Terrestrial and Aquatic Systems *19*

2 Ecosafety of Nanomaterials in the Aquatic Environment *21*
Maria J. Bebianno, Thiago L. Rocha, José P. Pinheiro, Margarida Ribau Teixeira, and Fernanda Cassio
2.1 Introduction *21*
2.2 Inputs of NMs to the Aquatic Environment *23*
2.3 How Wastewater Treatment Processes Act in the Removal of Nanomaterials? *23*
2.4 So What Is Expected to Occur in WWTPs Processes? *24*
2.5 The Importance of Understanding Speciation of NMs *27*
2.6 Ecotoxicological Effects of NMs in Freshwater Organisms *30*

2.7 Ecotoxicological Effects of NMs in Marine Organisms *34*
2.8 Interactive Effects of NMs with Other Contaminants *38*
2.9 Environmental Risk Assessment (ERA) of NMs *42*
Acknowledgments *43*
References *44*

3 Changes in Toxicant Physicochemistry and Bioavailability During Sorption/Desorption Processes with TiO_2 Nanoparticles in the Aqueous Phase *59*
Danae Patsiou, Martin R. S. McCoustra, Teresa F. Fernandes, and Theodore B. Henry
3.1 Introduction *59*
3.2 Properties of TiO_2 NPs in the Aqueous Phase *61*
3.2.1 Agglomeration *61*
3.2.2 Oxidation of TiO_2 NPs by UV Radiation *62*
3.3 Sorption of Organic Substances to TiO_2 NPs *64*
3.3.1 Influence of Organic Matter on Sorption *64*
3.3.2 Influence of TiO_2 NP Surface Area on Sorption *64*
3.3.3 Use of Bioavailability to Inform on Sorption of Organic Compounds on TiO_2 NPs *66*
3.4 Conclusions *72*
References *73*

4 Behavior, Fate, and Toxicity of Engineered Nanoparticles in Estuarine and Coastal Environments *79*
Daniel M. Lyons and Petra Burić
4.1 Introduction *79*
4.2 Types of Nanoparticles: Sources/Products/Release Routes *80*
4.3 Behavior of Nanoparticles in the Water Column *82*
4.4 Biota, Trophic Transfer, Toxicity, and Mechanisms *84*
4.5 Measurement Issues and Regulatory Environment *94*
4.6 Modeling *94*
4.7 Knowledge Gaps and Research Prospects *95*
References *96*

5 Interactive Effects of Nanomaterials with Other Contaminants on Aquatic Organisms: $nTiO_2$ as a Case Study *101*
Laura Canesi, Camilla Della Torre, Teresa Balbi, and Ilaria Corsi
5.1 Introduction *101*
5.2 Interactive Effects of NPs with Other Contaminants in Aquatic Organisms: $nTiO_2$ as a Case Study *104*
5.3 Interactions Between $nTiO_2$ and Other Contaminants in Marine Invertebrates: The Example of the Bivalve *Mytilus* *106*
5.3.1 Effects of $nTiO_2$ and Cd^{2+} *106*
5.3.2 Effects of $nTiO_2$ and TCDD *109*
5.4 Interactions Between $nTiO_2$ and Other Contaminants in Marine Fish: The Example of the European Sea Bass (*Dicentrarchus labrax*) *111*
5.5 Interactive Effects of NPs with Other Contaminants in Marine Species: Importance of Exposure Media *114*
5.6 Concluding Remarks *115*
Acknowledgments *115*
References *115*

6 Soil Nano-ecotoxicology: What Have We Learned from Standard Tests and What May We Be Missing? *121*
David J. Spurgeon, Elma Lahive, Carolin Schultz, and Claus Svendsen
6.1 Introduction *121*
6.2 Development of Standard Test Methods and Their Application to Nanomaterials *122*
6.3 From Soil Ecotoxicological Tests to Risk Assessment *127*
6.4 Looking Beyond Standardized Tests Toward Effects in Ecosystems *128*
6.4.1 Choice of Test Species *129*
6.4.2 Short-Term and Long-Term Effects of Particle "Aging" on Toxicity in Natural Environments *131*
6.4.3 How Soil Properties Interact with Nanomaterial Properties to Determine Bioavailability *133*
6.4.4 Nanomaterial Bioaccumulation and Food Chain Transfer *135*

6.4.5 Short-Term Tests Predict Long-Term Effects *136*
6.5 Standard Ecotoxicological Tests: A Blessing and A Curse? *138*
Acknowledgments *139*
References *140*

7 Impacts of Magnetic Iron Oxide Nanoparticles in Terrestrial and Aquatic Environments *147*
Renato Grillo and Leonardo F. Fraceto
7.1 Introduction *147*
7.1.1 Magnetic Nanoparticles and Their Properties *147*
7.1.2 Commercial Importance and Applications of IONPs *149*
7.1.3 Potential Toxic Effects of Magnetic Iron Oxide Nanoparticles *151*
7.2 Gaps and Obstacles *155*
7.3 Conclusions *158*
Acknowledgments *158*
References *158*

8 Carbon Nanotubes: Sublethal Effects and Unique Mechanisms of Toxicity in Aquatic Species *165*
Tara Sabo-Attwood, Christine Ngan, Candice Lavelle, Jaime Plazas-Tuttle, and Navid B. Saleh
8.1 Carbon Nanotubes in Aquatic Environments *165*
8.2 Classical Toxicity: What We Have Learned *167*
8.3 Unique Mechanisms and Effects *168*
8.3.1 Nutrient Depletion *168*
8.3.2 Immune Modulation *170*
8.3.3 Influence on Co-contaminants *172*
8.4 Next-Generation Nanomaterials: Nanohybrids *174*
8.4.1 Variation in Nanohybrid Composition and Environmentally Relevant Properties *174*
8.4.2 Toxic Responses Demonstrated by NHs *175*
8.5 Future Perspectives *176*
Acknowledgments *176*
References *176*

9 Surface Reactivity of Anatase and Rutile Samples: Relationship with Toxicity on Aquatic Organisms *187*
Charlotte Hurel, Norbert Jordan, Ulrike Gerber, Stephan Weiss, Bernd Kubier, and Reinhard Kleeberg
9.1 Introduction *187*
9.2 TiO_2 Solid Phase Characterization *190*
9.3 Potentiometric Titrations *194*
9.4 Electrophoresis Measurements *196*
9.4.1 In $NaNO_3$ *196*
9.4.2 In Synthetic Freshwater (SFW) *198*
9.5 Size Measurements of the Agglomerates *198*
9.5.1 In $NaNO_3$ *199*
9.5.2 In Synthetic Freshwater (SFW) *199*
9.6 Ecotoxicity Tests *201*
9.6.1 Rotifer Toxicity Test *201*
9.6.2 Microcrustacean Toxicity Test *203*
9.6.3 Diatoms Toxicity Test *204*
9.7 Discussion *206*
9.8 Conclusions *208*
Acknowledgments *208*
References *209*

10 Cardiorespiratory Toxicity of Nanoparticles in Aquatic Environments *213*
Christopher A. Dieni and Tyson J. MacCormack
10.1 Introduction *213*
10.2 Cellular and Molecular Mechanisms of Engineered Nanomaterial Toxicity *214*
10.2.1 Uptake-Independent Mechanisms *215*
10.2.1.1 Accumulation on Cell Surfaces and Interference with Membrane and Transport Functions *216*
10.2.1.2 Activation of Cell Surface Inflammatory Receptors *216*
10.2.1.3 Uptake-Independent Generation of Reactive Oxygen Species *219*
10.2.2 Uptake-Dependent Mechanisms *221*

10.2.2.1 Disruption of Ion Transporters by Intact Nanostructures and Ion Products and Physiological Regulation 223
10.2.2.2 Activation of Systemic Immunity 225
10.3 Complement 225
10.4 Phagocytosis 226
10.5 Conclusions and Ecological Perspectives 227
References 228

Section III Human Systems 237

11 Air Pollution and Neurodevelopmental Disorders 239
Joshua L. Allen, Carolyn Klocke, Keith Morris-Schaffer, Katherine Conrad, Marissa Sobolewski, and Deborah A. Cory-Slechta
11.1 Air Pollution and the Brain 239
11.1.1 The Brain as a Target of Air Pollution 240
11.2 Air Pollution and Neurodevelopmental Disorders 241
11.2.1 Shared Co-morbidities of Neurodevelopmental Disorders 242
11.2.2 Potential Mechanisms of Air Pollution Associations with Neurodevelopmental Disorders 243
11.2.2.1 Microglial Activation and Inflammation 243
11.2.2.2 Ventriculomegaly, White Matter Damage, and Consequent Interhemispheric Dysconnectivity 244
11.2.2.3 Altered Glutamate and Dopamine 245
11.3 An Animal Model of UFP-Induced Developmental Neuropathology and Behavioral Disorders 245
11.3.1 Developmental CAPS Exposures of Mice Produce Male-Specific Microglial Activation 246
11.3.2 Developmental CAPS Exposures of Mice Produce Male-Specific Ventriculomegaly 247
11.3.3 Developmental CAPS Exposures of Mice Produce Male-Specific White Matter Tract Disruption 247
11.3.3.1 Corpus Callosum Size 248
11.3.3.2 Corpus Callosum Myelination 250
11.3.4 Developmental CAPS Exposures of Mice Elevate Glutamate Levels and Result in Male-Specific Excitatory–Inhibitory Imbalance 250
11.3.5 Developmental CAPS Exposures of Mice Are Associated with Impulsive-Like Behavior 251

11.4 Summary and Conclusions *253*
References *256*

12 Toxicity of Nanomaterials to the Gastrointestinal Tract *277*
Penelope A. Rice
12.1 Introduction *277*
12.2 GI Physiology and Toxicity Testing *279*
12.3 Nanomaterial Toxicity Assessment: Challenges *286*
12.4 Toxicity of Specific Nanomaterial Types *289*
12.4.1 Titanium Dioxide *289*
12.4.2 Silica *305*
12.4.3 Nanosilver *311*
12.4.4 ZnO Nanoparticles *319*
12.4.5 Carbon Nanotubes and Fullerenes *327*
12.5 Miscellaneous Nanomaterials *332*
12.6 Analysis and Conclusions *337*
References *338*

13 The Mucosal Microbiome: Impact of Nanoparticles and Nanomaterials *353*
Katherine M. Williams, Kuppan Gokulan, and Sangeeta Khare
13.1 Introduction *353*
13.2 Types of Nanoparticles and Human Exposure *354*
13.3 Factors Influencing Nanomaterial/Microbiota Interactions in the Intestinal Mucosal Environment *356*
13.3.1 Nanomaterial-Specific Factors *356*
13.3.2 Gut Environment-Specific Factors *357*
13.3.3 Protein Corona *358*
13.4 Nanomaterial Effects on Bacterial Microbiota *358*
13.4.1 Metallic NM *361*
13.4.1.1 Antibacterial Activity *361*
13.4.1.2 Impact in Gut Models *362*
13.4.2 Metal Oxide NM *363*
13.4.2.1 Antibacterial Activity *363*
13.4.2.2 Impact in Gut Models *364*
13.4.3 Carbon-Based NM *364*
13.4.3.1 Antibacterial Activity *364*
13.4.3.2 Impact in Gut Models *366*

13.5 Nanomaterial Effects on Viral and Fungal Microbiota *366*
13.6 Conclusions *367*
13.6.1 Antibacterial Activity of NMs *367*
13.6.2 Evidence for NM Effects on Gut Mucosal Microbiota *367*
13.6.3 Strategies for Assessing NM–Microbiota Effects *368*
13.6.4 Direction for Future Research *370*
13.6.5 Conclusion *372*
References *372*

14 Human Health Impacts and Immunotoxicology of Metal Nanoparticles and Nanomaterials – An Overview *383*
Gregory P. Nichols and Jason Davis
14.1 Introduction and Background *383*
14.2 Welding Fumes as Surrogates for Metal Nanoparticles *384*
14.3 Immune-Related Health Effects *385*
14.4 Oxidative Stress and Immunologic Effects *386*
14.5 Conclusion *387*
14.6 Immune-Triggered Human Health Effects of Metal and Metal Oxide Nanoparticles *387*
14.7 Immune Interaction *387*
14.8 Cellular Mechanisms of Injury *389*
14.9 Oxidative Stress *389*
14.10 Interaction with Cellular Membranes and Proteins *391*
14.11 Disruption of Signaling Pathways *391*
14.12 Immune Response *392*
14.13 Immunosuppression *392*
14.14 Inflammation and Autoimmunity *393*
14.15 Sensitivity/Hypersensitivity *394*
14.16 Summary *396*
References *396*

15 Vasculotoxicity of Metal-Based Nanoparticles *401*
Maria S. Sepúlveda and Jiejun Gao
15.1 Nanoparticles in the Environment *401*
15.2 Vascular Development *402*
15.3 Critical Issues in Assessing the Toxicity of NPs *403*
15.4 Vascular Toxicity of NPs *In Vitro* *404*
15.5 Vascular Toxicity of NPs *In Vivo* *409*

15.6 Movement of NPs Through the Blood Brain Barrier (BBB) *412*
15.7 Conclusions *413*
List of Abbreviations *413*
References *415*

Section IV Future Directions and Gaps in the Knowledge *423*

16 Knowledge Gaps, Future Directions, and the Emergence of Nanoplastics as an Environmental Threat Pollutant *425*
Marc A. Williams and Desmond I. Bannon
16.1 Current Concerns and Scope of the Problem *425*
16.2 A survey of the Identified Knowledge Gaps and Needs for Future Research *427*
16.3 Knowledge Gaps – Aquatic and Terrestrial Ecotoxicology *428*
16.4 Knowledge Gaps – Adverse Health Effects in Humans *435*
16.5 Emerging Threats and Future Directions *439*
16.5.1 Nanoplastics – An Emergent Environmental Threat Pollutant *440*
16.6 Conclusions and Other Considerations *444*
References *446*

Index *453*

List of Contributors

Joshua L. Allen
Sarepta
Columbus
OH, USA

Teresa Balbi
Department of Earth, Environmental and Life Sciences
University of Genoa
Genoa
Italy

Desmond I. Bannon
Health Effects Division, Directorate of Toxicology
U.S. Army Public Health Center
Aberdeen Proving Ground
MD, USA

Maria J. Bebianno
Centre for Marine and Environmental Research (CIMA)
University of Algarve Campus de Gambelas
Faro
Portugal

Petra Burić
Center for Marine Research
Ruđer Bošković Institute
Rovinj
Croatia

Laura Canesi
Department of Earth, Environmental and Life Sciences
University of Genoa
Genoa
Italy

Fernanda Cassio
Centre of Molecular and Environmental Biology (CBMA)
Department of Biology
University of Minho
Braga
Portugal

and

Institute of Science and Innovation for Biosustainability
University of Minho
Campus de Gualtar
Braga
Portugal

Katherine Conrad
Department of Environmental Medicine
University of Rochester Medical Center
Rochester, NY
USA

Ilaria Corsi
Department of Physical, Earth and Environmental Sciences
University of Siena
Siena
Italy

Deborah A. Cory-Slechta
Department of Environmental Medicine
University of Rochester Medical Center
Rochester, NY
USA

and

Department of Environmental Medicine
University of Rochester Medical Center
Rochester, NY
USA

Jason Davis
Oak Ridge Associated Universities
Oak Ridge, TN
USA

Christopher A. Dieni
Department of Chemistry and Biochemistry
Mount Allison University
Sackville, New Brunswick
Canada

Teresa F. Fernandes
Institute of Life and Earth Sciences
School of Energy, Geoscience
Infrastructure and Society
Heriot-Watt University
Edinburgh
UK

Leonardo F. Fraceto
São Paulo State University – UNESP
Institute of Science and Technology of Sorocaba
Brazil

Jiejun Gao
Department of Forestry and Natural Resources, Bindley Biosciences Center
Purdue University
West Lafayette, IN
USA

Ulrike Gerber
Institute of Resource Ecology
Helmholtz-Zentrum
Dresden-Rossendorf
Dresden
Germany

Renato Grillo
Center of Natural and Human Sciences
Federal University of ABC
Santo André, SP, Brazil

Kuppan Gokulan
National Center for Toxicological Research, US Food and Drug Administration
Division of Microbiology
Jefferson, AR
USA

Theodore B. Henry
Institute of Life and Earth Sciences
School of Energy, Geoscience Infrastructure and Society
Heriot-Watt University
Edinburgh
UK

and

Center for Environmental Biotechnology, Department of Forestry Wildlife and Fisheries
The University of Tennessee
Knoxville, TN
USA

Charlotte Hurel
CNRS, Institut de Physique de Nice (INPHYNI)
Université Côte d'Azur
Nice Cedex 2
France

Norbert Jordan
Institute of Resource Ecology
Helmholtz-Zentrum
Dresden-Rossendorf
Dresden
Germany

Sangeeta Khare
National Center for Toxicological Research, US Food and Drug Administration
Division of Microbiology
Jefferson, AR
USA

Reinhard Kleeberg
Institute of Mineralogy
TU Bergakademie Freiberg
Freiberg
Germany

Carolyn Klocke
Bausch Health
Rochester
NY
USA

Bernd Kubier
Institute of Mechanical Process Engineering and Minerals Processing
TU Bergakademie Freiberg
Freiberg
Germany

Elma Lahive
UK Centre for Ecology and Hydrology
Oxon
UK

Candice Lavelle
Department of Environmental & Global Health, Center for Environmental and Human Toxicology
University of Florida
Gainesville, FL
USA

Daniel M. Lyons
Center for Marine Research, Ruđer Bošković Institute
Rovinj
Croatia

Tyson J. MacCormack
Department of Chemistry and Biochemistry
Mount Allison University
Sackville, New Brunswick
Canada

Martin R. S. McCoustra
Institute of Chemical Sciences, School of Engineering and Physical Sciences
Heriot-Watt University
Edinburgh
UK

Keith Morris-Schaffer
Exponent
Sacramento
CA, USA

Christine Ngan
Department of Civil, Architectural and Environmental Engineering
University of Texas at Austin
Austin, TX
USA

Gregory P. Nichols
GP Nichols & Company
Knoxville, TN
USA

Danae Patsiou
Institute of Life and Earth Sciences
School of Energy, Geoscience
Infrastructure and Society
Heriot-Watt University
Edinburgh
UK

and

Institute of Oceanography, Hellenic Centre for Marine Research
Anavyssos
Greece

José P. Pinheiro
CNRS/Université de Lorraine, LIEC
Vandoeuvre-lèes-Nancy
France

Jaime Plazas-Tuttle
Department of Civil, Architectural and Environmental Engineering
University of Texas at Austin
Austin, TX
USA

Gunda Reddy
Health Effects Division, Directorate of Toxicology
U.S. Army Public Health Center
Aberdeen Proving Ground
MD, USA

Penelope A. Rice
Division of Food Contact Notifications
Office of Food Additive Safety, Center for Food Safety and Nutrition, Food and Drug Administration
College Park, MD
USA

Thiago L. Rocha
Centre for Marine and Environmental Research (CIMA)
University of Algarve
Campus de Gambelas
Faro
Portugal

Tara Sabo-Attwood
Department of Environmental & Global Health, Center for Environmental and Human Toxicology
University of Florida
Gainesville, FL
USA

Navid B. Saleh
Department of Civil, Architectural and Environmental Engineering
University of Texas at Austin
Austin, TX
USA

Carolin Schultz
UK Centre for Ecology and Hydrology
Oxon
UK

Maria S. Sepúlveda
Department of Forestry and Natural Resources, Bindley Biosciences Center
Purdue University
West Lafayette, IN
USA

Marissa Sobolewski
Department of Environmental Medicine
University of Rochester Medical Center
Rochester, NY
USA

David J. Spurgeon
UK Centre for Ecology and Hydrology
Oxon
UK

Claus Svendsen
UK Centre for Ecology and Hydrology
Oxon
UK

Margarida Ribau Teixeira
CENSE, Center for Environmental and Sustainability Research
University of Algarve
Campus de Gambelas
Faro
Portugal

Camilla Della Torre
Department of Bioscience
University of Milan
Milano
Italy

Stephan Weiss
Institute of Resource Ecology
Helmholtz-Zentrum Dresden-Rossendorf
Dresden
Germany

Katherine M. Williams
National Center for Toxicological Research, US Food and Drug Administration
Division of Microbiology
Jefferson, AR
USA

Marc A. Williams
Health Effects Division, Directorate of Toxicology
U.S. Army Public Health Center
Aberdeen Proving Ground
MD, USA

Foreword

This publication was prepared for educational use. The views and assertions expressed in this book title are the private and personal views of all contributing authors and do not necessarily reflect, nor should they be construed as reflecting the views, and official policy of the U.S. Department of Defense, the Department of the Army, the U.S. Army Medical Department, or the U.S. Federal Government. Furthermore, the use of trademarked name(s) does not imply endorsement whatsoever by the U.S. Department of Defense, the Department of the Army, the U.S. Army Medical Department, or the U.S. Federal Government but is intended only to assist in the identification of a specific product, device, or test article.

Editor Biography

Marc A. Williams, Ph.D., Fellow AAAAI

Biologist, Directorate of Toxicology, US Army Public Health Center, Aberdeen, Proving Ground, MD, USA

Dr. Williams is a Biologist (Immunotoxicologist) and serves as the Project Manager for the Ecotoxicity Assessment Program for the Health Effects Division, of the Directorate of Toxicology at the U.S. Army Public Health Center. Dr. Williams provides leadership and subject matter expertise in applied toxicology and immunology, and environmental toxicity assessments of Military unique chemicals of interest that include PFAS, Nanomaterials, Particulate Matter and Complex Chemical Mixtures and informatics. Dr. Williams has previously served as a Research Biologist at the U.S. Environmental Protection Agency (U.S. EPA), with research interests in human health and toxicological exposures that affect wellness and protective immune competence. Prior to joining the U.S. EPA, Dr. Williams was an Assistant Professor of Medicine and Environmental Medicine at the University of Rochester School of Medicine and Dentistry, New York, and an Instructor of Medicine at the Johns Hopkins University School of Medicine where he led research programs in human immunology, applied toxicology and immunological models of human disease. Dr. Williams holds primary degrees in Molecular Biotechnology and Process Bioengineering, and second in Molecular Cell Biology and Immunology. Dr. Williams earned his Ph.D. in Hematological Oncology and Immunobiology from Queen Mary College, University of London, U.K.

Dr. Williams has authored over 100 peer-reviewed articles, review papers, book chapters, book titles, and formal scientific reports, which include published work during his time on the Special Emphasis Working Group of the American Academy of Allergy, Asthma and Immunology (AAAAI) to highlight the importance of nanoparticle toxicology in human disease pathways. Dr Williams serves on several ad hoc grant review panels for NASA aligned to their Atmospheric, Earth and Environmental Health Research Panels, and has served as the non-VA Federal Government Chair of the Intergenerational Health Effects of Military Exposures Working Group and Committee. Dr. Williams is a member of the Society of Environmental Toxicology and Chemistry (SETAC) North America (SNA) and the Society of Toxicology (SOT). For the past five years, Dr. Williams has served as the Editor in Chief of *Drug and Chemical Toxicology* and for 10 years as a Regular Editorial Advisory Board Member of *Toxicology Letters*.

Section I

Introduction

1

A One Health Perspective and Introduction

Marc A. Williams and Gunda Reddy

Health Effects Division, Directorate of Toxicology, U.S. Army Public Health Center, Aberdeen Proving Ground, MD 21010, USA

Disclaimer: This chapter, and all sections contained herein, was prepared for educational use. The views and assertions expressed in this article are the private views of the authors and do not necessarily reflect, nor should they be construed as reflecting the views, and official policy of the Department of Defense, the Department of the Army, the U.S. Army Medical Department or the U.S. Federal Government. Further, use of trademarked name(s) does not imply endorsement by the U.S. Army but is intended only to assist in identification of a specific product.

1.1 Background

Nanoparticles constitute a diversity of naturally occurring and anthropogenic source-derived particles with unique spatial-geometric dimensions, surface chemistry, biological and chemical properties, and geographic origin that are widely distributed in the environment, often persistent, and present diverse ecotoxicological effects [1–3]. Indeed, the diverse properties of naturally occurring and anthropogenic-derived nanoparticles, including those found in everyday consumer products (e.g. cosmetics, sunscreens, and cancer-targeting therapeutics), present applied environmental toxicologists with formidable challenges in documenting and understanding the toxicology and risk characterization and assessment of nanoparticles. These considerations include a need to advance understanding of their environmental fate, transport and transformation, and their potential for complex environmental exposure. This further requires an understanding of their dose-dependent toxicity and the impact of nanoparticles on aquatic, marine, and terrestrial life, including adverse human health effects [1].

Toxicology of Nanoparticles and Nanomaterials in Human, Terrestrial and Aquatic Systems, First Edition. Edited by Marc A. Williams.

At the most basic level, attempts to characterize nanoparticles and understand their potential for adverse effects to aquatic/marine and terrestrial wildlife, including human health, have benefited from attempts at defining the origins and the complex nature of nanoparticles. A useful way to appreciate nanoparticle complexity is to identify the diverse origins and sources of these materials [1–3]. Thus, we can segregate nano-particles into those that are either naturally occurring; for example, particles that are formed by gas to particulate condensation; airborne particulates derived from volcanic activity or seasonal brush/forest fires; biological nanomaterials including viral particles [1–4]; or those that are derived from anthropogenic activities, including inadvertent release and formation or intentional industrial synthesis [2, 3]. Moreover, anthropogenic-derived nanoparticles can be further and conveniently segregated into so-called ultrafine particulate matter (PM) species or ultrafine particulates UFPs, and engineered or manufactured nanoparticles (ENPs) [1–4].

Both naturally occurring and anthropogenic-derived nano-particulate matter can vary by geographical origin, and, when present in the natural environment, by chemical, biological, and physical composition and characteristics [1]. Despite this diversity, nanoparticles are defined as nanoscale PM with dimensions of about 1–100 nm, which, in the context of ultrafine ambient particulate matter (APM), allows us to distinguish them from their coarse and fine particulate counterparts with aerodynamic diameters of less than 10 μm (PM_{10}, coarse PM) and less than 2.5 μm ($PM_{2.5}$, fine PM), respectively [1]. Moreover, despite considerable advances in assay development, detection methods, standard phased approaches to toxicological testing, and the advent of new approach methodologies (NAMs) to include computational and mathematical modeling and predictive approaches, relatively little is known about the environmental, eco-toxicological, and human health effects of particles less than 100 nm in size as compared with our understanding of fine ($PM_{2.5}$) and coarse (PM_{10}) particulate matter; thus, a considerable data and knowledge gap remains [1].

It was mentioned above that nanoparticles comprise a diverse group that are conveniently segregated into two major subgroups including UFPs and ENPs [1–3].

Particles defined as UFPs broadly refer to particles that are released to environmental media, including the ambient air, water columns, effluent discharge, and soil. This occurs either by inadvertent generation and release, such as emissions from vehicular tailpipes following combustible degradation of diesel or gasoline, and as by-products from the combustion of other fossil fuels, condensation and re-condensation cycles of semi-volatile organic compounds, and industrial funnel stack emissions [1]. By contrast, particles referred to as "nanoparticles" are predominantly those produced by controlled manufacturing and engineering approaches and thus defined as ENPs [1, 4].

We and others have previously argued for the need to advance our understanding of the health impacts of UFPs, and, by extension, the biology and applied toxicology of nanoparticles in the broader sense; and the need to strengthen approaches whereby the derived data are aggregated, organized, and shared among collaborating and decision-making stakeholders [1, 5, 6]. A key reason for emphasizing the need for increased research on UFP and ENP toxicology is in part derived from observations of the physical and chemical properties of nanoparticles that are so clearly different from their coarse and fine PM counterparts [1, 5, 6]. These properties include the increased surface area to mass ratio, the greater organic carbon (as compared elemental carbon) content of nanoparticles, the highly variable surface chemistry of nanoparticles, and other properties [1, 5, 6]. In regard to human health, it is thought that the disparate properties of nanoparticles as compared their larger counterparts ($PM_{2.5}$ and PM_{10}) might translate to promoting new disease onset and the exacerbation of pre-existing diseases such as poor cardiovascular health outcomes and diseases of the lung, including allergic asthma and allergic sensitization of the airways to commonly encountered allergens, albeit by differential mechanisms [1].

Several experimental studies, complemented by quality (albeit limited), epidemiological analyses, have strengthened the weight of evidence that UFPs might be of greater concern than their $PM_{2.5}$ and PM_{10} counterparts (discussed in detail in [1, 4]). This is largely due to crucial differences in their chemical composition and/or surface chemistries, their uniquely small size, increased retention rate, deposition to the apical airways and alveolar spaces, and a subsequently wider physiological distribution [1, 6–14]. The ability of UFPs to migrate to the apical airways and disseminate physiologically is accompanied by their capacity to rapidly stimulate reactive oxygen species and free radical secretion where pro-oxidative and pro-inflammatory cellular damage contributes to adverse health effects [1, 4, 14, 15]. The observed adverse health effects are due in part to anti-oxidant defense pathways, including those mediated by the Nrf2 transcription factor, becoming progressively overwhelmed, leading to cellular damage, mitochondrial dysfunction, and programmed cell death [1, 4, 14, 15].

Many features displayed by UFPs and ENPs help explain their capacity to promote adverse health effects not only in humans and other mammals, but also in other vertebrate and aquatic species, which will be discussed in more detail across those chapters included in this book. Unlike their larger $PM_{2.5}$ counterparts, UFPs and ENPs are not easily recognized by the innate host immune response. Thus, the inhaled and deposited particles are not opsonized by host immunoglobulins and targeted for phagocytosis in the same way that $PM_{2.5}$ and PM_{10} are recognized as "dangerous" and cleared by endocytic or phagocytic mechanisms [1, 6]. This means that nanoparticles display a high retention rate, particularly in the apical regions of the lungs, which is accompanied by prolonged stimulation of a pulmonary pro-inflammatory environment, and an ability of the nanoparticles to traffic systemically to other organ systems, including the regional lymph nodes, cardiovascular and central nervous (including the brain) systems. In doing so, the nanoparticles provoke toxicological effects that are unique to UFPs or engineered nanomaterials (ENMs) and not necessarily seen on inhalational exposure to PM_{10} or $PM_{2.5}$ [1, 6–16].

A potential mechanism accounting for the persistent inflammation seen in individuals on exposure to a single inhalation of ambient UFP (and presumably other pro-inflammatory nanomaterials) has been reported [16]. Persistent inflammation following inhalational exposure to ambient UFP might originate from a so-called quarantining process that involves deposition of excreted immobilized residues of inhaled nanoparticulate–biomolecular complexes located on the cell surface. These organized complexes were hypothesized to recycle among different cells of the pulmonary airways including respiratory epithelial cells, thereby sustaining chronic inflammation [16]. This process would also benefit from an effective mucociliary and phagocytic cell-mediated clearance process, which is required for particulate clearance from the airways [17]. Whether or not such a mechanism is found in aquatic or terrestrial wildlife across taxa is unknown and likely worth exploring to advance understanding of nanoparticulate clearance and inflammatory pathways in aquatic and terrestrial wildlife in addition to known and postulated pathways seen in human health.

Additional features that are arguably unique to environmental exposures to UFPs or ENPs and can assist our understanding of their subsequent adverse health effects include observations of localized ambient "hotspots" of exposure (e.g. neighborhoods adjacent to busy vehicular traffic highways or geographic areas particularly impacted by bush or wildfire smoke plumes) or persistent "depots" of environmentally deposited nanoparticulate matter [14]. Thus, unlike $PM_{2.5}$ and PM_{10}, whose ambient atmospheric distribution is considered to be quite homogenous, the same might not be true of UFPs and ENPs due in large part to their nanoscale size and dimensions. This variance is particularly concerning because it has not only hampered efforts to move forward in defining UFPs as criteria air pollutants but has also dramatically limited efforts to conduct public health oriented epidemiological studies. Few if any epidemiological studies have employed dose metrics typically used for nanoparticles including relative surface area or particle number since the most commonly used exposure metrics for PM are in mass units, which do not apply to nanoparticles [14].

In addition to concerns regarding the adverse health impacts from exposure to UFPs, it is now recognized that the intended industrial synthesis of ENPs is a rapidly developing and expanding business in which almost 1500 diverse nano-technology-derived consumer products are now available

[4, 17–23]. Thus, extensive occupational and/or environmental exposure to ENPs or composite ENMs has increased our concerns regarding their detailed toxicological effects and safety characterization. Such exposures represent a particular concern in occupational settings where individuals can be exposed by inhalation of the fine powders during manufacturing, processing, and handling [4, 17–23].

The known and suspected health effects of ENPs/ENMs and UFPs are well reported and described, and include exacerbations of pre-existing asthma, COPD, cardiovascular disease, metabolic disorders and diabetes, respiratory infections in children, and other adverse health effects [1, 4, 8–15, 25]. Exposure routes available to a variety of nanomaterials and UFPs are not limited to inhalational deposition to the lung; they might also include gastrointestinal and dermal exposure pathways depending on the commercial products used and their intended application [1, 4, 19, 21, 22, 24, 25]. Adverse human health effects have also been reported by a limited set of epidemiological association studies that have implicated diverse pro-inflammatory, pro-oxidative, DNA-damaging effects, as well as immunotoxicological and epigenetic mechanisms serving to subvert physiological homeostasis and provoke or exacerbate adverse health outcomes [26–28]. One study highlighting the emerging concern with UFPs as compared their larger $PM_{2.5}$ counterparts showed that in the setting of diabetes there was an association between the relative levels of UFPs and a comparative decrease in lung function (e.g. FEV_1 and forced vital capacity) and heightened levels of HbA1c, which is marker of type 2 diabetes [29, 30]. Further highlighting the concern in regard inhalational exposure to UFPs (and presumably other complex metallic nanoparticle exposures) was a 6-year prospective multi-ethnic cohort-study [11]. This study reported mortality caused by ischemic heart disease following chronic exposure to UFPs using the statewide cohort of >100 000 women from the California Teachers Study who were followed from 2001 through 2007 [11]. Not only was mortality from heart disease associated with UFPs, it was found to also be associated with the content of elemental carbon, transition/heavy metals, and mobile sources of particulates [11].

Significant progress has been made in striving to understand the adverse health effects from exposure to ambient and engineered nanoparticles, and potential mechanisms of action. The chapters contained within this book will provide more detailed discussion and critical analysis of these issues, which are briefly surveyed below. Not only are the effects of UFPs and ENPs on human health an emerging concern, but their broader environmental or ecotoxicological impact are equally concerning as well. In separate chapters of this book (see Figure 1.1), the effects of nanoparticles on aquatic/marine and terrestrial organisms are also discussed by leading investigators in the field.

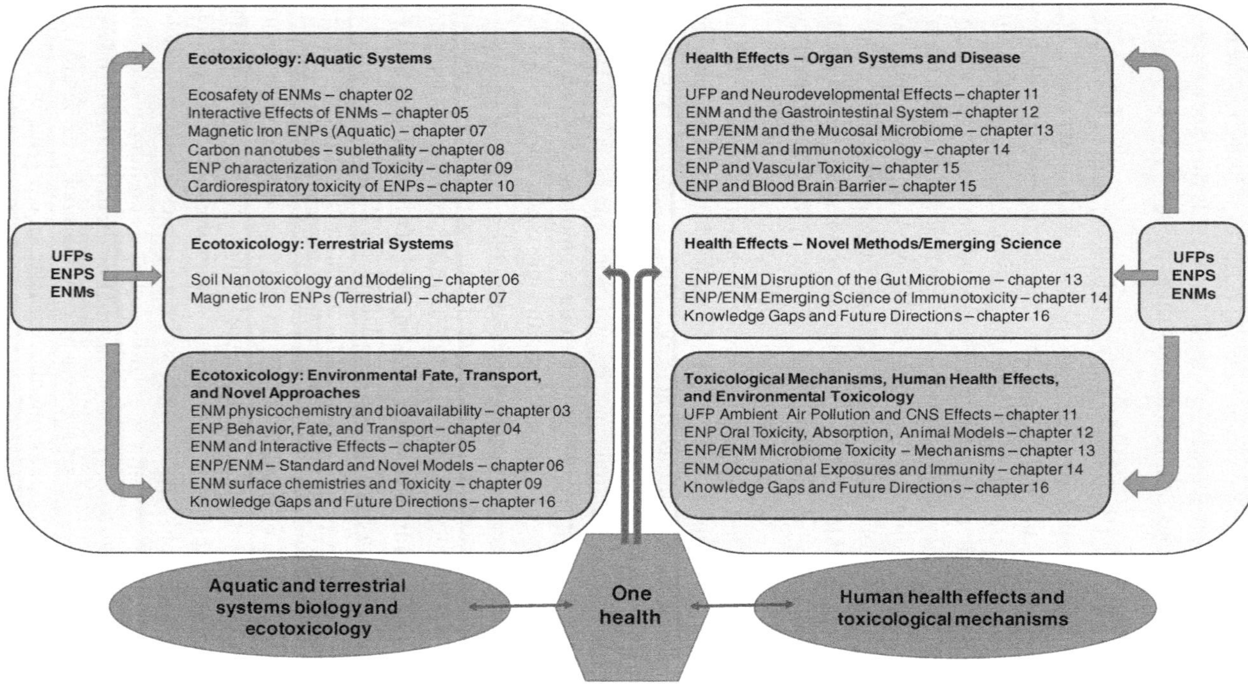

Figure 1.1 Illustrating the structural content to this book as it aligns to the applied toxicology of engineered ultrafine particulates (UFP) nanoparticles (ENP) and nanomaterials (ENM) in terrestrial and aquatic systems, and human health. Constituent chapters of this book critically discuss broadly two major themes. To the left, major themes include aquatic and terrestrial systems biology and ecotoxicology. To the right, major themes focus on human health effects and toxicological mechanisms. Both themes collectively align to the "One Health" paradigm where aquatic, terrestrial, and human health systems should be considered as complementary areas of focus. Ideally, the contents of this book should be considered as fitting that paradigm – at least in part. Source: Based on Ref. [31].

1.2 Structural and Logical Organization of the Book

An emerging and major concern in the field of toxicology is the broader environmental, ecotoxicological, and human health impact of an ever-growing number and diversity of environmentally persistent ENPs and ENMs. Of equal concern is the presence of a chemically complex and equally environmentally persistent array of UFPs, which the U.S. Environmental Protection Agency or EPA maintains cannot be regulated until additional data on their ecological and environmental toxicology become available [1, 4, 5, 19–23, 25]. Detracting from the argument in favor of including UFPs (and perhaps occupational exposures to ENPs/ENMs) in regulatory policy frameworks was a 2013 report by a review panel of the Health Effects Institute (HEI), an entity funded by the EPA and, privately, by the motor vehicle industry [1]. From an analysis of the weight of available data at the time, the HEI erroneously concluded in 2013 that there was insufficient evidence demonstrating any significant differences in the adverse health effects of UFPs as compared fine PM2.5, although ENPs/ENMs were not included in that analysis [1]. This conclusion was not only erroneous but also failed to demonstrate any awareness of in-progress research that challenged the HEI's conclusions. This highly cited clinical and epidemiological research data were published shortly after HEI made its opinions public [7, 8, 10, 11, 29, 30, 32]. Additionally, as discussed elsewhere in this chapter, it is the unique chemical composition, surface chemistry, and other physico-chemical properties of UFPs (and ENPs) that endow them with the capacity to promote greater adverse health effects than PM_{10} and $PM_{2.5}$ as well as ENPs such as silver nanoparticles and others [1, 9, 25].

In addition to the well-reported systems biological effects of UFPs and ENPs and adverse mammalian and human health effects, this book also focuses on other terrestrial and aquatic/marine species and their biological responses on exposure to ultrafine or nanoparticulate species. In separate sections of this book, the effects of nanoparticles on aquatic/marine and terrestrial organisms will be surveyed across several case examples, to include putative modes or mechanisms of action, and emerging paradigms in the field of environmental and ecological nanotoxicology.

In this context, the net impact of nanoparticle toxicants or pollutants on human health and the ecological environment fits the current paradigm of "One Health" that was first posited by the One Health Initiative Task Force, or OHITF [31]. Thus, the structure of this book (Figure 1.1) aligns with the "One Health" system of human and environmental health and does so in the context of discussing the health effects of ultrafine and engineered nanoparticles, their modification by environmental media, and subsequent health impacts seen in human, aquatic, and terrestrial life by means of case examples of observed exposures, toxicological effects, and technological advances. The intent of this book is to provide the reader with specific examples of the biology and toxicological outcomes following exposure to nanoparticles across 16 authoritatively authored chapters compartmentalized into four sections (see Figure 1.1).

From the four sections of this book, the reader will derive a strengthened appreciation of the environmentally dangerous and adverse health impact from exposure to nanoparticles. Section 1 – this chapter – provides an introduction to the emerging environmental and health impacts of nanoparticles and provides a concise survey of the content and organizational structure of this book.

Section 2 (Chapters 2–10) provides case examples of terrestrial and aquatic systems that are impacted by nanoparticles, including a critical discussion of their ecological safety in aquatic systems (Chapter 2), and the complex physicochemical mechanisms that critically influence ENP bioavailability. A case example (Chapter 3) presents a particular focus on titanium dioxide nanoparticle behavior in the aqueous phase. Complementing the ecotoxicological theme of Chapters 2 and 3, Chapter 4 provides a concise discussion of the behavior, fate, and transport of ENPs in the context of estuarine and coastal environments. The chapter highlights the importance of distinguishing two broad types of nanoparticles found in the environment: those that are naturally derived from geological and climatological processes and those that have been manufactured/engineered. Complementing the aquatic and environmental focus of Chapter 4, Chapter 5 provides a detailed and compelling analysis of the ways by which ENPs/ENMs interact with and subsequently alter the characteristics of other contaminants found in the aqueous phase and sediment. Additional analysis of ENPs subsequently interacting with aquatic life is achieved by focusing on the Mediterranean mussel (*Mytilus galloprovincialis*) and the European sea bass (*Dicentrarchus labrax*) as model marine species affected by exposure to ENPs.

Continuing with the theme of environmental and ecotoxicology, Chapter 6 critically discusses the relevance and importance of soils as potential substrates for nanomaterials, along with the possible impact of nanomaterials on soil-inhabiting organisms. The chapter also critically analyzes the effectiveness of current soil toxicity tests that have been optimized for standard toxicological assessment of conventional chemicals and their relevance for the hazard assessment of novel nanomaterials. Based on their surveys of currently available approaches, the authors critically discuss the limitations of current standardized tests systems in the context of a comprehensive soil ecosystem effect assessment for nanomaterials. This assessment recognizes that novel testing approaches will need to take into account the various soil types contaminated by given nanomaterials and will need to occur within settings and timescales that are aligned to ecotoxicologically relevant frameworks for risk assessment and management in soil ecosystems.

Recognizing that many different species of nanoparticle and nanomaterials might impact the aquatic and terrestrial ecosystems, the author[s] of Chapter 7 has[have] focused specifically on the toxicology of iron oxide nanoparticles or

IONPs, which are a specific class of widely used magnetic nanoparticle (MNP) with applications in diverse industrial and manufactured products including polymers, surfactants, important therapeutic agents in cancer therapy and contrast agents in clinical diagnostics; MNPs are found in a variety of inorganic materials, including silica. This comprehensive chapter discusses the synthesis, characteristics, and commercial utility of IONPs, a critical analysis of their potential aquatic and terrestrial toxicity, environmental fate and transport, and a detailed analysis of current knowledge gaps and approaches for overcoming roadblocks to advancing our understanding of IONP toxicology, particularly in terms of their potential for bioaccumulation and biomagnification.

Continuing the theme of focusing on specific classes of nanomaterials, Chapter 8 provides a critical analysis of the unique toxicological effects of carbon nanotubes (CNTs) in aquatic ecosystems, recognizing the paucity of data in this setting. In addition to discussing measured endpoints in the classical toxicological assessment of CNTs in aquatic organisms, the chapter presents a unique discussion of novel mechanisms of CNT toxicity. Included are discussions of nutrient depletion by CNTs and subsequent CNT-mediated cytotoxicity, immune modulation by CNTs in aquatic organisms (that might be associated with the limited nutrient availability), the reported influence of CNTs in altering the toxicity of co-contaminants with which they interact, and the subsequent increased or decreased toxicity of select co-contaminants in the presence of CNTs.

Chapter 9 provides a holistic approach to understanding and evaluating the impact of the anatase and rutile forms of titanium dioxide (TiO_2) nanoparticles on aquatic organisms. The chapter very carefully details the approaches and methodological limitations in assessing the aquatic ecotoxicity of TiO_2 nanoparticles with a particular focus on three organisms: marine rotifers (*Brachionus plicatilis*), marine microalgae (*Phaeodactylum tricornutum*), and the freshwater crustacean (*Daphnia magna*). The overarching objective of Chapter 9 is to provide a systematic study of how the environmental fate of aquatic organisms is influenced by TiO_2 nanoparticle size, surface site concentration and surface charge, agglomeration state, and the commercially available allotropic forms of anatase and rutile TiO_2 nanoparticles. The chapter will appeal to those with interests in the physical and biological chemistry of nanoparticles as well as those in the field of ecotoxicity and applied aquatic toxicology.

The final article (Chapter 10) in this section provides a mechanistic analysis of the cardiorespiratory toxicology of suspended nanoparticles in aquatic organisms. While inhalation of airborne ENMs provokes oxidative stress and inflammation in terrestrial vertebrates with subsequent onset of morbidity and dysfunction in cardiorespiratory systems, the chapter critically analyzes whether or not oxidative stress pathways might provoke cardiorespiratory toxicity in aquatic organisms. In detailed analyses of the available toxicological data in fish species, Chapter 10 focuses on the potential toxicity of ENMs with a specific mechanistic focus on the cardiorespiratory system since, as the authors point out, cardiac failure in fish species is in part considered limiting in environmental stress tolerance tests, including ENM exposure and sensitization of fish to hypoxia. Consistent with the holistic eco-toxicological approach described in Chapter 9, Chapter 10 also places into context the relevance of climate and environmental change that might increase the frequency and severity of environmental stressors like ENMs, and indeed other environmental toxicants and complex aquatic pollutants.

Section 3 (Chapters 11–15) provides the reader with a compendium of five chapters dedicated to the effects and mechanisms of action of ambient airborne ultrafine particulate matter and neurodevelopmental disorders (Chapter 11), the toxicity of nanomaterials in the settings of the gastrointestinal microbiome (Chapter 12), the mucosal microbiome in a broader context (Chapter 13), the immunotoxicity of metal nanoparticles and ENMs (Chapter 14), and lastly, a discussion of the vascular toxicity of engineered metal-based nanoparticles (Chapter 15).

The first chapter in this section (Chapter 11) systematically discusses the interaction of inhaled respirable UFP with the brain, and its association with neurodevelopmental disorders that include schizophrenia, the globally prevalent autism spectrum disorder (ASD), and attention hyperactivity disorder, among other conditions to include shared co-morbidities of particular neurodevelopmental disorders. This chapter also surveys the potential mechanisms that are thought to provoke neurodevelopmental disorders on exposure to UFPs. These include systemic inflammatory pathways on activation of microglia and vagal nerve afferents, white matter damage, and other observed central nervous system (CNS) neuropathologies commonly seen in schizophrenia and ASD. Chapter 11 also reviews the importance of altered glutamate activation on exposure to urban air nanoparticles and the putative role of the hyperglutamate theory of ASD onset and development. Lastly, this important chapter discusses critical research findings from murine models of UFP-induced CNS neuropathologies. These findings complement the theoretical and observational studies discussed earlier.

Continuing the theme of systems biology and organ systems affected by exposure to nanoparticles and nanomaterials, both Chapters 12 and 13 provide complementary discussion of the role of the gastrointestinal (GI) tract, barrier integrity, and the host microbiome in providing protection from, or susceptibility to, the effects of oral exposure to nanoparticles. The first of these chapters (Chapter 12) provides a comprehensive overview of GI physiology and approaches to toxicological testing across a variety of *in vitro* mechanistic and *in vivo* animal modeling studies, and critically highlights the unique physiological GI system and the ways by which nanoparticles might be modified following oral exposure. Also included are GI disease models, including inflammatory bowel disease (IBD), that have collectively determined the effects of nanomaterials in such disease states; and their limitations as compared normal physiological GI responses. Critically, this chapter comprehensively surveys the challenges of using complex cell culture models of the GI system and the technological difficulties in working with nanoparticles and materials. It also posits that many of the confounding factors that impact an assessment of the nanomaterial toxicity are not adequately considered by many studies in this area. Chapter 12, therefore, provides not only a comprehensive overview of GI toxicity testing of nanoparticles/nanomaterials but also a critical analysis of the experimental and technological challenges in doing so. Also included are specific case exemplars that critically survey the specific toxicity of titanium dioxide, silica, nanosilver, zinc oxide nanoparticles, and CNTs. These cases are complemented by a final section that concisely reviews the functional interplay of a variety of miscellaneous nanomaterials, including metal oxide and calcium carbonate nanomaterials and the host microbiome.

The complementary subject material of Chapter 13 discusses the importance of the gut microbiome, and the manner by which select ENMs are potentially toxic (i.e. display antibacterial activities) to gut microbiota and the mucosal microbiome in a more generalized context. Critically, this chapter provides a detailed overview of the mucosal microbiome, its composition, and the critical importance of maintaining a balanced, stable, and diverse microbiome. Gastrointestinal integrity is critical in maintaining effective and protective host immunity and intestinal barrier function. Failure to maintain a healthy homeostatic integrity of the mucosal microbiome, referred to as dysbiosis, could provoke adverse health consequences, due in large part to a loss in the beneficial microbiota that constitute the mucosal microbiome, a competent immune system, and elicitation of adverse gastrointestinal health conditions to include ulcerative colitis, Crohn's disease and colorectal cancer – all of which are surveyed in the chapter. Additionally, there is accumulating evidence linking disruption of the gut microbiome and an association with the development of diabetes, heart disease, allergic asthma, and, of relevance to Chapter 11, the development of neurodevelopmental disorders.

Continuing a discussion of the importance of host immunity and exposure to ENPs and ENMs, Chapter 14 provides a comprehensive analysis of the immunotoxicity of surveyed nanoparticles. Many sections of this chapter provide novel insights into the occupational health and safety relevancy of nanoparticles, their toxicity, and the disruption of normal host immunity on exposure to a variety of nanoparticulate species. In this chapter, authors focus on the adverse health effects of welding fumes as a surrogate of metal nanoparticle exposure. Given the complexity of welding fumes, which are thought to comprise a variety of metals and metal oxides, including aluminum (Al), cadmium (Cd), chromium (Cr), Iron (Fe), manganese (Mn), nickel (Ni), and zinc (Zn), there is significant concern that many welders (approximately one-fifth of those studied) were occupationally exposed to

Fe via the inhalational exposure pathway at levels that exceeded the recommended threshold limit value or TLV. Almost two-thirds of the welders studied were occupationally exposed above the TLV for Mn, as described in detail in the chapter.

Discussion is then given to the immunotoxicological effects of metal nanoparticles on prolonged inhalational exposure to welding fumes. These include heart disease, neuro-cognitive disorders, lung cancer, asthma, COPD, and increased susceptibility to community-acquired pneumonia and other lower respiratory-tract infections. Elsewhere in the chapter, the authors critically analyze the biochemical and immunological mechanisms of action of metal nanoparticles that provoke adverse health effects. A detailed discussion of the cellular and tissue compartments most affected is also provided, including the functional role played by host innate immunity, mechanisms of cell injury and inflammation, free radical generation and oxidative stress, and disrupted pathways of cell-mediated signaling and disease onset as a consequence of relative immune suppression, autoimmunity, and sensitivity and hypersensitivity. This chapter will appeal to those with interests in the toxicology and immunotoxicology of metal nanoparticles.

The penultimate chapter of this book, Chapter 15, also explores the toxicity of engineered metal and metal oxide nanoparticles in vertebrates. This chapter begins by briefly reviewing the omnipotent prevalence of nanoparticles in the environment and then very carefully describes the overall mechanisms that are involved in the development of the vertebrate vascular system and the manner by which metal nanoparticles interact with the vascular system in provoking toxicity. The toxicity of metal nanoparticles on the vascular system is described in the context of separately discussing *in vitro* and *in vivo* studies, and analyzing the different molecular mechanisms reported in the literature. In particular, the interaction of metal nanoparticles with the blood brain barrier is critically analyzed from the few animal studies available and discussed, indicating that both the integrity and function of the blood brain barrier is affected, with decreased viability of human brain microvascular endothelial cells being a particular manifestation of metal nanoparticle toxicity.

Finally, Chapter 16 provides a brief overview of current challenges in nanoparticulate toxicology, the known data gaps, and the novel approaches that are hoped will advance our understanding of both engineered nanoparticles and ambient airborne ultrafine particulate matter in human health as well as in terrestrial and aquatic/marine systems. The emergence and application of new technologies and alternative or complementary approaches will enable both the tracking of accumulating nanotoxicological research data and the analysis of that data. Novel methods will include the emergence of new approach methodologies or NAMs, physiologically based pharmacokinetic (PBPK) modeling tools, other computational approaches, and the need for sound data management tools to enable subsequent analysis of carefully archived, and publicly available quality data. This concluding chapter surveys some of the advances already made in these emerging domains of applied toxicology.

Acknowledgments

Authors would like to express their sincere thanks to Dr. Mark S. Johnson, Dr. Michael S. Quinn, and Dr. Desmond I. Bannon, Directorate of Toxicology, U.S. Army Public Health Center, for their technical review of the scientific content of this work.

References

1 Li, N., Georas, S.N., Alexis, N. et al. (2016). A work group report on ultrafine particles (American Academy of Allergy, Asthma and Immunology): why ambient ultrafine and engineered nanoparticles should receive special attention for possible adverse health outcomes in human subjects (review). *J. Allergy Clin. Immunol.* 138 (2): 386–396.
2 Oberdörster, G., Oberdörster, E., and Oberdörster, J. (2007a). Concepts of nanoparticle dose metric and response metric. *Environ. Health Perspect.* 115 (6): A290.
3 Oberdörster, G., Stone, V., and Donaldson, K. (2007b). Toxicology of nanoparticles: a historical perspective. *Nanotoxicology* 1 (1): 2–25.
4 Nel, A., Xia, T., Madler, L., and Li, N. (2006). Toxic potential of materials at the nano-level. *Science* 311: 622–627.
5 Andrea Haase and Fred Klaessig. (2018). EU-US Roadmap Nanoinformatics 2030. https://www.nanosafetycluster.eu/eu-us-nanoinformatics-2030-roadmap/ (17 February 2022).
6 Papadiamantis, A.G., Klaessig, F., Exner, T.E. et al. (2020). Metadata stewardship in nanosafety research: community-driven organisation of metadata schemas to support FAIR nanoscience data. *Nanomaterials* 10 (10): 2033.
7 Schaumann, F., Fromke, C., Dijkstra, D. et al. (2014). Effects of ultrafine particles on the allergic inflammation in the lung of asthmatics: results of a double-blinded randomized cross-over clinical pilot study. *Part. Fibre Toxicol.* 11: 39.
8 Devlin, R.B., Smith, C.B., Schmitt, M.T. et al. (2014). Controlled exposure of humans with metabolic syndrome to concentrated ultrafine ambient particulate matter causes cardiovascular effects. *Toxicol. Sci.* 140: 61–72.
9 Bezemer, G.F., Bauer, S., Oberdörster, G. et al. (2011). Activation of pulmonary dendritic cells and Th2-type inflammatory responses on instillation of engineered, environmental diesel emission source or ambient air pollutant particles in vivo. *J. Innate Immun.* 3: 150–166.
10 Karottki, D.G., Beko, G., Clausen, G. et al. (2014). Cardiovascular and lung function in relation to outdoor and indoor exposure to fine and ultrafine particulate matter in middle-aged subjects. *Environ. Int.* 73: 372–381.

11 Ostro, B., Hu, J., Goldberg, D. et al. (2015). Associations of mortality with long-term exposures to fine and ultrafine particles, species and sources: results from the California Teachers Study Cohort. *Environ. Health Perspect.* 123: 549–556.

12 Williams, M.A., Porter, M., Roman, J. et al. (2007). Ambient Baltimore particulate matter directs non-classical immune activation of human myeloid dendritic cells. *J. Allergy Clin. Immunol.* 119: 488–497.

13 Porter, M., Karp, M., Killedar, S. et al. (2007). Diesel exhaust particles induce a novel pattern of pro-allergic dendritic cell activation. *Am. J. Respir. Cell Mol. Biol.* 37: 706–719.

14 Li, N., Wang, M., Barajas, B. et al. (2013). Nrf2 deficiency in dendritic cells enhances the adjuvant effect of ambient ultrafine particles on allergic sensitization. *J. Innate Immun.* 5: 543–544.

15 Williams, M.A., Rangasamy, T.E., Bauer, S.M. et al. (2008). Disruption of the transcription factor Nrf2 promotes pro-oxidative dendritic cells that stimulate Th2-like immuno-responsiveness on activation by ambient particulate matter. *J. Immunol.* 181: 4545–4559.

16 Kokot, H., Kokot, B., Sebastijanović, A. et al. (2020). Prediction of chronic inflammation for inhaled particles: the impact of material cycling and quarantining in the lung epithelium. *Adv. Mater.* 32 (47): 2003913.

17 Navarrette, C.R., Sisson, J.H., Nance, E. et al. (2012). Particulate matter in cigarette smoke increases ciliary axoneme beating through mechanical stimulation. *J. Aerosol Med. Pulm. Drug Deliv.* 25 (3): 159–168.

18 Jain, P.K., Huang, X., El-Sayed, I.H., and El-Sayed, M.A. (2008). Noble metals on the nanoscale: optical and photothermal properties and some applications in imaging, sensing, biology, and medicine. *Acc. Chem. Res.* 41: 1578–1586.

19 Glover, R.D., Miller, J.M., and Hutchison, J.E. (2011). Generation of metal nanoparticles from silver and copper objects: nanoparticle dynamics on surfaces and potential sources of nanoparticles in the environment. *ACS Nano* 5: 8950–8957.

20 De Volder, M.F., Tawfick, S.H., Baughman, R.H., and Hart, A.J. (2013). Carbon nanotubes: present and future commercial applications. *Science* 339: 535–539.

21 Shi, H., Magaye, R., Castranova, V., and Zhao, J. (2013). Titanium dioxide nanoparticles: a review of current toxicological data. *Part. Fibre Toxicol.* 10: 15.

22 Wang, X., Mansukhani, N.D., Guiney, L.M. et al. (2015). Differences in the toxicological potential of 2D versus aggregated molybdenum disulfide in the lung. *Small* 11: 5079–5087.

23 Rudramurthy, G.R. and Swamy, M.K. (2018). Potential applications of engineered nanoparticles in medicine and biology: an update. *J. Biol. Inorg. Chem.* 23 (8): 1185–1204.

24 Liu, J., Sonshine, D.A., Shervani, S., and Hurt, R.H. (2010). Controlled release of biologically active silver from nanosilver surfaces. *ACS Nano* 4: 6903–6913.
25 Williams, M.A. (ed.) (2012). *Allergens and Respiratory Pollutants: The role of innate immunity*, Series on Medical and Health Science, Technology and Policy. Oxford, UK: Biohealthcare Publishing Ltd.
26 Peters, A., Wichmann, H.E., Tuch, T. et al. (1997). Respiratory effects are associated with number of ultrafine particles. *Am. J. Respir. Crit. Care Med.* 155: 1376–1383.
27 Diaz-Robles, L.A., Fu, J.S., Vergara-Fernandez, A. et al. (2014). Health risks caused by short term exposure to ultrafine particles generated by residential wood combustion: a case study of Temuco, Chile. *Environ. Int.* 66: 174–181.
28 Song, S., Paek, D., Park, C. et al. (2013). Exposure to ambient ultrafine particles and urinary 8-hydroxyl-2-deoxyguanosine in children with and without eczema. *Sci. Total Environ.* 458–460: 408–413.
29 Olsen, Y., Karottki, D.G., Jensen, D.M. et al. (2014). Vascular and lung function related to ultrafine and fine particles exposure assessed by personal and indoor monitoring: a cross-sectional study. *Environ. Health* 13: 112.
30 Evans, K.A., Halterman, J.S., Hopke, P.K. et al. (2014). Increased ultrafine particles and carbon monoxide concentrations are associated with asthma exacerbation among urban children. *Environ. Res.* 129: 11–19.
31 One Health Initiative Task Force (2008). *One Health: A New Professional Imperative*, 1–76. American Veterinary Medical Association.
32 Vora, R., Zareba, W., Utell, M.J. et al. (2014). Inhalation of ultrafine carbon particles alters heart rate and heart rate variability in people with type 2 diabetes. *Part. Fibre Toxicol.* 11: 31.

Section II

Terrestrial and Aquatic Systems

2

Ecosafety of Nanomaterials in the Aquatic Environment

Maria J. Bebianno[1], Thiago L. Rocha[1], José P. Pinheiro[2], Margarida Ribau Teixeira[3], and Fernanda Cassio[4,5]

[1]Centre for Marine and Environmental Research (CIMA), University of Algarve Campus de Gambelas, Faro, 8005-139, Portugal
[2]CNRS/Université de Lorraine, LIEC, UMR 7360, Vandoeuvre-lèes-Nancy F-54501, France
[3]CENSE, Center for Environmental and Sustainability Research, University of Algarve, Campus de Gambelas, Faro, 8005-139, Portugal
[4]Centre of Molecular and Environmental Biology (CBMA), Department of Biology, University of Minho, Campus de Gualtar, Braga, 4710-057, Portugal
[5]Institute of Science and Innovation for Biosustainability, University of Minho, IB-S, Campus de Gualtar, Braga, 4710-057, Portugal

2.1 Introduction

Nanomaterials (NMs) are novel materials with nanoscale dimensions (1–100 nm) that have attracted increasing attention due to their capacity to alter the physical and chemical properties of conventional materials. They have a wide range of applications into innovative technological uses (nanotechnology). Although a clear definition to characterize these NMs has not been agreed yet, they need to have at least one dimension on the 1–100 nm range. If they have three dimensions within this range, they play a central role in the advance of nanotechnology [1]. Because the definition behind these materials remains controversial, the European Commission defined NMs as "a natural, incidental or manufactured material containing particles, in an unbound state or as an aggregate or as an agglomerate and where, for 50% or more of the particles in the number size distribution, one or more external dimensions is in the size range 1–100 nm" [2].

Since their discovery in the 1980s, NMs production and variety increased enormously, being produced today in many different organic and inorganic chemical forms, such as carbon-based NMs (carbon nanotubes (CNTs) and carbon spheres), metal and metal oxides NMs, quantum dots (QDs), and nanoceramics [3, 4].

Toxicology of Nanoparticles and Nanomaterials in Human, Terrestrial and Aquatic Systems, First Edition. Edited by Marc A. Williams.

These materials have a wide variety of innovative technological and industrial applications, such as food products, textiles, systems for medical diagnostics (imaging, drug delivery, tumors treatments, and gene therapy), pharmaceuticals, personal care products (PCPs) (cosmetics and sunscreens), renewable energies, sports, electronic devices, telecommunications, paints, self-cleaning surfaces, and environmental applications [5–8]. Recently, a suite of new marine nanotechnologies was developed including antifouling paints (especially containing CNTs) designed to prevent bio-fouling or biofilms on vessel hulls [9, 10] and pollution remediation systems to protect the environment through pollution prevention, treatment, and cleanup of hazardous wastes sites [7, 11] or even in environmental remediation known as *nanoremediation* [7].

These intensive development and applications will produce an increase use of NMs, hence nanowastes will inevitably be released in the aquatic environment and might cause human, ecological, socioeconomics, and environmental impacts, particularly in the marine environment [11]. The uncertainty about their ecosafety and sustainability needs to be urgently addressed [12, 13].

NMs are considered a new class of emerging contaminants, thus there is a pressing need for risk assessment studies in order to introduce ecosafety measures to prevent potential adverse effects to human health and to the aquatic environment.

Due to the need for innovation and potential to synthesize, manipulate, and create, a wide range of nanoproducts/NMs with different chemical compositions, sizes, shapes, coatings, and functional groups are currently being produced. Materials in the nanoform behave quite differently from their bulk counterparts. Their specific physical and chemical characteristics (especially surface and quantum effects) and toxicity make them technologically more attractive [14]. In addition, they can be coated or capped with other chemical compounds to change their chemical properties such as dispersibility and conductivity and to prevent aggregation/agglomeration in aquatic solutions [15]. Due to their surface area/volume ratio even within a single chemical, a higher proportion of atoms are present at their surface, and this increase in surface area partly determines their reactivity. This will result in different chemical reactivity, bioavailability, and ecotoxicity to aquatic organisms, but their impact and ecosafety in the aquatic environment is difficult to predict [11, 16].

In this chapter we review the inputs of NMs in the aquatic environment, the wastewater treatment processes available to remove NMs, the importance of understanding speciation of NMs in the aquatic environment, the toxicological effects on fresh and marine species, the interactive effects of NMs with other contaminants, and the applicability of the existing framework of environmental risk assessment (ERA) to NMs.

2.2 Inputs of NMs to the Aquatic Environment

The amount of NMs in the aquatic environment is difficult to predict, so to ensure their ecosafety, proper tools and methodologies for their detection in the aquatic environment need to be developed to accurately assess the ecological risk of these materials [17–19].

Around 50 000 kg/year of nanosized materials are being produced. Some will enter the aquatic environment during the product's life cycle (e.g. by erosion of the materials, accidental spills, or use of products that contain NMs, such as paints, fabrics, and personal health care products), from production facilities and wastewater treatment plants (WWTPs) [12, 20, 21]. As the ultimate sink of most conventional contaminants, the aquatic environment is highly susceptible to NMs exposure either from point (sewage, effluents, and river influx) or from diffuse sources (atmospheric deposition, dumping, and runoff). The first evidence of water contamination of these NMs came from TiO_2 NMs present in runoff water released from self-cleaning exterior facades [22] and in raw sewage and effluents [23]. All these routes allow NMs to disperse through the aquatic environment in quantities that are likely to dramatically increase in the near future but the fate of each of them will vary [24].

2.3 How Wastewater Treatment Processes Act in the Removal of Nanomaterials?

The increase use of NMs in consumer products and industrial processes or materials introduces these materials intentionally or unintentionally into the waste streams and in wastewater treatment facilities. The capacity of WWTPs to remove these NMs is controversial. Recent studies suggested that some NMs escape from the wastewater treatment [23, 25, 26]. The fraction of NMs that is not removed has the potential to contaminate surface waters and soils and pose a risk to the biota or affect drinking water resources [27–29]. In addition, sewage sludge applied to agricultural soil may cause exposure of NMs to soil organisms and leach NMs to groundwater [30–33]. Therefore, via wastewater, potential environmental risks are linked to possible bioaccumulation of NMs in natural systems.

The fate of nanosized particles in WWTPs is not well characterized. Wastewater treatment consists of a combination of physical, chemical, and biological processes to remove solids, organic matter, and nutrients from wastewater. Depending on the characteristics of the wastewater and the countries standards, different degrees of treatment can be applied to increase the treatment level (preliminary, primary, secondary, and tertiary and/or advanced treatment, and sometimes disinfection).

However, WWTPs were not designed to remove NMs, thus the study of wastewater processes efficiency has been the focus of current research.

Several studies on NMs removal and release from WWTPs were reported. In Germany, Li et al. [34] concluded that just 5% of AgNPs from the wastewater effluent was discharged rising to low concentrations of AgNPs in treated water (<12 ng/L). In a WWTP effluent containing 100–3000 μg/L of Ti, 10–100 μg/L of Ti still remained in the effluent after treatment [23]. Westerhoff et al. [26] analyzed ten WWTPs with raw sewage whose TiO_xNPs concentrations ranged from 181 to 1233 μg/L. The WWTP removed more than 96% of the Ti from the influent, and all the effluents had Ti concentrations lower than 25 μg/L. In a pilot WWTP, Ma et al. [35] concluded that both Ag and ZnO NPs were transformed during the WWTP process into Ag_2S, Zn–Fe oxy/hydroxide, $Zn_3(PO_4)_2$, and ZnS, so these compounds appear to be the relevant species for toxicity and fate testing. Similar results for AgNPs were also obtained by Kaegi et al. [36]. These studies showed that WWTPs act as the gateway that control the release of NMs from domestic and/or industrial sources to the aquatic environment via treated effluents discharged into surface waters [34].

2.4 So What Is Expected to Occur in WWTPs Processes?

The ability of WWTPs processes to immobilize or destroy NMs depends on the chemical and physical nature of the particles and the residence time in the relevant compartments of the treatment plant. NMs are affected by sorption processes (for example, in primary clarifiers) and chemical reactions [37]. But settling, diffusion, and aggregation according to their size, density, and surface physical chemistry will also affect NPs removal [30]. Figure 2.1 presents a scheme of the NMs route after discharge in the wastewater and evidence of the NMs removal in each phase.

NMs are released to the wastewater collection system where transformation starts. The NMs transformation will influence their subsequent removal in the WWTPs and their bioavailability when released into the aquatic environment. The formation and fate of NMs in wastewater collection systems were investigated by Brunetti et al. [38] and Kaegi et al. [39]. ZnONPs and AgNPs underwent significant transformations during their transport through the sewerage network, being sulfur species the most important endpoint for these NPs [38, 39]. Sulfidation of AgNPs increased with adsorption of humic acids [46] and microbial processes play a direct role in speciation by driving changes in redox conditions and producing materials such as cysteine and histidine [38].

Several researchers investigated the removal of NMs in pre- and primary treatments [30, 34, 40–43] and concluded that sorption and agglomeration are the mechanisms responsible for the NPs removals in these two WWTP phases. NMs

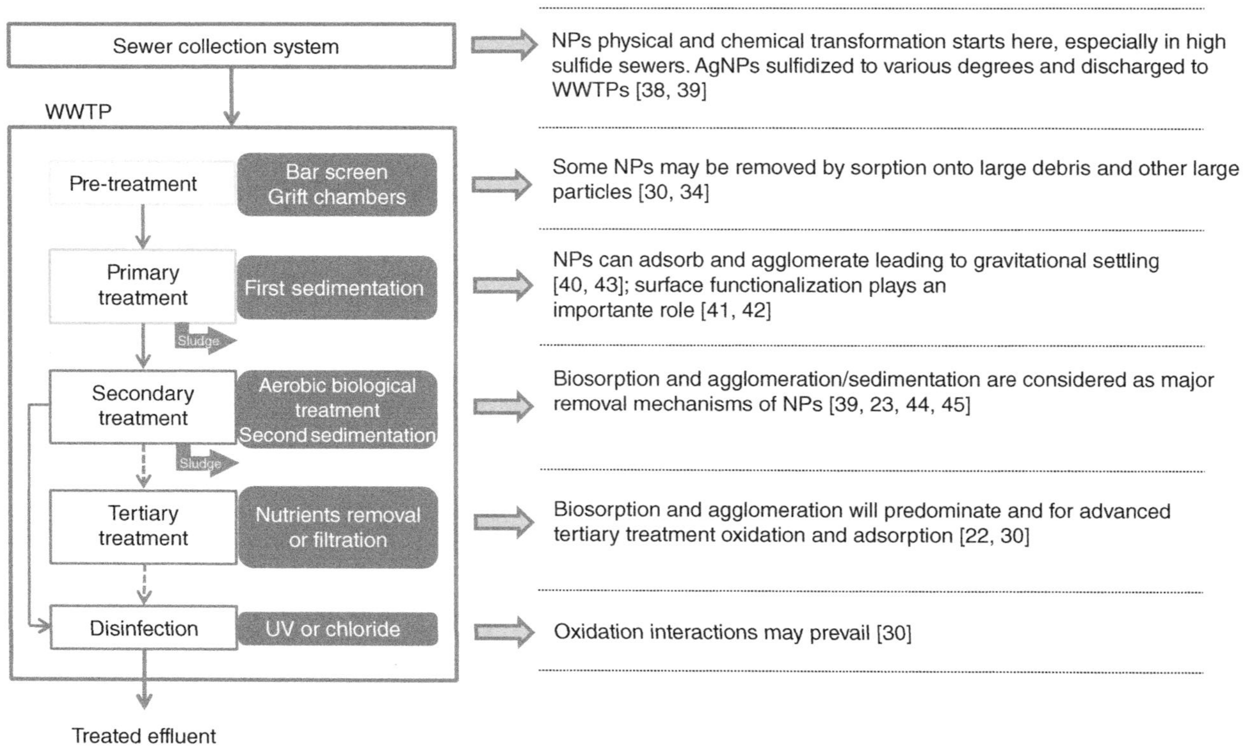

Figure 2.1 NMs route after discharge in wastewater and removal processes.

can adsorb and agglomerate with other particles present in wastewaters, leading to gravitational settling. Agglomeration involves the adherence of single or cluster of particles into larger masses, due to attractive forces or chemical or mechanical binding. In addition, sorption, agglomeration, and mobility of mineral colloids are strongly affected by pH, ions, and organic matter, so water characteristics may affect sorption and settling of NMs [30, 37]. Li et al. [34] concluded that Ag (ions and NPs) was partly attached to larger particles like suspended organic matter, resulting in removal of approximately 35% AgNPs. In a laboratory study using CeO_2NPs, the majority of NPs were captured through adhesion to clear sludge, and up to 6 wt% of CeO_2NPs was found in the exit stream [43]. Hou et al. [40] showed that over 90% of the AgNPs (23 nm) remained in the wastewater even using high suspended solid content during simulated primary clarification experiments. However, NPs surface functionalization plays an important role. King et al. [42] demonstrated that AgNPs (22 nm) coated with a non-ionic stabilizer (PVP) undergo rapid settling in wastewater and will be removed to sewage sludge by hetero-aggregation, rather than continuing to the secondary treatment stage. Results from Jarvie et al. [41] indicated that surface-functionalized SiO_2NPs (56 nm) were removed by sedimentation to sewage sludge, whereas uncoated SiO_2NPs continued through the effluent stream. Coated SiO_2NPs were adsorbed on Tween® molecules and sewage organic matter.

In the biological secondary treatment of WWTPs, where microorganisms are present, NMs can adhere or interact with them, promoting removal by agglomeration and settling of NMs. Interactions between NMs and bacteria appear to involve electrostatic attraction and to be size dependent [47]. NMs removal in activated sludge (A/S) systems seems to be a function of their composition, surface functionalization, size, and zeta potential [23]. In addition, higher biomass (total suspended solids [TSSs]) concentrations improved NMs removal [29, 44]. Gómez-Rivera et al. [45] concluded that CeO_2NPs (50 nm) were removed in a laboratory scale activated sludge (A/S) system mainly by aggregation and settling, promoted by neutral pH and interactions with organic and/or inorganic wastewater constituents. Batch bioassays demonstrated that CeO_2NPs only exerted inhibition of O_2 uptake by A/S at concentrations exceeding those in the bioreactor feed (50% inhibition at 950 mg CeO_2/L). Removals of 39–62% were obtained by Kiser et al. [23] for a group of NMs that includes surface functionalization (citrate-coated Ag, polyvinylpyrrolidone-coated Ag and Au, carboxylate-functionalized Ag, and gum arabic-coated Ag) and of 92–94% for a group of NMs that has properties that favor association with fresh activated sludge (nonfunctionalized fullerene, carboxylate- and sulfate-functionalized yellow-green fluorescent microspheres, and tannic acid-capped nanogold (AuNPs)). These authors concluded that zeta potential or size may not be sufficient for making *a priori* estimates of the removal of NMs in fresh activated sludge. Similar results were obtained by Kaegi et al. [39] for AgNPs (up to 10 nm) and AuNPs (up to 10 nm), where removals of approximately 99% were observed irrespective of size, coating, or type (elemental composition) of the core material by attachment to sludge flocs (hetero-aggregation).

Finally, in the tertiary treatment, the removal of the portion of the remaining organic and inorganic solids and pathogenic microorganisms occurs through a filtration step and/or the removal of nutrients. This treatment may be followed by chemical disinfection. The impact of this treatment in the removal of NMs is still not well addressed. A study was conducted in a WWTP in central Arizona (USA) that uses activated sludge process and tertiary filtration to treat a raw sewage containing 100–nearly 3000 μg Ti/L [23]. Results showed Ti concentrations between secondary and tertiary effluents of 20 and 12 ± 2 μg/L, respectively, and the same concentration (36 μg/L) for non-filtered samples. These authors concluded that the majority of Ti is associated with solids in wastewater, so as tertiary effluent has low solid levels (4 mg/L TSS), removals were higher in previous WWTP phases. Brar et al. [30] considered that in theses phases (tertiary and disinfection), the NMs removal processes previously discussed would predominate. However, there are many studies on the effect of NMs on the removal of nitrogen and phosphorus and on the biological communities responsible for nutrient removal. It seems that NMs have a negative impact on the removal of these nutrients from wastewater, especially for high NMs concentrations [48–51].

As a result of the different treatment phases, sludge from primary and secondary settling must be digested and dewatered and sent to landfill or reused for agricultural application. However, some of the NMs will persist under various biosolids processing procedures [35, 52]. Therefore, the fate and ecosafety of NMs in soils and water and interactions with plants coming for threat WWTPs sludge should be carefully evaluated.

2.5 The Importance of Understanding Speciation of NMs

Trace metal toxicity in natural systems is difficult to decipher due to the variety of metal ions and metal complexing agents present in the matrix. Thus, it is fundamental to know the specific dynamic metal speciation, also termed chemodynamics [53] to understand their toxicity. From the physicochemical point of view, the traditional approach is to simplify the problem considering that only the free metal ion (the hydrated cation) is toxic to the organisms, i.e. the well-known Free Ion Activity Model (FIAM) [54] and its evolution to the Biotic Ligand Model (BLM) [55].

The introduction of metal containing NMs adds a level of complexity to these already challenging studies. One must always consider the dissolution of NMs functioning as a source of metal ions [56], as well as the possibility of exchange of metal in the particle by another metal previously in solution and the interaction of the metal ions with the charged polymer layer often used to stabilize metal containing NMs [57]. In this context it is fundamental to quantify the amount of metal in the NMs and the amount of metal in solution during an exposure experiment.

The measurement of total metal in solution, comprising the metal contained in the NMs, the free metal ion, and the metal complexed by the matrix ligands is routinely performed by spectroscopic techniques, normally by inductively coupled plasma-mass spectrometry (ICP-MS).

The quantification of the concentration and/or number of NMs in solution is an involved problem normally requiring the physical separation of the NMs from the dissolved components of the matrix, usually by ultrafiltration [58], dialysis [59], or ultracentrifugation [60], followed by metal quantification by ICP-MS. Recently, the development of single particle ICP-MS [61, 62] provides a powerful, albeit expensive, technique to quantify very low NMs concentrations in complex matrices without previous separation.

Regarding the free metal ion activity, the traditional approach in ecotoxicological studies is not to measured but rather to buffer the free metal using a large total metal concentration and a metal buffer chelating agent (like EDTA). In experiments involving metal containing NMs, this approach is impossible since the chelating agent will affect the stability of the NMs and the metal speciation in the matrix [63]. The absence of metal buffer and the uncertainty in the amounts of free metal ions and metal in the NM makes it necessary to directly measure the free metal ion activity.

The quantification of free metal ion activity in solution, in principle, is not affected by the presence of NMs; nevertheless for the environmentally relevant concentrations, there are few available techniques [64], being the most adequate is the Donnan membrane technique (DMT) [65, 66] and the absence of gradients and Nernstian equilibrium stripping technique (AGNES) [67].

DMT is based on the transport of metal ions from solution, through a negatively charged semipermeable cation exchange membrane, to an acceptor solution. It has the advantage of being able to perform simultaneous measurements of several elements. When the solution is added, the cations will permeate the membrane until equilibrium is attained (usually 24–48 h), whereas the transport of anionic species is blocked by the negative charge of the membrane. The measurement of the total metal in the acceptor after reaching equilibrium yields the free metal ion concentration. The membranes used are calcium-based, implying that the background electrolyte will have calcium ions, which can be a major disadvantage if the matrix is calcium poor.

DMT is probably the best technique to determine several free metal ions in solution as long as the matrixes have already significant calcium content and were extensively applied to quantify free metal ions in soils [68, 69], natural waters [70–72], and ecotoxicological assessment [73–75]. In studies involving NMs, DMT was used to study the behavior of capped (by an organic coating composed of a mixture of C12–C15 alkyl benzoate, polyhydroxystearic acid, and triethoxycaprylylsilane) and uncapped ZnO (17.5 and 21 nm) in synthetic seawater [76] and the UV-induced photochemical transformations of citrate-capped AgNPs (20, 40, 60, and 80 nm) suspensions [77].

AGNES is an electrochemical stripping technique designed to quantify low free metal ion concentrations. It consists of two steps: a deposition step (where the metal ion is amalgamated in the working mercury electrode) and a stripping step (where the deposited metal is reoxidized and quantified). The major drawback of AGNES is the limited number of metals that it can study (Zn, Cd, Pb, Cu) when using a mercury electrode, but recently its extension to solid electrodes [78] opens good perspectives for its application to other metal ions. AGNES has been used in NPs studies involving the dissolution of CdTe/CdS QDs (3–4 nm) [63], the dissolution kinetics and solubility of ZnONPs (6, 20, and 71 nm) [79], the toxicity of ZnONPs (20–40 nm by TEM and 66.5–153.1 by DLS in water) on *Daphnia magna* [80], the bioavailability of Cd in presence of n-TiO_2 (5–20 nm) to *Corbicula fluminea* [81], and the toxic kinetics and tissue distribution of COOH-CdTe QDs (6 nm) in *Mytilus galloprovincialis* [82].

Another significant difference in the NMs' case is the aggregation/disaggregation phenomena that might happen when introducing the NMs in natural systems due to the changes of matrix conditions (like pH and ionic strength) and their interaction with other components of the matrix, like the natural organic matter (NOM) [83, 84].

The fate of NMs in the environment has been an object of great interest and many studies have been published [85], and it is important that this knowledge is used to understand the changes to the NMs and their interaction with the metal ions in solution during an ecotoxicological assessment, since they affect the solubility of the NMs and their ability to bind metal ions in solution.

Techniques and recommendations on how to follow the physicochemical changes of NMs in solution are abundant in the literature and many analytical techniques are accessible for size measurements comprising microscopy techniques, dynamic light scattering (DLS), fluorescence correlation spectroscopy (FCS), nanoparticle tracking analysis (NTA), and separation-based techniques, such as field flow fractionation (FFF), hydrodynamic chromatography (HDC), or analytical ultracentrifugation [86–88].

Within this framework it is important to recognize that an ecotoxicological assessment using NMs cannot be performed under the same hypothesis as for metal ions. The design and execution of an environmentally relevant toxicological experiment should take into consideration the following aspects [89]:

(a) In the exposure solution the physicochemical conditions of the medium should emulate those of the environmental matrix, and the following parameters need to be known and/or changed in a controlled way: temperature, pH, ionic strength, light, concentration of other metals, and complexing agents (especially NOM).
(b) In the ecotoxicological experiments in addition to the control and the metal-based NM contaminated solution, it is necessary to perform an additional control containing the same total metal concentration added in soluble form. To be able to analyze the toxicity results, it is necessary to follow, in the exposure media during the time of the experiment, the (i) metal containing NMs dissolution and aggregation/disaggregation behavior, (ii) free metal concentration, and (iii) metal containing NMs interaction with the other components of the matrix.

The ecotoxicity modeling of metal containing NMs also presents some interesting challenges, not only due to the possibility of direct internalization of NMs but also due to the possible dissolution of NMs and corresponding temporal variation of both total dissolved and free metal ion in solution. Therefore, the use of BLM becomes impossible since one of its main assumptions is that both total dissolved metal and free metal are constant during exposure.

Another interesting development in terms of modeling is the application of dynamic energy budget (DEB) theory to understand NP toxicity, namely, CdSe QDs (5 nm) toxicity [90, 91] and ZnONPs (20–30 nm) to *M. galloprovincialis* [92].

The ideal development would be the integration of physicochemical bioavailability models, considering the possible direct internalization of the NMs and their fate inside the organism, with the DEB model to provide a realistic modeling of the effect of NMs in the environment.

2.6 Ecotoxicological Effects of NMs in Freshwater Organisms

Most of the assessment of ecotoxicological effects of NMs in the aquatic environment was mainly focused on four out of five classes of NMs based on their chemical composition: carbon-based NMs (e.g. CNTs and fullerenes) [93, 94]; metal NPs and metal-oxide NPs (e.g. TiO_2NPs, AgNPs, and CuONPs) (e.g. [95–101]); and QDs [102]. Results so far suggested that NMs can be toxic to freshwater organisms, and effects are particle-type dependent. Furthermore, data from the field is scarce or inexistent in the aquatic environment.

Yan et al. [103] showed that the exposure to hydroxylated (OH-MWCNT; 4 ± 1.5 nm; 0.5 mg/L) and carboxylated multi-wall (COOH-MWCNTs; 4 ± 1.5 nm; 0.5 mg/L) for 4 and 18 days induced more stress to the fish *Carassius auratus* than raw MWCNTs (12 ± 2 nm; 0.5 mg/L). Furthermore, single-walled carbon nanotubes (SWCNTs) and multi-walled carbon nanotubes (MWCNTs), OH-MWCNT, and COOH-MWCNTs enhanced Cd toxicity to *D. magna* [94], suggesting that functionalization plays an important role in nanotoxicity. The exposure to SiO_2NPs (80–100 nm; 1–5 mg/L; 96 h) altered plasma electrolytes (Na^+, K^+, and Cl^-) levels, hematological parameters (e.g. hemoglobin and red and white blood cells), and Na^+/K^+ ATPase activity in gills of the teleost fish *Labeo rohita* [104].

The exposure of several algal species, such as *Scenedesmus quadricauda*, *Chlamydomonas moewusii*, and *Chlorella vulgaris*, to TiO_2NPs (27 nm; up to 300 ppm, 22 days) led to a decrease in their growth and metabolic functions [105]. The toxicity of TiO_2NPs was higher to the freshwater algae *Pseudokirchneriella subcapitata* followed by *Ceriodaphnia dubia* and fathead minnow *Pimephales promelas* [106] and induced genotoxicity and cytotoxicity in *C. dubia* [107]. The increased bioavailability of metal ions with the decrease in particle size contributed to explain the toxicity levels of each metal NP [98, 108, 109]. Several metal NPs are reported to be toxic to the microalgae *P. subcapitata*, and EC_{50} (72 h) indicated that toxicity decreased as follows: ZnONPs (50–70 nm) > CuONPs (30 nm) > TiO_2NPs (25–70 nm) [109]. Zhu et al. [110] showed that the chronic exposure to TiO_2NPs (100 nm; up to 50 mg/L; 96 h) resulted in bioaccumulation of NPs interfering with the daphnid feeding and inhibiting its growth and reproduction. In a short-term exposure, no lethal effects of TiO_2 NPs (approximately 24 and 210 nm; 150 mg/L; 5 days) were found on the fish *Poecilia reticulata* although NPs were able to penetrate into the fish blood through the gills and intestine [111]. The toxicity of TiO_2NPs (approximately 24 nm; up to 1.0 mg/L; 14 days) to the rainbow trout was associated with oxidative stress, organ pathologies, and the induction of antioxidant defenses [112]. Dietary exposure of this fish species to TiO_2-NPs (<100 nm; 100 mg/kg; 8 weeks) caused changes in Cu and Zn levels in the brain, with biochemical alterations in the gills and gut [113]. TiO_2NPs were reported to decrease the Na^+/K^+-ATPase activity in the gills and intestine, alteration of total glutathione levels (e.g. gills and liver) and the presence of apoptotic bodies in some hepatocytes of the rainbow trout [112]. Dietary exposure of rainbow trout to TiO_2NPs (100 nm; 100 mg/kg; 8 weeks) reduced 50% of Na^+/K^+-ATPase activity in the brain and thiobarbituric acid reactive substances (TBARS) in the gill and intestine, while Ti accumulation was detected in several fish organs (gills, gut, liver, brain, and spleen) [113]. TiO_2NPs (20 nm; 5.3 mg/L; 48 h) under natural UV radiation were toxic to aquatic microbes and generation of ROS and cell membrane damage were observed [114].

CeO_2NPs (15 and 30 nm; 1 mg/L; 96 h) and SiO_2NPs (7 and 10 nm; 1 mg/L; 96 h) induced toxicity against *D. magna* and to the larvae of *Chironomus riparius* [115]. CeO_2NPs promoted DNA strand breaks in both species, whereas neither exposure to SiO_2 nor TiO_2 had a genotoxic effect on either species. Moreover, the DNA damage was directly related with the mortality of *C. riparius* exposed to CeO_2NPs [115].

Lethal effects after 24 h of exposure to CuONPs (30 nm) and ZnONPs (70 nm) were reported against *Daphnia* and *Tetrahymena thermophila*, and the toxicity was mainly attributed to the solubilized ions [116, 117]. However, ionic metals released from metal NPs (TiO_2, 25–70 nm; ZnO, 50–70 nm; CuO, 30 nm) only partly explain the nanotoxicity against the yeast *Saccharomyces cerevisiae* [118]. Chronic exposure of the gastropod *Potamopyrgus antipodarum* to CuONPs for 9 weeks resulted in the accumulation of Cu and their toxicity varied with NP shape: rod (8 nm width, approximately 40 nm length), spheres (approximately 7 nm), or platelet (approximately 1 mm length, approximately 270 nm width) [119].

AgNPs (44 nm; 0.3 mg/L) and CuONPs (33 nm; 10 mg/L) showed bactericidal effects, while ZnONPs (<1000 nm; 10 mg/L) showed bacteriostatic effects against *Pseudomonas putida* after 60 min exposure [120]. In lake waters, the dissolution of ZnONPs (4.5 and 27 nm), carbon-coated CuNPs (7.5 and 27 nm), and CuONPs (7.5 and 45 nm) was higher than that of AgNPs (10 and 100 nm) and increase with the decrease of pH, ionic strength, and dissolved organic matter [56]. The latter can stabilize AgNPs, while sunlight can induce the transformation of AgNPs by causing particle-fusion or self-assembly to form aggregation of NPs in aquatic environment [121], indicating that the effects of NPs in aquatic environments depend also on light availability.

The exposure to CuONPs (30 nm; up to 175 mg/L; up to 48 h) led to ultrastructural changes in the midgut epithelium of daphnids and the presence of NPs in circular structures, analogous to membrane vesicles from holocrine secretion [122].

Exposure to sublethal concentrations of CuONPs (<50 nm; up to 1000 mg/L; 10 days) led to adverse effects on the growth and feeding behavior of invertebrate detritivore *Allogamus ligonifer* [101]; however, the effects on microbes [123] and invertebrate detritivores [124] decreased with the increase in NP size and presence of NOM. Histological and biochemical analyses revealed that the gills of zebrafish could be the primary target for CuNPs [125]. CuNPs decreased gill Na^+/K^+-ATPase activity up to 58%, and the transcription of stress response genes increased, such as the hypoxia-inducible factor 1, heat-shock protein, and Cu transport regulatory protein [125]. In fungal populations, CuONPs (<50 nm; up to 200 mg/L; up to 10 days) induced oxidative stress and increased the activity of superoxide dismutase (SOD), glutathione peroxidase (GPx), glutathione reductase (GR), and the extracellular enzyme laccases, which requires Cu as cofactor [126]. Also, CuONPs (12, 50, and 80 nm; up to 100 mg/L; 5 days) led to plasma membrane disruption, intracellular accumulation of ROS, and DNA strand breaks in aquatic fungi [126]. Waterborne and dietborne exposures of the snail *Lymnaea stagnalis* to (^{65}CuO)NPs (17 nm; up to 72 nM; 24 h) resulted in the bioaccumulation of the NPs most probably due to NP uptake [127].

The toxicity of AgNPs (20, 40, and 100 nm; 10 μM; 72 h) in *Synechococcus* sp. is explained by both the NP form and ionic Ag released by dissolution from the NPs [128]. Moreover, the exposure to AgNPs (5–25 nm; 48 h) and the ionic counterpart led to similar effects on mortality and abnormal swimming of *D. magna* [129]. The toxicity of AgNPs (50 nm; up to 10 mg/L; 24 h) in microalgae *C. vulgaris* was associated with a strong decrease in chlorophyll content and algal cell density, increased intracellular accumulation of reactive oxygen species (ROS), and lipid peroxidation (LPO) indicating that NPs induced oxidative stress [130]. The exposure of the mussel *Elliptio complanata* to AgNPs (20–80 nm; 0.8–20 μg/L; 24 h) increased the levels of MTs and LPO and induced changes in the protein ubiquitin, and DNA strand breaks in the digestive gland [131]. Because effects promoted by Ag^+ differed from those of AgNPs, a significant role in the toxicity was attributed to the NP form [131]. Results were consistent with those found when examining the proteomic profiles of *D. magna* exposed to AgNPs (3.0 and 11 nm; 20 mg/L; 6 and 10 months) and Ag^+ that showed an increase of carbonyl and hemoglobin levels only in the presence of AgNPs [132].

AgNP exposure (42 nm; up to 166 μg/L; 120 h) resulted in lethal and sublethal effects in zebrafish with various phenotypic deformities, including distorted yolk sac, degenerated body parts, twisted notochord, impaired blood circulation, low heart rate, and pericardial oedema [133]. AgNPs (5–20 nm; 100 μg/L; 72 h) induced apoptosis or programmed cell death in zebrafish embryos [134]. The exposure of the fish *Japanese medaka* to AgNPs (49.6 nm; up to 25 μg/L; up 10 days) led to DNA damage, oxidative stress, and upregulation of genes related to metal detoxification, metabolic regulation, and free radical scavenging activity [135]. AgNPs (5–20 nm; up to 100 μg/mL; 72 h) reduce hatching of zebrafish embryos [134] potentially compromising fish populations at long time.

Lethal toxicity of metal NPs is generally higher for organisms at lower trophic levels (filter-feeding invertebrates) compared with higher trophic levels (e.g. secondary consumers). Despite the few information available on NP effects at the level of communities and ecosystem processes, the knowledge on structural and functional responses of aquatic biota to NPs is required to predict the risk of these NMs. Most studies were done on freshwater bacterial communities or periphyton. For instance, the short-term exposure to AgNPs (<5 nm; 12.5 and 100 μM; 2 h) inhibited the respiration and photosynthesis and the activities of extracellular enzymes of periphyton [136]. However, phosphorus interactions with AgNPs attenuated the effects of NPs on the growth and stoichiometry of periphyton [137]. On the other hand, the extracellular polymeric substrate of periphytic communities altered the stability and surface of 25 nm of AgNPs and CeO_2NPs [138]. The exposure of benthic bacterial communities to TiO_2NPs (22.9 nm; 1 mg/L; up to 30 days) resulted in a decrease in bacterial abundance in artificial stream sediments [139]. TiO_2NPs led to a decrease in planktonic bacterial diversity and to an increase in sessile bacterial diversity [140]. TiO_2NPs (10–20 nm; 5.3 mg/L; up to 48 h) were toxic to planktonic and biofilm communities under natural levels of UV radiation and low TiO_2 concentrations, and effects were explained by the generation of ROS [114]. However, the exposure to low concentrations of AgNPs (<100 nm; up to 1000 μg/L; up to 30 days) induced minor changes in genetic diversity of bacterial communities from estuarine sediments [141]. Metal NPs (Ag and CuO <100 nm) led to changes in the structure and activity of fungal and bacterial decomposer communities [100, 142] with impacts to plant litter decomposition in streams [100].

2.7 Ecotoxicological Effects of NMs in Marine Organisms

NMs behave differently in fresh and marine systems. In seawater, NMs undergo complex interactions and transformations (e.g. aggregation/agglomeration, dissolution, NOM stabilization) [143–145], so their chemical nature and behavior is crucial to assess their bioavailability, bioaccumulation, and toxicity to marine organisms [145]. Aggregation/agglomeration kinetics, changes of the surface charge, and sedimentation rates of commercial NMs are salinity/ionic strength and pH dependent, as reported for CuONPs (33 nm) [146] and COOH-CdTe QDs (6 nm) [82, 102]. In aqueous medium, the ionized carboxylic groups confer negative surface charge for CdTe QDs aggregates, but salinity increase induces lower negative surface charge, higher hydrodynamic diameter (d_h), and sedimentation rate of QDs in seawater when compared with Milli-Q water, confirming the tendency of commercial NMs to aggregate/agglomerate in seawater, specially due to low repulsive electrostatic forces [102].

Therefore, ecotoxicological effects of NMs in seawater are different from freshwater especially because in seawater agglomeration/aggregation, dissolution and precipitation affect their bioavailability and ecotoxicity [147]. Figure 2.2 shows the different sources and interactions of NMs to the marine environment. Behavior and fate of NMs in marine environment are highly dependent on environmental conditions (e.g. ionic strength, pH, temperature, presence of NOM, and suspended particles), as well as on their nano-specific properties (e.g. size, shape, particle surface charge, functionalization, and coating) [144, 146, 148]. The interaction of NMs with surrounding components also affects their fate in seawater and sediments and

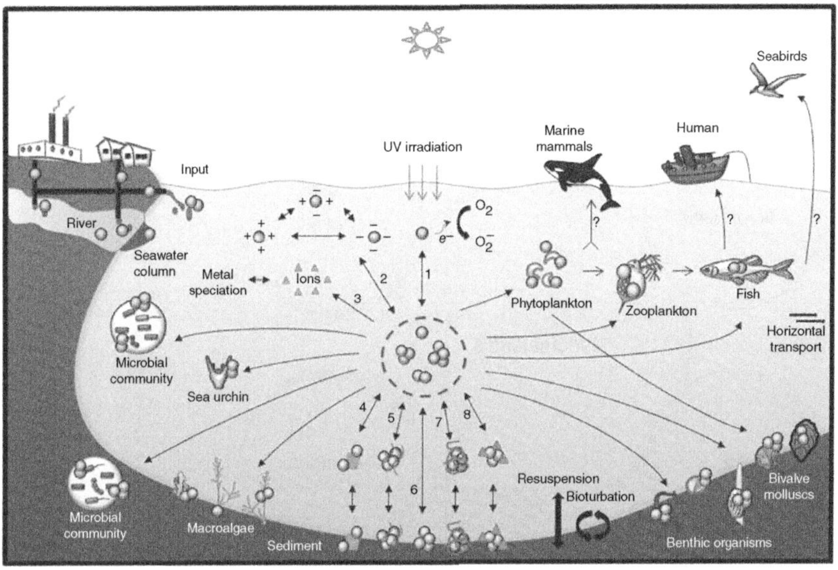

Figure 2.2 General scheme illustrating the potential pathways for the behavior and fate of nanomaterials (NMs) in the marine environment and their biological interactions at different trophic levels: (1) photo-oxidation, (2) surface charge change, (3) dissolution, (4) adsorption to suspended particles, (5) NOM stabilization, (6) sedimentation, (7) adsorption to biomolecules, and (8) interaction with other contaminants (*Trojan horse* effect).

thus determines their biokinetics and toxicity in benthic and pelagic organisms (Figure 2.2).

In natural environmental scenarios, different states of NMs may occur and interact with marine organisms: individual NMs; homo-aggregates; individual and NMs aggregates/agglomerates stabilized by NOM; individual and NMs aggregates adsorbed to suspended particle and/or biomolecules (e.g. clays, exopolymers and mucus); and individual and NMs aggregates adsorbed to organisms (e.g. phytoplankton and zooplankton), which interact with other organic and inorganic contaminants, as well as metal with ions released from metal-based NMs, metal complex formed after NM dissolution and ROS formed by NMs photo-activation (i.e. $O_2^{-\cdot}$ and $OH^{\cdot}$) [11, 145, 148, 149] (Figure 2.3).

The literature about the impact of NMs on aquatic environment increased since 2006, when the first nanoecotoxicological research was published [28]. Since then, a considerable effort was made by the scientific community to assess several ecotoxicological end points to quantify sublethal effects in marine organisms exposed to different classes of NMs, such as inorganic and organic NMs (reviewed in [145, 150–153]). Even so, there is more information regarding

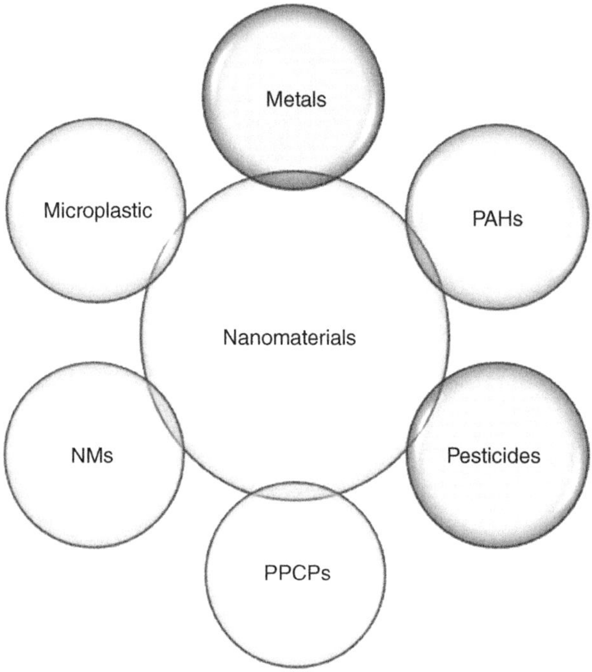

Figure 2.3 Scheme illustrating the potential interaction of NMs with other NMs and different types of pollutant in the aquatic environment. PAHs: polycyclic aromatic hydrocarbons; PPCPs: pharmaceutical and personal care products.

the effects of NMs in freshwater species when compared with seawater ones, especially for TiO_2 NPs, metal, and metal oxides NMs, while for ecotoxicity tests using bivalve species, more information exist on seawater species compared with freshwater ones [145]. Among marine invertebrates, bivalve mollusks were the most studied group (about 50%), followed by polychaetes and echinoderms, while sediment-dwelling invertebrates were less studied (about 20%) (reviewed by [153]). Among marine species, the blue mussel *Mytilus* spp. is the biological model most used to assess the effects and mode of action (MoA) of NMs in the marine environment [145]. Based on laboratory experiments there is information regarding NPs accumulation and ecotoxicological effects and MoA in marine organisms [144, 150–152]. Adsorption and accumulation of NMs were reported for a wide array of marine organisms at different trophic levels, such as bacteria, phytoplankton, zooplankton, macroalgae, echinoderm, polychaete, bivalve mollusks, and fish (Figure 2.2), while less information is available for marine mammals and seabirds [152, 154–157]. Furthermore, sea urchin was recently indicated as a suitable model for evaluating the effects of NMs on the early life stage and on the immune system [153, 158].

Despite NMs accumulation and trophic transfer in reconstructed marine trophic chain, limited biomagnification was confirmed, indicating the urgent need to assess the ecotoxicological effects of NMs in environmental relevant conditions, such as mesocosms and multispecies exposures (reviewed in [159]). Furthermore, to assess the ecosafety of NMs, further studies are needed to investigate the possible impact on human health due to the consumption of NMs-contaminated seafood.

Differential bioaccumulation and tissue-specific responses were detected in marine invertebrate tissues after exposure to NMs, such as in oysters *Crassostrea virginica* exposed to AgNPs (20–30 nm; 0.02–20 μg/L; 48 h; [160]), marine mussels *M. galloprovincialis* exposed to CuONPs (31 nm; 10 μg/L; 15 days; [96]), AgNPs (42 nm; 10 μg/L; 15 days; [161]), and CdTe QDs (6 nm; 10 μg/L; 15 days; [78]). Inorganic NPs (e.g. 42 nm Ag, 10–100 nm CuO, 22 nm TiO_2, and 6 nm QDs) are accumulated by mussel gills, translocated to the hemolymph and transported to the digestive gland where it induces changes in lysosomal membrane stability (LMS) and antioxidant enzymes activities, immunotoxicity, neurotoxicity, genotoxicity and pre-apoptotic processes [82, 95–97, 102, 162, 163], suggesting that *in vitro* and *in vivo* exposure support the hypothesis that several toxic effects of NMs on marine organisms are directly or indirectly mediated by oxidative stress. However, mechanisms of NMs-mediated toxicity are not well established and depend on nano-specific properties, exposure conditions, species, tissue, and end point analyzed.

Oxidative stress, changes in membrane proteins, and DNA damage are the main MoA of NMs in bivalve tissues (reviewed in [145]), while toxic effects of metal-based NMs were also related to extra- and intracellular dissolution and release of metal ions from NMs (reviewed in [152]). Alterations of antioxidant enzymes activities, ROS generation, and oxidative damage induced by different QDs were observed in the marine diatom *Phaeodactylum tricornutum* exposed to CdSe and CdSe/ZnS QDs (3.2–3.5 nm; 72 h; [164]), polychaete *Hediste diversicolor* exposed to CdSe/ZnS QDs (5.9 nm; 0.001–1 ng/(g w). sediment; [165]), marine mussel *M. galloprovincialis* exposed to CdTe QDs (6 nm; 10 μg/L; 14 days; [82, 102]), and fish *Gasterosteus aculeatus* exposed to CdS QDs (4.2 nm; 5–500 μg/L; 21 days, [166]) and showed that the modification of NM surface charges, functionalization, and coating have an effect on their uptake, accumulation and toxicity in marine organisms, indicating that the progress in the synthesis and modification of NM surface contribute to greater ecosafety of NMs in the aquatic environment.

Moreover, early life stages are a sensitive target for metal-based NMs. Embryo development is an important target for different types of NMs due to metal ion solubilization. Toxicity may be related to other mechanisms that are particle specific as the formation of a NM protein corona that might affect uptake routes and toxicity in different types of embryos [153].

2.8 Interactive Effects of NMs with Other Contaminants

Ecosafety and risk assessment of NMs in the marine environment should consider not only their inherent nanotoxicity but also the potential interaction of NMs with other pre-existing contaminants [167]. According to the *Trojan horse* concept, NMs can act as a carrier for other contaminants, which can change the bioavailability, uptake, accumulation, and toxicity of both NMs and contaminants in marine organisms [168].

In Figure 2.3 the possible interactions of NMs with other contaminants in the marine environment are described. Although the physicochemical interactions of NMs with several contaminants were investigated, the interactive effects of NMs with other contaminants in marine organisms are scarce. Regarding species, bivalve mollusks were the most studied group (64.3%) (with the mussel *M. galloprovincialis* as the main biological model used (50%)), while fish and sea urchin represent 28.6 (the marine fish *Dicentrarchus labrax* (21.4%)) and 7.1%, respectively (Table 2.1). They only exist for TiO_2NPs (71.4%), AuNPs (14.3%), C_{60} (7.1%), and SWCNTs (7.1%), while Cd^{2+} (28.6%) and 2,3,7,8-tetrachlorodibenzo-p-dioxins (TCDD; 28.6%) were contaminants whose interaction with NMs were the most investigated, followed by polycyclic aromatic hydrocarbons (PAHs: benzo(a)pyrene, fluoranthene, and phenanthrene), pro-oxidant compound menadione, and two emergent contaminants (1–5 μm microplastics and 15 nm AgNPs) (Table 2.1). TCDD accumulation increased in whole soft tissues of mussels *M. galloprovincialis* exposed to a mixture of TiO_2NPs (27 nm; 100 μg/L) and TCDD (0.25 μg/L) for 96 h [173], but a decreased in TCDD accumulation occurred in the digestive gland [175], indicating that the *T. horse* effect in marine mussels is tissue dependent, such as observed for the exposure to NMs alone [145, 160]. On the other hand, no Cd accumulation occurred in the digestive gland of *M. galloprovincialis* after the co-exposure to TiO_2 NPs (27 nm; 100 μg/L) and Cd^{2+} (100 μg/L) [172], but TiO_2 NPs accumulation increased in the gills [155]. NPs-induced chromosomal damage and partial recovery of genome template stability in the marine fish *D. labrax* and mussel *M. galloprovincialis* after co-exposure to TiO_2NPs (24 nm; 100 μg/L) and Cd (100 μg/L) [155, 177], indicating that changes in the NMs behavior during co-exposure lead to a decrease on nano-specific toxicity. Reduced C_{60} accumulation was observed in *M. galloprovincialis* after the co-exposure to C_{60} (100–200 nm; 0.1–1 mg/L) and

Table 2.1 Interactive effects of nanomaterials (NMs) with other contaminants in marine organisms.

	NM			Pollutant					
Species	**Type[a)]**	**Size (nm)**	**Concentration (μg/L)**	**Type[a)]**	**Concentration (μg/L)**	**Exposure time**	**Cell/ tissue[b)]**	**Interactive effects[c)]**	**References**
Bivalve mollusks									
Mytilus edulis	TiO_2	62	200, 2000	Benzo(a) pyrene	20	96 h	DG, H, W	Decreased B(a)P accumulation. No effects on SOD and GPx. Increased chromosomal damage and CAT	Farkas et al. [169]
Mytilus galloprovincialis	Au	15.6	750	Menadione	1 mM	24 h	DG	Reduced Au accumulation. No effects on oxidative stress parameters (TrxR; GSH/GSSG ratio; protein thiol group)	Tedesco et al. [170]
	C_{60}	100–200	100, 1000	Fluoranthene	32	72 h	DG, G, H, M	Reduced C_{60} accumulation. Additive (DNA damage) and synergic effects (GSH content)	Al-Subiai et al. [171]
	TiO_2	27	100	Cd	100	96 h	DG, E, H	No effect on Cd accumulation. Synergic effects (lysozyme, TLR-i genes). Decreased retarded development induced by Cd^{2+}	Balbi et al. [172]
	TiO_2	27	100	TCDD	0.25	96 h	DG, G, H	Increased TCDD accumulation. Antagonistic (hemocyte phagocytosis, ABC transport) more than synergic effects (neutral lipid content and lysosome/cytoplasm volume ratio, transcription of estrogens receptors)	Canesi et al. [173]

(Continued)

Table 2.1 (Continued)

	NM			Pollutant					
Species	Type[a)]	Size (nm)	Concentration (μg/L)	Type[a)]	Concentration (μg/L)	Exposure time	Cell/tissue[b)]	Interactive effects[c)]	References
	TiO_2	24	100	Cd	100	96 h	G	Decreased Cd accumulation. Antagonistic (*abcb* 1 gene, GST, DNA damage) more than additive effects (NO production).	Della Torre et al. [174]
	TiO_2	24	100	Cd	100	96 h		No effect on chromosomal damage. Partial recovery of genome template stability	Rocco et al. [13]
	TiO_2	27	100	TCDD	0.25	96 h	DG	Decreased TCDD accumulation. Up regulated 28 DEGs and down-regulated 34 DEGs. Decreased tissue damage	Banni et al. [175]
Scapharca subcrenata	TiO_2	≤10	500	Phenanthrene	20.9 ng	35 d	G, M, Pf	Increased phenanthrene accumulation	Tian et al. [176]
Fish									
Dicentrarchus labrax	TiO_2	27	1000	TCDD	46 pg/L	7 d	L, G, Er, S, M	No effect on TCDD accumulation	Della Torre et al. [174]

	TiO_2	24	1000	Cd	100	7 d	Er, M	Decreased NPs induced chromosomal damage. Partial recovery of genome template stability	Nigro et al. [177]
	TiO_2	27	1000	TCDD	46 pg/L	7 d	L	Down-regulated *ahrr*, *erβ2*, *Abcb1*, and *Abcb2* genes. No effects on *Cyp1a*, *gst*, and *elmod2* genes	Vannuccini et al. [178]
Pomatoschistus microps	Au	5	200	Microplastic (1–5μm)	184	96	He, O	No effects on NP accumulation and NP-induced toxic effects	Ferreira et al. [179]
Sea urchin									
Strongylocentrotus droebachiensis	COOH-SWCNT	1.4 μm × 2–20 nm	200–5000	PAAm-Ag NPs (15 nm)	20–500	96	E	Decreased toxic effects on early development stages	Magesky and Pelletier [180]

a) 2,3,7,8-Tetrachlorodibenzo-*p*-dioxins (TCDD); poly(allylamine)-coated silver nanoparticles (PAAm-AgNPs); single-walled carbon nanotubes (SWCNT).

b) Digestive gland (DG); embryo (E); erythrocytes (Er); gills (G); head (He); hemolymph/hemocyte (H); liver (L); muscle (M); pseudofeces (Pf); skin (S); whole tissues (W).

c) 2,3,7,8-Tetrachlorodibenzo-*p*-dioxins (TCDD); benzo(a)pyrene [B(a)P]; catalase (CAT); differentially expressed genes (DEGs); glutathione (GSH); glutathione disulfide (GSSG); glutathione peroxidase (GPx); glutathione s-transferase (GST); nanoparticles (NPs); nitric oxide (NO); superoxide dismutase (SOD); thioredoxin reductase (TrxR) toll-like receptor (TLR).

fluoranthene (32 μg/L) for 72 h [171]. Data showed that more antagonistic effects than synergic or additive effects occur in marine organisms after co-exposure to NMs and other contaminants (Table 2.1). In light of this, further studies are needed on the interactive effects of NMs with other types of NMs and other traditional and emerging contaminants, such as pharmaceuticals and PCPs in phytoplankton, zooplankton, macroalgae, marine vertebrates, and marine mammals, as well as sediment-dwelling species.

2.9 Environmental Risk Assessment (ERA) of NMs

In order to establish if the use of NMs is ecosafe, there is a need to establish guidelines for ERA. ERA for NMs was first based on the same criteria applied for other traditional chemical compounds [181, 182] and consists of three phases: problem formulation, planning and conducting risk assessment, and risk management. Furthermore, NMs are among the chemical compounds selected to be monitored by the European Organization for the Registration, Evaluation, Authorization and Restriction of Chemicals (REACH) of the European Union, which requires a chemical safety assessment for chemicals whose annual production is higher than 10 tonnes. The chemical safety assessment also includes four phases: registration, evaluation, authorization, and restrictions in the use [183]. However, soon it was clear that there was a need to adapt ERA to the specific nano-characteristics of NMs not included in the current ERA framework. Once the areas of uncertainty of NMs are identified, ERA can include hazard identification, dose–response assessment, exposure assessment, and risk characterization [182, 184].

ERA is function of hazard identification and dose exposure assessment. However, there are a lot of limitations and uncertainties mainly related to hazard identification and to define the most sensitive and relevant endpoints for exposure assessment particularly for inputs and environmental fate and behavior [183]. Koelmans [185] suggested that due to the uncertainties of hazard and exposure assessment, ERA of NMs should rely on prognostic approaches mainly due to the difficulty of measuring chemical concentrations and speciation of NMs. Furthermore, *in situ* exposure assessment has to rely on models to predict environmental concentrations (PECs) that are still very deficient because they do not take into account NMs dissolution by-products and their speciation in the aquatic environment [148]. Models used to estimate PECs of conventional contaminants are calibrated and validated by field and/or laboratory observations. As mentioned before, NMs may have different size, shape, and surface coatings that will affect NM behavior along with environmental characteristics (pH, NOM, dissolved oxygen, and ionic strength) [148]. However, models only predict exposure, and to determine risk, both hazard identification and exposure toxicological data are needed to consider all NMs transformations [148] and field and mesocosm data are lacking to calibrate and validate these models.

In what concerns hazard identification, a lot of gaps still exist related to the lack of scientific information and to the inadequacy of the results obtained from standard toxicity tests end points on nanotoxic effects, so it is questioned which are most suitable descriptors, and a clear interface between exposure and assessment is difficult to establish. For exposure assessment there is a debate whether at present it is possible to establish reliable dose–response relationships for NMs due to a number of factors. NMs are present in several environmental compartments (water, sediment, and biota) with a suite of forms (individuals or aggregates/agglomerates), different sizes, shape, surface area, functionalization, and coatings that can vary with time and space along with a number of biotic and abiotic factors, such as pH, salinity, exposure to sunlight, flow, and the presence of NOM [33]. Therefore, Notter et al. [186] suggested that the toxicity of metals NMs (based on EC_{50} values) is lower than the dissolved counterparts for Ag, Cu, and Zn (93.8%, 100%, and 81%, respectively). Although there were exceptions for Ag and ZnNPs, where toxicity (1.1% and 2.8%) was less than twofold higher to that of the ionic metal, it was suggested that applying a safety factor of two for these dissolved metals would protect water and sediments against the toxicity of these NMs. Although no official NMs test guidelines exist for aquatic species, similar results point out that in marine filter-feeder mollusks exposed to CuO NMs (31 nm) and QDs (6 nm) in seawater, NMs bioavailability is lower than their soluble counterparts [82, 97, 102]. Although in a more environmental realistic scenario, size and aggregation/agglomeration might be a key factor for NPs bioavailability, but biological effects are different and need to be taken into consideration in ERA. Because biomarkers seem to provide more useful information about the toxicity effects of NMs, there is an urgent need to integrate this information and use it as a relevant ecotoxicological metrics in ERA, although care should be taking in the interpretation of results. Moreover, there is a gap of information on sediments assessment of NMs; therefore, there is an urgent need to establish a relationship between the effects obtained from laboratory experiments to those occurring in the real environment.

Acknowledgments

This work was supported by the Emergemix project (PTDC/BIABMA/30922/2017) funded by the Portuguese Foundation for Science and Technology (FCT) I.P. and by the ERDF through the COMPETE2020 – Programa Operacional Competitividade e Internacionalização. We further thank FCT for the funds attributed to CIMA, University of Algarve UID/00350/2020 and to CBMA, University of Minho UID/BIA/04050/2020 (POCI-01-0145-FEDER- 007569).

References

1 Ju-Nam, Y. and Lead, J.R. (2008). Manufactured nanoparticles: an overview of their chemistry, interactions and potential environmental problems. *Sci. Total Environ.* 400: 396–414.

2 European Union (2011). Commission recommendation on the definition of nanomaterial of 18 October 2011 (2011/696/EU). *Off. J. Eur. Union* 275: 38–40.

3 Boxall, A.B.A., Chaudhry, Q., Sinclair, C. et al. (2008). Current and future predicted environmental exposure to engineered nanoparticles. In: *Report by Central Science Laboratory for Department for Environment, Food and Rural Affairs*, 1–89. UK: Her Majesty's, Government.

4 Stone, V., Nowack, B., Baun, A. et al. (2010). Nanomaterials for environmental studies: classification, reference material issues, and strategies for physico-chemical characterisation. *Sci. Total Environ.* 408: 1745–1754.

5 Sinha, R., Kim, G.J., Nie, S., and Shin, D.M. (2006). Nanotechnology in cancer therapeutics: bioconjugated nanoparticles for drug delivery. *Mol. Cancer Ther.* 5: 1909–1917.

6 Cormode, D.P., Skajaa, T., Fayad, Z.A., and Mulder, W.J.M. (2009). Nanotechnology in medical imaging: probe design and applications. *Arterioscler. Thromb. Vasc. Biol.* 29: 992–1000.

7 Karn, B., Kuiken, T., and Otto, M. (2009). Nanotechnology and in situ remediation: a review of the benefits and potential risks. *Environ. Health Perspect.* 117: 1813–1831.

8 Fabrega, J., Luoma, S.M., Tyler, C.R. et al. (2011). Silver nanoparticles: behaviour and effects in the aquatic environment. *Environ. Int.* 37: 517–531.

9 Dineshram, R., Subasri, R., Somaraju, K.R.C. et al. (2009). Biofouling studies on nanoparticles-based metal oxide coatings on glass coupons exposed to marine environment. *Colloids Surf., B* 74: 75–83.

10 Upadhyayula, V.K.K. and Gaghamshetty, V. (2010). Appreciating the role of carbon nanotube composites in preventing biofouling and promoting biofilms on material surfaces in environmental engineering. *Biotechnol. Adv.* 28: 802–816.

11 Corsi, I., Cherr, G., Lenihan, H., and Labille, J. (2014). Common strategies and technologies for the ecosafety assessment and design of nanomaterials entering the marine environment. *ACS Nano* 8: 9694–9709.

12 Nowak, B. and Bucheli, T.D. (2007). Occurrence, behaviour and effects of nanoparticles in the environment. *Environ. Pollut.* 150: 5–22.

13 Rocco, L., Santonastaso, M., Nigro, M. et al. (2015). Genomic and chromosomal damage in the marine mussel *Mytilus galloprovincialis*: effects of the combined exposure to titanium dioxide nanoparticles and cadmium chloride. *Mar. Environ. Res.* 111: 144–148.

14 Dowling, A.P. (2004). Development of nanotechnologies. *Mater. Today* 7: 30–35.

15 Peralta-Videa, J.R., Zhao, L., Lopez-Moreno, M.L. et al. (2011). Nanomaterials and the environment: a review for the biennium 2008–2010. *J. Hazard. Mater.* 186: 1–15.

16 Handy, R.D., Cornelis, G., Fernandes, T. et al. (2012). Ecotoxicity test methods for engineered nanomaterials: practical experiences and recommendations from the bench. *Environ. Toxicol. Chem.* 31: 15–31.

17 Hassellöv, M., Readman, J.W., Ranville, J.F., and Tiede, K. (2008). Nanoparticle analysis and characterization methodologies in environmental risk assessment of engineered nanoparticles. *Ecotoxicology* 17: 344–361.

18 Klaine, S.J., Alvarez, P.J.J., Batley, G.E. et al. (2008). Nanomaterials in the environment: behavior, fate, bioavailability and effects. *Environ. Toxicol. Chem.* 27 (9): 1825–1851.

19 Maynard, A. (2006). *Nanotechnology: A Research Strategy for Addressing Risk*, Project on Emerging Nanotechnologies. Publication 3. Washington, DC: Woodrow Wilson International Centre for Scholars and PEW Charitable Trusts.

20 Oberdörster, G., Oberdörster, E., and Oberdörster, J. (2005). Nanotoxicology: an emerging discipline evolving from studies of ultrafine particles. *Environ. Health Perspect.* 113 (7): 823–839.

21 Xia, T., Li, N., and Nel, A.E. (2009). Potential health impact of nanoparticles. *Ann. Rev. Public Health* 30: 137–150.

22 Kaegi, R., Ulrich, A., Sinnet, B. et al. (2008). Synthetic TiO_2 nanoparticles emission from exterior facades into the aquatic environment. *Environ. Pollut.* 156: 233–239.

23 Kiser, M.A., Westerhoff, P., Benn, T. et al. (2009). Titanium nanomaterial removal and release from wastewater treatment plants. *Environ. Sci. Technol.* 43: 6757–6763.

24 Klaine, S.J., Koelmans, A.A., Horne, N. et al. (2012). Paradigms to assess the environmental impact of manufactured nanomaterials. *Environ. Toxicol. Chem.* 31 (1): 3–14.

25 Kiser, M.A., Ryu, H., Jang, H. et al. (2010). Biosorption of nanoparticles to heterotrophic wastewater biomass. *Water Res.* 44: 4105–4114.

26 Westerhoff, P., Song, G., Hristovski, K., and Kiser, M.A. (2011). Occurrence and removal of titanium at full scale wastewater treatment plants: implications for TiO_2 nanomaterials. *J. Environ. Monit.* 13: 1195–1203.

27 Chalew, T.E.A., Ajmani, G.S., Huang, H., and Schwab, K.J. (2013). Evaluating nanoparticle breakthrough during drinking water treatment. *Environ. Health Perspect.* 121: 1161–1166.

28 Moore, M.N. (2006). Do nanoparticles present ecotoxicological risks for the health of the aquatic environment? *Environ. Int.* 32: 967–976.

29 Simeonidis, K., Mourdikoudis, S., Kaprara, E. et al. (2016). Inorganic engineered nanoparticles in drinking water treatment: a critical review. *Environ. Sci. Water Res. Technol.* 2: 43–70.

30 Brar, S.K., Verma, M., Tyagi, R.D., and Surampalli, R.Y. (2010). Engineered nanoparticles in wastewater and wastewater sludge – Evidence and impacts. *Waste Manage. (Oxford)* 30: 504–520.

31 Kim, B., Park, C.S., Murayama, M., and Hochella, M.F. (2010). Discovery and characterization of silver sulfide nanoparticles in final sewage sludge products. *Environ. Sci. Technol.* 44: 7509–7514.

32 Nowack, B. and Bucheli, T.D. (2007). Occurrence, behavior and effects of nanoparticles in the environment. *Environ. Pollut.* 150: 5–22.

33 Nowack, B., Ranville, J.F., Diamond, S. et al. (2012). Potential scenarios for nanomaterial release and subsequent alteration in the environment. *Environ. Toxicol. Chem.* 31: 50–59.

34 Li, L., Hartmann, G., Döblinger, M., and Schuster, M. (2013). Quantification of nanoscale silver particles removal and release from municipal wastewater treatment plants in Germany. *Environ. Sci. Technol.* 47: 7317–7323.

35 Ma, R., Levard, C., Judy, J.D. et al. (2014). Fate of zinc oxide and silver nanoparticles in a pilot wastewater treatment plant and in processed biosolids. *Environ. Sci. Technol.* 48: 104–112.

36 Kaegi, R., Voegelin, A., Sinnet, B. et al. (2011). Behavior of metallic silver manoparticles in a pilot wastewater treatment plant. *Environ. Sci. Technol.* 45: 3902–3908.

37 Nanotechnology EPA's Group (2007). Nanotechnology White Paper. EPA 100/B-07/001. EPA's Science Policy Council. U.S. Environmental Protection Agency (EPA), USA.

38 Brunetti, G., Donner, E., Laera, G. et al. (2015). Fate of zinc and silver engineered nanoparticles in sewerage networks. *Water Res.* 77: 72–84.

39 Kaegi, R., Voegelin, A., Ort, C. et al. (2013). Fate and transformation of silver nanoparticles in urban wastewater systems. *Water Res.* 47: 3866–3877.

40 Hou, L., Li, K., Ding, Y. et al. (2012). Removal of silver nanoparticles in simulated wastewater treatment processes and its impact on COD and NH_4 reduction. *Chemosphere* 87: 248–252.

41 Jarvie, H.P., Al-Obaidi, H., King, S.M. et al. (2009). Fate of silica nanoparticles in simulated primary wastewater treatment. *Environ. Sci. Technol.* 43: 8622–8628.

42 King, S.M., Jarvie, H.P., Bowes, M.J. et al. (2015). Exploring controls on the fate of PVP-capped silver nanoparticles in primary wastewater treatment. *Environ. Sci. Nanotechnol.* 2: 177–190.

43 Limbach, L., Bereiter, R., Muller, E. et al. (2008). Removal of oxide nanoparticles in a model wastewater treatment plant: Influence of agglomeration and surfactants on clearing efficiency. *Environ. Sci. Technol.* 42: 5828–5833.

44 Wang, Y., Westerhoff, P., and Hristovski, K.D. (2012). Fate and biological effects of silver, titanium dioxide, and C60 (fullerene) nanomaterials during simulated wastewater treatment processes. *J. Hazard. Mater.* 201-202: 16–22.

45 Gómez-Rivera, F., Field, J.A., Brown, D., and Sierra-Alvarez, R. (2012). Fate of cerium dioxide (CeO_2) nanoparticles in municipal wastewater during activated sludge treatment. *Bioresour. Technol.* 108: 300–304.

46 Thalmann, B., Voegelin, A., Morgenroth, E., and Kaegi, R. (2016). Effect of humic acid on the kinetics of silver nanoparticle sulfidation. *Environ. Sci. Nano* 3: 203–212.

47 Zhang, W., Stack, A.G., and Cheng, Y. (2011). Interaction force measurement between *E. coli* cells and nanoparticles immobilized surfaces by using AFM. *Colloids Surf., B* 82: 316–324.

48 Chen, J., Tang, Y.-Q., Li, Y. et al. (2014). Impacts of different nanoparticles on functional bacterial community in activated sludge. *Chemosphere* 104: 141–148.

49 Chen, Y., Su, Y., Zheng, X. et al. (2012). Alumina nanoparticles-induced effects on wastewater nitrogen and phosphorus removal after short-term and long-term exposure. *Water Res.* 46: 4379–4386.

50 Puay, N.-Q., Qiu, G., and Ting, Y.-P. (2015). Effect of Zinc oxide nanoparticles on biological wastewater treatment in a sequencing batch reactor. *J. Clean. Product* 88: 139–145.

51 Wu, D., Shen, Y., Ding, A. et al. (2013). Effects of nanoscale zero-valent iron particles on biological nitrogen and phosphorus removal and microorganisms in activated sludge. *J. Hazard. Mater.* 262: 649–655.

52 Lombi, E., Donner, E., Tavakkoli, E. et al. (2012). Fate of zinc oxide nanoparticles during anaerobic digestion of wastewater and post-treatment processing of sewage sludge. *Environ. Sci. Technol.* 46: 9089–9096.

53 Buffle, J., Wilkinson, K.J., and van Leeuwen, H.P. (2009). Chemodynamics and bioavailability in natural waters. *Environ. Sci. Technol.* 43: 7170–7174.

54 Morel, F.M.M. and Hering, J. (1993). *Principles and Applications of Aquatic Chemistry*, 1ste. New York: Wiley-Interscience.

55 Di Toro, D.M., Allen, H.E., Bergman, H.L. et al. (2001). Biotic ligand model of the acute toxicity of metals. 1. Technical basis. *Environ. Toxicol. Chem.* 20: 2383–2396.

56 Odzak, N., Kistler, D., Behra, R., and Sigg, L. (2014). Dissolution of metal and metal oxide nanoparticles in aqueous media. *Environ. Pollut.* 191: 132–138.

57 Domingos, R.F., Franco, C., and Pinheiro, J.P. (2015). The role of charged polymer coatings of nanoparticles on the speciation and fate of metal ions in the environment. *Environ. Sci. Pollut. Res.* 22: 2900–2906.

58 Schmidt, J. and Vogelsberger, W. (2006). Dissolution kinetics of titanium dioxide nanoparticles: the observation of an unusual kinetic size effect. *J. Phys. Chem. B* 110: 3955–3963.

59 Misra, S.K., Dybowska, A., Berhanu, D. et al. (2012). Isotopically modified nanoparticles for enhanced detection in bioaccumulation studies. *Environ. Sci. Technol.* 46: 1216–1222.

60 Ma, R., Levard, C., Marinakos, S.M. et al. (2012). Size-controlled dissolution of organic-coated silver nanoparticles. *Environ. Sci. Technol.* 46: 752–759.

61 Degueldre, C., Favarger, P.Y., and Wold, S. (2006). Gold colloid analysis by inductively coupled plasma-mass spectrometry in a single particle mode. *Anal. Chim. Acta* 555: 263–268.

62 Bustos, A.R.M., Petersen, E.J., Possolo, A., and Winchester, M.R. (2015). Post hoc interlaboratory comparison of single particle ICP-MS size measurements of NIST gold nanoparticle reference materials. *Anal. Chem.* 87: 8809–8817.

63 Domingos, R.F., Franco, C., and Pinheiro, J.P. (2013). Stability of core/shell quantum dots-role of pH and small organic ligands. *Environ. Sci. Pollut. Res.* 20: 4872–4880.

64 Pesavento, M., Alberti, G., and Biesuz, R. (2009). Analytical methods for determination of free metal ion concentration, labile species fraction and metal complexation capacity of environmental waters: a review. *Anal. Chim. Acta* 631: 129–141.

65 Temminghoff, E.J.M., Plette, A.C.C., Van Eck, R., and Van Riemsdijk, W.H. (2000). Determination of the chemical speciation of trace metals in aqueous systems by the Wageningen Donnan membrane technique. *Anal. Chim. Acta* 417: 149–157.

66 Kalis, E.J.J., Weng, L., Temminghoff, E.J.M., and van Riemsdijk, W.H. (2007). Measuring free metal ion concentrations in multicomponent solutions using the Donnan membrane technique. *Anal. Chem.* 79: 1555–1563.

67 Galceran, J., Companys, E., Puy, J. et al. (2004). AGNES: a new electroanalytical technique for measuring free metal ion concentration. *J. Electroanal. Chem.* 566: 95–109.

68 Fest, E.P.M.J., Temminghoff, E.J.M., Comans, R.A.J., and van Riemsdijk, W.H. (2008). Partitioning of organic matter and heavy metals in a sandy soil: effects of extracting solution, solid to liquid ratio and pH. *Geoderma* 146: 66–74.

69 Duffner, A., Weng, L., Hoffland, E., and van der Zee, S.E.A.T.M. (2014). Multi-surface modeling to predict free zinc ion concentrations in low-zinc soils. *Environ. Sci. Technol.* 48: 5700–5708.

70 Kalis, E.J.J., Weng, L.P., Dousma, F. et al. (2006). Measuring free metal ion concentrations in situ in natural waters using the Donnan membrane technique. *Environ. Sci. Technol.* 40: 955–961.

71 Chito, D., Weng, L., Galceran, J. et al. (2012). Determination of free Zn^{2+} concentration in synthetic and natural samples with AGNES (Absence of Gradients and Nernstian Equilibrium Stripping) and DMT (Donnan Membrane Technique). *Sci. Total Environ.* 421: 238–244.

72 Vega, F.A. and Weng, L. (2013). Speciation of heavy metals in River Rhine. *Water Res.* 47: 363–372.

73 Weng, L.P., Lexmond, T.M., Wolthoorn, A. et al. (2003). Phytotoxicity and bioavailability of nickel: chemical speciation and bioaccumulation. *Environ. Toxicol. Chem.* 22: 2180–2187.

74 Kalis, E.J.J., Temminghoff, E.J.M., Visser, A., and van Riemsdijk, W.H. (2007). Metal uptake by Lolium perenne in contaminated soils using a four-step approach. *Environ. Toxicol. Chem.* 26: 335–345.

75 Semenzin, E., Temminghoff, E.J.M., and Marcomini, A. (2007). Improving ecological risk assessment by including bioavailability into species sensitivity distributions: an example for plants exposed to nickel in soil. *Environ. Pollut.* 148: 642–647.

76 Gelabert, A., Sivry, Y., Ferrari, R. et al. (2014). Uncoated and coated Zno nanoparticle life cycle in synthetic seawater. *Environ. Toxicol. Chem.* 33: 341–349.

77 Gorham, J.M.R., MacCuspie, I., Klein, K.L. et al. (2012). UV-induced photochemical transformations of citrate-capped silver nanoparticle suspensions. *J. Nanopart. Res.* 14: 1139.

78 Rocha, L.S., Galceran, J., Puy, J., and Pinheiro, J.P. (2015). Determination of the free metal ion concentration using AGNES implemented with environmentally friendly bismuth film electrodes. *Anal. Chem.* 87: 6071–6078.

79 David, C.A., Galceran, J., Rey-Castro, C. et al. (2012). Dissolution kinetics and solubility of ZnO nanoparticles followed by AGNES. *J. Phys. Chem. C* 116: 11758–11767.

80 Adam, N., Schmitt, C., Galceran, J. et al. (2014). The chronic toxicity of ZnO nanoparticles and $ZnCl_2$ to *Daphnia magna* and the use of different methods to assess nanoparticle aggregation and dissolution. *Nanotoxicology* 8: 709–717.

81 Vale, G., Franco, C., Diniz, M.S. et al. (2014). Bioavailability of cadmium and biochemical responses on the freshwater bivalve *Corbicula fluminea* - the role of TiO_2 nanoparticles. *Ecotoxicol. Environ. Saf.* 109: 161–168.

82 Rocha, T.L., Gomes, T., Pinheiro, J.P. et al. (2015). Toxicokinetics and tissue distribution of cadmium-based quantum dots in the marine mussel *Mytilus galloprovincialis*. *Environ. Pollut.* 204: 207–214.

83 Aiken, G.R., Hsu-Kim, H., and Ryan, J.N. (2011). Influence of dissolved organic matter on the environmental fate of metals, nanoparticles, and colloids. *Environ. Sci. Technol.* 45: 3196–3201.

84 Philippe, A. and Schaumann, G.E. (2014). Interactions of dissolved organic matter with natural and engineered inorganic colloids: a review. *Environ. Sci. Technol.* 48: 8946–8962.

85 Peijnenburg, W.J.G.M., Baalousha, M., Chen, J. et al. (2015). A review of the properties and processes determining the fate of engineered nanomaterials in the aquatic environment. *Crit. Rev. Env. Sci. Technol.* 45: 2084–2134.

86 Domingos, R.F., Baalousha, M.A., Ju-Nam, Y. et al. (2009). Characterizing manufactured nanoparticles in the environment: Multimethod determination of particle sizes. *Environ. Sci. Technol.* 43: 7277–7284.

87 Jiang, J., Oberdorster, G., and Biswas, P. (2009). Characterization of size, surface charge, and agglomeration state of nanoparticle dispersions for toxicological studies. *J. Nanopart. Res.* 11: 77–89.

88 Diaz, L., Peyrot, C., and Wilkinson, K.J. (2015). Characterization of polymeric nanomaterials using analytical ultracentrifugation. *Environ. Sci. Technol.* 49: 7302–7309.

89 Domingos, R.F. and Pinheiro, J.P. (2014). Challenges in designing environmentally relevant toxicological experiments. In: *Implications of the Use of Nanomaterials for Environmental Protection* (ed. B.I. Kharisov, O.V. Kharissova and H.V.R. Dias), 535–548. John Wiley & Sons Inc.

90 Klanjscek, T., Nisbet, R.M., Priester, J.H., and Holden, P.A. (2013). Dynamic energy budget approach to modeling mechanisms of CdSe quantum dot toxicity. *Ecotoxicology* 22 (2): 319–330.

91 Priester, J.H., Stoimenov, P.K., Mielke, R.E. et al. (2009). Effects of soluble cadmium salts versus CdSe quantum dots on the growth of planktonic *Pseudomonas aeruginosa*. *Environ. Sci. Technol.* 43: 2589–2594.

92 Muller, E.B., Hanna, S.K., Lenihan, H. et al. (2014). Impact of engineered zinc oxide nanoparticles on the energy budgets of *Mytilus galloprovincialis*. *J. Sea Res.* 94: 29–36.

93 Oberdörster, E. (2004). Manufactured nanomaterials (fullerenes, C60) induce oxidative stress in the brain of juvenile largemouth bass. *Environ. Health Perspect.* 112: 1058–1062.

94 Wang, X., Qu, R., Liu, J. et al. (2016). Effect of different carbon nanotubes on cadmium toxicity to *Daphnia magna*: the role of catalyst impurities and adsorption capacity. *Environ. Pollut.* 208: 732–738.

95 Buffet, P.E., Tankoua, O.F., Pan, J.F. et al. (2011). Behavioural and biochemical responses of two marine invertebrates *Scrobicularia plana* and *Hediste diversicolor* to copper oxide nanoparticles. *Chemosphere* 84: 166–174.

96 Gomes, T., Pereira, C.G., Cardoso, C. et al. (2012). Accumulation and toxicity of copper oxide nanoparticles in the digestive gland of *Mytilus galloprovincialis*. *Aquat. Toxicol.* 118–119: 72–79.

97 Gomes, T., Pereira, C.G., Cardoso, C., and amd Bebianno, M.J. (2013). Differential protein expression in mussels *Mytilus galloprovincialis* exposed to nano and ionic Ag. *Aquat. Toxicol.* 136–137: 79–90.

98 Heinlaan, M., Ivask, A., Blinova, I. et al. (2008). Toxicity of nanosized and bulk ZnO, CuO and TiO_2 to bacteria *Vibrio fischeri* and crustaceans *Daphnia magna* and *Thamnocephalus platyurus*. *Chemosphere* 71: 1308–1316.

99 Ringwood, A.H., McCarthy, M., Levi-Polyachenko, N., and Carroll, D.L. (2010). Nanoparticles in natural aquatic systems and cellular toxicity. *Comp. Biochem. Physiol. A: Mol. Integr. Physiol.* 157: S53.

100 Pradhan, A., Seena, S., Pascoal, C., and Cássio, F. (2011). Can metal nanoparticles be a threat to microbial decomposers of plant litter in streams? *Microb. Ecol.* 62: 58–68.

101 Pradhan, A., Seena, S., Pascoal, C., and Cássio, F. (2012). Copper oxide nanoparticles can induce toxicity to the freshwater shredder *Allogamus ligonifer*. *Chemosphere* 89: 1142–1150.

102 Rocha, T.L., Gomes, T., Cardoso, C. et al. (2014). Immunocytotoxicity, cytogenotoxicity and genotoxicity of cadmium-based quantum dots in the marine mussel *Mytilus galloprovincialis*. *Mar. Environ. Res.* 101C: 29–37.

103 Yan, L., Feng, M., Liu, J. et al. (2016). Antioxidant defences and histological changes in *Carassius auratus* after combined exposure to zinc and three multi-walled carbon nanotubes. *Ecotoxicol. Environ. Saf.* 125: 61–71.

104 Priya, K.K., Ramesh, M., Saravanan, M., and Ponpandian, N. (2015). Ecological risk assessment of silicon dioxide nanoparticles in a fresh-water fish *Labeo rohita*: hematology, ionoregulation and gill Na^+/K^+-ATPase activity. *Ecotoxicol. Environ. Saf.* 120: 295–302.

105 Cardinale, B.J., Bier, R., and Kwan, C. (2012). Effects of TiO_2 nanoparticles on the growth and metabolism of three species of freshwater algae. *J. Nanopart. Res.* 14: 913–921.

106 Hall, S., Bradley, T., Moore, J.T. et al. (2009). Acute and chronic toxicity of nano-scale TiO_2 particles to freshwater fish, cladocerans, and green algae, and effects of organic and inorganic substrate on TiO_2 toxicity. *Nanotoxicology* 3: 91–97.

107 Dalai, S., Pakrashi, S., Chandrasekaran, N., and Mukherjee, A. (2013). Acute toxicity of TiO_2 manoparticles to *Ceriodaphnia dubia* under visible light and dark conditions in a freshwater system. *PLoS One* 8 (4): e62970.

108 Franklin, N.M., Rogers, N.J., Apte, S.C. et al. (2007). Comparative toxicity of nanoparticulate ZnO, bulk ZnO, and $ZnCl_2$ to a freshwater microalga

(*Pseudokirchneriella subcapitata*): the importance of particle solubility. *Environ. Sci. Technol.* 41: 8484–8490.

109 Aruoja, V., Dubourguier, H.-C., Kasemets, K., and Kahru, A. (2009). Toxicity of nanoparticles of CuO, ZnO and TiO_2 to microalgae *Pseudokirchneriella subcapitata*. *Sci. Total Environ.* 407: 1461–1468.

110 Zhu, X., Wang, J., Zhang, X. et al. (2010). Trophic transfer of TiO(2) nanoparticles from *Daphnia* to zebrafish in a simplified freshwater food chain. *Chemosphere* 79: 928–933.

111 Krysanov, E.Y., Demidova, T.B., Pel'gunova, L.A. et al. (2009). Effect of hydrated tin dioxide ($SnO_2 \cdot xH_2O$) nanoparticles on guppy (*Poecilia reticulate* Peters, 1860). *Dokl. Biol. Sci.* 426: 288–289.

112 Federici, G., Shaw, B.J., and Handy, R.D. (2007). Toxicity of titanium dioxide nanoparticles to rainbow trout (*Oncorhynchus mykiss*): gill injury, oxidative stress, and other physiological effects. *Aquat. Toxicol.* 84: 415–430.

113 Ramsden, C.S., Smith, T.J., Shaw, B.J., and Handy, R.D. (2009). Dietary exposure to titanium dioxide nanoparticles in rainbow trout (*Oncorhynchus mykiss*): no effect on growth, but subtle biochemical disturbances in the brain. *Ecotoxicology* 18: 939–951.

114 Battin, T.J., Kammer, F.V.D., Weilhartner, A. et al. (2009). Nanostructured TiO_2: transport behavior and effects on aquatic microbial communities under environmental conditions. *Environ. Sci. Technol.* 43: 8098–8104.

115 Lee, S.-W., Kim, S.-M., and Jinhee Choi, J. (2009). Genotoxicity and ecotoxicity assays using the freshwater crustacean *Daphnia magna* and the larva of the aquatic midge *Chironomus riparius* to screen the ecological risks of nanoparticle exposure. *Environ. Toxicol. Pharmacol.* 28: 86–91.

116 Blinova, I., Ivask, A., Heinlaan, M. et al. (2010). Ecotoxicity of nanoparticles of CuO and ZnO in natural water. *Environ. Pollut.* 158: 41–47.

117 Mortimer, M., Kasemets, K., and Kahru, A. (2010). Toxicity of ZnO and CuO nanoparticles to ciliated protozoa *Tetrahymena thermophila*. *Forensic Toxicol.* 269: 182–189.

118 Kasemets, K., Ivask, A., Dubourguier, H.C., and Kahru, A. (2009). Toxicity of nanoparticles of ZnO, CuO and TiO_2 to yeast *Saccharomyces cerevisiae*. *Toxicol. in Vitro* 23: 1116–1122.

119 Ramskov, T., Croteau, M.-N., Forbes, V.E., and Selck, H. (2015). Biokinetics of different-shaped copper oxide nanoparticles in the freshwater gastropod, *Potamopyrgus antipodarum*. *Aquat. Toxicol.* 163: 71–80.

120 Gajjar, P., Pettee, B., Britt, D.W. et al. (2009). Antimicrobial activities of commercial nanoparticles against an environmental soil microbe, *Pseudomonas putida* KT2440. *J. Biol. Eng.* 3: 3–13.

121 Yu, S.-J., Yin, Y.-G., Jing-bo Chao, J.-B. et al. (2010). Highly dynamic PVP-coated silver nanoparticles in aquatic environments: chemical and morphology change induced by oxidation of Ag0 and reduction of Ag^+ toxicity and bioaccumulation of TiO_2 nanoparticle aggregates in *Daphnia magna*. *Chemosphere* 78: 209–215.

122 Heinlaan, M., Kahru, A., Kasemets, K. et al. (2011). Changes in the *Daphnia magna* midgut upon ingestion of copper oxide nanoparticles: a transmission electron microscopy study. *Water Res.* 45: 179–190.

123 Pradhan, A., Geraldes, P., Seena, S. et al. (2016). Humic acid can mitigate the toxicity of small copper oxide nanoparticles to microbial decomposers and leaf decomposition in streams. *Freshwater Biol.* 61 (12): 2197–2210.

124 Pradhan, A., Geraldes, P., Seena, S. et al. (2015). Natural organic matter alters size-dependent effects of nanoCuO on the feeding behaviour of freshwater shredders. *Sci. Total Environ.* 535: 94–101.

125 Griffitt, R.J., Weil, R., Hyndman, K.A. et al. (2007). Exposure to copper nanoparticles causes gill injury and acute lethality in Zebrafish (*Danio rerio*). *Environ. Sci. Technol.* 41: 8178–8186.

126 Pradhan, A., Seena, S., Dobritzsch, D. et al. (2014). Physiological responses to nanoCuO in fungi from non-polluted and metal-polluted streams. *Sci. Total Environ.* 466–467: 556–563.

127 Croteau, M.-N., Superb, M.K., Luoma, S.N., and Valsami-Jones, E. (2014). Bioaccumulation and toxicity of CuO nanoparticles by a freshwater invertebrate after waterborne and dietborne exposures. *Environ. Sci. Technol.* 48: 10929–10937.

128 Burchardt, A.D., Carvalho, R.N., Valente, A. et al. (2012). Effects of silver nanoparticles in diatom *Thalassiosira pseudonana* and cyanobacterium *Synechococcus* sp. *Environ. Sci. Technol.* 46: 11336–11344.

129 Asghari, S., Johari, S.A., Lee, J.H. et al. (2012). Toxicity of various silver nanoparticles compared to silver ions in *Daphnia magna*. *J. Nanobiotechnol.* 10: 14–25.

130 Oukarroum, A., Bras, S., Perreault, F., and Popovic, R. (2012). Inhibitory effects of silver nanoparticles in two green algae, *Chlorella vulgaris* and *Dunaliella tertiolecta*. *Ecotoxicol. Environ. Saf.* 78: 80–85.

131 Gagné, F., Auclair, J., Fortier, M. et al. (2013). Bioavailability and immunotoxicity of silver nanoparticles to the freshwater mussel *Elliptio complanata*. *J. Toxicol. Environ. Health Part A* 76: 767–777.

132 Rainville, L.-C., Carolan, D., Varela, A.C. et al. (2014). Proteomic evaluation of citrate-coated silver nanoparticles toxicity in *Daphnia magna*. *Analyst* 139: 1678–1686.

133 Lee, K.J., Nallathamby, P.D., Browning, L.M. et al. (2007). *In vivo* imaging of transport and biocompatibility of single silver nanoparticles in early development of zebrafish embryos. *ACS Nano* 1: 133–143.

134 Asharani, P.V., LianWu, Y., Gong, Z., and Valiyaveettil, S. (2008). Toxicity of silver nanoparticles in zebrafish models. *Nanotechnology* 19: 255102–255110.

135 Chae, Y.J., Pham, C.H., Lee, J. et al. (2009). Evaluation of the toxic impact of silver nanoparticles on Japanese medaka (*Oryzias latipes*). *Aquat. Toxicol.* 94: 320–327.

136 Gil-Allué, C., Kristin Schirmer, K., Tlili, A. et al. (2015). Silver nanoparticle effects on stream periphyton during short-term exposures. *Environ. Sci. Technol.* 49: 1165–1172.

137 Norman, B.C., Xenopoulos, M.A., Braun, D., and Frost, P.C. (2015). Phosphorus availability alters the effects of silver nanoparticles on periphyton growth and stoichiometry. *PLoS One* 10 (6): e0129328.

138 Kroll, A., Behra, R., Kaegi, R., and Sigg, L. (2014). Extracellular polymeric substances (EPS) of freshwater biofilms stabilize and modify CeO_2 and Ag nanoparticles. *PLoS One* 9 (10): e110709.

139 Ozaki, A., Adams, E., Binh, C.T.T. et al. (2016). One-time addition of nano-TiO_2 triggers short-term responses in benthic bacterial communities in artificial streams. *Microb. Ecol.* 71: 266–275.

140 Jomini, S., Clivot, H., Bauda, P., and Pagnout, C. (2015). Impact of manufactured TiO_2 nanoparticles on planktonic and sessile bacterial communities. *Environ. Pollut.* 202: 196–204.

141 Bradford, A., Handy, R.D., Redman, J.W. et al. (2009). Impact of silver nanoparticle contamination on the genetic diversity of natural bacterial assemblages in estuarine sediments. *Environ. Sci. Technol.* 43: 4530–4536.

142 Tlili, A., Cornut, J., Behra, R. et al. (2016). Harmful effects of silver nanoparticles on a complex detrital model system. *Nanotoxicology* 26: 1–8.

143 Misra, S.K., Dybowska, A., Berhanu, D. et al. (2012). The complexity of nanoparticle dissolution and its importance in nanotoxicological studies. *Sci. Total Environ.* 438: 225–232.

144 Lowry, G.V., Gregory, K.B., Apte, S.C., and Lead, J.R. (2012). Transformations of nanomaterials in the environment. *Environ. Sci. Technol.* 46: 6893–6899.

145 Rocha, T.L., Gomes, T., Sousa, V.S. et al. (2015). Ecotoxicological impact of engineered nanomaterials in bivalve molluscs: an overview. *Mar. Environ. Res.* 111: 74–88.

146 Sousa, V.S. and Teixeira, M.R. (2013). Aggregation kinetics and surface charge of CuO nanoparticles: the influence of pH, ionic strength and humic acids. *Environ. Chem.* 10: 313–322.

147 Stolpe, B. and Hassellov, M. (2007). Changes in size distribution of fresh water nanoscale colloidal matter and associated elements on mixing in seawater. *Geochim. Cosmochim. Acta* 71: 3292–3301.

148 Dale, A.L., Casman, E., Lowry, G.V. et al. (2015). Modeling nanomaterial environmental fate in aquatic systems. *Environ. Sci. Technol.* 49: 2587–2593.

149 Ipe, B.I., Lehnig, M., and Niemeyer, C.M. (2005). On the generation of free radical species from quantum dots. *Small* 1: 706–709.

150 Matranga, V. and Corsi, I. (2012). Toxic effects of engineered nanoparticles in the marine environment: model organisms and molecular approaches. *Mar. Environ. Res.* 76: 32–40.

151 Lapresta-Fernández, A., Fernández, A., and Blasco, J. (2012). Nanoecotoxicity effects of engineered silver and gold nanoparticles in aquatic organisms. *Trends Anal. Chem.* 32: 40–59.

152 Baker, T.J., Tyler, C.R., and Galloway, T.S. (2014). Impacts of metal and metal oxide nanoparticles on marine organisms. *Environ. Pollut.* 186: 257–271.

153 Canesi, L. and Corsi, I. (2016). Effects of nanomaterials on marine invertebrates. *Sci. Total Environ.* 1–8. https://doi.org/10.1016/j.scitotenv.2016.01.085.

154 Paterson, G., Ataria, J.M., Hoque, M.E. et al. (2011). The toxicity of titanium dioxide nanopowder to early life stages of the Japanese medaka (*Oryzias latipes*). *Chemosphere* 82: 1002–1009.

155 Della Torre, C., Buonocore, F., Frenzilli, G. et al. (2015). Influence of titanium dioxide nanoparticles on 2,3,7,8-tetrachlorodibenzo-p-dioxin bioconcentration and toxicity in the marine fish European sea bass (*Dicentrarchus labrax*). *Environ. Pollut. C* 196: 185–193.

156 Bernardeschi, M., Guidi, P., Scarcelli, V. et al. (2010). Genotoxic potential of TiO_2 on bottlenose dolphin leukocytes. *Anal. Bioanal.Chem.* 396: 619–623.

157 Frenzilli, G., Bernardeschi, M., Guidi, P. et al. (2014). Effects of *in vitro* exposure to titanium dioxide on DNA integrity of bottlenose (*Tursiops truncatus*) dolphin fibroblasts and leukocytes. *Mar. Environ. Res.* 100: 68–73.

158 Gambardella, C., Morgana, S., Bari, G.D. et al. (2015). Multidisciplinary screening of toxicity induced by silica nanoparticles during sea urchin development. *Chemosphere* 139: 486–495.

159 Bour, A., Mouchet, F., Verneul, L. et al. (2015). Toxicity of CeO_2 nanoparticles at different trophic levels – effects on diatoms, chironomids and amphibians. *Chemosphere* 120: 230–236.

160 McCarthy, M.P., Carroll, D.L., and Ringwood, A.H. (2013). Tissue specific responses of oysters, *Crassostrea virginica*, to silver nanoparticles. *Aquat. Toxicol.* 138-139: 123–128.

161 Gomes, T., Pereira, C.G., Cardoso, C. et al. (2014). Effects of silver nanoparticles exposure in the mussel *Mytilus galloprovincialis*. *Mar. Environ. Res.* 101: 208–214.

162 Canesi, L., Ciacci, C., Vallotto, D. et al. (2010). *In vitro* effects of suspensions of selected nanoparticles (C_{60} fullerene, TiO_2, SiO_2) on *Mytilus* hemocytes. *Aquat. Toxicol.* 96: 151–158.

163 Gomes, T., Pinheiro, J.P., Cancio, I. et al. (2011). Effects of copper nanoparticles exposure in the mussel *Mytilus galloprovincialis*. *Environ. Sci. Technol.* 45: 9356–9362.

164 Morelli, E., Salvadori, E., Bizzarri, R. et al. (2013). Interaction of CdSe/ZnS quantum dots with the marine diatom *Phaeodactylum tricornutum* and the green alga *Dunaliella tertiolecta*: a biophysical approach. *Biophys. Chem.* 182: 4–10.

165 Saez, G., Aye, M., De Meo, M. et al. (2015). Genotoxic and oxidative responses in coelomocytes of *Eisenia fetida* and *Hediste diversicolor* exposed to lipid-coated CdSe/ZnS quantum dots and $CdCl_2$. *Environ. Toxicol.* 30: 918–926.

166 Sanders, M.B., Sebire, M., Sturve, J. et al. (2008). Exposure of sticklebacks (*Gasterosteus aculeatus*) to cadmium sulfide nanoparticles: biological effects and the importance of experimental design. *Mar. Environ. Res.* 66: 161–163.

167 Baun, A., Sørensen, S.N., Rasmussen, R.F. et al. (2008). Toxicity and bioaccumulation of xenobiotic organic compounds in the presence of aqueous suspensions of aggregates of nano-C60. *Aquat. Toxicol.* 86: 379–387.

168 Limbach, L.K., Wick, P., Manser, P. et al. (2007). Exposure of engineered nanoparticles to human lung epithelial cells: influence of chemical composition and catalytic activity on oxidative stress. *Environ. Sci. Technol.* 41: 4158–4163.

169 Farkas, J., Bergum, S., Nilsen, E.W. et al. (2015). The impact of TiO_2 nanoparticles on uptake and toxicity of benzo(a)pyrene in the blue mussel (*Mytilus edulis*). *Sci. Total Environ.* 511C: 469–476.

170 Tedesco, S., Doyle, H., Blasco, J. et al. (2010). Exposure of the blue mussel, *Mytilus edulis*, to gold nanoparticles and the pro-oxidant menadione. *Comp. Biochem. Physiol. C* 151: 167–174.

171 Al-Subiai, S.N., Arlt, V.M., Frickers, P.E. et al. (2012). Merging nano-genotoxicology with eco-genotoxicology: an integrated approach to determine interactive genotoxic and sub-lethal toxic effects of C_{60} fullerenes and fluoranthene in marine mussels, *Mytilus* sp. *Mutat. Res.* 745: 92–103.

172 Balbi, T., Smerilli, A., Fabbri, R. et al. (2014). Co-exposure to n-TiO_2 and Cd^{2+} results in interactive effects on biomarker responses but not in increased toxicity in the marine bivalve *M. galloprovincialis*. *Sci. Total Environ.* 493C: 355–364.

173 Canesi, L., Frenzilli, G., Balbi, T. et al. (2014). Interactive effects of n-TiO_2 and 2,3,7,8-TCDD on the marine bivalve *Mytilus galloprovincialis*. *Aquat. Toxicol.* 153: 53–65.

174 Della Torre, C., Balbi, T., Grassi, G. et al. (2015). Titanium dioxide nanoparticles modulate the toxicological response to cadmium in the gills of *Mytilus galloprovincialis*. *J. Hazard. Mater.* 297: 92–100.

175 Banni, M., Sforzini, S., Balbi, T. et al. (2016). Combined effects of n-TiO_2 and 2,3,7,8-TCDD in *Mytilus galloprovincialis* digestive gland: a transcriptomic and immunohistochemical study. *Environ. Res.* 145: 135–144.

176 Tian, S., Zhang, Y., Song, C. et al. (2014). Titanium dioxide nanoparticles as carrier facilitate bioaccumulation of phenanthrene in marine bivalve, ark shell (*Scapharca subcrenata*). *Environ. Pollut.* 192: 59–64.

177 Nigro, M., Bernardeschi, M., Costagliola, D. et al. (2015). n-TiO_2 and $CdCl_2$ co-exposure to titanium dioxide nanoparticles and cadmium: genomic, DNA and chromosomal damage evaluation in the marine fish European sea bass (*Dicentrarchus labrax*). *Aquat. Toxicol.* 168: 72–77.

178 Vannuccini, M.L., Grassi, G., Leaver, M.J., and Corsi, I. (2016). Combination effects of nano-TiO_2 and 2,3,7,8-tetrachlorodibenzo-p-dioxin (TCDD) on biotransformation gene expression in the liver of European sea bass *Dicentrarchus labrax. Comp. Biochem. Physiol. C: Pharmacol. Toxicol. Endocrinol.* 176-177: 71–78.

179 Ferreira, P., Fonte, E., Soares, M.E. et al. (2016). Effects of multi-stressors on juveniles of the marine fish *Pomatoschistus microps*: gold nanoparticles, microplastics and temperature. *Aquat. Toxicol.* 170: 89–103.

180 Magesky, A. and Pelletier, É. (2015). Toxicity mechanisms of ionic silver and polymer-coated silver nanoparticles with interactions of functionalized carbon nanotubes on early development stages of sea urchin. *Aquat. Toxicol.* 167: 106–123.

181 OECD (2012). Chemicals Committee of the OECD. Important issues on risk assessment of manufactured nanomaterials. Series on the Safety of Manufactured Nanomaterials No. 33.

182 NRC (2009). Committee on improving risk analysis approaches used by the U.S. EPA, National Research.

183 Syberg, K. and Hansen, S.F. (2016). Environmental risk assistance of chemicals and nanomaterials — The best foundation for regulatory decision-making? *Sci. Total Environ.* 541: 784–794.

184 Mouneyrac, C., Syberg, K., and Selck, H. (2015). Ecotoxicological risk of nanomaterials. In: *Aquatic Toxicology. Advancing Tools for Dealing with Emerging Risks* (ed. C. Amiard-Triquet, J.-C. Amiard and C. Mouneyrac), 417–437. Academic Press.

185 Koelmans, A.A., Diepens, N.J.M., Velzeboer, I. et al. (2015). Guidance for the prognostic risk assessment of nanomaterials in aquatic ecosystems. *Sci. Total Environ.* 535: 141–149.

186 Notter, D.A., Mitrano, D.M., and Nowack, B. (2014). Are nanosized or dissolved metals more toxic in the environment? A meta-analysis. *Environ. Toxicol. Chem.* 33: 2733–2739.

3

Changes in Toxicant Physicochemistry and Bioavailability During Sorption/Desorption Processes with TiO_2 Nanoparticles in the Aqueous Phase

Danae Patsiou[1,2], Martin R. S. McCoustra[3], Teresa F. Fernandes[1], and Theodore B. Henry[1,4]

[1]*Institute of Life and Earth Sciences, School of Energy, Geoscience, Infrastructure and Society, Heriot-Watt University, Edinburgh, EH14 4AS, UK*
[2]*Institute of Oceanography, Hellenic Centre for Marine Research, 46.7 km Athens-Sounio Ave, Anavyssos, 19013, Greece*
[3]*Institute of Chemical Sciences, School of Engineering and Physical Sciences, Heriot-Watt University, Edinburgh, EH14 4AS, UK*
[4]*Center for Environmental Biotechnology, Department of Forestry Wildlife and Fisheries, The University of Tennessee, Knoxville, TN 37996, USA*

3.1 Introduction

TiO_2 nanomaterials (NMs) are the most widely produced engineered nanoparticles (NPs) worldwide with an estimated current production of more than 200 000 tons/year [1, 2]. The term NM is used collectively to describe all materials with at least one dimension in the nanoscale (1–100 nm), and NPs are particles with all three dimensions in the nanoscale [3]. More than 60% of TiO_2 NMs are used in cosmetics and sunscreens, and due to their white color, TiO_2 NMs are widely used as a pigment in paints, plastics, and in the food industry [2]. The most frequently used TiO_2 NMs are the Aeroxide P25 NPs (Evonik Degussa), and numerous studies on P25 have been conducted since this NP was first mentioned in the literature in 1985 [4]. Specifically, a search on "TiO_2" and "P25" in Scopus produces 4425 document results (from 1980 since March 2017), 40% in the field of chemistry and 21.3% in the field of environmental sciences.

Their low cost and high chemical stability have made TiO_2 NMs attractive to support remediation of contaminated environments, and TiO_2 NMs have been used in decontamination of air, water, and soil [5]. TiO_2 is a common semiconductor and an excellent photocatalyst and with potential for application to the photo-degradation of organic compounds in the aqueous phase.

Toxicology of Nanoparticles and Nanomaterials in Human, Terrestrial and Aquatic Systems, First Edition. Edited by Marc A. Williams.

The photocatalytic reaction starts with photo-excitation of a valence electron and the subsequent generation of an electron–hole pair that leads to creation of hydrogen peroxide and consequent photo-degradation of organic compounds in the presence of water [6]. Anatase and rutile, the two most common crystalline structures of TiO_2, can be photo-excited in the ultraviolet A (UVA) and visible light spectrum, respectively [7]. The nanoscale properties of TiO_2 NPs increase the catalytic functionalities of the material and the efficiency of photo-degradation of unwanted substances (e.g. contaminants).

The increased production and use of engineered NPs has led to increased occurrence of TiO_2 NPs in the environment. Release of TiO_2 NPs in the aquatic environment can be natural, for NPs that occur naturally in the terrestrial environment; accidental, during industrial processes and fabrication of engineered NPs; or intentional, during and after use of products containing TiO_2 NPs. Studies based on the large amount of information available for production of economically important engineered NPs have been used to model and predict TiO_2 NP concentrations in the aquatic environment [8, 9]. Taken under consideration the production volumes, the importance of product allocation, and the life cycle, the predicted environmental concentrations of TiO_2 NPs was the highest in sewage effluent, surface waters, sediments, and soils and was calculated to be less than 1 μg/L on surface waters [2].

Once in the aquatic environment, TiO_2 NPs can undergo transformations that can alter their physicochemistry compared to the pristine state. After TiO_2 NP release into the environment, ozone, UV light, and humidity alter the NP surfaces and can affect surface coatings [10, 11]. The process of oxidation is affected by environmental conditions and the duration in the environment. The oxidation that occurs on the surface of the particle after the release and weathering of commercially coated TiO_2 NPs in the aquatic environment can erode the coating and expose the core TiO_2 NPs [12]. The ionic strength and changes in ionic strength of surface waters (e.g. transitions between freshwater and marine waters) can induce changes in physicochemistry of NPs including alterations in surface properties that can lead to larger masses of NPs. High calcium levels can neutralize the negative surface charge of TiO_2 NPs in fresh water, and the presence of divalent ions tends to promote agglomeration of NPs [13]. Changes of the surface charge and surface area can in turn lead to alteration of surface physicochemistry of the particles and complicate the interactions and sorption of other compounds present in the aquatic environment (i.e. natural organic matter (NOM), organic contaminants, metals). For instance, thorium sorption on TiO_2 NPs is independent of the presence of NaCl or KCl in aqueous solution [14, 15], but the presence of K^+, Li^+, or Na^+ can alter the surface of TiO_2 NPs and compete with thorium ions for sorption sites on TiO_2 NPs, therefore changing the sorption capacity on TiO_2 NPs [14].

The interactions between TiO_2 NPs and organic contaminants present in the aquatic environment can have implications on the environmental fate and transport of the compounds involved. Numerous studies have investigated the complicated processes of adsorption of organic compounds on the surface of TiO_2 NPs. Sorption refers to the collective processes of adsorption, the process of adhesion of a molecule to a surface due to surface forces [16], and absorption, the process of molecules crossing the surface and retained by the material [16]. The physicochemical properties of both adsorbent and adsorbate (e.g. size and the charge of the TiO_2 NPs and the organic compound) can influence sorption of substances to NPs. The chemistry of the sorption of organic contaminants on TiO_2 NPs has been investigated in the context of photo-degradation of organic contaminants in the aquatic environment [17–20], but these studies have not been critically reviewed. Further, sorption of contaminants to TiO_2 NPs and effects on bioavailability and toxicity in aquatic organisms have been researched, [21–23], and these studies must be critically evaluated to advance the collective understanding of the environmental implications of toxicant sorption to TiO_2 NPs.

The objective of this chapter is to review the aqueous-phase behavior of TiO_2 NPs with a focus on sorption/desorption processes with organic contaminants. The chapter will review the physicochemistry of the TiO_2 in the aqueous phase, factors that affect sorption of organic contaminants to TiO_2 NPs, the influence of UV photo-activation of TiO_2 NPs and other chemical substances, and the use of chemicals bioavailability as a tool for the evaluation of sorption/desorption processes. New data from our research on the sorption of anthracene with anatase TiO_2 NPs under UVA irradiation will demonstrate utility of detection methods based on bioavailability of decomposition products of organic contaminants in zebrafish larvae (Danio rerio).

3.2 Properties of TiO_2 NPs in the Aqueous Phase

3.2.1 Agglomeration

Titanium dioxide NPs agglomerate, and agglomeration is influenced by NP surface charge, size, and presence of other substances in the aqueous phase. The term "agglomerate" is defined as a cluster of NPs and other substances that is loosely held (i.e. a cluster of NPs that readily return to original nanoscale size by high frequency sonication), whereas the term aggregate is used to refer to a cluster of strongly held NPs that are not returned to original nanoscale dispersion by high frequency sonication [24]. It has been documented that surface charge of oxide NPs is an important property responsible for the agglomeration of the particles [25, 26].

At the nanoscale, attractive forces between particles (van der Waals forces) can cause NPs to agglomerate. The isoelectric point of anatase TiO_2 NPs in the aqueous phase is at pH 6 [24] and the NPs acquire a negative surface charge in waters of higher pH (down to approximately −40 mV; [17, 19, 24, 27–29]). A pH closer to the isoelectric point of the NPs enhances NP agglomeration, while at higher pH values, the absolute value of the surface charge can increase repulsive forces between NPs and reduce agglomeration [24].

The NP surface area available for interactions with surrounding environment and the chemical structure of the surface are additional factors that affect NP agglomeration. The crystalline structure of TiO_2 NPs appears to influence the agglomeration and sedimentation for the NPs in the aqueous phase. More rapid sedimentation has been observed for rutile TiO_2 NPs compared with anatase TiO_2 NPs in the water column. Specifically, for rutile TiO_2 NPs (20–80 nm primary size) 48 h after being introduced to a freshwater medium, more than 80% of the TiO_2 concentration was decreased in the upper water layer, and in turn large agglomerates of 800 nm size were found in the bottom layer based on measurements by dynamic light scattering (DLS; [23]). On the contrary, the same study observed that the numbers of anatase TiO_2 NPs of a primary size 20–50 nm reduced by 52% in the upper water layer at the same time [23]. The particle size and surface area have been reported to be more important than the crystalline structure in terms of the propensity of the TiO_2 NPs to agglomerate in the aqueous phase [19, 30].

The ionic strength of the medium can induce the formation of TiO_2 NP agglomerates in the aqueous phase. Divalent cations adsorb to specific sites on the surface of TiO_2 NPs and neutralize the negative charges on the surface, thereby preventing repellent forces between NPs and facilitating agglomeration [31]. Calcium ions of a freshwater medium neutralize the surface charge of NPs and cause agglomeration. The presence of NOM can stabilize aqueous dispersions of NPs when concentrations of divalent cations are relatively low (e.g. 0.3 mM of $CaCl_2$; [31]), and this is likely because calcium ions bind to the NOM and are less available to sorb to TiO_2 NPs. The TiO_2 NP agglomerates tend to sediment rapidly in freshwater and sea water media. Specifically, 80% of the TiO_2 NPs were removed from water column in less than 100 min and formed larger agglomerates in seawater compared with fresh water [26]. The hydrodynamic diameter (i.e. the average size) of TiO_2 NPs in suspension increased from 793 to 2833 nm as the concentration of TiO_2 NPs increased from 5 to 50 mg/L, after 3 h in Roswell Park Memorial Institute (RPMI) culture medium [27].

3.2.2 Oxidation of TiO_2 NPs by UV Radiation

Ultraviolet illumination can cause changes in the surface chemistry of TiO_2 NPs and other metal oxide NPs such as ZnO and CeO_2. TiO_2 NPs under UV light produce electron–hole pairs (e^-–h^+), and the oxidation of H_2O or OH^- generates hydroxyl radicals ($OH^\bullet$; [32]). Additionally, the presence of O_2 acts as an electron trap that leads to generation of superoxide ($O_2^{-\bullet}$; [6, 32]). The preceding observations indicate that UV light can induce the production of reactive oxygen species (ROS) by TiO_2 NPs. Pigeot-Remy et al. [33] observed high levels of hydroxyl radicals at the surface of TiO_2 P25 NPs by spin trapping after 15 s of UVA irradiation. Anatase band gap energy of 3.2 eV corresponds to 385 nm UVA wavelength, while rutile TiO_2 has a smaller band gap energy (3.0 eV) with excitation at 410 nm wavelength [7], and therefore, oxidation of TiO_2 NPs by UV light can potentially occur after their release to the aquatic environment and exposure to sunlight. No free radicals were observed with P25 TiO_2 NPs under visible light; however, the intensity of 5,5-dimethyl-1-pyrroline N-oxide (DMPO)-$OH^\cdot$ was significantly increased after the incorporation of boron and nickel in TiO_2 NPs, indicating that chemical composition of TiO_2 NPs can influence production of free radicals [34]. Brame et al. [35] indirectly demonstrated the presence of ROS during degradation of 4-chlorophenol in the presence of sunlight and food-grade TiO_2 NPs, when upon addition of the hydroxyl radical ($^\cdot OH$) scavenger, isopropanol, the degradation ended. Weathering of TiO_2 NPs, used in cosmetic products, by artificial daylight degraded the polydimethylsiloxane coating of the TiO_2 NPs [36]. Silicon as part of the coating of the TiO_2 NPs was leached in solution (up to 90% of the initial weight), and the remaining coating layer was oxidized [36]. This alteration process of the surface of TiO_2 NPs after release in the aquatic environment can alter the physicochemical properties and reactivity of the NPs.

Photo-degradation of organic compounds by metal oxide NPs is a promising method for the elimination of organic contaminants from the water. In particular, TiO_2 P25 NPs under UV light facilitated photo-degradation and led to a reduction in half of the amount of paracetamol present within 300 min [37]. Further, Peterson et al. [20] observed total degradation of ampicillin after 500 min in the presence of anatase TiO_2 NPs under 254 nm UV light. For the TiO_2 materials, the NPs provide larger surface area that can lead to higher photocatalytic activity compared to bulk [38]. Specifically, a TiO_2 NP of 30 nm diameter achieved total degradation of methylene blue in 10 min under 300 nm UV light in aqueous solution, while TiO_2 particles of 49 μm diameter needed 60 min to facilitate the same amount of degradation [39]. Estrone (E1) and 17β-estradiol (E2; ng/L range) were totally degraded under light (238–579 nm) in the presence of TiO_2 P25 (1 g/L) within 4 h, and, when the light was fixed at 253 nm, the degradation time reduced to less than 1 h [40]. Investigation of ofloxacin degradation by TiO_2 NPs in dark conditions showed that sorption on TiO_2 NPs leads to breakage of organic compounds and degradation of ofloxacin occurred after 100 h [41] compared with UV photo-degradation of the same compound in the range of minutes to 1 h [42, 43].

3.3 Sorption of Organic Substances to TiO_2 NPs

3.3.1 Influence of Organic Matter on Sorption

Natural organic matter (NOM) can associate with TiO_2-NPs and influence the NP physicochemistry and environmental fate. NOM is a mixture of organic compounds naturally occurring in the aquatic environment [44]; and NOM includes carboxylic and amino acids, proteins, and substances that include molecules of high molecular weight such as humic and fulvic acids [45]; and the amount and the properties of NOM may differ according to area and season. As mentioned earlier, NOM can sorb on the surface of TiO_2 NPs in the aqueous phase and stabilize the NPs in dispersion. In the presence of NOM, the agglomeration of particles is inhibited; however, higher concentrations of Ca ions (>1.5 mM of $CaCl_2$) promote neutralization of the surface charge and increase of the hydrodynamic diameter [46].

The presence of NOM in aquatic environments can influence sorption of organic contaminants on TiO_2 NPs. Fries et al. [47] investigated sorption of the antibiotic ciprofloxacin (CIP) on TiO_2 NPs, in the presence of NOM, and concluded that NOM and CIP competed for sorption sites on TiO_2 NPs, and in the presence of NOM, CIP sorption on TiO_2 NPs was reduced. The study used 500 mg/L TiO_2 NPs and achieved sorption of 50% of the primary CIP concentration. Wang et al. [48] found that the presence of dissolved organic matter (DOM, specifically tannic acid) as a coating on TiO_2 NPs, enhanced the sorption of phenanthrene on TiO_2 NPs. The DOM coating of TiO_2 NPs promoted agglomeration of NPs reducing the surface area available for sorption; therefore, the sorption of phenanthrene between NPs and bulk TiO_2 was not different. Zhang et al. [40] observed that in the presence of humic acid, the photo-degradation of E1 and E2 by TiO_2 P25 was enhanced and that was attributed to the additional photosensitivity of NOM itself (Table 3.1).

3.3.2 Influence of TiO_2 NP Surface Area on Sorption

The surface area of TiO_2 NPs available is important for sorption, and the total surface area in the preparation is influenced by the number of particles and the surface-to-volume ratio of the particles present. More perfluorooctanesulfonic acid (PFOS) was sorbed to small anatase particles of the same mass compared with larger rutile TiO_2 NPs as detected by measurement of the PFOS concentration in the top layer of water column after centrifugation of PFOS–TiO_2 NPs and analysis by ultra-performance liquid chromatography-tandem mass spectrometer [23]. The conclusion that the size of the particles is more important than the crystalline structure was supported by another study that investigated sorption of algal toxin, microcystin-LR: microcystin with leucine and arginine as L-amino acids, on TiO_2 NPs [19]. In particular, TiO_2 P25 (80% anatase, 20% rutile) displayed the

Table 3.1 Evaluation of sorption of organic contaminants on TiO_2 NPs, based on analytical chemistry.

TiO_2 NPs specifics	Adsorbate	Factors assessed	Results	References
P25	Orange II	Agglomeration and photocatalytic performance	Less agglomeration of NPs, efficient photo-degradation of orange II	[18]
P25	Trichloroethylene	Activated carbon and humic acid	Humic acid reduced sorption of trichloroethylene on TiO_2 NPs	[29]
Anatase, rutile, rutile hydrophilic, rutile hydrophobic	Phenanthrene	Organic matter: different types of TiO_2 NPs	Sorption enhancement due to organic matter	[48]
P25, anatase (<25 nm) and rutile (10–30 nm) NPs	Microcystin LR	Different crystalline phase of TiO_2 NPs: NPs compared to bulk	Sorption of MLR enhanced with smaller size of TiO_2 NPs	[19]
Anatase NPs (32 nm)	CIP	Natural organic matter	NOM competitive against CIP for sorption on TiO2 NPs	[47]
Rutile NPs 10 × 40 nm average particle size	17β-Estradiol	Effect of 17β-estradiol on aggregation and NP characteristics	Quick sorption Changes in NPs size after sorption	[17]
P25	Dissolved organic matter	Photo-degradation of DOM with TiO_2 and UVA	Removal of 73–90% DOM	[49]
25 nm diam. anatase	Ofloxacin	Adsorption vs. degradation	TiO_2 NPs remove ofloxacin by adsorption and degradation	[50]
25 nm diam. anatase	Ampicillin	Effect of pH	TiO_2 NPs remove penicillin by adsorption and degradation	[20]
25 nm diam. anatase	Ofloxacin	Dark conditions	Ofloxacin sorption to TiO_2, degradation occurred in dark conditions	[41]

highest capacity for microcystin adsorption measured in supernatant by liquid chromatography – mass spectrometry, followed by pure anatase NPs and pure rutile NPs; however, P25 NPs did have the smallest hydrodynamic diameter. All TiO_2 NPs displayed higher capacity for adsorption of the cyanobacterial toxin compared with bulk TiO_2 [19].

Larger surface area can be achieved with higher concentrations of NPs, but greater numbers of particles can increase the potential for agglomeration and thereby reduce surface area available for sorption. Benzo(a)pyrene concentration in the aqueous phase decreased to 61% of the starting concentration in the presence of 0.2 mg/L TiO_2 NPs after sorption on the surface of the particles. Sedimentation of the TiO_2 NPs with sorbed BaP decreased further (37% of the starting BaP concentration) when more NPs were added in the preparations (2 mg/L; [22]). At higher TiO_2 NP concentrations (>150 mg/L), the effectiveness of the particle as a photocatalyst was decreased due to agglomeration of the particles [18]. At higher concentrations, the agglomeration rate was faster due to increased probability of contact between particles. Increasing pH of solution from 2.29 to 6.15 led to the formation of larger agglomerates and the observed sorption/degradation of orange II dye was decreased [18]. Summarized and additional information regarding organic compound adsorption on TiO_2 NPs can be found in Table 3.1.

3.3.3 Use of Bioavailability to Inform on Sorption of Organic Compounds on TiO_2 NPs

Aquatic toxicology of TiO_2-NPs has demonstrated that aqueous-phase agglomerates of TiO_2 NPs are of minimal toxicity at environmentally relevant concentrations. Aqueous concentrations of 0.1–1 mg/L (2–3 orders of magnitude above estimated TiO_2 NPs concentration on surface waters; [2]) can interfere with respiratory surfaces in fish [51, 52], and subsequent effects of TiO_2-NPs on fish behavior [51], reproduction [53], and tissues biochemistry [51] have been reported. A critical comparison among studies conducted with identical TiO_2 NPs in rainbow trout *Oncorhynchus mykiss* exposed via diet, aqueous phase, and injection into cardiovascular system demonstrated that Ti did not accumulate in internal tissues of rainbow trout after waterborne or dietary exposure. Minor sublethal toxicity indications (e.g. elevated TBARS [i.e. thiobarbituric acid reactive substances] and GSH [i.e. glutathione reduced form]) were attributed to physical disruption of fish respiration (accumulation of TiO_2 NPs in gills) after waterborne exposure [54]. Although also at concentrations that are orders of magnitude higher than environmental relevance, TiO_2 NPs under simulated solar radiation with environmentally relevant UVA intensity of 1700 $\mu W/cm^2$ enhanced mortality on smaller organisms after daily renewal of exposure solutions: zebrafish free swimming larvae with 96-h LC_{50} of 134.6 mg/L TiO_2 NPs P25 [55], 96-h LC_{50} of 2.2 mg/L in Japanese medaka larvae, and 48-h LC_{50} of 29.8 μg/L in *Daphnia magna* [56]; no investigation of ROS has been conducted.

Sorption/desorption of substances with TiO_2 NPs has been effectively investigated within the aqueous phase by evaluation of changes in substance bioavailability in organisms. The bioavailability of a substance can be defined as the amount of a substance that has entered the body and interacted with organs/tissues to generate a measurable response in the organism [57, 58]. The interpretation of a biomarker relies on the concentration–response principle, assuming a demonstrated relationship between the response of the biomarker and the concentration of the contaminant. For some organic substances, bioavailability can be influenced by the presence or absence of TiO_2 NPs, and, presumably, the change in substance bioavailability occurs because of sorption to TiO_2 NPs. The investigated organic substance should cause a well-defined concentration–response pattern; thus any changes in the biomarker response can be attributed to the presence of TiO_2 NPs.

Organisms such as fish and bivalves have been used to evaluate the sorption of organic contaminants on TiO_2 NPs in the aqueous phase (Table 3.2). The sorption of the organic compound on TiO_2 NPs is shown as changes in the response (e.g. toxicity) of the organism to the compound. Sorption depends on the nature of the organic compound and the characteristics of TiO_2 NPs (size and crystal phase) as it has been mentioned earlier, and additionally it can depend on the organism used for the assessment of bioavailability and the route of exposure (e.g. waterborne or dietary exposure). Direct comparison between studies is difficult because of differences in methods that include types of TiO_2 NPs, the organic compounds tested, and the organisms and end points used for measurement of bioavailability. Several different biomarkers and aquatic organisms have been used to assess bioavailability.

TiO_2 NP agglomerates with sorbed contaminants can be ingested by aquatic organisms, and the bioavailability of the sorbed contaminant can increase after ingestion. Marine bivalves such as ark shell *Scapharca subcrenata* have been used to assess sorption of phenanthrene [58] and polybrominated diphenyl ether (BDE; [63]), *Mytilus galloprovincialis* to evaluate sorption of 2,3,7,8-tetrachlorodibenzo-*p*-dioxin (TCDD; [61]), and *Haliotis diversicolor* to assess sorption of tributyltin on TiO_2 NPs [67]. The previously mentioned studies observed enhanced bioavailability of organic compounds in the presence of TiO_2 NPs suggesting that TiO_2 NPs facilitated transport of sorbed organic compounds into the studied organisms. One study observed reduced uptake (in μg/L) of benzo(a)pyrene in the presence of TiO_2 NPs in *Mytilus edulis*. No changes were observed in the activity of glutathione peroxidase and superoxide dismutase proteins; however, the catalase protein activity of the digestive glands was increased significantly in the presence of TiO_2 NPs indicating the presence

Table 3.2 Evaluation of sorption of organic contaminants on TiO_2 NPs by assessment of bioavailability in aquatic organisms.

TiO_2 NPs specifics	Adsorbate	Factors assessed	Target organism	Results	References
P25	2,3,7,8-Tetrachlorodibenzo-p-dioxin (TCDD)	Sorption	Dicentrarchus labrax	TiO_2 NPs showed no effect in TCDD biotransformation gene expression	[59]
25 nm diam.	Pentachlorophenol (PCP)	PCP range of concentrations sorption	Danio rerio larvae	TiO_2 reduced uptake of PCP	[21]
62 nm diam.	Benzo-a-Pyrene (BaP)	Sorption on TiO_2 range of conc.	*Mytilus edulis*	TiO_2 reduced uptake of BaP	[22]
<10 nm diam. anatase NPs	Phenanthrene	Sorption	*Scapharca subcrenata*	TiO_2 enhanced uptake	[58]
20–50 nm diam. anatase NPs, 20–80 nm diam. rutile NPs	Perfluorooctanesulfonate (PFOS)	TiO_2 anatase and TiO_2 rutile NPs sorption	Danio rerio	TiO_2 NPs enhanced uptake	[23]
P25	TCDD	Sorption	Dicentrarchus labrax	TiO_2 NPs did not affect biotransformation or bio-concentration of TCDD	[60]
P25	TCDD	*In vitro* and *in vivo* exposure	*Mytilus galloprovincialis*	TiO_2 NPs enhanced bioavailability of TCDD	[61]

7 nm diam.	Polybrominated diphenyl ether (BDE)	Range of BDE conc. Range of TiO_2 NPs conc.	Danio rerio larvae	TiO_2 enhanced uptake	[62]
<10 nm diam. anatase	Tributyltin (TBT)	Range of conc. of TBT	*Haliotis diversicolor* embryos	TiO_2 enhanced TBT toxicity	[59]
<10 nm diam.	BDE	Range of conc. of BDE	*Scapharca subcrenata*	TiO_2 enhanced uptake of BDE	[63]
20–30 nm diam. anatase	PFOS	Different fish species	Danio rerio, *Ctenopharyngodon idella*, *Hypostomus plecostomus*	TiO_2 NPs increased bioaccumulation of PFOS in all three species	[64]
P25	Ciprofloxacin (CIP)	UVA	*Vibrio fischeri*	Initial decrease in toxicity after sorption of CIP to TiO_2 NPs, high toxicity of CIP by-products	[65]
P25	Anthracene, phenanthrene	UVA	*Danio rerio*	PAH time-dependent photoproduct formation in the presence of TiO_2 NPs	[66]

of hydrogen peroxide, and an increased frequency of micronucleus in *M. edulis* hemocytes was also observed [22]. As bivalves are organisms of low mobility and feed by drawing large quantities of water via the inhalant siphon and filtering the water through the gill, waterborne TiO_2 NP agglomerates and potential sorbents can end up inside the organisms.

The bioavailability of a sorbed organic toxicant on TiO_2 NP agglomerates may differ when investigated in different developmental stages of the same organism. TiO_2 NPs facilitated bioaccumulation of PFOS in adult zebrafish, and bioaccumulation of pentachlorophenol (PCP) was reduced in zebrafish embryos. Qiang et al. [23] observed increased bioavailability of PFOS, as mass of PFOS in adult zebrafish wet weight, in the presence of TiO_2 NPs. Differences were found between two crystalline phases tested: anatase TiO_2 NPs increased PFOS bioavailability by 55.9% and rutile TiO_2 NPs increased PFOS by 25.4%. Zebrafish ingested TiO_2 agglomerates with sorbed PFOS and the TiO_2 NPs acted as a carrier increasing the concentration of hydrophobic PFOS in the whole fish body, although no information is provided on whether PFOS was accumulated or present on fish skin or in the lumen of the gut. Fang et al. [21] investigated the bioavailability of PCP) in larval zebrafish to assess sorption of PCP on TiO_2 NPs and observed that the mass of PCP accumulated by the organism was reduced in the presence of TiO_2 NPs by 33.7%. Further experimentation showed that the presence of TiO_2 NPs in the PCP exposure increased ROS measured by florescence and increased superoxide dismutase gene expression, suggesting increased bioavailability of PCP [21].

The bioavailability of the organic compound sorbed on TiO_2 NPs can be different between biomarkers and tissues of the same organism. The presence of TiO_2 P25 NPs had no effect on TCDD bioavailability in European sea bass [59, 60]. The presence of TiO_2 NPs did not change the bioavailability of TCDD as assessed by measurement of upregulation of cytochrome P4501A (*cyp1A*) and glutathione s-transferase gene expression or ethoxyresorufin-*O*-deethylase (EROD) enzyme activity [59, 60]; however the presence of TiO_2 NPs and adsorbate suppressed significantly suppressed the expression of adenosine triphosphate-binding cassette genes, *abcb1* and *abcc1* [59]. Further analysis of expression of gene related to immunomodulation showed upregulation in gills and spleen of fish exposed in TiO_2 NPs or TCDD alone, but the presence of NPs reduced the induction caused by TCDD when fish were exposed to the mixture [60]. This response, in combination with the reduced body burden of TCDD in the presence of TiO_2 NPs, suggests that the sorption of TCDD on TiO_2 NPs can cause alteration of physicochemical characteristics of agglomerates and sorbent and result in lower TCDD bioavailability. One explanation for the reduction of the concentration of primary substance can be the decomposition of the substance after sorption on the surface of TiO_2 NPs. Sorption on TiO_2 NPs leads to photo-induced degradation or slower degradation of organic compounds in the dark conditions. TiO_2 NPs

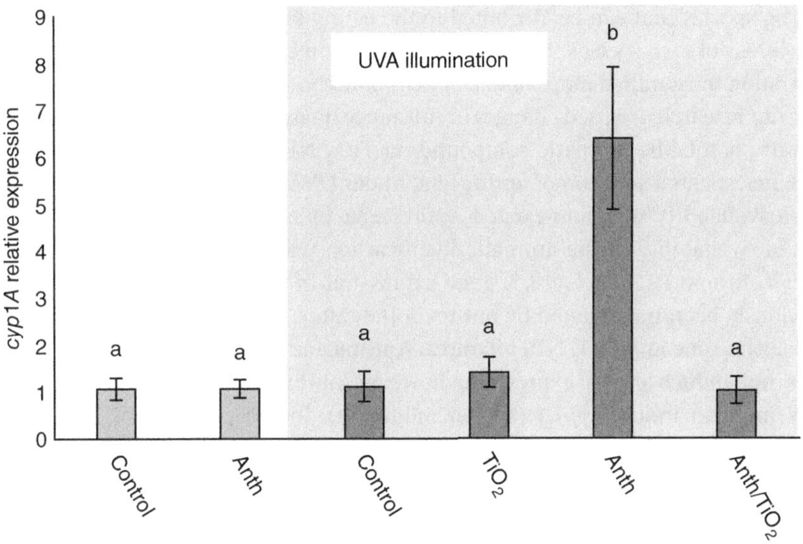

Figure 3.1 Relative expression of *cyp1A* in zebrafish larvae (72 hpf), after normalization to *β-actin*. When larvae (20 per sample) were exposed to 15 μg/L of anthracene (anth; $n = 3$), no induction of *cyp1A* was observed. On the contrary, larvae exposed to anthracene under UVA illumination (80 kJ/m^2; $n = 6$) led to a 6.3-fold induction of *cyp1A*, suggesting that anthracene after UVA exposure creates photo-activated compound that is biodegraded by cyp1A. Exposure of TiO_2 NPs (2 mg/L) under UVA ($n = 3$) do not induce *cyp1A* expression. The mixture of anthracene and TiO_2 NPs under UVA do not induce *cyp1A* expression, indicating that photo-activated anthracene is not bioavailable to the larvae. The control exposure is synthesized freshwater medium (79, 38, 12, 17, and 2 mg/L of Ca^{2+}, Mg^{2+}, Na^+, and K^+, respectively). Mean ± Sem, ANOVA, $p < 0.05$.

may induce decomposition of sorbed parent substances to by-products that can have biomarker response. For instance, as it is shown in our work (Figure 3.1), parent polycyclic aromatic hydrocarbons (PAHs) and degradation by-products may not have the same bioactivity, thus cause different biomarker response in the organism and inform on sorption of compound on TiO_2 NPs differently.

Qiang et al. [64] conducted a comparative study to assess sorption of PFOS on TiO_2 NPs in three species of fish that naturally occupy different levels of water column. Zebrafish resided in the top layer of exposure chamber, *Ctenopharyngodon idella*, occupied the middle layer, and a benthic fish species, *Hypostomus plecostomus*, was in the bottom layer of the exposure chamber. The PFOS whole body burden was increased in the presence of TiO_2 NPs, 59% in zebrafish, 67.6% in *C. idella*, and 16.4% in *H. plecostomus*. The highest concentration of PFOS was found in the intestines, gills, and skin of all three fish species after exposure to TiO_2–PFOS [64]. The results showed differences in the bioavailability of PFOS among species that can be attributed to the eating behavior, natural habitat, and physiology of each species. The PFOS–TiO_2 NPs combination increased PFOS concentration in zebrafish more than *H. plecostomus*.

In our research, we used changes in substance bioavailability to investigate sorption of photolabile aromatic compound on TiO_2 NPs under UV irradiation. We have investigated sorption of anthracene under UVA on TiO_2 NPs by assessment of bioavailability of photo-excited anthracene in zebrafish larvae (72–96 hpf). The bioavailability of the aromatic hydrocarbon was assessed through changes in cytochrome P4501A (*cyp1A*) gene expression in zebrafish embryos as it has previously been investigated by our team [68] after 24-h exposure to anthracene and anthracene and TiO_2 NPs mixtures. Anthracene exposure in zebrafish larvae does not induce *cyp1A* expression; however, anthracene exposed under UVA (5 W/m^2) can induce *cyp1A* (sixfold induction). In the presence of TiO_2 NPs, *cyp1A* remained at low levels, reducing the effect of photo-excited anthracene on *cyp1A* expression (Figure 3.1). The results indicate that anthracene under UVA irradiation alters chemical structure and metabolism of photo-excited anthracene-induced *cyp1A* gene expression. The presence of TiO_2 NPs reduced the bioavailable anthracene by either providing enhanced surface area for anthracene to sorb leaving less/no amount of anthracene available to larvae or by degrading anthracene into by-products that do not involve aryl hydrocarbon pathway biodegradation. A recent study by our team by [66] investigated the photoproduct formation and transformation of anthracene following UVA in the presence of TiO_2 NPs and demonstrated a time dependent degradation of anthracene after analyses of gene expression profiles of PAH-specific metabolic enzymes in larval zebrafish.

3.4 Conclusions

To summarize, our evaluation indicates several implications of the sorption of organic contaminants on TiO_2 NPs and agglomerates. TiO_2 NPs are likely to interact with organic contaminants present in the aquatic environment during and after the NP release. After manufacturing and during use and release in the aquatic environment, TiO_2 NPs can go through changes in their physicochemistry. Interactions between NPs or with ionic compounds, NOM, and solar irradiation can change the properties of the NP surface, making laboratory-based experiments more complicated to assess and interpret. Sorption of organic contaminants on TiO_2 NPs can occur and promote transport of sorbent to aquatic organisms, but it can be influenced by the changes in the physicochemistry of the surface of NPs and sorbent in the aqueous phase. TiO_2 NPs can facilitate photo-degradation with potential for use in environmental remediation; however, there is also potential that biologically active decomposition products are produced after interaction with UV light and organic substances sorbed to TiO_2 NPs. Evaluation of organic substance bioavailability to investigate the sorption/desorption processes with NPs is an environmentally relevant and sensitive approach that contributes important new information toward understanding sorption phenomena.

References

1 Robichaud, C.O., Uyar, A.E., Darby, M.R. et al. (2009). *Estimates of Upper Bounds and Trends in Nano-TiO_2 Production as a Basis for Exposure Assessment*. ACS Publications.

2 Sun, T.Y., Gottschalk, F., Hungerbühler, K., and Nowack, B. (2014). Comprehensive probabilistic modelling of environmental emissions of engineered nanomaterials. *Environ. Pollut.* 185: 69–76.

3 ISO (International Organization for Standardization), (2008). Nanotechnologies—Terminology and definitions for nano-objects—Nanoparticle, Nanofibre and Nanoplate. Geneva, Switzerland.

4 Cãrdoba, A. and Luque, J.J. (1985). Mechanism of surface dehydration of anatase (TiO_2). *Phys. Rev. B* 31: 8111–8118.

5 Choi, H., Stathatos, E., and Dionysiou, D.D. (2006). Sol–gel preparation of mesoporous photocatalytic TiO_2 films and TiO_2/Al_2O_3 composite membranes for environmental applications. *Appl. Catal. B* 63 (1): 60–67.

6 Gaya, U.I. and Abdullah, A.H. (2008). Heterogeneous photocatalytic degradation of organic contaminants over titanium dioxide: a review of fundamentals, progress and problems. *J. Photochem. Photobiol. C* 9 (1): 1–12.

7 Hurum, D.C., Agrios, A.G., Gray, K.A. et al. (2003). Explaining the enhanced photocatalytic activity of Degussa P25 mixed-phase TiO_2 using EPR. *J. Phys. Chem. B* 107 (19): 4545–4549.

8 Gottschalk, F., Sonderer, T., Scholz, R.W., and Nowack, B. (2009). Modeled environmental concentrations of engineered nanomaterials (TiO_2, ZnO, Ag, CNT, Fullerenes) for different regions. *Environ. Sci. Technol.* 43 (24): 9216–9222.

9 Gottschalk, F., Sun, T.Y., and Nowack, B. (2013). Environmental concentrations of engineered nanomaterials: review of modeling and analytical studies. *Environ. Pollut.* 181: 287–300.

10 Labille, J., Feng, J., Botta, C. et al. (2010). Aging of TiO_2 nanocomposites used in sunscreen. Dispersion and fate of the degradation products in aqueous environment. *Environ. Pollut.* 158 (12): 3482–3489.

11 Mitrano, D.M., Motellier, S., Clavaguera, S., and Nowack, B. (2015). Review of nanomaterial aging and transformations through the life cycle of nano-enhanced products. *Environ. Int.* 77: 132–147.

12 Kaegi, R., Ulrich, A., Sinnet, B. et al. (2008). Synthetic $TiO_{(2)}$ nanoparticle emission from exterior facades into the aquatic environment. *Environ. Pollut.* 156 (2): 233–239.

13 Shih, Y.-h., Zhuang, C.-m., Peng, Y.-H. et al. (2012). The effect of inorganic ions on the aggregation kinetics of lab-made TiO_2 nanoparticles in water. *Sci. Total Environ.* 435: 446–452.

14 Tan, X., Wang, X., Chen, C., and Sun, A. (2007). Effect of soil humic and fulvic acids, pH and ionic strength on Th(IV) sorption to TiO_2 nanoparticles. *Appl. Radiat. Isot.* 65 (4): 375–381.

15 Zhijun, G., Lijun, N., and Zuyi, T. (2005). Sorption of Th(IV) ions onto TiO_2: effects of contact time, ionic strength, thorium concentration and phosphate. *J. Radioanal. Nucl. Chem.* 266 (2): 333–338.

16 Calvert, J.G. (1990). Glossary of atmospheric chemistry terms (Recommendations 1990). *Pure Appl. Chem.* 62 (11): 2167–2219.

17 Lee, J., Bartelt-Hunt, S.L., Li, Y., and Morton, M. (2015). Effect of 17β-estradiol on stability and mobility of TiO_2 rutile nanoparticles. *Sci. Total Environ.* 511: 195–202.

18 Li, G., Lv, L., Fan, H. et al. (2010). Effect of the agglomeration of TiO_2 nanoparticles on their photocatalytic performance in the aqueous phase. *J. Colloid Interface Sci.* 348 (2): 342–347.

19 Okupnik, A., Contardo-Jara, V., and Pflugmacher, S. (2015). Potential role of engineered nanoparticles as contaminant carriers in aquatic ecosystems: estimating sorption processes of the cyanobacterial toxin microcystin-LR by TiO_2 nanoparticles. *Colloids Surf. A* 481: 460–467.

20 Peterson, J.W., Petrasky, L.J., Seymour, M.D. et al. (2012). Adsorption and breakdown of penicillin antibiotic in the presence of titanium oxide nanoparticles in water. *Chemosphere* 87 (8): 911–917.

21 Fang, Q., Shi, X., Zhang, L. et al. (2015). Effect of titanium dioxide nanoparticles on the bioavailability, metabolism, and toxicity of pentachlorophenol in zebrafish larvae. *J. Hazard. Mater.* 283: 897–904.

22 Farkas, J., Bergum, S., Nilsen, E.W. et al. (2015). The impact of TiO_2 nanoparticles on uptake and toxicity of benzo(a)pyrene in the blue mussel (*Mytilus edulis*). *Sci. Total Environ.* 511: 469–476.

23 Qiang, L., Shi, X., Pan, X. et al. (2015). Facilitated bioaccumulation of perfluorooctanesulfonate in zebrafish by nano-TiO_2 in two crystalline phases. *Environ. Pollut.* 206: 644–651.

24 Jiang, J., Oberdörster, G., and Biswas, P. (2009). Characterization of size, surface charge, and agglomeration state of nanoparticle dispersions for toxicological studies. *J. Nanopart. Res.* 11 (1): 77–89.

25 Bian, S.-W., Mudunkotuwa, I.A., Rupasinghe, T., and Grassian, V.H. (2011). Aggregation and dissolution of 4 nm ZnO nanoparticles in aqueous environments: influence of pH, ionic strength, size, and adsorption of humic acid. *Langmuir* 27 (10): 6059–6068.

26 Keller, A.A., Wang, H., Zhou, D. et al. (2010). Stability and aggregation of metal oxide nanoparticles in natural aqueous matrices. *Environ. Sci. Technol.* 44 (6): 1962–1967.

27 Allouni, Z.E., Cimpan, M.R., Høl, P.J. et al. (2009). Agglomeration and sedimentation of TiO_2 nanoparticles in cell culture medium. *Colloids Surf. B* 68 (1): 83–87.

28 Loosli, F., Vitorazi, L., Berret, J.F., and Stoll, S. (2015). Isothermal titration calorimetry as a powerful tool to quantify and better understand agglomeration mechanisms during interaction processes between TiO_2 nanoparticles and humic acids. *Environ. Sci. Nano* 2 (5): 541–550.

29 Salih, H.H., Sorial, G.A., Patterson, C.L., and Speth, T.F. (2015). The implications of Fe_2O_3 and TiO_2 nanoparticles on the removal of trichloroethylene by activated carbon in the presence and absence of humic acid. *Water Air Soil Pollut.* 226 (2): 1–10.

30 Suttiponparnit, K., Jiang, J., Sahu, M. et al. (2010). Role of surface area, primary particle size, and crystal phase on titanium dioxide nanoparticle dispersion properties. *Nanoscale Res. Lett.* 6 (1): 27.

31 Romanello, M.B. and de Cortalezzi, M.M.F. (2013). An experimental study on the aggregation of TiO_2 nanoparticles under environmentally relevant conditions. *Water Res.* 47 (12): 3887–3898.

32 Li, J., Chen, C., Zhao, J. et al. (2002). Photodegradation of dye pollutants on TiO_2 nanoparticles dispersed in silicate under UV–VIS irradiation. *Appl. Catal. B: Environ.* 37 (4): 331–338.

33 Pigeot-Remy, S., Dufour, F., Herissan, A. et al. (2017). Bipyramidal anatase TiO_2 nanoparticles, a highly efficient photocatalyst? Towards a better understanding of the reactivity. *Appl. Catal. B* 203: 324–334.

34 Zhao, W., Ma, W., Chen, C. et al. (2004). Efficient degradation of toxic organic pollutants with $Ni_2O_3/TiO_{2-x}B_x$ under visible irradiation. *J. Am. Chem. Soc.* 126 (15): 4782–4783.

35 Brame, J.A., Hong, S.W., Lee, J. et al. (2013). Photocatalytic pre-treatment with food-grade TiO_2 increases the bioavailability and bioremediation potential of weathered oil from the Deepwater Horizon oil spill in the Gulf of Mexico. *Chemosphere* 90 (8): 2315–2319.

36 Auffan, M., Pedeutour, M., Rose, J. et al. (2010). Structural degradation at the surface of a TiO_2-based nanomaterial used in cosmetics. *Environ. Sci. Technol.* 44 (7): 2689–2694.

37 Yang, L., Liya, E.Y., and Ray, M.B. (2008). Degradation of paracetamol in aqueous solutions by TiO_2 photocatalysis. *Water Res.* 42 (13): 3480–3488.

38 Belessi, V., Lambropoulou, D., Konstantinou, I. et al. (2007). Structure and photocatalytic performance of TiO_2/clay nanocomposites for the degradation of dimethachlor. *Appl. Catal. B: Environ.* 73 (3–4): 292–299.

39 Xu, N., Shi, Z., Fan, Y. et al. (1999). Effects of particle size of TiO_2 on photocatalytic degradation of methylene blue in aqueous suspensions. *Ind. Eng. Chem. Res.* 38 (2): 373–379.

40 Zhang, Y., Zhou, J., and Ning, B. (2007). Photodegradation of estrone and 17β-estradiol in water. *Water Res.* 41 (1): 19–26.

41 Peterson, J.W., Gu, B.H., and Seymour, M.D. (2015). Surface interactions and degradation of a fluoroquinolone antibiotic in the dark in aqueous TiO_2 suspensions. *Sci. Total Environ.* 532: 398–403.

42 Paul, T., Miller, P.L., and Strathmann, T.J. (2007). Visible-light-mediated TiO_2 photocatalysis of fluoroquinolone antibacterial agents. *Environ. Sci. Technol.* 41 (13): 4720–4727.

43 An, T., Yang, H., Song, W. et al. (2010). Mechanistic considerations for the advanced oxidation treatment of fluoroquinolone pharmaceutical compounds using TiO_2 heterogeneous catalysis. *J. Phys. Chem. A.* 114 (7): 2569–2575.

44 Sillanpää, M. (2014). *Natural Organic Matter in Water: Characterization and Treatment Methods*. Butterworth-Heinemann.

45 Belin, C., Quellec, C., Lamotte, M. et al. (1993). Characterization by fluorescence of the dissolved organic matter in natural water. application to fractions obtained by tangential ultrafiltration and XAD resin isolation. *Environ. Technol.* 14 (12): 1131–1144.

46 Zhang, Y., Chen, Y., Westerhoff, P., and Crittenden, J. (2009). Impact of natural organic matter and divalent cations on the stability of aqueous nanoparticles. *Water Res.* 43 (17): 4249–4257.

47 Fries, E., Crouzet, C., Michel, C., and Togola, A. (2016). Interactions of ciprofloxacin (CIP), titanium dioxide (TiO_2) nanoparticles and natural organic matter (NOM) in aqueous suspensions. *Sci. Total Environ.* 563: 971–976.

48 Wang, X.L., Ma, E.X., Shen, X.F. et al. (2014). Effect of model dissolved organic matter coating on sorption of phenanthrene by TiO_2 nanoparticles. *Environ. Pollut.* 194: 31–37.

49 Phong, D.D. and Hur, J. (2015). Insight into photocatalytic degradation of dissolved organic matter in UVA/TiO_2 systems revealed by fluorescence EEM-PARAFAC. *Water Res.* 87: 119–126.

50 Van Wieren, E.M., Seymour, M.D., and Peterson, J.W. (2012). Interaction of the fluoroquinolone antibiotic, ofloxacin, with titanium oxide nanoparticles in water: adsorption and breakdown. *Sci. Total Environ.* 441: 1–9.

51 Boyle, D., Al-Bairuty, G.A., Ramsden, C.S. et al. (2013b). Subtle alterations in swimming speed distributions of rainbow trout exposed to titanium dioxide nanoparticles are associated with gill rather than brain injury. *Aquat. Toxicol.* 126: 116–127.

52 Federici, G., Shaw, B.J., and Handy, R.D. (2007). Toxicity of titanium dioxide nanoparticles to rainbow trout (*Oncorhynchus mykiss*): gill injury, oxidative stress, and other physiological effects. *Aquat. Toxicol.* 84 (4): 415–430.

53 Ramsden, C., Henry, T., and Handy, R. (2013). Sub-lethal effects of titanium dioxide nanoparticles on the physiology and reproduction of zebrafish. *Aquat. Toxicol.* 126: 404–413.

54 Boyle, D., Al-Bairuty, G.A., Henry, T.B., and Handy, R.D. (2013a). Critical comparison of intravenous injection of TiO_2 nanoparticles with waterborne and dietary exposures concludes minimal environmentally-relevant toxicity in juvenile rainbow trout *Oncorhynchus mykiss*. *Environ. Pollut.* 182: 70–79.

55 Ma, H.B. and Diamond, S.A. (2013). Phototoxicity of TiO_2 nanoparticles to zebrafish (*Danio rerio*) is dependent on life stage. *Environ. Toxicol. Chem.* 32 (9): 2139–2143.

56 Ma, H.B., Brennan, A., and Diamond, S.A. (2012). Phototoxicity of TiO_2 nanoparticles under solar radiation to two aquatic species: *Daphnia magna* and *Japanese medaka*. *Environ. Toxicol. Chem.* 31 (7): 1621–1629.

57 Semple, K.T., Doick, K.J., Jones, K.C. et al. (2004). *Peer reviewed: Defining Bioavailability and Bioaccessibility of Contaminated Soil and Sediment is Complicated.* ACS Publications.

58 Tian, S., Zhang, Y., Song, C. et al. (2014). Titanium dioxide nanoparticles as carrier facilitate bioaccumulation of phenanthrene in marine bivalve, ark shell (*Scapharca subcrenata*). *Environ. Pollut.* 192: 59–64.

59 Vannuccini, M.L., Grassi, G., Leaver, M.J., and Corsi, I. (2015). Combination effects of nano-TiO_2 and 2,3,7,8-tetrachlorodibenzo-*p*-dioxin (TCDD) on biotransformation gene expression in the liver of European sea bass *Dicentrarchus labrax*. *Comp. Biochem. Physiol. C Toxicol. Pharmacol.* 176: 71–78.

60 Della Torre, C., Buonocore, F., Frenzilli, G. et al. (2015). Influence of titanium dioxide nanoparticles on 2, 3, 7, 8-tetrachlorodibenzo-*p*-dioxin bioconcentration and toxicity in the marine fish European sea bass (Dicentrarchus labrax). *Environ. Pollut.* 196: 185–193.

61 Canesi, L., Frenzilli, G., Balbi, T. et al. (2014). Interactive effects of n-TiO_2 and 2, 3, 7, 8-TCDD on the marine bivalve *Mytilus galloprovincialis*. *Aquat. Toxicol.* 153: 53–65.

62 Wang, Q.W., Chen, Q., Zhou, P. et al. (2014). Bioconcentration and metabolism of BDE-209 in the presence of titanium dioxide nanoparticles and impact on the thyroid endocrine system and neuronal development in zebrafish larvae. *Nanotoxicology* 8: 196–207.

63 Tian, S.Y., Zhang, Y.D., Song, C.Z. et al. (2015). Bioaccumulation and biotransformation of polybrominated diphenyl ethers in the marine bivalve (*Scapharca subcrenata*): influence of titanium dioxide nanoparticles. *Marine Pollut. Bull.* 90 (1–2): 48–53.

64 Qiang, L., Pan, X., Zhu, L. et al. (2016). Effects of nano-TiO_2 on perfluorooctanesulfonate bioaccumulation in fishes living in different water layers: implications for enhanced risk of perfluorooctanesulfonate. *Nanotoxicology* 10 (4): 471–479.

65 Silva, A.R., Martins, P.M., Teixeira, S. et al. (2016). Ciprofloxacin wastewater treated by UVA photocatalysis: contribution of irradiated TiO_2 and ZnO nanoparticles on the final toxicity as assessed by *Vibrio fischeri*. *RSC Adv.* 6 (98): 95494–95503.

66 St. Mary, L., Trine, L.S., Roper, C. et al. (2021). Time-related alteration of aqueous-phase anthracene and phenanthrene photoproducts in the presence of TiO_2 nanoparticles. *Environ. Sci. Technol.* 55 (6): 3727–3735.

67 Zhu, X., Zhou, J., and Cai, Z. (2011). TiO_2 nanoparticles in the marine environment: impact on the toxicity of tributyltin to abalone (haliotis diversicolor supertexta) embryos. *Environ. Sci. Technol.* 45 (8): 3753–3758.

68 Sleight, V.A., Bakir, A., Thompson, R.C., and Henry, T.B. (2017). Assessment of microplastic-sorbed contaminant bioavailability through analysis of biomarker gene expression in larval zebrafish. *Mar. Pollut. Bull.* 116 (1): 291–297.

4

Behavior, Fate, and Toxicity of Engineered Nanoparticles in Estuarine and Coastal Environments

Daniel M. Lyons and Petra Burić

Center for Marine Research, Ruđer Bošković Institute, Rovinj 52210, Croatia

4.1 Introduction

The field of nanotechnology, encompassing research on the creation, tailoring, and application of nanoscale materials for technological purposes, has undergone tremendous growth over the past decade. It is the new and unusual properties of materials at the nanoscale compared to their bulk analogs that has provided the impetus to research in this area. In parallel with the ever-increasing number of applications of nanoparticles in consumer and industrial products, the likelihood for their incidental escape into the environment is also growing. With this realization, significant research efforts have been brought to bear over the past number of years in determining the behavior and fate of nanoparticles in the environment. However, the vast majority of that research has focused on terrestrial and freshwater aquatic systems, and determining nanoparticle toxicity toward living organisms has equally been focused in great part on terrestrial and freshwater organisms. Somewhat belatedly there is an increasing awareness that freshwater systems do not necessarily represent the final sink for nanoparticles, but these nanoparticles may also find their way into estuarine and coastal environments, either by transport from freshwater systems or by direct input from areas of heavy human activity. Indeed, as more than 50% of the world's population live in near-coastal areas, both estuarine and coastal waters are prime candidates for coming under significant environmental pressure from the increasing use of nanoparticles over the coming years.

It is important at this point to draw a distinction between the two broad classes of nanoparticles that may be present in the environment – natural nanoparticles,

Toxicology of Nanoparticles and Nanomaterials in Human, Terrestrial and Aquatic Systems,
First Edition. Edited by Marc A. Williams.

which arise from processes such as erosion and weathering of geological formations, and engineered nanoparticles, which are purposefully formed by man with specific technological applications in mind. It is the latter that are of interest here not only as they derive from anthropogenic activity, with the consequent implication that a certain level of control may be exercised over their release, but also because they may be considerably different in character and hence behave differently, from nanoparticles formed naturally.

This chapter will focus on providing a broad overview of efforts to date on understanding the behavior of engineered nanoparticles in non-freshwater systems and their potential impact on organisms that live in those environmental compartments and will conclude with a perspective on future research needs in this area.

4.2 Types of Nanoparticles: Sources/Products/Release Routes

A broad range of nanomaterials have found application in many consumer products and industrial applications over the past decade. However, only a relatively small subset of them are presently manufactured in significant production volumes, which include, for example, titanium dioxide, carbon nanotubes, cerium dioxide, and silver nanoparticles with estimated production volumes in the United States alone of up to 38 000, 1100, 700, and 20 tons per annum, respectively [1]. It is the nature of the application in which these nanoparticles are included that has facilitated their release into the environment in ever-increasing quantities. The use of titanium dioxide nanoparticles in paints and concrete products make them susceptible to weathering and runoff to surface waters; silver nanoparticles as a biocide in clothing leach during washing cycles and enter wastewater streams, while cerium oxide nanoparticles are added as an aid to combustion in diesel fuels and after escape in exhaust eventually settle from atmospheric deposition and may also enter surface waters. For population centers located at estuaries and coastal areas, the quantity of nanomaterials reaching these environmental compartments has already been shown to be significant. For example, just from the use of sunscreen alone, estimated quantities of titanium dioxide nanoparticles released at a single beach may be up to 4 kg/day [2]. Moreover, not all wastewaters are treated, and, taken in combination with rainwater runoff to surface waters, they represent significant sources of nanoparticles reaching brackish and marine waters (Figure 4.1).

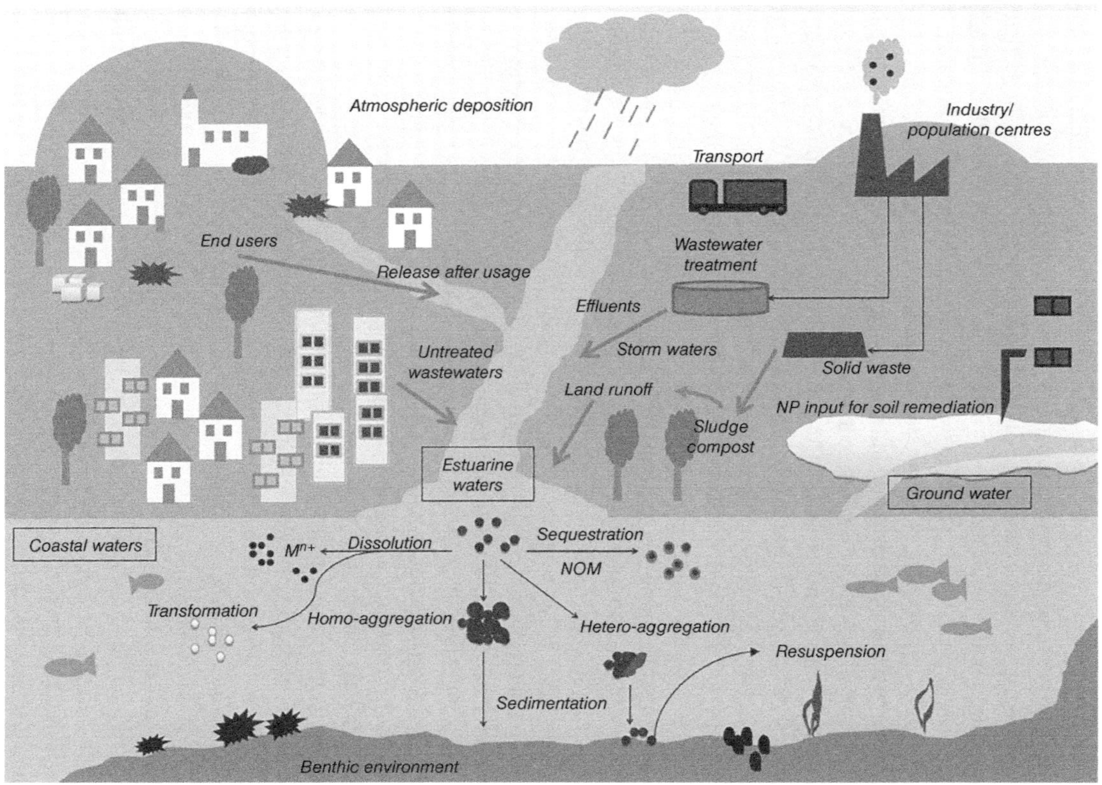

Figure 4.1 Nanomaterial pathways to estuarine and coastal aquatic environments and subsequent transformation processes.

4.3 Behavior of Nanoparticles in the Water Column

One of the key issues regarding the transport of nanoparticles in environmental compartments is their agglomeration or aggregation state (Figure 4.1). While nanoparticles in simple electrolyte-free aqueous solutions may be able to remain as stable dispersions over significant periods of time due to electrostatic repulsion of surface charges or steric hindrance due to bulky surface ligands, nanoparticles in the environment encounter more complex aqueous solutions that contain a range of salts and organic matter, which dramatically affect the stability of nanoparticles' dispersions. The presence of electrolytes disrupts the electrical double layer around nanoparticles, and with the diminishing electrostatic repulsion, nanoparticles begin to agglomerate. As nanoparticles pass from freshwater systems to estuaries, and eventually the sea, there is an enormous increase in electrolyte concentration that would ensure that nanoparticles do not remain in their unagglomerated state. However, apart from salts, a second component present in these aqueous matrices, which has a critical bearing on the stability of nanoparticles and their agglomeration behavior, is dissolved organic matter [3]. Dissolved organic matter, sometimes referred to as natural organic matter, represents a continuum of poorly defined organic material, ranging from humic substances in soil and rivers to alginates, polysaccharides, and phytoplankton exudates in brackish and marine waters. This organic material has been shown to stabilize nanoparticles in high strength electrolytes to some degree, with individual nanoparticles still present alongside large aggregates, albeit for relatively short periods of time. It is likely that some of this organic matter can coat the nanoparticles due to a combination of electrostatic attraction and hydrophobicity of parts of the organic matter, and the resulting steric hindrance acts to slow nanoparticles agglomeration processes.

A second factor that is important for the fate of nanoparticles in the environment is their chemical stability. Environmental conditions are extremely harsh for many types of nanoparticles, and the combination of salts and presence of various oxidizing or reducing species results in dissolution or reformation of the nanoparticles over time. It is well known that for metal nanoparticles such as silver, the presence of sulfur generates metal sulfides on the nanoparticles surfaces, which may then be released resulting in a gradual attrition of the surface [4]. As the surface to bulk atom ratio of nanoparticles is large, complete nanoparticle dissolution can occur over a relatively short period of time. To complicate this picture, recent research has shown that some dissolved organic matter is reducing in nature and so, in the presence of metal ions, encourages the formation of metal nanoparticles. Thus there are two competing processes, and while nanoparticles dissolution ultimately dominates, dissolved organic matter slows the kinetics of this process, allowing nanoparticles to surprisingly persist for some time in saltwater solutions [3].

Apart from overall ionic strength, other factors that impact on chemical stability are the nature of the ions present, oxygen content, pH and temperature of the aqueous matrix, and light intensity. Not all ions have the same effect on the electrical double layer around nanoparticles, and it has been shown that monovalent alkali cations including Na^+ and K^+ exert less of an influence than divalent alkaline earth cations such as Mg^{2+} and Ca^{2+}, the latter causing more rapid agglomeration of nanoparticles. Nanoparticles in alkaline solutions such as coastal waters, e.g. pH 7.8, and further from their isoelectric point may experience some increased stability related to electrostatic repulsion compared with their analogs in pH neutral water. However, in reality, this may be only a small effect when taking into consideration the overall impact of ionic strength on collapsing the electric double layer around the nanoparticles. Light intensity in the environment is another factor to be considered with the stability of nanoparticles, e.g. silver nanoparticles, decreasing with increasing light intensity. Thus, in marine environments nanoparticles in near-surface waters are likely to be less stable than those in deeper waters due to higher light intensity at the surface, particularly in oligotrophic waters. However, in eutrophic waters, there is greater primary production in the photic zone and higher organic matter content that may tend to counter nanoparticle agglomeration. For comparison, in estuarine environments light attenuation is significant due to the extremely high organic matter and suspended sediment content; hence the sensitivity of nanoparticle agglomeration processes to light becomes a minor component compared to other processes [5, 6].

While individual nanoparticles have been shown to be present in high ionic strength aqueous solutions representing brackish and marine waters, the bulk of the nanoparticles remain as agglomerates. Such agglomerates, while comprising of individual nanoparticles, represent essentially larger particles that may have dimensions in the micrometer range. As a consequence of increased mass and reduced surface area compared with its individual nanoparticle components, the transport and fate of such agglomerates in the environment is thus likely different from single nanoparticles. The kinetics of dissolution are slower due to a change in the surface-to-bulk ratio and hence reduced surface area, while the nature of interaction of dissolved organic matter on agglomerates has not been thoroughly addressed to date [7]. In brackish and coastal marine waters, particularly those near large urban centers, the dissolved organic matter content may be high and, through abiotic routes such as mixing of the water column and resuspension of organic matter or biotic processes such as the formation of marine mucilage, nanoparticle agglomerates may be more easily incorporated into particulate organic matter and hence persist for longer time periods. In addition to nanoparticles remaining suspended in the water column, a portion would be expected to eventually reach the sediment and become incorporated in this matrix. For nanoparticles such as those of silver, assuming an oxic environment, oxidation processes will continue until the nanoparticles are eventually dissolved, while other less soluble nanoparticles including titanium dioxide will likely persist in discrete form for longer periods. However, porosity and sediment organic matter may play a role in stabilizing nanoparticles and, where encapsulating nanoparticles, may allow for detachment from the surface and resuspension in the water column. However, this area has not been researched for estuarine and coastal sediments in significant detail to date.

4.4 Biota, Trophic Transfer, Toxicity, and Mechanisms

A key concern that has arisen over the past number of years is that engineered nanoparticles, by virtue of their size and surface chemistry, and hence chemical identity, may present a threat to living organisms in the environment. In cases these may represent essentially new materials to which organisms may not have evolved a tolerance or sufficiently efficient detoxification pathways to protect themselves upon encountering such materials in the environment.

While the toxicity of a broad range of nanoparticles toward a wide variety of freshwater aquatic organisms has been investigated, much less data is available regarding toxic or deleterious effects in organisms whose habitats include estuarine and coastal environments [8]. There are several reasons for this. Firstly, because of nanoparticles high surface area and reactivity, it is expected that they will be quickly transformed upon reaching the environment and will either undergo oxidative dissolution, transition to a more stable, less reactive form, or become combined with natural organic matter before being transported very far. The second expectation is that should nanoparticles be transported from freshwater systems to estuaries and eventually coastal waters, increasing dilution will result in nanoparticles presenting less of a toxic risk to any organism. A third reason is that, with increasing ionic strength, agglomeration processes will ensure that nanoparticles do not remain well dispersed but will form larger agglomerates or aggregates that then present reduced risk of toxicity to organisms. However, it is increasingly being recognized that such expectations are not entirely accurate in that nanoparticles have been shown to persist for longer time periods in both freshwater and saline environments and be present as both individual nanoparticles and agglomerates [9, 10]. Further, even in the form of agglomerates or aggregates, nanoparticles show significant toxicity toward organisms living in those environmental compartments.

To date little consensus has been reached on how to group nanoparticles with respect to their impact on organisms. Mechanisms of toxicity such as disruption of cell membranes or ion transport, or endocytosis followed by disruption to intracellular processes, are essentially particle specific. That notwithstanding, nanoparticles may be broadly classified into two categories with respect to their mode of toxicity to biota: those such as silver, copper, or zinc oxide, which undergo dissolution or essentially non-soluble nanoparticles such as titanium dioxide and cerium oxide.

Algae, occupying a vital position at the base of the food chain as primary producers, have shown specific responses to the presence of nanoparticles, which in large part may be related to the adherence of nanoparticles to cell surfaces and uptake of released ions. Investigation of the impact of silver nanoparticles (20–100 nm, 10 μM) on the diatom *Thalassiosira pseudonana* has found growth inhibition characteristics similar to that of silver ions, suggesting that toxicity may arise by a combination of both nanoparticles and released ions although the contribution of each was not resolved [11]. Similar results were found for the interaction of zinc oxide nanoparticles (100 nm, 0.01–10 mg/L), and the alga *Dunaliella tertiolecta* [12] where inhibition of growth was significant after exposure to nanoparticles and which could not be explained only on the basis of released zinc ions and would thus indicate a nanoparticle size effect.

Echinoderms have proven increasingly popular for testing the impact of nanoparticles on living organisms. For example, the echinoderm *Lytechinus pictus* was exposed to titanium dioxide and cerium oxide nanoparticles (27 nm and 67 × 8 nm, respectively), and no adverse effects were noted in terms of developmental abnormalities at concentrations of up to 10 mg/L [13]. This would suggest that relatively insoluble nanoparticles may not present a significant risk. However, a subsequent study on cerium oxide (50–105 nm), tin oxide (61 nm), and iron oxide (20–30 nm) nanoparticles showed significant toxicity to the echinoderm *Paracentrotus lividus* at the same concentrations of 10 mg/L [14]. For nanoparticles that undergo dissolution, it is generally considered that the released metal ions are the main drivers of toxicity although there is evidence that a size effect in cases causes toxicity at levels exceeding that of the corresponding ions. For example, silver nanoparticles (60 nm) have been found more toxic to several species of echinoderm during embryonal development than the corresponding mass of silver ions, with EC_{50} values lying in the range of 10–50 μg/L [15]. Further, the development stage of those embryos at the moment when they first encountered nanoparticles was also found to be an important factor that modulated toxicity with respect to retarded embryonal development or deformation of the skeletal rods in the larvae. For comparison, an EC_{50} value of 99.5 μg/L for toxicity of zinc oxide nanoparticles (24 nm) to *L. pictus* was noted [13]. Thus, while decreasing nanoparticle size would be expected to enhance toxicity due to faster ion release kinetics, it is clear that size is not the only factor but rather toxicity depends on the interplay between chemical identity of the nanoparticle, its size, and ion release characteristics and also on the species being investigated. Interestingly, apart from nanoparticles' negative effects on embryonal development, 13 nm carbon nanoparticles have been found to provoke anomalous aragonite biomineralization in *P. lividus* embryos through induced expression of the Sp-CyP-1 gene [16], indicating that nanoparticles and/or their released ions such as Zn^+ may impact on a broad range of fundamental genetic cellular and metabolic processes [17, 18].

While echinoderms have traditionally proven important as test organisms, particularly as models for developmental biology or immune system research, the most investigated marine organisms in terms of nanoparticle toxicity are bivalves, as members of the phyla Mollusca [19]. This is related to a number of factors including being widely distributed in coastal zones and their filtering of prodigious volumes of water, hence enhanced possibility for nanoparticle uptake and potential for bioaccumulation of nanoparticles in their tissues. It has been noted that while individual nanoparticles are less well trapped by the gills, larger aggregates, likely aided by encapsulation in muco-polysaccharide exudates of the bivalve, are more readily taken up [20]. However, extensive accumulation of nanoparticles has not been noted in the gills but rather in the digestive gland with, for example, up to 95% of gold nanoparticles that had been taken up (5 nm, *Mytilus edulis*; [21]) transferred to the latter. Further, for cases where a range of nanoparticle sizes are taken up, bivalves have shown the ability to retain smaller nanoparticles with more rapid egestion of larger particles [20]. Those nanoparticles that remain for longer periods diffuse (nanoparticles <5 nm) to various cellular compartments including the nucleus and mitochondria, while larger nanoparticles (5–25 nm) may undergo endocytosis, particularly by digestive cells.

A variety of *in vitro* biochemical tests have been applied to bivalves to investigate the impact of nanoparticles and focus on a wide range of end points including modulation of enzymatic activity levels such as carbonic anhydrase or acetylcholinesterase and induction of metal-binding proteins such as metallothionein and heat shock proteins. A key focus has been on determining the nanoparticle-induced oxidative stress that derives from excessive formation of reactive oxygen species that is not mitigated by cells antioxidant capacity, ultimately resulting in cellular damage. A number of antioxidant enzymatic scavengers have been identified as useful markers including glutathione-S-transferase, superoxide dismutase, and catalase. For example, copper oxide nanoparticles (31 nm; 10 μg/L) increased catalase and glutathione peroxidase levels while suppressing superoxide dismutase activities in the digestive gland of *M. galloprovincialis* [22]. In parallel, an increase in lipid peroxidation was noted, and metallothionein was induced with levels increasing linearly with increasing exposure time to the nanoparticles. *M. galloprovincialis* was also exposed to a range of other nanoparticles including those of C_{60}-fullerene (0.7 nm), carbon black (30 nm), silica (12 nm), and titanium dioxide (22 nm) at concentrations in the range of 0.05–5 mg/L/mussel, and alterations to lysosomal membrane stability in hemocytes, induction of oxidative stress, and also significant lysosomal accumulation of neutral lipids and lipofuscin in digestive gland tissue were noted [23]. Protein expression profiles of the gills and digestive gland in *M. gallo provincialis* have shown differential responses to nanoparticles such as those of silver (<100 nm; 10 μg/L) and the corresponding mass of silver ions, indicating that toxicity is not necessarily driven by ions released by nanoparticles but that a size effect may also be important [24].

Gastropods have shown similar effects to bivalves, e.g. titanium dioxide nanoparticles (≤10 nm; ≥1 mg/L) have resulted in a decrease of glutathione levels and increased of lipid peroxidation and nitric oxide generation [25]. However, data on bioaccumulation are equivocal with reports showing that nanoparticles may be taken up although are not retained for significant amounts of time in other gastropods, e.g. for silver nanoparticles (58 nm; 20 μg/L) in periwinkle *Littorina littorea* [26]. Equally, *L. littorea* did not show accumulation of silver when feeding on the marine macroalga *Ulva lactuca* pretreated with silver nanoparticles. However, this may be related to poor uptake of the nanoparticles by *U. lactuca* where nanoparticles were found to be only weakly associated with the macroalga's surface [27]. Further, the nanoparticles did not cause phytotoxicity as determined through changes in photosynthesis rate, and it is likely that toxicity may arise only indirectly by the absorption of released ions from nanoparticles attached to the algal surface.

Considering organisms higher in the food chain, fish have received far less attention than invertebrates even though fish, by consuming lower trophic level organisms that may have accumulated nanoparticles, may be particularly at risk. Many experiments to investigate nanoparticle impact in fish have relied on direct exposures and in many cases have found sublethal effects such as thickening of gill epithelia and nanoparticles distributed in tissue likely as a result of ingestion. For example, in the brackish water sheepshead minnow (*Cyprinodon variegates*), silver nanoparticles (35 nm, 10 μg/L) caused thickening in the gills of adults and some hemorrhaging in those of juveniles over several days but which then gradually diminished over several weeks [28]. In broad terms, loss of water through osmosis requires marine fish to compensate by drinking saltwater, suggesting that ingestion of nanoparticles may be the most important uptake route and that the tissue of the gastrointestinal tract may be the first at risk. Apart from macroscopic changes to tissue, subcellular effects have also been shown to be important. For example, exposure of larvae of medaka *Oryzias melastigma* to zinc oxide nanoparticles (26 nm, 4 mg/L) results in significant induction of heat shock protein HSP70, indicating genotoxic effects although it is possible that zinc ions from dissolution of the nanoparticles may be responsible rather than a specific size effect [29].

In spite of the growing use of biochemical tests, it is becoming increasingly clear that the nanoparticles themselves may interfere with the assays, and it has been reported that in the 2010–2012 period, 90–95% of papers using fluorescence or colorimetric assays in nanotoxicity assessment did not take into consideration potential interference in the assay from nanoparticles. Indeed, a wide range of assays including protein quantification, cell viability and proliferation, catalase, and lactate dehydrogenase were shown to be significantly affected by the presence of nanoparticles at concentrations exceeding 10 mg/L [30]. Therefore, not only must greater caution be exercised in the concentrations used in these tests but also care should be taken regarding assumptions on the degree of reliability of data gained so far on toxicity tests of nanoparticles. In broader terms, potentially significant artifacts may be introduced even during the experimental procedure due to factors such as the agglomeration or aggregation behavior of the nanoparticles in the test media, leading to overestimation of true dose, changes to nanoparticle surface chemistry due to growth of an eco-/bio-coating, etc. [31].

Overall, there are an ever-increasing number of reports available in the literature focused toward determining the impacts of a wide range of nanoparticles of different sizes, shapes, surface chemistry, and concentrations in brackish and marine organisms and which induce a myriad of responses that are too wide in breadth to address here. However, a representative selection of papers on the impact of nanoparticles on marine invertebrates has been given in Table 4.1.

An important aspect of the fate of nanoparticles in brackish and coastal waters is their movement through the food web by trophic transfer. The association of nanoparticles with low trophic level organisms, including bacteria and algae, or bioaccumulation in filter feeders such as bivalves provide important entry points for nanoparticles into the food web. As tracking transfer of nanoparticles in the food web is difficult due to the complex interrelationships of all of the organisms in an ecosystem, research has generally focused on determining effects in simplified food webs. In a two level food web, silver nanoparticles (20–80 nm, 25–1000 μg/L) were taken up by brine shrimp *Artemia salina* nauplii, which were subsequently fed to medaka *Oryzias melastigma* [56]. While the nanoparticles were readily taken up by *A. salina*, low transfer efficiency to *O. melastigma* over a period of 28 days was noted. However, a wide range of significant effects was noted in the fish including reduction of total body length and inhibition of superoxide dismutase activity and Na^+/K^+-ATPase. Another report has shown zinc oxide nanoparticles (10–1000 μg/L) being trialed for trophic transfer from diatom *Thalassiosira weissflogii*, which were cultured in the presence of the nanoparticles, to copepod *Acartia tonsa* [57]. Apart from dose-dependent growth reduction in the diatom, transfer to *A. tonsa* resulted in reduced copepod reproduction and survival. A recent study has focused on assessing the trophic transfer of silica, tin oxide, iron oxide, and cerium oxide nanoparticles (4–40, 61, 20–30, 50–105 nm, respectively; 5 mg/L) from microalgae *Cricosphaera elongata* to urchin *P. lividus*

Table 4.1 Selected recent research on the impact of nanomaterials on marine invertebrates.

Phylum	Species	NP type[a)]	Size (nm)	Concentration (μg/L)	Exposure (h)	Effects[b)]	References
Plankton	*Chlorella vulgaris*	ZnO	40–48	$50–300 \times 10^3$	24, 72	↓Viability, ↓LDH, ↓GSH, ↑SOD, ↑LPO	[32]
	Natural plankton community	Ag	10	5	24	meta-transcriptome approach: significant difference compared to the control for $AgNO_3$, no effect for AgNP	[33]
	Dunaliella tertiolectaIsochrysis galbanaPhaeodactylum tricornu-tumRhino-monas reticulata	CdSe/ZnS	17 ± 3	0.5–4 nM	3 d	↓Growth rate	[34]
	Isochrysis galbana *Phaeodactylum tricornutum* *Tetraselmis suecica*	PVP/PEI Ag Ag	5 47	$0.00001–100 \times 10^3$ $1–10 \times 10^3$	72	PVP/PEI AgNP + algal cells → hetero-aggregates, PVP/PEI AgNP differences in algae sensitivity (*T. suecica* > *I. galbana* > *P. tricornutum*). PVP/PEI toxic to algae	[35]
	Dunaliella tertiolecta *Skeletonema costatum*	Ag	1–10	$0.1–100 \times 10^3$	72	↓Growth rate	[35]
Cnidaria	*Aurelia aurita*	Ag	1–10	$0.1–100 \times 10^3$	24, 48	Dose-dependent effect on ephyra immobilization and frequency of pulsation	[36]
	Exaiptasia pallida	CuO	<100	10–100	21 d	↑CAT, ↑GPx, ↑GR, ↓CA	[37]

(Continued)

Table 4.1 (Continued)

Phylum	Species	NP type[a)]	Size (nm)	Concentration (μg/L)	Exposure (h)	Effects[b)]	References
Crustacea	*Acartia tonsa*	CdSe/ZnS	17 ± 3	0.15–2.5 nM	48 (acute), 4 d (chronic)	↑Mortality of larvae after 48 h (EC_{50} = 0.7 nM), chronic 4 d exposure: no effect on egg production (except for 2.5 nM), no reduction of egg hatching, ↑*HSP 70* gene expression	[34]
	Amphibalanus amphitrite *Artemia salina*	Ag	1–10	0.1–100 × 10^3	24, 48	↑Mortality and ↓swimming behavior	[36]
	Artemia franciscana	n-PS	1.4	5–100 × 10^3	48	No mortality, n-PS accumulated inside the gut and on the surface of sensorial antennules and appendages, hampering larvae motility	[38]
	Tigriopus japonicus	MWCNT	Diameter = 6–13; length = 10 000	6.25–100 × 10^3	96	↓ROS, ↓CAT, ↓GST, no change in GSH and GR, ERK activation	[39]
Echinodermata	*Arbacia lixulaParacentrotus lividusSphaerechinus granularis*	Ag	60	1–100	48: *A. lixula*, *P. lividus*72: *S. granularis*	AgNP effect is species-specific (*A. lixula* > *S. granularis* > *P. lividus*) and dependent upon moment of first exposure	[15]

developed. While the latter have proven useful for modeling toxicological and environmental risk, they should be applied in consort with population balance models, which consider the broader temporal features of aggregation processes of size classes of nanoparticles, and probabilistic models describing the flow of nanoparticles between different environmental compartments [58]. Probabilistic models may use as input the amount of nanoparticles entering specific environmental compartments. However a limiting factor is the uncertainty on actual nanoparticle concentrations in the environment that could be used to infer input quantities [59]. Ultimately, modeling nanoparticle behavior and transport in estuarine and coastal waters has not been well addressed to date, and data remain scarce [60].

4.7 Knowledge Gaps and Research Prospects

Comparing available research on the behavior, fate, and toxicity of nanoparticles in brackish and coastal marine environments to that on freshwater and terrestrial systems, the former have received relatively little attention to date. While there were specific reasons for this in the past, the increasing realization that population centers at estuaries or coasts may represent potentially enormous sources of nanoparticle release into the environment, and which is only increasing with time, emphasizes the need to refocus on these environmental compartments. In particular there are a number of aspects that must be addressed before a clear understanding of nanoparticles fate in brackish and marine waters can be gained. Research remains fragmented with a wide variety of nanoparticles, sizes, concentrations, and test organisms used. This does not easily enable drawing conclusions about how specific nanoparticles behave across a range of species, and this lack of ability for read-across or grouping of nanomaterials, based on their physicochemical characteristics and toxicity profiles, prevents comprehensive risk assessment. Investigations, in cases, have used high nanoparticle concentrations or low ionic strength media that are not environmentally relevant for brackish and coastal waters, and exposure scenarios have overly focused on acute effects while paying insufficient attention to chronic or repeated dose scenarios. Further, inadequate consideration has been given to determine real internal dose in toxicity studies compared to initial nominal dose, and this is a particularly important issue to resolve in view of nanoparticles agglomeration behavior in saltwater. Behavioral studies of organisms have been shown to provide relevant endpoints in toxicity testing over shorter times and with fewer test organisms than some classical tests and hence should be encouraged [61, 62]. While complex, more effort should be dedicated to studies on trophic transfer, and biomagnification across the food web must be addressed, as these will be to key in enabling vital estimates of long term ecotoxicological effects at the broader ecosystem level. Modeling of nanoparticles in these environmental compartments remains a poorly developed area, and there is a need for developing modules for these compartments and integrating with existing models for freshwater systems. In particular, there are few data on the physicochemical behavior of nanoparticles in estuarine and marine sediments, and this remains an area to receive attention in the future. Thus, there remain many important aspects of nanoparticles residence in estuaries and coastal waters that still remain to be resolved. Ultimately, it will only be by including these key compartments as potentially final sinks for engineered nanoparticles that will lead to a comprehensive understanding of the impact of nanoparticles as emerging contaminants in the environment as a whole.

References

1 Hendren, C.O., Mesnard, X., Droge, J., and Wiesner, M.R. (2011). Estimating production data for five engineered nanomaterials as a basis for exposure assessment. *Environ. Sci. Technol.* 45: 2562–2569.

2 Sanchez-Quiles, D. and Tovar-Sanchez, A. (2014). Sunscreens as a source of hydrogen peroxide production in coastal waters. *Environ. Sci. Technol.* 48: 9037–9042.

3 Grillo, R., Rosa, A.H., and Fraceto, L.F. (2015). Engineered nanoparticles and organic matter: a review of the state-of-the-art. *Chemosphere* 119: 608–619.

4 Liu, J. and Hurt, R.H. (2010). Ion release kinetics and particle persistence in aqueous nanosilver colloids. *Environ. Sci. Technol.* 44: 2169–2175.

5 António, D.C., Cascio, C., Jakšić, Ž. et al. (2015). Assessing silver nanoparticles behaviour in artificial seawater by mean of AF4 and spICP-MS. *Mar. Environ. Res.* 111: 162–169.

6 Čarapar, I., Jurković, L., Pavičić-Hamer, D. et al. (2022). Simultaneous influence of gradients in natural organic matter and abiotic parameters on the behavior of silver nanoparticles in the transition zone from freshwater to saltwater environments. *Nanomaterials* 12: 296.

7 Wang, H., Burgess, R.M., Cantwell, M.G. et al. (2014a). Stability and aggregation of silver and titanium dioxide nanoparticles in seawater: role of salinity and dissolved organic carbon. *Environ. Toxicol. Chem.* 33: 1023–1029.

8 Canesi, L. and Corsi, I. (2016). Effects of nanomaterials on marine invertebrates. *Sci. Total Environ.* 565: 933–940.

9 Chinnapongse, S.L., MacCuspie, R.I., and Hackley, V.A. (2011). Persistence of singly dispersed silver nanoparticles in natural freshwaters, synthetic seawater, and simulated estuarine waters. *Sci. Total Environ.* 409: 2443–2450.

10 Levak, M., Burić, P., Dutour Sikirić, M. et al. (2017). Effect of protein corona on silver nanoparticle stabilization and ion release kinetics in artificial seawater. *Environ. Sci. Technol.* 51: 1259–1266.

11 Burchardt, A.D., Carvalho, R.N., Valente, A. et al. (2012). Effects of silver nanoparticles in diatom *Thalassiosira pseudonana* and *Cyanobacterium synechococcus* sp. *Environ. Sci. Technol.* 46: 11336–11344.

12 Manzo, S., Miglietta, M.L., Rametta, G. et al. (2013). Toxic effects of ZnO nanoparticles towards marine algae *Dunaliella tertiolecta. Sci. Total Environ.* 445-446: 371–376.

13 Fairbairn, E.A., Keller, A.A., Madler, L. et al. (2011). Metal oxide nanomaterials in seawater: linking physicochemical characteristics with biological response in sea urchin development. *J. Hazard. Mater.* 192: 1565–1571.

14 Falugi, C., Aluigi, M.G., Chiantore, M.C. et al. (2012). Toxicity of metal oxide nanoparticles in immune cells of the sea urchin. *Mar. Environ. Res.* 76: 114–121.

15 Burić, P., Jakšić, Ž., Štajner, L. et al. (2015). Effect of silver nanoparticles on Mediterranean sea urchin embryonal development is species specific and depends on moment of first exposure. *Mar. Environ. Res.* 111: 50–59.

16 Manno, D., Carata, E., Tenuzzo, B.A. et al. (2012). High ordered biomineralization induced by carbon nanoparticles in the sea urchin *Paracentrotus lividus. Nanotechnology* 23: 495104.

17 Li, J., Schiavo, S., Xiangli, D. et al. (2018). Early ecotoxic effects of ZnO nanoparticle chronic exposure in *Mytilus galloprovincialis* revealed by transcription of apoptosis and antioxidant-related genes. *Ecotoxicology* 27: 369–384.

18 Oliviero, M., Schiavo, S., Dumontet, S., and Manzo, S. (2019). DNA damages and offspring quality in sea urchin *Paracentrotus lividus* sperms exposed to ZnO nanoparticles. *Sci. Total Environ.* 651: 756–765.

19 Canesi, L., Ciacci, C., Fabbri, R. et al. (2012). Bivalve molluscs as a unique target group for nanoparticle toxicity. *Mar. Environ. Res.* 76: 16–21.

20 Ward, J.E. and Kach, D.J. (2009). Marine aggregates facilitate ingestion of nanoparticles by suspension-feeding bivalves. *Mar. Environ. Res.* 68: 137–142.

21 Tedesco, S., Doyle, H., Blasco, J. et al. (2010). Oxidative stress and toxicity of gold nanoparticles in *Mytilus edulis. Aquat. Toxicol.* 100: 178–186.

22 Gomes, T., Pereira, C.G., Cardoso, C. et al. (2012). Accumulation and toxicity of copper oxide nanoparticles in the digestive gland of *Mytilus galloprovincialis. Aquat. Toxicol.* 118-119: 72–79.

23 Canesi, L., Fabbri, R., Gallo, G. et al. (2010). Biomarkers in *Mytilus galloprovincialis* exposed to suspensions of selected nanoparticles (Nano carbon black, C60 fullerene, Nano-TiO_2, Nano-SiO_2). *Aquat. Toxicol.* 100: 168–177.

24 Gomes, T., Pereira, C.G., Cardoso, C., and Bebianno, M.J. (2013). Differential protein expression in mussels *Mytilus galloprovincialis* exposed to nano and ionic Ag. *Aquat. Toxicol.* 136-137: 79–90.

25 Zhu, X., Zhou, J., and Cai, Z. (2011). The toxicity and oxidative stress of TiO_2 nanoparticles in marine abalone (*Haliotis diversicolor supertexta*). *Mar. Pollut. Bull.* 63: 334–338.

26 Li, H., Turner, A., and Brown, M.T. (2012). Accumulation of aqueous and nanoparticulate silver by the marine gastropod *Littorina littorea*. *Water Air Soil Pollut.* 224: 1354.

27 Turner, A., Brice, D., and Brown, M.T. (2012). Interactions of silver nanoparticles with the marine macroalga *Ulva lactuca*. *Ecotoxicology* 21: 148–154.

28 Griffitt, R.J., Brown-Peterson, N.J., Savin, D.A. et al. (2012). Effects of chronic nanoparticulate silver exposure to adult and juvenile sheepshead minnows (*Cyprinodon variegates*). *Environ. Toxicol. Chem.* 31: 160–167.

29 Wong, S.W.Y., Leung, P.T.Y., Djurisic, A.B., and Leung, K.M.Y. (2010). Toxicities of nano zinc oxide to five marine organisms: influence s of aggregate size and ion solubility. *Anal. Bioanal. Chem.* 396: 609–618.

30 Ong, K.J., MacCormack, T.J., Clark, R.J. et al. (2014). Widespread nanoparticle-assay interference: implications for nanotoxicity testing. *PLoS One* 9: e90650.

31 Petersen, E.J., Henry, T.B., Zhao, J. et al. (2014). Identification and avoidance of potential artefacts and misinterpretations in nanomaterial ecotoxicity measurements. *Environ. Sci. Technol.* 48: 4226–4246.

32 Suman, T.Y., Radhika Rajasree, S.R., and Kirubagaran, R. (2015). Evaluation of zinc oxide nanoparticles toxicity on marine algae Chlorella vulgaris through flow cytometric, cytotoxicity and oxidative stress analysis. *Ecotoxicol. Environ. Saf.* 113: 23–30.

33 Boenigk, J., Beisser, D., Zimmermann, S. et al. (2014). Effects of silver nitrate and silver nanoparticles on a planktonic community: general trends after short-term exposure. *PLoS One* 9: e95340.

34 Zhou, C., Vitiello, V., Pellegrini, D. et al. (2016). Toxicological effects of CdSe/ZnS quantum dots on marine planktonic organisms. *Ecotoxicol. Environ. Saf.* 123: 26–31.

35 Schiavo, S., Duroudier, N., Bilbao, E. et al. (2017). Effects of PVP/PEI coated and uncoated silver NPs and PVP/PEI coating agent on three species of marine microalgae. *Sci. Total Environ.* 577: 45–53.

36 Gambardella, C., Costa, E., Piazza, V. et al. (2015). Effect of silver nanoparticles on marine organisms belonging to different trophic levels. *Mar. Environ. Res.* 111: 41–49.

37 Siddiqui, S., Goddard, R.H., and Bielmyer-Fraser, G.K. (2015). Comparative effects of dissolved copper and copper oxide nanoparticle exposure to the sea anemone, Exaiptasia pallida. *Aquat. Toxicol.* 160: 205–213.

38 Bergami, E., Bocci, E., Vannuccini, M.L. et al. (2016). Nano-sized polystyrene affects feeding, behavior and physiology of brine shrimp Artemia franciscana larvae. *Ecotoxicol. Environ. Saf.* 123: 18–25.

39 Lee, J.W., Won, E.-J., Kang, H.-M. et al. (2016). Effects of multi-walled carbon nanotube (MWCNT) on antioxidant depletion, the ERK signaling pathway, and copper bioavailability in the copepod (*Tigriopus japonicus*). *Aquat. Toxicol.* 171: 9–19.

40 Gambardella, C., Gallus, L., Gatti, A.M. et al. (2014). Toxicity and transfer of metal oxide nanoparticles from microalgae to sea urchin larvae. *Chem. Ecol.* 30: 308–316.

41 Mesarič, T., Sepčić, K., Drobne, D. et al. (2015). Sperm exposure to carbon-based nanomaterials causes abnormalities in early development of purple sea urchin (*Paracentrotus lividus*). *Aquat. Toxicol.* 163: 158–166.

42 Magesky, A. and Pelletier, É. (2015). Toxicity mechanisms of ionic silver and polymer-coated silver nanoparticles with interactions of functionalized carbon nanotubes on early development stages of sea urchin. *Aquat. Toxicol.* 167: 106–123.

43 Saggese, I., Sarà, G., and Dondero, F. (2016). Silver nanoparticles affect functional bioenergetic traits in the invasive red sea mussel *Brachidontes pharaonis. Biomed. Res. Int.* 2016: 1872351.

44 Vale, G., Franco, C., Diniz, M.S. et al. (2014). Bioavailability of cadmium and biochemical responses on the freshwater bivalve *Corbicula fluminea* – the role of TiO_2 nanoparticles. *Ecotoxicol. Environ. Saf.* 109: 161–168.

45 Trevisan, R., Delapedra, G., Mello, D.F. et al. (2014). Gills are an initial target of zinc oxide nanoparticles in oysters *Crassostrea gigas*, leading to mitochondrial disruption and oxidative stress. *Aquat. Toxicol.* 153: 27–38.

46 Doyle, J.J., Ward, J.E., and Mason, R. (2015). An examination of the ingestion, bioaccumulation, and depuration of titanium dioxide nanoparticles by the blue mussel (*Mytilus edulis*) and the eastern oyster (*Crassostrea virginica*). *Mar. Environ. Res.* 110: 45–52.

47 Huang, X., Lin, D., Ning, K. et al. (2016). Hemocyte responses of the thick shell mussel *Mytilus coruscus* exposed to nano-TiO_2 and seawater acidification. *Aquat. Toxicol.* 180: 1–10.

48 Wang, Y., Hu, M., Li, Q. et al. (2014b). Immune toxicity of TiO_2 under hypoxia in the green-lipped mussel *Perna viridis* based on flow cytometric analysis of hemocyte parameters. *Sci. Total Environ.* 470–471: 791–799.

49 Gomes, T., Pereira, C.G., Cardoso, C. et al. (2014a). Effects of silver nanoparticles exposure in the mussel *Mytilus galloprovincialis. Mar. Environ. Res.* 101: 208–214.

50 Gomes, T., Chora, S., Pereira, C.G. et al. (2014b). Proteomic response of mussels *Mytilus galloprovincialis* exposed to CuO NPs and Cu^{2+}: an exploratory biomarker discovery. *Aquat. Toxicol.* 155: 327–336.

51 Hanna, S.K., Miller, R.J., and Lenihan, H.S. (2014). Deposition of carbon nanotubes by a marine suspension feeder revealed by chemical and isotopic tracers. *J. Hazard. Mater.* 279: 32–37.

52 Canesi, L., Ciacci, C., Bergami, E. et al. (2015). Evidence for immunomodulation and apoptotic processes induced by cationic polystyrene nanoparticles in the hemocytes of the marine bivalve *Mytilus*. *Mar. Environ. Res.* 111: 34–40.

53 Muller, E.B., Hanna, S.K., Lenihan, H.S. et al. (2014). Impact of engineered zinc oxide nanoparticles on the energy budgets of *Mytilus galloprovincialis*. *J. Sea Res.* 94: 29–36.

54 Munari, M., Sturve, J., Frenzilli, G. et al. (2014). Genotoxic effects of Ag_2S and CdS nanoparticles in blue mussel (*Mytilus edulis*) haemocytes. *Chem. Ecol.* 30: 719–725.

55 Tian, S., Zhang, Y., Song, C. et al. (2014). Titanium dioxide nanoparticles as carrier facilitate bioaccumulation of phenanthrene in marine bivalve, ark shell (*Scapharca subcrenata*). *Environ. Pollut.* 192: 59–64.

56 Wang, J. and Wang, W.-X. (2014). Low bioavailability of silver nanoparticles presents trophic toxicity to marine medaka (*Oryzias melastigma*). *Environ. Sci. Technol.* 48: 8152–8161.

57 Jarvis, T.A., Miller, R.J., Lenihan, H.S., and Bielmyer, G.K. (2013). Toxicity of ZnO nanoparticles to the copepod *Acartia tonsa*, exposed through a phytoplankton diet. *Environ. Toxicol. Chem.* 32: 1264–1269.

58 Gottschalk, F., Scholz, R.W., and Nowack, B. (2010). Probabilistic material flow modeling for assessing the environmental exposure to compounds: methodology and an application to engineered nano-TiO_2 particles. *Environ. Model Softw.* 25: 320–332.

59 Gottschalk, F., Sun, T.Y., and Nowack, B. (2013). Environmental concentrations of engineered nanomaterials: review of modeling and analytical studies. *Environ. Pollut.* 181: 287–300.

60 Corsi, I., Cherr, G.N., Lenihan, H.S. et al. (2014). Common strategies and technologies for the ecosafety assessment and design of nanomaterials entering the marine environment. *ACS Nano* 8: 9694–9709.

61 Buffet, P.-E., Fossi Tankoua, O., Pan, J.-F. et al. (2011). Behavioural and biochemical responses of two marine invertebrates *Scrobicularia plana* and *Hediste diversicolor* to copper oxide nanoparticles. *Chemosphere* 84: 166–174.

62 Melvin, S.D. and Wilson, S.P. (2013). The utility of behavioural studies for aquatic toxicity testing: a meta-analysis. *Chemosphere* 93: 2217–2223.

5

Interactive Effects of Nanomaterials with Other Contaminants on Aquatic Organisms: $nTiO_2$ as a Case Study

Laura Canesi[1], Camilla Della Torre[2], Teresa Balbi[1], and Ilaria Corsi[3]

[1]*Department of Earth, Environmental and Life Sciences, University of Genoa, Corso Europa 26, 16132, Genoa, Italy*
[2]*Department of Bioscience, University of Milan, via Celoria 26, 20133, Milano, Italy*
[3]*Department of Physical, Earth and Environmental Sciences, University of Siena, via Mattioli, 4-53100, Siena, Italy*

5.1 Introduction

The increasing production and usage in various fields of different types of manufactured nanomaterials (NMs) and nanoparticles (NPs), estimated to grow to over half a million tons by 2020, would lead to their release in substantial amounts in the environment. Being used, disposed of, and degraded, NMs are expected to enter soils and waterways and eventually reach the marine environment as nanowaste [1]. The possible impact of nanowaste on human and environmental health can have significant socioeconomic consequences [2]. Considerable concern is growing on environmental behavior of NMs (aggregation/agglomeration, redox reactions, dissolution, exchange of surface moieties, and reactions with biomacromolecules) and consequent impact on aquatic organisms [1–5]. This in particular applies to those types of NPs produced in higher amounts, such as metal oxides (silver, titanium dioxide, zinc oxide, silica) and carbon-based NMs (fullerenes, including single- and multi-walled carbon nanotubes). Due to their extremely small size, NPs exhibit unique physicochemical properties, such as high specific surface area and increased reactivity, which account for their widespread use in a number of industrial and biomedical applications. A better understanding of their fate and behavior in the natural environment will come from a prediction of NM exposure.

Major scientific gaps on fate models and tools (analytical and biological) for *ecosafety* assessment and design of NMs entering the natural environment need

Toxicology of Nanoparticles and Nanomaterials in Human, Terrestrial and Aquatic Systems, First Edition. Edited by Marc A. Williams.

to be filled [2]. The development, for instance, of quantitative approaches for integrating more realistic NP exposure scenarios and predicting ecosystem impacts are urgently needed for a proper risk assessment [4]. Particular attention should be given to those NPs intended for environmental applications as, for instance, wastewater treatment and, more generally, remediation that can cause a significant release in environmental matrices of much higher concentrations than those expected to be released by other anthropogenic sources [6].

In this respect, the inherent properties of NPs can represent a double-edged sword in an environmental context and in particular when released into the aquatic compartment. Both in the water phase and in the sediments, NPs will mix and interact with other components, not only with dissolved and particulate organic matter but also with hydrophobic organic contaminants and inorganic ions, including heavy metals [1, 5, 7, 8] (Figure 5.1). How these interactions may affect the biological impact of NPs *per se* or that of other pre-existing contaminants is largely unknown and is therefore the subject of current considerable debate.

Limbach et al. [9] first postulated the "*Trojan horse*" effect, which implies facilitated entry of toxic molecules adsorbed to NPs into the cells. Soon later Baun et al. [10] extended this concept in an environmental context. Risk assessment of NPs should not only focus on their inherent toxicity but also consider any possible interactions with existing environmental contaminants that may lead to changes in their bioavailability/bioconcentration/toxicity.

The assumption that NPs may interact with pre-existing environmental contaminants implies the need of studies aimed at evaluating these interactions in laboratory exposure media and natural environmental matrices. Although a great deal of effort is being made in evaluating NP behavior in aquatic media

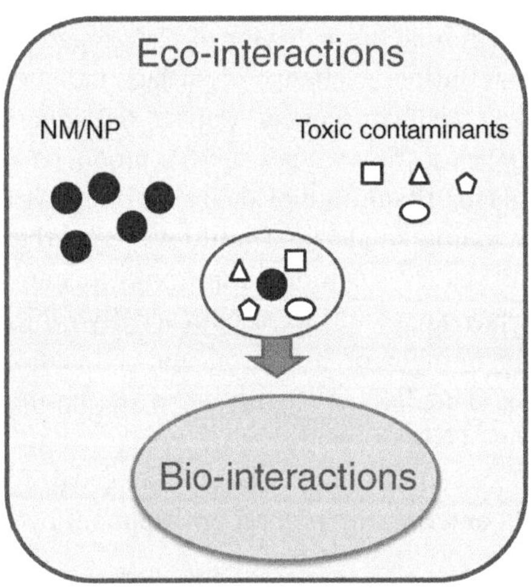

Figure 5.1 Schematic representation of the possible interactions of NPs with existing contaminants in water column or sediments (eco-interactions) possibly affecting bio-interactions within the organisms and with other molecules present in internal body fluids of the organisms (bio-interactions). These interactions will lead to the formation of complexes as NP–contaminant or as NP–biomolecule with new peculiar behavior and toxicity.

characterized by different physicochemical parameters [11–13], information is still scanty on absorption of other contaminants to different types of NPs. This particularly applies to marine waters, where the high ionic strength may represent an additional factor influencing NP–contaminant interactions.

Interactions with other contaminants will involve not only different types of manufactured and natural NPs but also nanoplastics derived by fragmentation of plastic wastes, which represent a further environmental concern [14]. Adsorption of hydrophobic aquatic pollutants by nanosized plastic debris has been shown to increase their bioavailability and consequent cellular uptake, once accumulated in the gut of different organisms [15, 16]. Interactions could take place at different levels starting from their release in the environmental matrices (water, sediment, soil), in the extracellular media (i.e. mucus, and algal extracellular polysaccharides [EPS]), and inside the organisms' body within biological fluids (i.e. blood, hemolymph, etc.) [17] (Figure 5.1). In particular, NP surface charges, size, and shape, irrespective of core composition, have been recently shown to drive their behavior toward different types of biological molecules (proteins, colloids, polysaccharides, etc.) released by organisms in environmental media, thus affecting particle recognition and uptake by different biota [17].

Once within the organism, both NPs and chemical contaminants may give rise to different types of interactive effects at different levels of biological organization, from molecular to the organism level, to induce a wide range of responses (i.e. additive, synergistic, or antagonistic) [18]. The biological effects of mixtures of pollutants are usually hard to predict. This is even more challenging when dealing with NP peculiar properties, behavior in different media, and possible interactions with other contaminants under realistic environmental conditions [19]. However, in recent years, many attempts have been made in order to understand whether the predicted increase of NP discharge into the aquatic environment in the presence of other chemicals, including "legacy" or "emerging" contaminants, may pose a further risk to organism and ecosystem health.

The purpose of this chapter is to summarize available data obtained from co-exposure experiments with titanium dioxide NPs ($nTiO_2$) and chemical contaminants, utilizing experimental models of both freshwater and marine organisms. In particular, data gathered in the framework of a project aimed at understanding the interactive effects of $nTiO_2$ with cadmium (Cd^{2+}) and dioxin (2,3,7,8-tetrachlorodibenzo-*p*-dioxin, 2,3,7,8-TCDD) in marine organisms in laboratory exposure experiments (Figure 5.2) will be critically discussed. The Mediterranean bivalve *Mytilus galloprovincialis* and the European sea bass *Dicentrarchus labrax* were chosen as models of marine invertebrates and vertebrates, respectively. Among invertebrates, bivalve mollusks, and in particular the marine *Mytilus* spp., long been utilized for the evaluation of biological impact of contaminants in coastal areas, have been shown to represent a significant target for

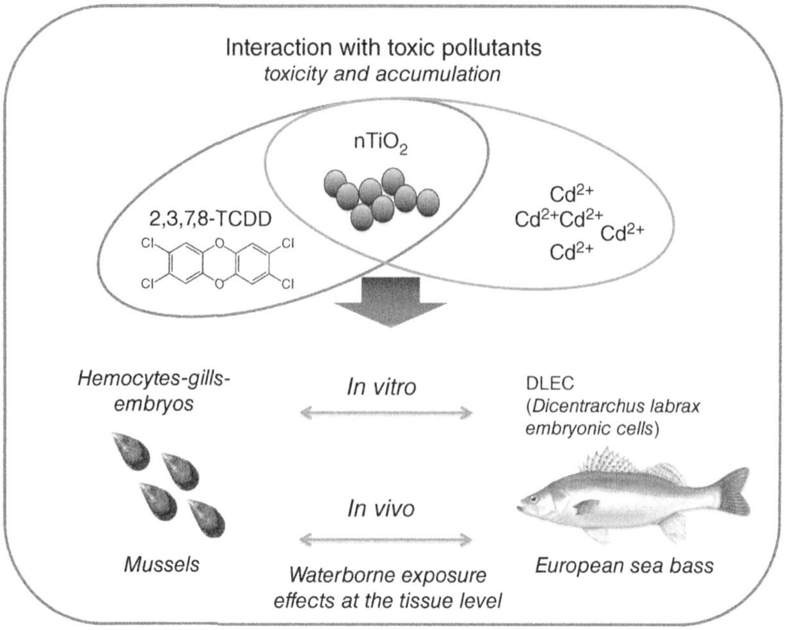

Figure 5.2 Schematic representation of the case study on the interactive effects of $nTiO_2$ with Cd^{2+} and 2,3,7,8-tetrachlorodibenzo-p-dioxin (TCDD) as models of inorganic and organic contaminants in marine mussels (*Mytilus galloprovincialis*) and fish (*Dicentrarchus labrax*).

different types of NPs, alone [20] and in mixtures with other pollutants [17, 18]. The teleost *D. labrax* has been widely used as a bioindicator in Mediterranean marine coastal areas. The species has been proposed as a useful ecotoxicological model for acute toxicity tests, biomarkers analysis, and bioaccumulation studies [21], and it has been has recently proposed for inclusion among species recommended in the Organization for Economic Co-operation and Development (OECD) guidelines for the toxicity testing of chemicals [22].

5.2 Interactive Effects of NPs with Other Contaminants in Aquatic Organisms: $nTiO_2$ as a Case Study

$nTiO_2$ is one of the most widespread NPs in use [23, 24]. Due to its large production and growing application, $nTiO_2$ is released in huge amounts in urban and industrial sewages and occurs in the aquatic environment and eventually in the sea [24, 25], where it may represent a risk to the health of marine organisms [1].

Photocatalytic interactions of $nTiO_2$ with different organic compounds and metals have been investigated, in view of its application in water treatment [26, 27].

The most studied example of the interactive effects of $nTiO_2$ and heavy metals – in particular Cd^{2+} – is on freshwater species. Available information shows that bioavailability, bioaccumulation, and toxicity of the metal are affected by the presence of $nTiO_2$, at mg/L levels, but with conflicting results [18].

In the green algae *Chlamydomonas reinhardtii*, Cd^{2+} toxicity as well as accumulation seemed to be reduced by $nTiO_2$ [28]. In the microcrustacean *Daphnia magna*, increased Cd^{2+} bioavailability has been reported in the presence of $nTiO_2$ [29]. By contrast, no influence of $nTiO_2$ was observed on Cd^{2+} uptake in the earthworm *Lumbriculus variegatus* [29]. In the freshwater bivalve *Corbicula fluminea*, Cd^{2+} uptake and antioxidant biomarker responses were unaffected by the presence of $nTiO_2$, despite $nTiO_2$ significantly reducing the free Cd^{2+} content in the medium [30].

Although these studies indicate changes in bioavailability/uptake/bioaccumulation of Cd^{2+} in the presence of $nTiO_2$, probably due to sorption of the metal to $nTiO_2$ in freshwater media, no clear evidence of increased toxicity was provided. These results were obtained with different $nTiO_2$ types, model organisms, and experimental conditions, utilizing concentrations of $nTiO_2 \geq 1$ mg/L. These concentrations, although higher than the predicted environmental concentrations (PECs) for $nTiO_2$ in surface waters (μg/L) [4], might be informative to characterize the potential toxicological impact of TiO_2 NPs [18].

With regard to freshwater fish, enhanced bioaccumulation of Cd^{2+} was reported in the carp (*Cyprinus carpio*) co-exposed with $nTiO_2$ [31]. Similarly, in zebrafish (*Danio rerio*), $nTiO_2$ enhanced Cd^{2+} bioaccumulation, but this effect was reduced in the presence of humic acid [32]. $nTiO_2$ increased the bioconcentration of Pb in zebrafish larvae, enhancing the metal toxic effect on thyroid endocrine and neuronal systems [33].

The combined effects of $nTiO_2$ and organic contaminants have been investigated using as biological model the developing zebrafish. Co-exposure to $nTiO_2$ (0.1 mg/L) and the polybrominated diphenyl ether (PBDE) polybrominated diphenyl ether (BDE)-209 increased BDE-209 bioavailability and metabolism and altered thyroid hormone levels, gene and protein expression levels, and larval locomotor behavior, indicating thyroid endocrine disruption and developmental neurotoxicity [34]. A significant sorption of bisphenol A on $nTiO_2$ has been reported, leading to enhanced toxic effects in zebrafish embryos and adults [35, 36]. In embryos, co-exposure reduced survival, increased morphological aberrations, and delay in hatching [35]. In adults, the adsorptive interactions lead to increased bisphenol A (BPA) body burden and altered concentration of several sex-related hormones and higher vitellogenin concentration in males

[36]. $nTiO_2$ did not increase pentachlorophenol (PCP) content in zebrafish larvae, but enhanced the metabolism of PCP to tetrachlorohydroquinone [37]. Besides, co-exposure of $nTiO_2$ with PCP enhanced the oxidative stress and genotoxicity with respect to each single pollutant, such as increased production of reactive oxygen species (ROS), upregulation of *nrf2* and *sod* gene transcription, increase in lipid peroxidation, and DNA damage.

Limited information is available on the combined effects of NMs, including $nTiO_2$ and other contaminants, in marine organisms. The presence of $nTiO_2$ significantly increased the embryotoxicity of tributyltin (TBT) to the abalone (*Haliotis diversicolor*) [38]. In the ark shell *Scapharca subcrenata*, enhanced uptake of phenanthrene by clams was recently observed in the presence of $nTiO_2$ [39]. On the contrary, co-exposure with $nTiO_2$ did not alter the bioaccumulation of several PBDE congeners in the same species [40].

5.3 Interactions Between $nTiO_2$ and Other Contaminants in Marine Invertebrates: The Example of the Bivalve *Mytilus*

Among aquatic organisms, suspension feeders have been shown to represent important test organisms in ecotoxicological research with respect to NPs [10], as well as in studies aimed at evaluating the possible interactive effects of NPs with other pollutants [29]. In the marine bivalve *M. galloprovincialis*, the effects of $nTiO_2$ alone have been investigated in a wide concentration range (from 1 μg/L to 5 mg/L). Significant effects from molecular to tissue level were observed, in particular those related to immune and digestive gland function, at concentrations (μg/L) much lower than those usually utilized in ecotoxicity tests on aquatic species (mg/L), and closer to PECs for $nTiO_2$ [20, 41]. These studies represented the basis for investigating the possibility that in *Mytilus* $nTiO_2$ may interact with other contaminants to induce responses at different levels of biological organization. Cd^{2+} and TCDD were chosen as models of inorganic and organic contaminants, respectively.

5.3.1 Effects of $nTiO_2$ and Cd^{2+}

The effects of $nTiO_2$ and Cd^{2+} were investigated in mussels exposed to $nTiO_2$ (100 μg/L) and to Cd^{2+} (100 μg/L), alone and in combination, for 96 h [42, 43]. The outcome of the interactive effects is summarized in Table 5.1. In the hemolymph of mussels exposed *in vivo* to the mixture, antagonistic effects were observed on functional immune parameters. Cd^{2+} abolished the increase in lysozyme activity and nitric oxide (NO) production induced by $nTiO_2$. Interestingly, short-term

Table 5.1 Interactive effects (antagonistic, additive, synergistic) of *in vivo* exposure to $nTiO_2$ and Cd^{2+} on different biomarkers measured in *Mytilus* hemolymph, gills and digestive gland (+, ++ increase; – , ––inhibition; /no effect).

Interactive effect/endpoint	$nTiO_2$	Cd^{2+}	Mixture
	Hemolymph		
ANTAGONISTIC			
Hemocyte NO production	+	/	/
Serum lysozyme activity	+	/	/
	Gills		
ADDITIVE			
NO production	+	+	++
ANTAGONISTIC			
Cd accumulation	/	++	+
GST activity	/	+	/
abcb1 expression	/	+	/
SYNERGISTIC			
Ti Accumulation	+	/	++
	Digestive gland		
ADDITIVE			
Lysosomal membrane stability	–	–	– –
ANTAGONISTIC			
Ti accumulation	++	–	+
Neutral Lipids	/	+	/
Lys/Cyt volume ratio	/	++	+
SYNERGISTIC			
GSTπ expression	/	/	–
Lysozyme Expression	+	/	++
TLR expression	/	/	–

exposure experiments (60 min) subsequently carried out in hemocytes *in vitro*, although at different concentrations, also showed antagonistic effects of Cd^{2+} toward $nTiO_2$ – Cd^{2+} protected from the $nTiO_2$ induced lysozyme release and reduced the $nTiO_2$-mediated increase in NO production. These data support the hypothesis that the two contaminants could rapidly act on common targets in the hemocyte.

In the digestive gland, distinct interactive effects of nTiO_2 and Cd^{2+} were observed *in vivo* on different lysosomal biomarkers. Despite the additive effects on lysosomal membrane stability (LMS), indicating stronger stressful conditions induced by the mixture, co-exposure with TiO_2 apparently protected from the effects of Cd^{2+} on other lysosomal parameters, such as increased neutral lipid accumulation and lysosome/cytoplasm volume ratio. Synergistic effects on transcription of the immune genes lysozyme and toll-like receptor (TLR) were observed, confirming interactive effects of the two contaminants on molecular targets related to the immune function.

The interactive effects of nTiO_2 and Cd^{2+} observed on the digestive gland could be due to differences in metal bioavailability/accumulation in the presence of nTiO_2. However, negligible adsorption of Cd^{2+} to nTiO_2 agglomerates was observed in artificial sea water (ASW) standard suspensions, and comparable concentrations of soluble Cd^{2+} were detected in exposure media in the absence and presence of nTiO_2 [42]. In addition, no differences in total Cd^{2+} tissue content were observed in the digestive gland of mussels exposed to Cd^{2+}, alone and in combination with nTiO_2. Moreover, nTiO_2 did not affect the Cd-induced increase in metallothionein at both transcriptional and protein level, and no interactive effects were observed on oxidative stress-related biomarkers (i.e. catalase and GSH transferase transcription and activities and lysosomal lipofuscin accumulation; Balbi, personal observations). Overall, these results clearly showed that the interactive effects of nTiO_2 and Cd^{2+} on digestive gland biomarkers could not be ascribed to changes in bioavailability and/or bioaccumulation of Cd^{2+} in the presence of nTiO_2 or in the heavy metal detoxication response.

The interactive effects of nTiO_2 and Cd^{2+} were further investigated in the gills, a tissue that in bivalves represent the first site of entry of soluble chemicals, including metals, from the surrounding environment [43]. Due to their role in the feeding process, the gills might represent the main site of disposal of NPs and their agglomerates. Several cellular biomarkers were evaluated at gene/protein and functional level: ATP-binding cassette (ABC) transport proteins and metallothioneins (*abcb1*, *abcc-like*, and *mt-20*), glutathione-S-transferase (GST) activity, and NO production. Significant responses to Cd^{2+} exposure were observed, such as upregulation of *abcb1* and *mt-20* gene transcription, increases in total MT content, P-gp efflux functionality and GST activity, DNA damage, and NO production. nTiO_2 alone increased P-gp efflux activity and NO production. When combined with Cd^{2+}, nTiO_2 reduced the metal-induced effects by significantly lowering *abcb1* gene transcription, GST activity, and DNA damage, whereas additive effects were observed on NO production. Accumulation of total Cd^{2+} and titanium in gills, as in whole soft tissue, was also investigated. Distinct from that observed in the digestive gland, a lower concentration of Cd^{2+} was observed in the gills upon co-exposure, suggesting a tissue-specific competitive effect in uptake/accumulation of nano-TiO_2 and Cd^{2+} in the gills.

With regard to Ti concentrations, no accumulation was observed in whole mussel tissues. This was in line with the observation that, in our experimental conditions, only a minor fraction (<10%) of the nominally added $nTiO_2$ is available to organism in the suspended form, due to both particle sedimentation and agglomeration/adsorption processes occurring in the presence organic particles produced by the mussels (mucus, fecal pellets, gametes) [42]. However, significant accumulation of Ti was observed both in the digestive gland and gills of mussels exposed to $nTiO_2$, alone and in combination with Cd^{2+} [42, 43]. Interestingly, in the digestive gland, lower tissue levels of Ti were measured in mussels co-exposed with Cd^{2+}, whereas in the gills co-exposure resulted in higher Ti levels.

Studies on different freshwater model species highlighted the interference of metals with NP uptake and *vice versa* [44, 45]. Nevertheless, these findings were often conflicting and did not consider tissue-specific accumulation. The results obtained on Ti and Cd^{2+} concentrations in mussel gills and digestive gland in comparison with whole soft tissues underline the importance of investigating tissue-specific accumulation when dealing with exposure to NPs in combination with other contaminants.

Finally, the possible impact of $nTiO_2/Cd^{2+}$ mixtures in *Mytilus* at the organism level was also evaluated in the 48 h embryo toxicity assay [43]. When fertilized eggs were exposed for 48 h to $nTiO_2$ and Cd^{2+}, at the same concentrations utilized in adult exposure experiments, significant effects were observed only with Cd^{2+}, in terms of both retarded development and malformed D-larvae. However, the effects of the metal were abolished in the presence of $nTiO_2$, further indicating antagonistic effects of the two contaminants. The observed effects of $nTiO_2$ and Cd^{2+} may result from interactions of either contaminant with both common and distinct targets/mechanisms of action at different levels of biological organization. Since developing larvae of aquatic organisms are most sensitive to environmental perturbations, further research is needed to investigate the interactive effects of NPs and other contaminants in the first developmental stages of mussels. Overall, the results obtained so far indicate that in *M. galloprovincialis* co-exposure to the $nTiO_2/Cd^{2+}$ mixture seemed to mitigate, rather than exacerbate, the adverse effect of each contaminant alone in different adult tissues, as well as in the developing embryo. In this light, the predominant antagonistic effects of the two contaminants seem to be confirmed at the whole organism level.

5.3.2 Effects of $nTiO_2$ and TCDD

Mussels were exposed *in vivo* to $nTiO_2$ (100 μg/L) and to TCDD (0.25 μg/L), alone and in combination, for 96 h [46]. A wide range of biomarkers, from molecular to

Table 5.2 Interactive effects of *in vivo* exposure to $nTiO_2$ and 2,3,7,8-TCDD, alone and in combination, on different biomarkers measured in *Mytilus* hemolymph, gills and digestive gland (+, ++ increase; – , ––inhibition; /no effect). Antagonistic effects were observed on all biomarkers.

Interactive effect/endpoint	$nTiO_2$	2,3,7,8-TCDD	Mixture
	Hemolymph		
Phagocytosis	–	–	/
	Gills		
MN frequency	+	/	/
	Digestive gland		
Total TCDD content	/	++	+
Neutral lipids	/	++	+
Lys/Cyt volume ratio	/	/	+

tissue level, were measured: LMS and phagocytosis in hemocytes, ATP-binding cassette efflux transporters in gills (gene transcription and efflux activity), several biomarkers as lysosomal biomarkers, and transcription of selected genes in the digestive gland. The results showed that co-exposure to $nTiO_2$ and TCDD affected the immune, gill, and digestive gland function. Both synergistic or antagonistic effects were observed, depending on experimental condition, cell/tissue, or type of measured response. However, antagonistic effects on different biomarkers seem to prevail (Table 5.2). In particular, *in vivo* as well as *in vitro* co-exposure with TCDD prevented the effects of $nTiO_2$ in terms of reduced phagocytic activity of the hemocytes and decreased multixenobiotic resistance (MXR) activities and increased genotoxicity in gills; moreover, $nTiO_2$ blunted TCDD-induced lysosomal lipid accumulation in the digestive gland. With regard to the effects at the molecular level, no interactive effects of $nTiO_2$/TCDD were observed on expression of antioxidant, heat shock response, abcc-like or immune-related genes in hemocytes, gills, or digestive gland.

In order to elucidate whether the interactive effects observed at the tissue level could be related to differences in bioaccumulation of TCDD in the presence of $nTiO_2$ and to identify possible molecular targets of $nTiO_2$ and TCDD, alone and in combination, the effects of different experimental conditions in the digestive gland were further investigated using an integrated approach transcriptomics/histopathology/immunohistochemistry [47]. Transcriptomic analysis identified 48, 49, and 62 differentially expressed genes (DEGs) in response to $nTiO_2$, TCDD, and $nTiO_2$/TCDD, respectively. Gene ontology (GO) term analysis revealed distinct biological processes affected in different experimental conditions. $nTiO_2$ mainly upregulated cytoskeletal genes, while TCDD upregulated endocrine and signal transduction-related processes. Co-exposure induced transcriptional changes common to individual treatments and identified a newly generated process, response to chemical stimulus. Transcriptomics data underlined changes in transcription of genes associated with cytoskeletal structures, revealing a general upregulation of microtubule-based movement-related genes in samples from $nTiO_2$ and mixture exposure groups. Among these, upregulation of tubulin at the mRNA and protein level was also confirmed by quantitative real time PCR (qRT-PCR) and tissue immunohistochemistry. These data indicated that the effects of $nTiO_2$ on cytoskeletal components prevailed over those of TCDD. In contrast, TCDD alone and in combination with $nTiO_2$ upregulated transcription of vitelline envelope-related genes, whereas no effects were induced by $nTiO_2$ alone. However, $nTiO_2$ reduced the effects of TCDD on processes related to signal transduction. Both transcriptional and histopathological data confirmed that exposure to the mixture did not increase overall stressful conditions in the tissue [47]. Tissue TCDD accumulation, evaluated by immunofluorescence with an anti-dioxins antibody, demonstrated that exposure to the mixture decreased, rather than increase, TCDD levels in the digestive gland. In contrast, higher TCDD accumulation was observed in whole mussel tissues in the presence of $nTiO_2$ [46]. Again, these data, which represent the first report on the effects of NPs and other contaminants on the transcriptome of marine invertebrates, underline the need to investigate contaminant accumulation in co-exposure experiments with other contaminants at the tissue level, in order to elucidate the occurrence of a *Trojan horse* effect.

Overall, the results obtained in the marine invertebrate model *Mytilus* represent the most extensive data available so far on the sublethal effects of NPs and other contaminants in aquatic organisms. Moreover, these data extend the knowledge on the molecular and cellular targets of NPs in bivalves. These data further underline the importance of evaluating accumulation of contaminants at the tissue level in co-exposure experiments with NPs and other contaminants [46].

5.4 Interactions Between $nTiO_2$ and Other Contaminants in Marine Fish: The Example of the European Sea Bass (*Dicentrarchus labrax*)

The interactive effects of NPs and pollutants to fish have been investigated on early life stages and on adults. In general, such studies showed that NPs are able to affect the bioavailability of either organic and inorganic contaminants. Nevertheless, the toxicological consequences of such interactions remain largely controversial – the influence of $nTiO_2$ and TCDD bioconcentration and sublethal toxicity (biotransformation) in the marine fish European sea bass *D. labrax*. Bioconcentration of TCDD in the presence of $nTiO_2$ was also investigated in the liver tissue as well as interaction of the pollutants in artificial sea water (ASW).

The cellular machinery activated in response to toxic contaminants exposure known as biotransformation represents the main defense mechanism devoted to protect cells against both endogenous and exogenous compounds [1]. Phase I of biotransformation has been well characterized in marine fish species and the CYP1A family recognized as induced by dioxin and dioxin-like compounds in fish liver [48, 49]. Phase 0 and III of biotransformation, played by ABC transport proteins belonging to the ABCB1 and ABCC1–2 families, respectively, far less known in marine fish, play a fundamental role in chemical disposition and toxicity [50–52]. MXR mechanism played by ABC genes and proteins has been proved to act as defense against toxicants also in marine fish species [53]. Nevertheless, up to now no information were available on the effects of NPs on fish ABC transporters and even more upon exposure to mixtures of different chemicals, as it happens in more realistic environmental scenarios.

Being the main organ involved in the biotransformation, fish liver is more susceptible to $nTiO_2$ waterborne exposure than other organs [54–56]. Recent studies showed that $nTiO_2$ induce oxidative stress in fish liver in waterborne exposure conditions, while other organs seem more susceptible upon dietary exposure [57, 58]. Therefore, there is an increasing need to address the impact of $nTiO_2$ on fish liver biotransformation (at different phase level as 0, I, II, and III) with the aim to identify potential interactions with known pathways of aquatic toxic pollutants [59–61].

An *in vivo* exposure study (7 day) was performed using $nTiO_2$ (1 mg/L) alone and in combination with TCDD (46 pg/L). Several genes were investigated in the liver as aryl hydrocarbon receptor repressor (*Ahrr*), estrogen receptor (*erβ2*), ABC transport proteins as *abcb1* and *abcc1-c2-g2*, cytochrome P450 (*cyp1a*), glutathione-*s*-transferase (*gst*), and glutathione reductase (*gr*). Two enzyme activities as 7-ethoxyresorufin-*O*-deethylase (EROD) activity and glutathione-*s*-transferase (GST) were also assayed in liver cytosolic fractions [62, 63]. Taking into consideration the recent findings on the cross-talk between aryl hydrocarbon receptor (AhR) and estrogen receptor (ER), the latter has been included as potentially affected by TCDD coupled with $nTiO_2$ [64].

No differences in liver bioconcentration/bioaccumulation of TCDD were observed in fish exposed to the mixture compared with single chemicals, although higher levels of total Ti were detected in the presence of TCDD [62]. These findings resulted in agreement of what reported in mussel co-exposed to Cd^{2+}, which showed higher levels of total Ti in gills of co-exposed specimens [43].

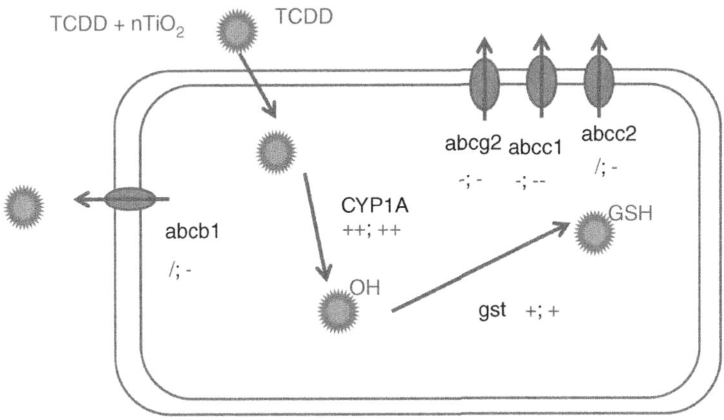

Figure 5.3 Schematic representation of modulation (+ upregulation, – downregulation, / no effect) of proteins involved in liver biotransformation of 2,3,7,8-TCDD and the influence of co-exposure with $nTiO_2$ in a marine fish, the European sea bass (*Dicentrarchus labrax*).

We hypothesized that, once in seawater, the TCDD partitioned directly into the fish without being affected by the presence of $nTiO_2$.

Concerning the gene expression profile in the liver of European sea bass, a confirmation of the ability of TCDD to induce *cyp1a* and *gst* genes was obtained. A significant upregulation of both *cyp1a* and *gst* was observed upon exposure to TCDD alone compared to control fish. Co-exposure with $nTiO_2$ seemed not to affect the upregulation induced by TCDD. $nTiO_2$ alone did not affect *cyp1a* and *gst* gene expression. Such findings seem to suggest the absence of interaction of $nTiO_2$ with TCDD on both phase I and II of biotransformation genes. Further evidence is given by the similar levels of TCDD detected in liver of fish both single and co-exposed to TCDD and $nTiO_2$ [62] (Figure 5.3).

Concerning phase 0 and III responses, ABC genes resulted downregulated with respect to controls in the liver of fish exposed to single TCDD and $nTiO_2$ and to the mixture [63]. Exposure to the mixture caused a further significant downregulation of *abcb1* and *abcc2* genes compared with single chemical exposure ($nTiO_2$ and TCDD alone), suggesting an interactive effect. *Abcc2* gene again resulted downregulated by $nTiO_2$ in single and in co-exposure experiments, but not by TCDD, which still caused a decrease in expression but not significant compared to controls. *Abcg2*, which overlaps substrate selectivity with both *abcb1* and *abcc1-2*, also showed a similar trend being downregulated by single and co-exposure conditions.

Such findings are in agreement with the documented less responsiveness of *abcb1* gene compared to *abcc1-2* and *abcg2* in the liver of European sea bass upon exposure to toxicants [64]. An interactive effect of TCDD and $nTiO_2$ was therefore hypothesized toward *abcb1* and *abcc2* genes expression but not for *abcc1* and *abcg2* in the liver of this species.

Abcb1 encodes for MDR1 or P-gp, which pumps out a broad range of uncharged or moderately basic amphipathic substrates [65], playing a central role in phase 0 of biotransformation. On the contrary, Abcc1 encodes for multidrug resistance protein 1 (MRP1), and together with *abcc2* they pump out of the cell, free glutathione (GSH), chemical conjugates with GSH, glucuronic acid, or sulfate as products of phase II of biotransformation known as phase III of biotransformation [66]. Based on these findings, ABC transporters have been identified as potential targets of $nTiO_2$ toxicity in the liver of a marine fish species, suggesting that further investigation are needed in order to find pathways/mechanism of interaction at molecular and cellular level and for risk assessment purposes.

Regarding phase I, TCDD, as expected, caused a significant induction of *cyp1a* and EROD activity of the same extent observed in fish exposed to mixture, while $nTiO_2$ alone did not affect neither gene expression nor enzyme activity [62]. Overall, the results indicate that $nTiO_2$ is unlikely to interfere with TCDD-dependent biotransformation gene expression in the liver of European sea bass, although the effects of co-exposure observed in ABC transport mRNAs might suggest an impact on xenobiotic metabolite disposition and transport in European sea bass liver.

5.5 Interactive Effects of NPs with Other Contaminants in Marine Species: Importance of Exposure Media

$nTiO_2$ behavior (sedimentation/agglomeration) in water media at different concentrations and ionic strengths had been previously characterized [67]. Our studies of laboratory exposure of marine mollusks and fish to $nTiO_2$ in combination with Cd^{2+} and TCDD were carried in ASW. In a first attempt to understand whether the interactive effects (or the lack of them) of different contaminants with $nTiO_2$, observed in different experimental conditions and model organisms, were due to interactions in exposure media, adsorption of either contaminants to $nTiO_2$ agglomerates in ASW was investigated.

Adsorption of Cd^{2+} to $nTiO_2$ in ASW standard suspensions was evaluated by sequential filtration, followed by acidic digestion and inductively coupled plasma mass spectrometry (ICP-MS) analysis of Ti and Cd content. The results indicated that the average amount of Cd adsorbed to the $nTiO_2$ retained by filtration was less than 10% of the nominal concentration of the metal. The small amount of adsorbed Cd detected could be explained by the natural tendency of Cd ions in sea water to form stable soluble complexes, both ionic and neutral, with chloride ions, depending on chloride concentration level, while only a minor portion is available as free Cd^{2+} to be adsorbed to $nTiO_2$ [42].

Interactions between $nTiO_2$ and TCDD were investigated by adsorption spectroscopy and by nuclear magnetic resonance (NMR) spectroscopy. The results suggested no interaction between $nTiO_2$ and TCDD in ASW. In addition, a further investigation with the same techniques highlighted that no interaction between $nTiO_2$ and naphthalene and 1,3-benzodioxole could be established in ASW [62].

5.6 Concluding Remarks

The results summarized in this chapter still represent to date one of the first and more thorough investigation on the interactive effects between NPs and other contaminants in marine organisms from the molecular to the organism level. From this experience, we could also learn that when dealing with seawater media, the interactions between NPs and other contaminants may be less relevant than generally thought in other aquatic media as freshwater to affect contaminant bioavailability/bioconcentration and toxicity in marine organisms.

Acknowledgments

This work was partly supported by the Italian Ministry of Research (PRIN 2009 FHHP2W "Marine ecotoxicology of manufactured nano-materials. Biological effects and bioaccumulation of combined exposure to nano-TiO_2, metals and dioxin on edible species").

References

1 Corsi, I., Cherr, G., Lenihan, H. et al. (2014). Common strategies and technologies for the ecosafety assessment and design of nanomaterials entering the marine environment. *ACS Nano* 8: 9694–9709.
2 Matranga, V. and Corsi, I. (2012). Toxic effects of engineered nanoparticles in the marine environment: model organisms and molecular approaches. *Mar. Environ. Res.* 76: 32–40.
3 Delay, M. and Frimmel, F.H. (2012). Nanoparticles in aquatic systems. *Anal. Bioanal. Chem.* 402: 583–592.
4 Gottschalk, F., Sun, T., and Nowack, B. (2013). Environmental concentrations of engineered nanomaterials: review of modeling and analytical studies. *Environ. Pollut.* 181: 287–300.
5 Baker, T.J., Tyler, C.R., and Galloway, T. (2014). Impacts of metal and metal oxide nanoparticles on marine organisms. *Environ. Pollut.* 186: 257–271.

6 Sanchez, A., Recillas, S., Font, X. et al. (2011). Ecotoxicity of, and remediation with, engineered inorganic nanoparticles in the environment. *Trends Anal. Chem.* 30: 507–516.
7 Maurer-Jones, M.A., Gunsolus, I.L., Murphy, C.J., and Haynes, C.L. (2013). Toxicity of engineered nanoparticles in the environment. *Anal. Chem.* 85: 3036–3049.
8 Grillo, R., Rosa, A.H., and Fraceto, L.F. (2014). Engineered nanoparticles and organic matter: a review of the state-of-the-art. *Chemosphere* 119C: 608–619.
9 Limbach, L.K., Wick, P., Manser, P. et al. (2007). Exposure of engineered nanoparticles to human lung epithelial cells: influence of chemical composition and catalytic activity on oxidative stress. *Environ. Sci. Technol.* 41: 4158–4163.
10 Baun, A., Sørensen, S.N., Rasmussen, R.F. et al. (2008). Toxicity and bioaccumulation of xenobiotic organic compounds in the presence of aqueous suspensions of aggregates of nano-C60. *Aquat. Toxicol.* 86: 379–387.
11 Christian, P., Von der Kammer, F., Baalousha, M., and Hofmann, T. (2008). Nanoparticles: structure, properties, preparation and behaviour in environmental media. *Ecotoxicology* 17: 326–343.
12 Philippe, A. and Schaumann, G.E. (2014). Interactions of dissolved organic matter with natural and engineered inorganic colloids: a review. *Environ. Sci. Technol.* 48: 8946–8962.
13 Sani-Kast, N., Scheringer, M., Slomberg, D. et al. (2015). Addressing the complexity of water chemistry in environmental fate modeling for engineered nanoparticles. *Sci. Tot. Environ.* 535: 150–159.
14 Lambert, S. and Wagner, M. (2016). Characterisation of nanoplastics during the degradation of polystyrene. *Chemosphere* 145: 265–268.
15 Rochman, C.M., Manzano, C., Hentschel, B.T. et al. (2013). Polystyrene plastic: a source and sink for polycyclic aromatic hydrocarbons in the marine environment. *Environ. Sci. Technol.* 47: 13976–13984.
16 Velzeboer, I., Kwadijk, C.J., and Koelmans, A.A. (2014). Strong sorption of PCBs to nanoplastics, microplastics, carbon nanotubes, and fullerenes. *Environ. Sci. Technol.* 48: 4869–4876.
17 Canesi, L. and Corsi, I. (2016). Effects of nanomaterials on marine invertebrates. *Sci. Total Environ.* 565: 933–940. doi: S0048-9697(16)30086-9.
18 Canesi, L., Ciacci, T., and Balbi, T. (2015). Interactive effects of nanoparticles with other contaminants in aquatic organisms: friend or foe? *Mar. Environ. Res.* 111: 128–134.
19 Hartmann, N.B. and Baun, A. (2010). The nano cocktail: ecotoxicological effects of engineered nanoparticles in chemical mixtures. *Integr. Environ. Assess. Manag.* 6: 311–314.

20 Canesi, L., Ciacci, C., Fabbri, R. et al. (2012). Bivalve molluscs as a unique target group for nanoparticle toxicity. *Mar. Environ. Res.* 76: 16–21.

21 Tornambe, A., Manfra, L., Canepa, S. et al. (2016). Adaptation of the fish juvenile growth test (OECD TG 215, 2000) to the marine species *Dicentrarchus labrax. Ecotoxicol. Environ. Saf.* 148: 1090–1095. doi:10.1016/j.ecoenv.2015.12.032.

22 Conti, D., Balzamo, S., Paina, A. et al. (2015). European sea bass (*Dicentrarchus labrax L.* 1758) as a sentinel species in Europe to study the effects of contaminants. *Annu. Res. Rev. Biol.* 8: 1–13.

23 Robichaud, C., Uyar, A.E., Darby, M.R. et al. (2009). Estimates of upper bounds and trends in nano-TiO_2 production as a basis for exposure assessment. *Environ. Sci. Technol.* 43: 4227–4233.

24 Menard, A., Drobne, D., and Jemec, A. (2011). Ecotoxicity of nanosized TiO_2. Review of *in vivo* data. *Environ. Pollut.* 159: 677–684.

25 Gondikas, A.P., von der Kammer, F., Reed, R.B. et al. (2014). Release of TiO_2 nanoparticles from sunscreens into surface waters: a one year survey at the old Danube recreational lake. *Environ. Sci. Technol.* 48: 5415–5422.

26 Huang, A., Wang, N., Lei, M. et al. (2013). Efficient oxidative debromination of decabromodiphenyl ether by TiO_2-mediated photocatalysis in aqueous environment. *Environ. Sci. Technol.* 47: 518–525.

27 Zou, X., Shi, J., and Zhang, H. (2014). Coexistence of silver and titanium dioxide nanoparticles: enhancing or reducing environmental risks? *Aquat. Toxicol.* 154: 168–175.

28 Yang, W.-W., Miao, A.-J., and Yang, L.-J. (2012). Cd^{2+} toxicity to a green alga *Chlamydomonas reinhardtii* as influenced by its adsorption on TiO_2 engineered nanoparticles. *PLoS One* 7, e32300–3.

29 Hartmann, N.B., Legros, S., Von der Kammer, F. et al. (2012). The potential of TiO_2 nanoparticles as carriers for cadmium uptake in *Lumbriculus variegatus* and *Daphnia magna*. *Aquat. Toxicol.* 118–119: 1–8.

30 Vale, G., Franco, C., Diniz, M.S. et al. (2014). Bioavailability of cadmium and biochemical responses on the freshwater bivalve *Corbicula fluminea*-the role of TiO_2 nanoparticles. *Ecotoxicol. Environ. Saf.* 109: 161–168.

31 Zhang, X.Z., Sun, H.W., Zhang, Z.Y. et al. (2007). Enhanced bioaccumulation of cadmium in carp in the presence of titanium dioxide nanoparticles. *Chemosphere* 67: 160–166.

32 Hu, X., Chen, Q., Jiang, L. et al. (2011). Combined effects of titanium dioxide and humic acid on the bioaccumulation of cadmium in zebrafish. *Environ. Pollut.* 159: 1151–1158.

33 Miao, W., Zhu, B., Xiao, X. et al. (2015). Effects of titanium dioxide nanoparticles on lead bioconcentration and toxicity on thyroid endocrine system and neuronal development in zebrafish larvae. *Aquat. Toxicol.* 161: 117–126.

34 Wang, Q., Chen, Q., Zhou, P. et al. (2014). Bioconcentration and metabolism of BDE-209 in the presence of titanium dioxide nanoparticles and impact on the thyroid endocrine system and neuronal development in zebrafish larvae. *Nanotoxicology* 8: 196–207.

35 Yan, J., Lin, B., Hu, C. et al. (2014). The combined toxicological effects of titanium dioxide nanoparticles and bisphenol A on zebrafish embryos. *Nanoscale Res. Lett.* 9: 406.

36 Fang, Q., Shi, Q., Guo, Y. et al. (2016). Enhanced bioconcentration of bisphenol A in the presence of nano-TiO_2 can lead to adverse reproductive outcomes in zebrafish. *Environ. Sci. Technol.* 50: 1005–1013.

37 Fang, Q., Shi, X.J., Zhang, L.P. et al. (2015). Effect of titanium dioxide nanoparticles on the bioavailability, metabolism, and toxicity of pentachlorophenol in zebrafish larvae. *J. Hazard. Mater.* 283: 897–904.

38 Zhu, X., Zhou, J., and Cai, Z. (2011). TiO_2 nanoparticles in the marine environment: impact on the toxicity of tributyltin to abalone (*Haliotis diversicolor* supertexta) embryos. *Environ. Sci. Technol.* 45: 3753–3758.

39 Tian, S., Zhang, Y., Song, C. et al. (2014). Titanium dioxide nanoparticles as carrier facilitate bioaccumulation of phenanthrene in marine bivalve, ark shell (*Scapharca subcrenata*). *Environ. Pollut.* 192: 59–64.

40 Tian, S., Zhang, Y., Song, C. et al. (2015). Bioaccumulation and biotransformation of polybrominated diphenyl ethers in the marine bivalve (*Scapharca subcrenata*): influence of titanium dioxide nanoparticles. *Mar. Polltu. Bull.* 90: 48–53.

41 Barmo, C., Ciacci, C., Canonico, B. et al. (2013). In vivo effects of n-TiO_2 on digestive gland and immune function of the marine bivalve *Mytilus galloprovincialis*. *Aquat. Toxicol.* 15: 132–133.

42 Balbi, T., Smerilli, A., Fabbri, R. et al. (2014). Co-exposure to n-TiO_2 and Cd^{2+} results in interactive effects on biomarker responses but not in increased toxicity in the marine bivalve *M. galloprovincialis*. *Sci. Tot. Environ.* 493: 355–364.

43 Della Torre, C., Balbi, T., Grassi, G. et al. (2015). Titanium dioxide nanoparticles modulate the toxicological response to cadmium in the gills of *Mytilus galloprovincialis*. *J. Haz. Mat.* 297: 92–100.

44 Tan, C. and Wang, W.-X. (2014). Modification of metal bioaccumulation and toxicity in *Daphnia magna* by titanium dioxide nanoparticles. *Environ. Pollut.* 186: 34–42.

45 Pavagadhi, S., Sathishkumar, M., and Balasubramanian, R. (2014). Uptake of Ag and TiO_2 nanoparticles by zebrafish embryos in the presence of other contaminants in the aquatic environment. *Wat. Res.* 55: 280–291.

46 Canesi, L., Frenzilli, G., Balbi, T. et al. (2014). Interactive effects of n-TiO_2 and 2,3,7,8-TCDD on the marine bivalve *Mytilus galloprovincialis*. *Aquat. Toxicol.* 153: 53–65.

47 Banni, M., Sforzini, S., Balbi, T. et al. (2015). Combined effects of n-TiO_2 and 2,3,7,8-TCDD in *Mytilus galloprovincialis* digestive gland: a transcriptomic and immunohistochemical study. *Environ. Res.* 145: 135–144.

48 van der Oost, R., Beyer, J., and Vermeulen, N.P.E. (2003). Fish bioaccumulation and biomarkers in environmental risk assessment: a review. *Environ. Toxicol. Pharmacol.* 13: 57–149.

49 Della Torre, C., Mariottini, M., Trisciani, A. et al. (2014). Induction of CYP1A and ABC transporters in European sea bass (*Dicentrarchus labrax*) upon 2,3,7,8-TCDD waterborne exposure. *Mar. Environ. Res.* 99: 218–222.

50 Kurelec, B. (1992). The multixenobiotic resistance mechanism in aquatic organisms. *Crit. Rev. Toxicol.* 22: 23–43.

51 Leslie, E.M., Deeley, R.G., and Cole, S.P. (2005). Multidrug resistance proteins: role of P-glycoprotein, MRP1, MRP2, and BCRP (ABCG2) in tissue defense. *Toxicol. Appl. Pharmacol.* 204: 216–237.

52 Xu, C.C., Yong-Tao, L., and Kong, A.N.T. (2005). Induction of phase I, II and III drug metabolism/transport by xenobiotics. *Arch. Pharm. Res.* 28: 249–268.

53 Luckenbach, T., Fischer, S., and Sturm, A. (2014). Current advances on ABC drug transporters in fish. *Comp. Biochem. Physiol. C* 165: 28–52.

54 Linhua, H., Zhenyu, W., and Baoshan, X. (2009). Effect of sub-acute exposure to TiO_2 nanoparticles on oxidative stress and histopathological changes in Juvenile Carp (*Cyprinus carpio*). *J. Environ. Sci.* 21: 1459–1466.

55 Chen, J.Y., Dong, X., Xin, Y.Y., and Zhao, M.R. (2011). Effects of titanium dioxide nanoparticles on growth and some histological parameters of zebrafish (*Danio rerio*) after a long-term exposure. *Aquat. Toxicol.* 101: 493–499.

56 Diniz, M.S., Alves de Matos, A.P., Lourenco, J. et al. (2013). Liver alterations in two freshwater fish species (*Carassius auratus* and *Danio rerio*) following exposure to different TiO_2 nanoparticles concentrations. *Microsc. Microanal.* 19: 1131–1140.

57 Ramsden, C., Smith, T., Shaw, B., and Handy, R. (2009). Dietary exposure to titanium dioxide nanoparticles in rainbow trout, *Oncorhynchus mykiss*: no effect on growth, but subtle biochemical disturbances in the brain. *Ecotoxicology* 18: 939–951.

58 Ramdsen, C.S., Henry, T.B., and Handy, R.D. (2013). Sub-lethal effects of titanium dioxide nanoparticles on the physiology and reproduction of zebrafish. *Aquat. Toxicol.* 126: 404–413.

59 Paetzold, S.C., Ross, N.W., Richards, R.C. et al. (2009). Upregulation of hepatic ABCC2, ABCG2, CYP1A1 and GST in multixenobiotic-resistant killifish (*Fundulus heteroclitus*) from Sydney Tar Ponds, Nova Scotia, Canada. *Mar. Environ. Res.* 68: 37–47.

60 Zucchi, S., Corsi, I., Luckenbach, T. et al. (2010). Identification of five partial ABC genes in the liver of the Antarctic fish *Trematomus bernacchii* and sensitivity of ABCB1 and ABCC2 to Cd exposure. *Environ. Pollut.* 158: 2746–2756.

61 Costa, J., Reis-Henriquesa, M.A., Wilson, M.J., and Ferreira, M. (2013). P-glycoprotein and CYP1A protein expression patterns in Nile tilapia (*Oreochromis niloticus*) tissues after waterborne exposure to benzo(a)pyrene (BaP). *Environ. Toxicol. Pharmacol.* 36: 611–625.

62 Della Torre, C., Buonocore, F., Frenzilli, G. et al. (2015). Influence of titanium dioxide nanoparticles on 2,3,7,8-tetrachlorodibenzo-*p*-dioxin bioconcentration and toxicity in the marine fish European sea bass (*Dicentrarchus labrax*). *Environ. Pollut.* 196: 185–193.

63 Vannuccini, M.L., Grassi, G., Leaver, M.J., and Corsi, I. (2015). Combination effects of nano-TiO_2 and 2,3,7,8-tetrachlorodibenzo-p-dioxin (TCDD) on biotransformation gene expression in the liver of European sea bass *Dicentrarchus labrax*. *Comp. Biochem. Physiol. C Toxicol. Pharmacol.* 176-177: 71–78.

64 Ferreira, M., Costa, J., and Reis-Henriques, M.A. (2014). ABC transporters in fish species: a review. *Front. Phys.* 5: 1–12.

65 Ambudkar, S.V., Cardarelli, C.O., Pashinsky, I., and Stein, W.D. (1997). Relation between the turnover number for vinblastine transport and for vinblastine-stimulated ATP hydrolysis by human P-glycoprotein. *J. Biol. Chem.* 272: 21160–21166.

66 Deeley, R.G., Westlake, C., and Cole, S.P. (2006). Transmembrane transport of endo-and xenobiotics by mammalian ATP-binding cassette multidrug resistance proteins. *Physiol. Rev.* 86: 849–899.

67 Brunelli, A., Poiana, G., Callegaro, S., and Marcomini, A. (2013). Agglomeration and sedimentation of titanium dioxide nanoparticles (n-TiO_2) in synthetic and real waters. *J. Nanopart. Res.* 15: 1–10.

6

Soil Nano-ecotoxicology: What Have We Learned from Standard Tests and What May We Be Missing?

David J. Spurgeon, Elma Lahive, Carolin Schultz, and Claus Svendsen

UK Centre for Ecology and Hydrology, MacLean Building, Benson Lane, Wallingford, Oxon, OX10 8BB, UK

6.1 Introduction

Nanotechnology can create a wide array of nanomaterials of different chemistries, sizes, shapes, coatings, surface reactivity, and surface properties. Such variations provide the essential characteristics needed for the range of different applications for which nanotechnologies are used and proposed. However, those same features can also influence the way nanomaterials behave in the environment, interact with living organisms, and perturb different biochemical, metabolic, and physiological pathways and processes. The potential for new or different fates, exposure routes, and modes of action of nanomaterials compared to bulk materials has raised specific concern that releases from nanotechnology products may result in exposure and toxicity in ways that may not be fully understood based on current understanding. This concern has been the basis for the growth in interest in understanding fate and effects of nanomaterials for humans and other species.

The growing use of nanomaterials in consumer and industrial products means that nanomaterials are currently already entering the environment. These releases can be expected to increase as the take-up of nanotechnology increases in different industrial and consumer sectors. Initial efforts to identify and characterize these emissions to the environment have identified a range of forms of release from different sources. Major potential routes include uncontrolled releases during production or from accidents (e.g. through effluent or solid waste releases); direct product use (e.g. nanopesticides or nanofertilizers); product weathering (e.g. release from paints and surface coatings); and transport through waste streams (e.g. cosmetics and household products). While solid waste releases to landfill and incineration may be relatively well contained, transport through sewage treatment systems can result in nanomaterials reaching surface waters via effluent and

Toxicology of Nanoparticles and Nanomaterials in Human, Terrestrial and Aquatic Systems, First Edition. Edited by Marc A. Williams.

soil through sewage sludge amendment. Studies of nanomaterial behavior during sewage treatment have generally indicated a high degree of sedimentation to the sludge compartment during primary and secondary sewage treatment [1]. The consequence of this sedimentation is that a large fraction of many nanomaterials will frequently end up in sewage sludge [2]. Hence where sludge application occurs, soils can be a major receiving compartment.

Recognizing that soils can be a major sink, there has been growing interest in assessing the toxicity of nanomaterials for soil species. Over the past decade, numerous studies have sought to identify hazards for a range of soil invertebrates, plants, and microbial species and their functions. This research on exposure and effect in soils has been both an academic and a regulatory science pursuit. For academics, questions to be answered in nano-ecotoxicology include (i) the relationship between the toxicities of nano- and non-nano-material forms, (ii) the mechanisms of toxicity, (iii) the environmental drivers of bioavailability, (iv) the traits that make species potentially vulnerable to exposure and effect, (v) the influence of nanomaterial characteristics on observed toxicity, and (vi) how chemical and physical transformations of nanomaterials in waste streams and in the environment affect exposure, uptake, and toxicity. The regulatory science perspective in nano-ecotoxicology has been focused on the development of standard methods appropriate for nanomaterial toxicity tests (mainly based on adaptation of existing test guidelines), the identification and certification of reference materials, the feasibility of quantitative structure/activity/property relationships for predictive toxicology, the development of grouping approaches, and the design of risk assessment, risk management, and governance frameworks.

6.2 Development of Standard Test Methods and Their Application to Nanomaterials

The regulatory drivers that underpin the development of standardized soil organism toxicity tests for nanomaterials match those for conventional chemical risk assessment practices. Within regulatory regimes such as the Europe's Registration, Evaluation, Authorization and Restriction of Chemicals (REACH) policy of the European Union and product-specific legislative frameworks for classes of chemicals such as pesticides, biocides, veterinary medicines, and pharmaceuticals, standardization of test procedures meet the requirement to be able to compare studies conducted across different laboratories on an equivalent basis. The overall process of standardization is a long-term and highly organized activity involving draft protocol development followed by ring-testing, amendment, and finally ratification. Ultimately the final product of a process that takes years, rather than months, is a detailed protocol setting out the overall approach to be used: the choice of test species (and strains) for study, experimental design and test conditions, reference chemicals to be used, performance criteria for the results, and techniques for statistical analysis. Efforts by researchers across the world from academic, industry, and regulatory backgrounds have, to date, resulted in the ratification of ecotoxicity test protocols for soil organisms by transnational bodies such as the Organisation for Economic Co-operation and Development (OECD), International Organization for Standardization (ISO), and European Centre for Standardisation. These highly controlled bioassays, or variant thereof, now form the backbone of current prospective and retrospective risk assessment policies.

Efforts to standardize toxicity test procedures generally seek to optimize desirable characteristics. Preferred features include being rapid, simple, replicable, economical, sensitive, to have discriminatory power, be ecological relevant, be relevant to field effects, and be useable for developing robust regulatory criteria [3]. While seeking to match as many criteria as possible, the nature of some characteristics can mean they are mutually exclusive. Hence no matter the experimental design, it is not possible to achieve all desired criteria within a given test. For example, controlling environmental variables within a system by maintaining constant temperature or allowing unrestricted access to food can often come at the cost of ecological relevance, since species in the field will rarely, if ever, be subject to such stable, often idealized, conditions. Similarly the use of short-term exposures, to allow rapid and cost-effective testing, may mean that chemicals with different toxicokinetics may be sampled at different windows within their toxicokinetic and toxicodynamic time course of effect [4, 5]. Such variability in sampling effects over time may limit the discriminatory power of a given assay, especially for chemical with slow toxicokinetics [5].

The standardization of toxicity testing has been a key factor that has supported the generation of an extensive database of ecotoxicity data for conventional chemicals (see www.epa.gov/ecotox; also [6, 7]). The large majority of this data is collected for aquatic species, most notably water flea (Cladocerans), different algal species, some insect taxa (e.g. chironomids, Ephemeroptera), and fish (e.g. rainbow trout, fathead minnow, zebra fish). Studies for soil organisms form only a relatively minor portion of the total reported studies. Current standardized soil organism toxicity tests include methods that cover terrestrial plants (OECD, ISO), earthworms (OECD, ISO), Enchytraeids (OECD), Collembola (OECD, ISO), predatory Acari (ISO), Dipteran dung flies (ISO), nematodes (ISO), soil nitrogen mineralization (OECD, ISO), soil carbon mineralization (OECD, ISO), and soil dehydrogenase enzyme activity (ISO). The toxicity data that is cataloged for different soil taxa within the ecotoxicological database reflects the availability of standardized test procedures, being dominated by earthworms, enchytraeids, springtails, and nematodes (Figure 6.1), as well as some studies for terrestrial (crop) plants and soil microbial functions [6, 8].

Figure 6.1 Invertebrate species current used for standardized toxicity tests for soil: (a) *Eisenia fetida*; (b) *Folsomia candida*; (c) *Enchytraeus albidus*; (d) *Caenorhabditis elegans*.

The rapid ramp-up of nanotechnology research, coupled with the identification of soils as a major receiving compartment, has resulted in a growing number of publications documenting nanomaterial behavior in soil and effects on soil organisms. This growing focus is illustrated by the burgeoning number of research papers addressing aspects of the topic. In the past 10 years, the number of articles published retrieved by a Web of Knowledge search that includes the terms "soil and nano* and *toxic*" has increased from less than 10 papers in 2006 to over 200 papers in later years. Only in 2015 and 2016 has this rapid rise shown signs of "leveling off," staying at a constant output of around 220 papers per year in 2015 and 2016 (see Figure 6.2). The potential interest in the field is further indicated by the rapid rise in citations, reaching over 5000 in 2016 alone.

Interest in quantifying nanomaterial impact on soil ecosystems naturally brings a need to quantify nano-ecotoxicological effects in relevant species. For this purpose, the toxicity tests that had been developed, standardized, and used for routine hazard screening of chemicals were the obvious choices through which nano-ecotoxicologists could approach such studies. The early application of existing standardized laboratory test protocols in nano-ecotoxicology is clearly reflected by the species bias of the current literature on nanomaterial effects for soil organisms. Not only are current studies dominated by laboratory-based assessment, with few semi-field and field studies reported, but also the species for which standard tests exist, namely, earthworms and Enchytraeids (*Eisenia fetida*,

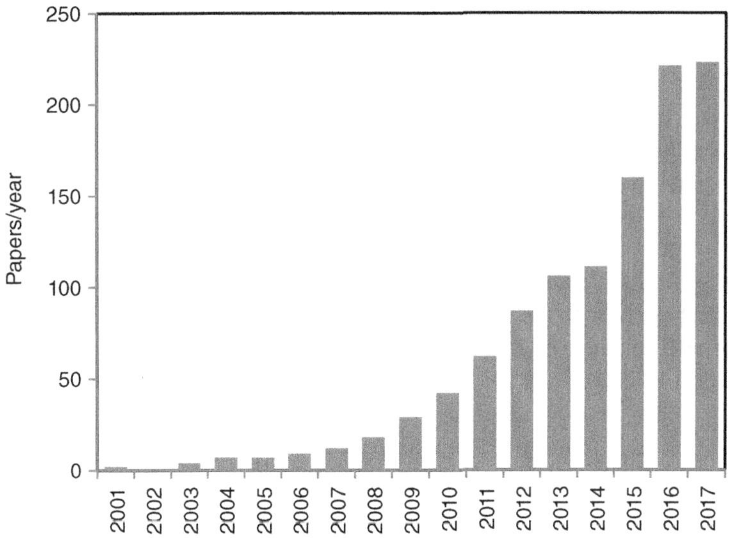

Figure 6.2 Number of papers in Web of Knowledge by year containing the search terms "soil and nano* and *toxic*" showing the rapid rise in publication relating to the field from a low base in 2004.

Enchytraeus crypticus), springtails (*Folsomia candida*), crop plants, and soil microorganisms, dominate the published literature. A further soil-dwelling species, the nematode *Caenorhabditis elegans* is perhaps the most extensively used species for nano-ecotoxicology studies (along with *Daphnia* species for freshwater studies). However, even though normally soil dwelling, the large majority of nano-ecotoxicology studies with this model nematode are conducted in aquatic media (e.g. M9 buffer, K-media). Such exposure conditions poorly represent true soil or even soil solution conditions [9]. Hence studies with this species, although valuable in other contexts (e.g. mechanistic toxicology, comparative ecotoxicology), may not be considered as "true" soil studies.

When undertaking soil organism nano-ecotoxicology tests, a key consideration was whether there were features of relevant test systems that may need specific adaptation to allow robust and reliable effect assessment for different nanomaterial types and forms. Reviews undertaken within a series of research initiatives, for example, the OECD Sponsorship Program for Representative Nanomaterials, have given considerable thought to the applicability of standard tests for nanomaterials [10]. These reviews have largely supported the choices of test species, conditions, media, and overall experimental approaches in respect of duration, measurement end points, and overall design. Where specific concerns have been raised, these have generally related to the method for spiking nanomaterials into the test medium and characterization needs, both at the start and during the test. Key focusses in both respects include understanding homo-agglomeration and subsequently tracking changes in core chemistry and speciation, understanding changes in coating integrity, and assessing surface properties over time. As a consequence, these aspects have been widely investigated within the context of soil nanotoxicology studies.

First consideration for standardized testing is creating homogeneous exposures that produce replicable results across different dosing schemes. Currently no single method is universally accepted for spiking nanoparticles into soil prior to toxicity testing. To date, three broad classes of techniques have mainly been used. These approaches cover studies in which nanomaterials are (i) added to soil as a powder prior to mixing, (ii) spiked into soil following nanoparticle dispersion in an aqueous or organic carrier, or (iii) added to soil following dispersion in solution designed to mimic pore water conditions, or a soil extract. In a study comparing the distribution and characteristics of ZnO nanoparticles added to soil using a separate technique from each of these classes, Miglietta et al. [11] found that each achieved a similar homogeneous distribution. However, analysis of soil extracts found higher concentrations of Zn in soils that were spiked with dry powder than with water or soil extract dispersions. Accordingly, the effects of exposure on root length growth of *Lepidium sativum* were also greatest in the dry spiked soil. Waalewijn-Kool et al. [12] also reported on a study to compare a dry spiking method to a soil extract dispersion approach for ZnO nanoparticles. Measurement of total metal levels in sub-samples of the spiked soil also found that each method produced a homogeneous distribution. In this case, however, toxicity to the springtail *F. candida* was similar between the two spiking methods. Hence while there is clearly a need to identify and standardize a preferred method for spiking, the available evidence points to the ability of all approaches to obtain a homogeneous soil distribution (at least at broader measurement scales). Further, method choices does not consistently lead to variations in nanoparticle behavior and toxicity.

A second identified requirement for soil nano-ecotoxicology studies is to better understand the fate and behavior in the test system. The need for high quality nanomaterial characterization, not just prior to the start, but also as the test progresses, is widely recognized in nano-ecotoxicology. In aquatic tests, there are a number of potential characterization options available covering analytical, spectroscopic, and imaging methods. In soils, the granular form of the medium itself, coupled to its opaque nature, limits the simple application of many characterization methods that are more easily applied in aquatic media, such as in dynamic light scattering, nanoparticle tracking analysis, atomic force microscopy, and some applications of electron microscopy. Frequently bulk chemical analysis following harsh extraction is the only method pursued for measuring nanomaterial concentrations in soil. Beyond simple total measurements, weak chemical extractions or direct collection of soil pore water (e.g. by centrifugation) can be used to study partitioning between the soil and pore water phases [13]. Methods such as field flow fractionation coupled with elemental detection using, for example, single particle inductively coupled plasma mass spectrometry, can offer the opportunity to further identify particle agglomeration status in these soil extracts [14]. This analysis can be supported by imaging using methods such as electron microscopy and other spatial techniques, although generally these can only visualize nanoparticles at the soil surface [15].

6.3 From Soil Ecotoxicological Tests to Risk Assessment

While valuable as phenomenological data in their own right, it is the use of "meta-analysis" of the growing nano-ecotoxicology database that has provided the coherent framework within which individual results can be placed to help support hazard and risk assessment. These "species sensitivity distributions" (SSDs) approaches were initially developed for conventional chemicals from principles outlined by Kooijman [16] and later refined by Van Straalen and Denneman [17] and Aldenberg and Slob [18]. To develop an SSD for a chemical or nanomaterial, a review of the published scientific and "gray" literature is conducted to collect all toxicity data available for taxa that are relevant to the environmental compartment in question (e.g. soil organisms, sediment dwellers, freshwater species, marine species). Data is collected as calculated toxicity metrics, such as no observed effect concentration (NOEC) or effect concentrations (LC_x/EC_x) for all relevant measurement parameters – usually life cycle-related end points such as survival, growth, or reproduction. Collected values are compiled into a cumulative frequency plot of all species, with the value for the toxicity metric on the *x*-axis and the cumulative probability on the *y*-axis. A statistical model is fitted to describe this ranked order distribution from which a concentration in the left hand tail of the curve can be identified that is predicted to result in an adverse effects for only a small proportion (usually 5%) of all species. This concentration designated as the HC*p*, "hazardous concentration for *p* percent of species" [19], can be taken as a protection goal above which environmental concentrations should not rise if exposed communities are to remain largely protected.

As the amount of nano-ecotoxicology data has grown, it has become possible to compile an initial set of species sensitivity distributions for well-studied nanomaterials [20–22]. The prevailing dominance of aquatic species effect data within the current literature has so far meant that SSDs have been generated only for the aquatic compartment. Further, even for aquatic ecosystems, the published SSDs cover only the most widely studied nanomaterials, including some metals and metal oxides, such as Ag, ZnO, and CuO, as well as carbon nanotubes. One of the key decisions that needs to be made prior to the compilation of nanomaterial SSDs is the degree to which the available ecotoxicology data for nanomaterials with the same primary chemistry should be further divided by differences in properties such as size ranges, shapes, or coatings. This question is naturally governed by the extent to which the particle properties in question are known to affect toxicity. Among published nanomaterial SSDs, current data separation has been made solely based on the chemical composition of the primary particle, not taking other properties into consideration.

Efforts have been made to relate the toxicity of metal nanomaterial forms to that of the matching dissolved ion. For example, in a meta-analysis of studies that measured metal and metal oxide nanomaterial and the relevant metal ion effects in the same exposure system, Notter et al. [23] found that nanomaterial toxicity was the same or below the toxicity of the ionic form in 93.8%, 100%, and 81% of cases for Ag, CuO, and ZnO paired studies. This comparative work highlights the potential value that ionic metal studies could play in the support of nanomaterial risk assessments in those cases where sufficient bespoke nanomaterial toxicity data is lacking. To specifically study how different properties affect toxicity, Garner et al. [20] analyzed single species toxicity data (LC_x, LD_x, EC_x, IC_x, LOEC, NOEC) from >300 published studies for aquatic species exposed to Ag, CuO, ZnO nanoparticles, and carbon nanotubes. Results suggested that size, formulation, and the presence of a coating can alter toxicity, indicating the potential need to separate data beyond core chemistry. This implies a need to further understand how variations in properties contribute to the hazard and risk assessment of nanomaterials. Such work, encompassing the development of read-across approaches, is a key focus of current research in the field.

6.4 Looking Beyond Standardized Tests Toward Effects in Ecosystems

Although of undoubted value for routine assessment in both regulatory and academic research, a dogma of soil nano-ecotoxicology research that is based exclusively on the use of standard tests alone cannot capture and detect a range of key nanoparticle effects relevant to possible impacts on ecosystems. These limitations arise because of well-recognized differences in the characterization of the exposure and effects represented in laboratory tests compared to those in ecosystems. Current standardized tests systems are generally (i) restricted in the range of species that are studied; (ii) consider toxicity of only the "pristine," as manufactured form; (iii) measure toxicity only in a single media; (iv) do not routinely consider accumulation; and (v) assess effects only over relatively short exposure times. Already in their original paper setting out the SSD approach, Van Straalen and Denneman [17] identified a range of factors that could both increase and decrease the effects of chemicals in the field compared to those anticipated based on laboratory test results. These factors are related to differences in environmental chemistry (e.g. bioavailability, mixtures), environmental conditions (optimal, variable), ecological processes (compensatory, over extended timescales), and population genetic structure (adaptation, meta-population dynamics). Similar factors are also likely to complicate field effect prediction for nanomaterials based on laboratory test data. These challenges for laboratory to field extrapolation are set within the recognized limitations of current standard tests and the data that they produce as discussed in the following text with specific reference to soil nano-ecotoxicology studies.

6.4.1 Choice of Test Species

Soils are known to contain a wealth of bacterial, fungal, and invertebrate biodiversity that, both alone and in combination, contribute to key ecosystem services such as controlling soil nutrient balance, regulating water flows, acting as a reservoir of natural resources, and providing catabolism potential for organic molecule degradation [24]. The dynamics of soil biogeochemical cycles have been shown to be impacted when stressors, such as pollution, reduce standing biomass and biodiversity [25, 26]. Hence, the challenge for maintaining soil functionality is to protect soil biodiversity in its broadest context [27].

The negative effects of any soil contaminant, including nanomaterials, can be driven through change to any one of the different taxa found in soils or to species interactions. The current range of standardized soil ecotoxicology tests covers only a relatively small portion of this range of taxonomic diversity. Nano-ecotoxicology tests for terrestrial plants can, in principle, be conducted with any species. However, the reality of the nano-ecotoxicology studies conducted to date is that studies with only a small number of monocotyledonous (e.g. maize, rice, wheat) and dicotyledous (e.g. *Brassica* sp., tomato) crop plants and the model plant *Arabidopsis thaliana* form the large majority of the literature. For invertebrates, the current set of standard tests covers only three phyla, namely, annelids, arthropods, and nematodes, and the effects on microorganisms are assessed only at the coarse functional level through assessments of nitrogen and carbon mineralization and the activity of one soil enzymes (dehydrogenase). Hence, the species coverage of a set of standard test studies is significantly limited.

The restricted taxonomic spectrum of standardized toxicity tests currently available poses specific problems for the robust hazard assessment of nanomaterials. Data requirements for the reliable derivation of SSDs necessitate that toxicity data for the specific chemical or nanomaterial under consideration is available for at least five and preferably more species. In contrast, the number of taxa for which standardized soil ecotoxicity tests are available is comparatively small, barely surpassing the minimum requirement. This potential limitation can mean that even for relatively well-studied nanomaterials, adherence to a standard test only paradigm, could mean that toxicity metrics for only a small number of species will be available. This issue is further exacerbated because, with the exception of the terrestrial plant tests, current soil organism tests make specific recommendations on the species to be tested. Reflecting this, the large majority of the soil nano-ecotoxicology literature currently covers those standard test species, most notably the earthworms *E. fetida* and *Eisenia andrei*, the springtail *F. candida*, and the nematode *C. elegans*. For nanomaterials, as for other chemicals, it is often those studies conducted with nonstandard tests species from a previously routinely studied taxa or, even more so, species from to date untested taxa, that are the most valued for future SSD derivation in support of hazard and risk assessment. Hence, studies with such nonstandard test species should be encouraged to increase the robustness of hazard assessments.

A further issue arising from the current range of standardized soil ecotoxicity tests is the potential for effects on vulnerable species, processes, or ecological interactions to be missed. Such oversight can lead to an underassessment of hazard in the field when compared to laboratory data. One approach by which to identify ecosystem level effects is the use of mesocosm and terrestrial model ecosystem studies. The value of these approaches for nanomaterials is illustrated from the insights that have already emerged from the small number of studies conducted to date. Judy et al. [28] used a simple leguminous plant and associated nitrifying bacterial tests system to compare the toxicity of metal and metal oxide (Ag, ZnO, TiO_2) nanomaterials and ionic and bulk forms spiked into sewage sludge generated from a pilot reactor on the plant host and symbiont. Observed effects of the nanomaterial on the growth of the plant *Medicago truncatula* were greater than those found for ionic/bulk treatments. This higher nanomaterial toxicity was associated both with an increased overall plant metal uptake and an associated reduction in root nodulation. Toxicogenomic analysis of root tissues identified a unique gene expression signature in the nanomaterial-exposed plants. Gene pathways specifically affected included nitrogen metabolism, nodule formation, metal homeostasis, and stress responses. Addition of nanomaterial to the system, therefore, appears to change metal bioavailability and uptake into the plant, with toxicological impacts affecting the interactions between the plant and its bacterial symbiont with consequences for plant health.

In a second more complex mesocosm study, Colman et al. [29] observed effects of nanoparticle addition on a multispecies plant community and soil microbial diversity and functions. Effects of exposure on individual plants species (e.g. *Microstegium vimineum*) were seen, even though overall aboveground primary production was not affected. Significant differences in bacterial community composition, N_2O flux, and soil microbial extracellular enzymes leucine amino peptidase and phosphatase also occurred. Characterization of the effect of nanomaterials on the plant species that were found to be affected in this study are unlikely to be included within the normal course of standardized plant species nano-ecotoxicology testing. Similarly, current standard tests for soil microbial communities measure only simple functions, meaning that the structural changes and effects on specific soil enzyme activities seen in this mesocosm study would routinely be missed. The need for researchers to go beyond standard tests to consider more complex and realistic exposure case for priority nanomaterial types is, therefore, evident. Soil mesocosm studies that cover different nanomaterials, soil types, and species assemblages would, therefore, be an important addition to the current literature.

6.4.2 Short-Term and Long-Term Effects of Particle "Aging" on Toxicity in Natural Environments

The majority of standardized toxicity tests have been conducted using "pristine" (i.e. as manufactured) nanomaterials spiked into soil followed by only a short stabilization period prior to the addition of test organisms. This exposure scenario lies counter to what is known about the most likely routes and forms of nanomaterials release to soil ecosystems. While direct release through accidents and spills and from specific products (e.g. biocides) may be important, a major route by which nanomaterials are likely to enter soils is in association with solid waste streams including sewage sludge [30]. During passage through waste reactors and sewage treatment facilities, nanomaterials have been shown to undergo a range of different physical and chemical transformations affecting both their chemistry and form. Such modifications include speciation change, dissolution, hetero- and homo-agglomeration, surface modification, and coating exchange. Over time, this range of possible transformations will affect pristine forms in different, yet predictable, ways, with resulting effects on bioavailability, uptake, and toxicity. Not all types of possible transformations will, however, necessarily be relevant for all nanomaterials. For example, some metal oxides (e.g. TiO_2) are chemically stable and so do not undergo extensive dissolution or chemical speciation changes, yet others such as Ag, ZnO, and CuO may be readily subject to such reactions. In contrast, some transformation processes are likely to be more commonly found across a range of different nanomaterial types. For example, the dissolved organic ligands that are ubiquitously present in soil pore water can be expected to bind particle surfaces, resulting in "eco-corona" formation across a range of particle types and forms.

The role of dissolution leading to the release of ions in nano-ecotoxicology has been a major focus of research in studies with (soil) species. The debate concerning the extent to which ion release is able to fully account for the toxicity in different cases remains an open question. Certainly it is fair to say that the majority of studies conducted to date have recognized that dissolution can be a key mechanism governing toxic effects for a range of different potentially soluble nanomaterials. For example, dissolution has been found to fully account for observed toxicity without the need to include any contribution from the remaining nanomaterial fraction in soil organism toxicity tests conducted with Ag [31, 32], ZnO [33, 34], and CuO [35]. Not all studies have, however, shown toxicity fully explained by ion release. For example, Dimkpa et al. [36] and Starnes et al. [37] both found a contribution of both released ions and also the remaining nanoparticles in studies conducted with CuO and Ag in a bacteria and nematode species, respectively.

Studies of nanomaterial dissolution in aquatic toxicity tests have shown the potential for rapid dissolution [38], dependent to an extent on nanomaterial properties [39]. In soils, however, particles can potentially become associated with the solid phase matrix, with such interactions can reduce dissolution rates. This association has the potential to extend the timescales over which intact nanoparticles may remain in the test system. In a study in which both Ag ions and Ag nanoparticles were incubated in a clay loam soil for 52 weeks and chemical partitioning and toxicity for earthworms periodically assessed, Diez-Ortiz et al. [40] found different temporal patterns of dissolution, bioavailability, and effects on reproduction between the Ag forms. For ionic Ag, a classic pattern of decreasing bioavailability and toxicity with longer soil incubation time was found. This is consistent with an established paradigm of metal ion "aging" in soil by which metal ions become increasingly strongly associated with the soil matrix over time, thereby reducing their bioavailability [41, 42]. In clear contrast, the toxicity of the Ag nanoparticles was shown to be increased as the incubation time in soil was extended. This pattern of change was driven by Ag ion release through dissolution over time resulting in increased concentrations of Ag in soil solution that ultimately in greater toxicity. These results clearly show that short-term exposures conducted without prior long-term nanoparticle incubation in soil, as would be routinely used in current standardized tests, may not fully represent the hazard of nanomaterials following extended incubation in, as may occur in field exposures.

Studies of the behavior of nanomaterials in waste streams have also identified the potential for changes in the chemical speciation of the primary particle. Reactions such as the sulfidation for Ag and CuO and phosphatization of ZnO have been identified following wastewater treatment system passage [43–45]. These changes in speciation have been shown to affect both the bioavailability and toxicity of different nanomaterials. For example, passivation of Ag nanomaterial surfaces by sulfidation of the surface has been shown to reduce toxicity compared with the untransformed form [46–48]. However, such reduction of effects for transformed materials have not always been shown. For example, the toxicity of CuS nanoparticles has been found to be higher than for pristine CuO forms in exposures conducted using medaka embryos [49]. It is certainly the case that wastewater processing to generate sewage sludge provides a highly reactive environment in which transformation reactions can be expected to also ubiquitously occur. Currently, the relationship between the physicochemical transformation and toxicity of nanomaterials with different starting properties has not been fully established. As soil species will frequently be exposed to these altered forms, further work is needed to investigate the relationships between different starting properties and resulting transformed states. Such analysis will help to identify the need to modify any "nano-specific" uncertainty factor included in a risk assessment tool to include the influences of environmental transformations on ecological risks.

6.4.3 How Soil Properties Interact with Nanomaterial Properties to Determine Bioavailability

One of the most important standardizations to allow for interlaboratory comparison is the use of a common test soil. The two media most widely used for routine ecotoxicological testing are either the natural sandy loam soil commonly known as "LUFA 2.2" or a wholly artificial soil that comprises of a pH-modified mixture of 70% sand, 20% kaolin clay, and 10% organic material as peat or peat substitute. The need to use a standard soil for toxicity testing reflects the known effects of soil properties, such as pH, organic matter content, clay fraction, cation exchange capacity and presence of iron oxides on chemical uptake and toxicity. Studies with trace elements have characterized in detail the effects that major soil properties can have on metal bioavailability and toxicity [42, 50]. Extensive work in this area has resulted in the development of modeling approaches like the terrestrial biotic ligand and WHAM-FTOX that incorporate major soil chemical processes into a quasi-mechanistic framework [51, 52]. Similar understanding will be needed if the site-specific risks of nanoparticles for different soil types are to be better understood and used in assessments. Such work covering studies of nanomaterial fate and effects in multiple, rather than a single soil, are starting to emerge.

The range of processes governing nanoparticle fate and exposure in soil and once assimilated into living organisms are notably complex covering aspects of colloidal and environmental chemistry and biological processes (see Figure 6.3). Because of its known key role in controlling metal partitioning and also nanoparticle environemntal behaviour, soil pH has been identified as one of the main

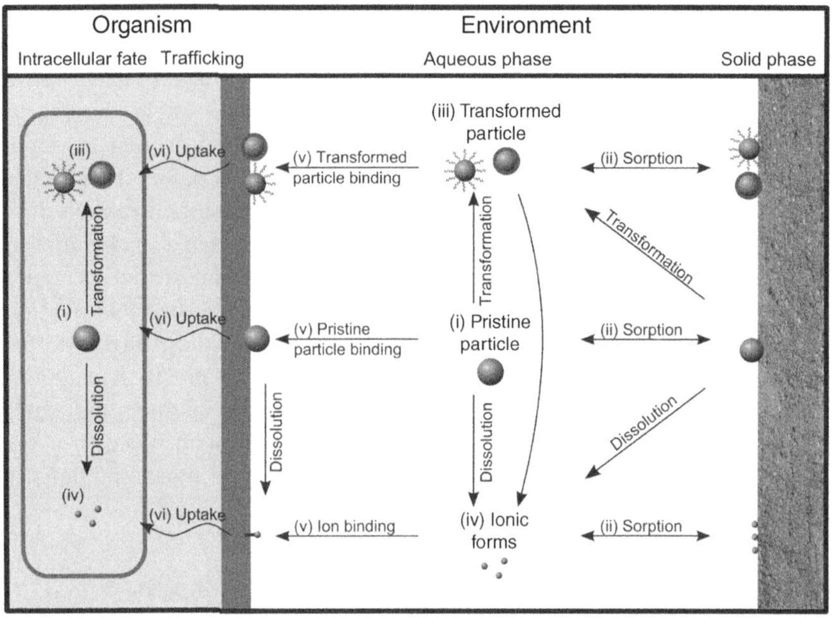

Figure 6.3 Major processes governing nanomaterial environmental fate, bioavailability, and toxicological availability in soil systems.

factors that could potentially influence nanomaterial fate, bioavailability, and toxicity in soil [48]. A series of studies have investigated the influence of soil pH on the bioavailability and toxicity of a 300 and 30 nm ZnO for the earthworm *E. fetida*, springtail *F. candida*, and woodlouse *Porcellionides pruinosus* [53–55]. In all studies, zinc sorption to the soil solid phase increased, and consequently Zn levels in the soil pore water decreased in progressively more neutral soils. This effect was independent of particle size. Toxicity for earthworm and Collembola reproduction was also clearly influenced by pH for both ZnO forms. In each case, increasing soil pH resulted in a decrease in toxicity that was associated with a reduction in the overall rate of dissolution [53, 55]. The influence of pH on toxicity was less clearly seen in the study with the woodlouse. Here the greatest effect on reproduction was seen at an intermediate pH [54]. These combined results point to a number of important ways by which pH changes dissolution rates and speciation of the dissolved ionic forms to modify exposure and toxicity. Such effects appear more pertinent to those species, such as earthworms and springtails that live in close association with the soil pore water than for soil surface-dwelling species.

Organic carbon content is a further physiochemical factor that can potentially influence nanomaterial bioavailability and toxicity in soil. Organic matter fragments present in the soil matrix present charged surfaces for nanomaterial attachment and hetero-agglomeration, while dissolved organic carbon species may coat nanomaterial surfaces with a corona layer that can affect fate parameters such as attachment efficiencies and dissolution rates. Waalewijn-Kool et al. [56] specifically studied the joint effects of soil organic matter content and pH on the bioavailability and toxicity of ZnO nanomaterials for *F. candida* in four natural soils. Zinc levels in pore water extracts were found to differ between soils dependent on both organic matter content and pH. The calculated 28-day EC_{50} values for the effect of ZnO on springtail reproduction indicated no relationship between observed toxicity and soil organic matter content. In agreement with the previously discussed results, a relationship between toxicity and soil pH was found. Thus, while organic matter content changed the partitioning of nanomaterial-derived Zn, no influence on bioavailability or toxicity was found. Whether the same pattern will hold across an extended range of soils and for different nanomaterials remains to be investigated.

6.4.4 Nanomaterial Bioaccumulation and Food Chain Transfer

Current standard laboratory tests focus on measuring toxicological end points at a single trophic level. While nanomaterial uptake can also be assessed in these tests, a more complete understanding of bioaccumulation can be gained using specifically designed bioaccumulation assays. The issue of chemical bioaccumulation in soil organisms is addressed in OECD Guideline No. 317: Bioaccumulation in Terrestrial Oligochaetes [57]. This test is designed for neutral organic chemicals with log Kow 3–6, superlipophilic substances and stable metallo-organic compounds and has not yet been approved for use with nanomaterials. The protocol describes a procedure in which earthworms are exposed to the test chemical spiked into a soil. Individuals are then periodically sampled over time during both an accumulation phase in the presence of the chemical and an elimination phase following transfer to clean soil. Following chemical analysis of the sampled earthworms, toxicokinetic parameters including uptake rate, elimination rate, and the bioaccumulation factor can be estimated by fitting a compartment model to both the accumulation and elimination phase measurements.

The potential for nanomaterials to accumulate in soil-dwelling species has been demonstrated in studies covering a number of the major soil taxa [58]. In some cases, there is emerging evidence that nanomaterial forms can sometimes be accumulated to higher concentrations than ionic forms. For example, Diez-Ortiz et al. [40] found that earthworms accumulated higher concentrations when exposed to Ag nanoparticles compared with the same concentration of Ag ions. However, these increased tissue concentrations do not necessarily result in greater toxicity. In a study in which earthworms were exposed to both CeO_2 and cerium ions, Lahive et al. [59] found that the nanoparticles accumulated higher internal cerium concentrations yet showed only subtle signs of toxicity, while the worms exposed to the ions suffered acute toxic effects at lower body burdens. Such findings suggest a need to consider the use of more complex multiple compartment models in studies of nanomaterial bioaccumulation in terrestrial species that may include a less toxicologically active nanomaterial fraction.

The need for standard tests able to quantify nanomaterial bioaccumulation is further emphasized by a number of studies that have identified trophic transfer within food chains. For example, Werlin et al. [60] showed that uncoated CdSe quantum dots accumulated in the bacteria *Pseudomonas aeruginosa* were transferred to and biomagnified in the predatory protozoa *Tetrahymena thermophile*, while Luo et al. [61] showed the tropic transfer of polyvinylpyrrolidone-coated Ag nanoparticles accumulated in *Escherichia coli* to the bacterivorous nematode *C. elegans*. In a plant–herbivore system, Judy et al. [62] showed that Au nanomaterials accumulated in leaf tissues of *Nicotiana tabacum* could be transferred into the caterpillar *Manduca sexta*. Biomagnification at 6.2, 11.6, and 9.6-fold was found from the primary producer to a primary consumer for primary particle of sizes 5, 10, and 15 nm, respectively. A similar tropic transfer of Au nanoparticles was also seen in a macroinvertebrate–predator system using the earthworm *E. fetida* and bullfrog *Rana catesbeiana* [63]. The Au nanoparticles transferred from earthworm prey to frog were found in the liver, kidney, spleen, muscle, stomach, and intestine. However in this system, no biomagnification was found. Indeed tissue concentrations measured in earthworms exceeded those in any frog tissue by approximately 100-fold. These results clearly demonstrate that in terrestrial ecosystems, materials accumulated from nanoparticles can both be taken up into organisms at the base of food chains and also be transferred to higher tier consumers. This potential for trophic transfer clearly emphasizes the need to assess bioaccumulation and, further, the need to monitor the possible presence of nanoparticles in higher trophic levels in field-based assessments.

6.4.5 Short-Term Tests Predict Long-Term Effects

The extent to which exposure time influences observed toxicity has important implications for the use of laboratory toxicity data to predict field effects. The current generation of standardized laboratory toxicity tests fails to capture this important dynamic element of toxicity. Thus, although generally longer than the 48 and 96 h durations used in some tests, such as *Daphnia* and honeybees, current standardized soil toxicity tests are still only up to 28 days in duration. Further, even over this limited timescale, effects are generally only quantified for each measurement end point at one time point placed at the end of the test. In contrast, when exposure to a persistent chemical occurs in the field, the duration may extend over the full life span of different test species. For springtails, this may be for up to one year and for earthworms up to five years. Hence field exposure durations may exceed those used in laboratory tests by over an order of magnitude.

The disconnection between standard test durations and realistic exposure times raises the possibility that short-duration tests may not fully reflect realized toxicity in the field. The extent of this difference can depend on a number of factors. These may include the toxicokinetics of the chemical, species traits, and the transgenerational nature of toxic effects. The time dependence of toxicity in hazard assessment can be represented through the use of toxicokinetic and toxicodynamic models. By capturing the key rate-dependent process of chemical uptake and interaction with target sites [64, 65], models such as those based on dynamic energy budget (DEB) theory and associated general unified threshold model for survival (GUTS) may have an important role to play in understanding the time courses of toxic effect for a range of end points. For parametrization, these types of models benefit from time series measurements of effects that go beyond the single time point assessment generally made in standardized tests. Hence to support use of such models, modifications to standard test methods are needed to increase the frequency of effect measurements.

A recent illustration of the effect of exposure time on observed sensitivity was provided by Hesketh et al. [5]. This study continuously monitored survival effects for seven chemicals in the honeybee *Apis mellifera* over a 10-day test and used a stochastic survival model to predict LC_{50}s for different exposure times from 24 h up to the 720 h and 2160 h life span of summer and overwintering worker bees. For all chemicals where toxic effects were found, the temporal nature of the estimated LC_{50} was clearly shown, with values reduced progressively over time until ultimately a constant value was reached. The period over which the LC_{50}s changed with time differed between the test chemicals. For chemicals with slow predicted elimination rates, higher time dependence of the LC_{50} was found than for chemicals with more rapid predicted elimination. Indeed for cadmium in *A. mellifera*, modeling predicted no elimination. Hence for this metal, no asymptote for the LC_{50} was found, and values were predicted to continuously reduce over the full life span of the organism [5]. Data from studies of the uptake and accumulation of nanomaterials in soil-dwelling organisms (e.g. [66–68]) can provide relevant information on their toxicokinetics, which can be used to assess the extent to which the short-term exposure times employed in standard tests may underestimate the effects resulting from long-term exposure.

The influences of exposure time on toxicity can become further nuanced when the persistence of the chemical extends exposure beyond a single generation. The exposure of parents to a toxicant is commonly observed to result in effects on offspring, such as reduced birth weight, developmental defects, and impacts on metabolism. These effects are consistent with a paradigm whereby maternal exposure affects gametes or the developing embryo. In addition to impacts on single generations, persistent toxicant effects can potentially also be transferred to later generations. To date there have only been a small number of studies that have assessed how multigenerational exposures to nanomaterials may affect populations. Such studies for soil-dwelling species have, to date, been conducted almost exclusively with the nematode *C. elegans*. Different nanomaterials tested including Ag [69], Au [70], and zero-valent Fe [71]. In all three studies, the main finding was of heightened sensitivity in the later-exposed generations. In the study of Schultz et al. [69], this greater effect was even maintained after the population had been transferred and reared in clean media for multiple generations. Such finding of greater effects in populations under continuous stress certainly favor further investigations of multigenerational impacts on observed toxicity. This is something that cannot be achieved with current standard tests, but rather requires specifically designed methods tailored to each species.

6.5 Standard Ecotoxicological Tests: A Blessing and A Curse?

The expected rise in the use of nanotechnology products means that in future, we will become reliant on the availability of robust risk assessment schemes to help balance decisions of risks against benefits for new materials and applications. Work conducted to date has, so far, largely shown that key methods for toxicity testing and biological end point assessment already established for conventional chemicals can be successfully applied in studies with nanomaterials, as long as attention is paid to the exposure setup. These existing tests provide tractable and useable tools for the rapid generation of initial toxicity data that are comparable between laboratories with only a few modifications needed for nanomaterials. These specific adjustments include clearer guidance on spiking methods and an increased focus on material characterization prior to and during the conducted study, in order to document and understand the degree to which it provided a true worst-case exposure.

While valuable for routine assessment, current standard tests are certainly not a panacea for all cases. By their nature, they encourage the use of only a very limited number of species, meaning that there will be many species and indeed whole phyla that will rarely, if ever, be tested. This both limits the value of the obtained data for generating SSDs, because the same species may be tested on multiple occasions and also raises the real possibility that vulnerable species may be missed. Further with only data from short-term studies in a single standard test medium often available, it is difficult to know how relevant that data generated will be to actual effects in different environments over extended exposure times and for all species.

When considering the use of standard tests for wider chemical risk assessment, Sumpter and Johnson [72] considered the lessons learned from studies of endocrine disrupting chemicals conducted in the aquatic environment. One of their most notable conclusions was that "One lesson [...] during the 10 years of research on endocrine disruption is that the current testing regime used to determine the toxicity of a chemical to aquatic organisms has "failed," in the sense that it has not detected the (biological) activity of many chemicals." Further these authors also noted that "even more surprisingly, it transpired that ethinyl estradiol (n.b. a potent endocrine disrupting chemical) is acutely toxic to fish, but this effect is delayed long enough that it is not detected in the acute toxicity test protocols." Such insights, derived on the basis of lessons learned after 20 years of work, certainly indicate that those involved in the hazard and risk assessment of any emerging chemical, should be careful not to become over-reliant on methods that lacks the flexibility to identify novel effects simple because standard tests do not measure the right endpoint or are too short for exposure and toxicokinetics to manifest into observable effects.

Where standard tests are to be used for developing exposure or hazard-based rankings or groupings for nanomaterials, it is essential that they are conducted in a manner that is sufficiently comparable even considering the complexities of some nano-properties to be investigated and compared [73]. For nano-enabled products going beyond routine tests by developing environmentally relevant-specific risk assessments for classes of pollutants (e.g. based on product type, use, and waste handling-driven release forms and patterns), studying their exposure dynamics, mechanistic toxicology, and long-term and transgenerational effects can provide methods needed for a more integrated and science-driven assessment of nanomaterial release into the environment [74]. To exploit and develop such approaches, it is important to establish a sound theoretical scientific basis for nanomaterial ecotoxicology, using specifically tailored test designs to address the most relevant and pressing issues. Such understanding can then inform a regulatory approach that can make use of, but is not inhibited by, standard tests. In short, as nano-ecotoxicologists we need to be sure that we use standard tests not because we can, but rather when we should!

Acknowledgments

The work was conducted in NanoFASE H2020 project under Grant Agreement Number 646002 (2015-2019) (H2020-NMP-2014), coordinated by C. Svendsen (NERC – Centre for Ecology & Hydrology, UK) (www.nanofase.eu).

References

1 Westerhoff, P.K., Kiser, A., and Hristovski, K. (2013). Nanomaterial removal and transformation during biological wastewater treatment. *Environ. Eng. Sci.* 30: 109–117.

2 Johnson, A.C., Bowes, M.J., Crossley, A. et al. (2011). An assessment of the fate, behaviour and environmental risk associated with sunscreen TiO_2 nanoparticles in UK field scenarios. *Sci. Total Environ.* 409: 2503–2510.

3 Giesy, J.P. and Hoke, R.A. (1989). Freshwater sediment toxicity bioassessment: rationale for species selection and test design. *J. Great Lakes Res.* 1: 539–569.

4 Heckmann, L.H., Baas, J., and Jager, T. (2010). Time is of the essence. *Environ. Toxicol. Chem.* 29: 1396–1398.

5 Hesketh, H., Lahive, E., Horton, A.A. et al. (2016). Toxicokinetic and toxicodynamic of the time course effects of chemicals in honeybees (*Apis mellifera mellifera*) analysed through DEBtox modelling. *Sci. Rep.* 6: 37655.

6 Henegar, A., Mombelli, E., Pandard, P., and Pery, A.R.R. (2011). What can be learnt from an ecotoxicity database in the framework of the REACh regulation? *Sci. Total Environ.* 409: 489–494.

7 Posthuma, L., Traas, T.P., and Suter, G.W. (2001). *The Use of Species Sensitivity Distributions in Ecotoxicology*. Boca Raton, FL: Lewis Publishers.

8 Spurgeon, D.J., Svendsen, C., Hankard, P.K. et al. (2002). *Review of Sublethal Ecotoxicological Tests for Measuring Harm in Terrestrial Ecosystems*. Bristol: Environment Agency.

9 Tyne, W., Lofts, S., Spurgeon, D.J. et al. (2013). A new medium for *Caenorhabditis elegans* toxicology and nanotoxicology studies designed to better reflect natural soil solution conditions. *Environ. Toxicol. Chem.* 32: 1711–1717.

10 Handy, R.D., Cornelis, G., Fernandes, T. et al. (2012). Ecotoxicity test methods for engineered nanomaterials: Practical experiences and recommendations from the bench. *Environ. Toxicol. Chem.* 31: 15–31.

11 Miglietta, M.L., Rametta, G., Manzo, S. et al. (2015). Methodological issues about techniques for the spiking of standard OECD soil with nanoparticles: evidence of different behaviours. *J. Nanopart. Res.* 17: 12.

12 Waalewijn-Kool, P.L., Ortiz, M.D., and van Gestel, C.A.M. (2012). Effect of different spiking procedures on the distribution and toxicity of ZnO nanoparticles in soil. *Ecotoxicology* 21: 1797–1804.

13 Lahive, E., Matzke, M., Durenkamp, M. et al. (2017). Sewage sludge from waste water treatment plant treated with metal nanomaterials inhibits earthworm reproduction more strongly than sludge treated with metals in bulk/salt forms. *Environ. Sci. Nano* 7: e34197.

14 Schultz, C., Powell, K., Crossley, A. et al. (2015). Analytical approaches to support current understanding of exposure, uptake and distributions of engineered nanoparticles by aquatic and terrestrial organisms. *Ecotoxicology* 24: 239–261.

15 Tuoriniemi, J., Gustafsson, S., Olsson, E., and Hassellov, M. (2014). In situ characterisation of physicochemical state and concentration of nanoparticles in soil ecotoxicity studies using environmental scanning electron microscopy. *Environ. Chem.* 11: 367–376.

16 Kooijman, S.A.L.M. (1987). A safety factor for LC_{50} values allowing for differences in the sensitivity among species. *Ecotoxicol. Environ. Saf.* 18: 241–252.

17 Van Straalen, N.M. and Denneman, C.A.J. (1989). Ecotoxicological evaluation of soil quality criteria. *Ecotoxicol. Environ. Saf.* 18: 241–251.

18 Aldenberg, T. and Slob, W. (1993). Confidence limits for hazardous concentrations based on logistically distributed NOEC toxicity data. *Ecotoxicol. Environ. Saf.* 25: 48–63.

19 Van Straalen, N.M. (2002). Threshold models for species sensitivity distributions applied to aquatic risk assessment for zinc. *Environ. Toxicol. Pharmacol.* 11: 167–172.

20 Garner, K.L., Suh, S., Lenihan, H.S., and Keller, A.A. (2015). Species sensitivity distributions for engineered nanomaterials. *Environ. Sci. Technol.* 49: 5753–5759.

21 Kwak, J.I., Cui, R.X., Nam, S.H. et al. (2016). Multispecies toxicity test for silver nanoparticles to derive hazardous concentration based on species sensitivity distribution for the protection of aquatic ecosystems. *Nanotoxicology* 10: 521–530.

22 Wang, Y., Deng, L., Caballero-Guzman, A., and Nowack, B. (2016). Are engineered nano iron oxide particles safe? an environmental risk assessment by probabilistic exposure, effects and risk modeling. *Nanotoxicology* 10: 1545–1554.

23 Notter, D., Mitrano, D.M., and Nowack, B. (2014). Are nanosized or dissolved metals more toxic? A meta-analysis. *Environ. Toxicol. Chem.* 33: 2733–2739.

24 Bardgett, R.D. and van der Putten, W.H. (2014). Belowground biodiversity and ecosystem functioning. *Nature* 515: 505–511.

25 Bradford, M.A., Jones, T.H., Bardgett, R.D. et al. (2002). Impacts of soil faunal community composition on model grassland ecosystems. *Science* 298: 615–618.

26 Hector, A., Beale, A.J., Minns, A. et al. (2000). Consequences of the reduction of plant diversity for litter decomposition: effects through litter quality and microenvironment. *Oikos* 90: 357–371.

27 De Ruiter, P.C., Wolters, V., Moore, J.C., and Winemiller, K.O. (2005). Food web ecology: playing Jenga and beyond. *Science* 309: 68.

28 Judy, J.D., McNear, D.H., Chen, C. et al. (2015). Nanomaterials in biosolids inhibit nodulation, shift microbial community composition, and result in increased metal uptake relative to bulk/dissolved metals. *Environ. Sci. Technol.* 49: 8751–8758.

29 Colman, B.P., Arnaout, C.L., Anciaux, S. et al. (2013). Low concentrations of silver nanoparticles in biosolids cause adverse ecosystem responses under realistic field scenario. *PLos One* 8: e57189.

30 Keller, A.A., McFerran, S., Lazareva, A., and Suh, S. (2013). Global life cycle releases of engineered nanomaterials. *J. Nanopart. Res.* 15: 1692.

31 Topuz, E. and van Gestel, C.A.M. (2015). Toxicokinetics and toxicodynamics of differently coated silver nanoparticles and silver nitrate in *Enchytraeus crypticus* upon aqueous exposure in an inert sand medium. *Environ. Toxicol. Chem.* 34: 2816–2823.

32 Waalewijn-Kool, P.L., Klein, K., Fornies, R.M., and van Gestel, C.A.M. (2014a). Bioaccumulation and toxicity of silver nanoparticles and silver nitrate to the soil arthropod *Folsomia candida*. *Ecotoxicology* 23: 1629–1637.

33 Hooper, H.L., Jurkschat, K., Bailey, J. et al. (2011). Comparative chronic toxicity of nanoparticulate and ionic zinc to the earthworm *Eisenia veneta* in a soil matrix. *Environ. Int.* 37: 1111–1117.

34 Liu, X.Q., Wang, F.Y., Shi, Z.Y. et al. (2015). Bioavailability of Zn in ZnO nanoparticle-spiked soil and the implications to maize plants. *J. Nanopart. Res.* 17: 11.

35 Rousk, J., Ackermann, K., Curling, S.F., and Jones, D.L. (2012). Comparative toxicity of nanoparticulate CuO and ZnO to soil bacterial communities. *PLos One* 7: 8.

36 Dimkpa, C.O., McLean, J.E., Britt, D.W. et al. (2012). Nanospecific inhibition of pyoverdine siderophore production in *Pseudomonas chlororaphis* O6 by CuO nanoparticles. *Chem. Res. Toxicol.* 25: 1066–1074.

37 Starnes, D.L., Unrine, J.M., Starnes, C.P. et al. (2015). Impact of sulfidation on the bioavailability and toxicity of silver nanoparticles to *Caenorhabditis elegans*. *Environ. Pollut.* 196: 239–246.

38 Franklin, N.M., Rogers, N.J., Apte, S.C. et al. (2007). Comparative toxicity of nanoparticulate ZnO, bulk ZnO, and $ZnCl_2$ to a freshwater microalga (*Pseudokirchneriella subcapitata*): The importance of particle solubility. *Environ. Sci. Technol.* 41: 8484–8490.

39 Dobias, J. and Bernier-Latmani, R. (2013). Silver release from silver nanoparticles in natural waters. *Environ. Sci. Technol.* 47: 4140–4146.

40 Diez-Ortiz, M., Lahive, E., George, S. et al. (2015a). Short-term soil bioassays may not reveal the full toxicity potential for nanomaterials; bioavailability and toxicity of silver ions ($AgNO_3$) and silver nanoparticles to earthworm *Eisenia fetida* in long-term aged soils. *Environ. Pollut.* 203: 191–198.

41 Smit, C.E., van Beelen, P., and Van Gestel, C.A.M. (1997). Development of zinc bioavailability and toxicity for the springtail *Folsomia candida* in an experimentally contaminated field plot. *Environ. Pollut.* 98: 73–80.

42 Smolders, E., Oorts, K., van Sprang, P. et al. (2009). Toxicity of trace metals in soil as affected by soil type and aging after contamination: using calibrated bioavailability models to set ecological soil standards. *Environ. Toxicol. Chem.* 28: 1633–1642.

43 Levard, C., Hotze, E.M., Lowry, G.V., and Brown, G.E. Jr., (2012). Environmental transformations of silver nanoparticles: impact on stability and toxicity. *Environ. Sci. Technol.* 46: 6900–6914.

44 Ma, R., Levard, C., Judy, J.D. et al. (2014). Fate of zinc oxide and silver nanoparticles in a pilot wastewater treatment plant and in processed biosolids. *Environ. Sci. Technol.* 48: 104–112.

45 Wang, Z.Y., von dem Bussche, A., Kabadi, P.K. et al. (2013). Biological and environmental transformations of copper-based nanomaterials. *ACS Nano* 7: 8715–8727.

46 Levard, C., Hotze, E.M., Colman, B.P. et al. (2013). Sulfidation of silver nanoparticles: natural antidote to their toxicity. *Environ. Sci. Technol.* 47: 13440–13448.

47 Sharma, V.K. (2009). Aggregation and toxicity of titanium dioxide nanoparticles in aquatic environment - A Review. *J. Environ. Sci. Health, Part A* 44: 1485–1495.

48 Tourinho, P.S., van Gestel, C.A.M., Lofts, S. et al. (2012). Metal-based nanoparticles in soil: Fate, behavior, and effects on soil invertebrates. *Environ. Toxicol. Chem.* 31: 1679–1692.

49 Li, L., Hu, L., Zhou, Q. et al. (2015). Sulfidation as a natural antidote to metallic nanoparticles is overestimated: CuO sulfidation yields CuS nanoparticles with increased toxicity in medaka (*Oryzias latipes*) embryos. *Environ. Sci. Technol.* 49: 2486–2495.

50 Lofts, S., Criel, P., Janssen, C.R. et al. (2013). Modelling the effects of copper on soil organisms and processes using the free ion approach: towards a multi-species toxicity model. *Environ. Pollut.* 178: 244–253.

51 He, E. and Van Gestel, C.A.M. (2015). Delineating the dynamic uptake and toxicity of Ni and Co mixtures in *Enchytraeus crypticus* using a WHAM-F-TOX approach. *Chemosphere* 139: 216–222.

52 Thakali, S., Allen, H.E., Di Toro, D.M. et al. (2006). Terrestrial biotic ligand model. 2. Application to Ni and Cu toxicities to plants, invertebrates, and microbes in soil. *Environ. Sci. Technol.* 40: 7094–7100.

53 Heggelund, L.R., Diez-Ortiz, M., Lofts, S. et al. (2014). Soil pH effects on the comparative toxicity of dissolved zinc, non-nano and nano ZnO to the earthworm *Eisenia fetida*. *Nanotoxicology* 8: 559–572.

54 Tourinho, P.S., van Gestel, C.A.M., Lofts, S. et al. (2013). Influence of soil pH on the toxicity of zinc oxide nanoparticles to the terrestrial isopod *Porcellionides pruinosus*. *Environ. Toxicol. Chem.* 32: 2808–2815.

55 Waalewijn-Kool, P.L., Ortiz, M.D., van Straalen, N.M., and van Gestel, C.A.M. (2013). Sorption, dissolution and pH determine the long-term equilibration and toxicity of coated and uncoated ZnO nanoparticles in soil. *Environ. Pollut.* 178: 59–64.

56 Waalewijn-Kool, P.L., Rupp, S., Lofts, S. et al. (2014b). Effect of soil organic matter content and pH on the toxicity of ZnO nanoparticles to *Folsomia candida*. *Ecotoxicol. Environ. Saf.* 108: 9–15.

57 OECD (2010). *Test No. 317: Bioaccumulation in Terrestrial Oligochaetes. OECD Guidelines for the Testing of Chemicals*. Paris: OECD.

58 Hou, W.C., Westerhoff, P., and Posner, J.D. (2013). Biological accumulation of engineered nanomaterials: a review of current knowledge. *Environ. Sci. Process. Impacts* 15: 103–122.

59 Lahive, E., Jurkschat, K., Shaw, B.J. et al. (2014). Toxicity of cerium oxide nanoparticles to the earthworm *Eisenia fetida*: subtle effects. *Environ. Chem.* 11: 268–278.

60 Werlin, R., Priester, J.H., Mielke, R.E. et al. (2011). Biomagnification of cadmium selenide quantum dots in a simple experimental microbial food chain. *Nat. Nanotechnol.* 6: 65–71.

61 Luo, X., Xu, S.M., Yang, Y.N. et al. (2016). Insights into the ecotoxicity of silver nanoparticles transferred from *Escherichia coli* to *Caenorhabditis elegans*. *Sci. Rep.* 6: 36465.

62 Judy, J.D., Unrine, J.M., and Bertsch, P.M. (2011). Evidence for biomagnification of gold nanoparticles within a terrestrial food chain. *Environ. Sci. Technol.* 45: 776–781.

63 Unrine, J.M., Shoults-Wilson, W.A., Zhurbich, O. et al. (2012). Trophic transfer of Au nanoparticles from soil along a simulated terrestrial food chain. *Environ. Sci. Technol.* 46: 9753–9760.

64 Ashauer, R., Albert, C., Augustine, S. et al. (2016). Modelling survival: exposure pattern, species sensitivity and uncertainty. *Sci. Rep.* 6: 29178.

65 Jager, T., Albert, C., Preuss, T.G., and Ashauer, R. (2011). General unified threshold model of survival – a toxicokinetic-toxicodynamic framework for ecotoxicology. *Environ. Sci. Technol.* 45: 2529–2540.

66 Diez-Ortiz, M., Lahive, E., Kille, P. et al. (2015b). Uptake routes and toxicokinetics of silver nanoparticles and silver ions in the earthworm *Lumbricus rubellus*. *Environ. Toxicol. Chem.* 34: 2263–2270.

67 Laycock, A., Diez-Ortiz, M., Larner, F. et al. (2016). Earthworm uptake routes and rates of Ionic Zn and ZnO nanoparticles at realistic concentrations, traced using stable isotope labeling. *Environ. Sci. Technol.* 50: 412–419.

68 Tourinho, P.S., van Gestel, C.A.M., Morgan, A.J. et al. (2016). Toxicokinetics of Ag in the terrestrial isopod *Porcellionides pruinosus* exposed to Ag NPs and $AgNO_3$ via soil and food. *Ecotoxicology* 25: 267–278.

69 Schultz, C., Unrine, J., Tsyusko, O. et al. (2016). The transgenerational toxicity of Ag ions, nanopaticles and sulphadised nanoparticles across 10 *Caenorhabditis elegans* generations. *Proc. R. Soc. B Biol. Sci.* 283: 20152911.

70 Moon, J., Kwak, J.I., Kim, S.W., and An, Y.J. (2017). Multigenerational effects of gold nanoparticles in *Caenorhabditis elegans*: continuous versus intermittent exposures. *Environ. Pollut.* 220: 46–52.

71 Yang, Y.F., Chen, P.J., and Liao, V.H.C. (2016). Nanoscale zerovalent iron (nZVI) at environmentally relevant concentrations induced multigenerational reproductive toxicity in *Caenorhabditis elegans*. *Chemosphere* 150: 615–623.

72 Sumpter, J.P. and Johnson, A.C. (2005). Lessons from endocrine disruption and their application to other issues concerning trace organics in the aquatic environment. *Environ. Sci. Technol.* 39: 4321–4332.

73 Petersen, E.J., Diamond, S.A., Kennedy, A.J. et al. (2015). Adapting OECD aquatic toxicity tests for use with manufactured nanomaterials: key issues and consensus recommendations. *Environ. Sci. Technol.* 49: 9532–9547.

74 Holden, P.A., Gardea-Torresdey, J.L., Klaessig, F. et al. (2016). Considerations of environmentally relevant test conditions for improved evaluation of ecological hazards of engineered nanomaterials. *Environ. Sci. Technol.* 50: 6124–6145.

7

Impacts of Magnetic Iron Oxide Nanoparticles in Terrestrial and Aquatic Environments

Renato Grillo[1] and Leonardo F. Fraceto[2]

[1] *Federal University of ABC, Center of Natural and Human Sciences, Santo André, SP, Brazil*
[2] *São Paulo State University – UNESP, Institute of Science and Technology of Sorocaba, Brazil*

7.1 Introduction

In recent years, the use of magnetic iron oxide nanoparticles (IONPs) has been explored in areas including biotechnology, materials science, biomedicine, catalysis, engineering, and environmental remediation, due to their small size, surface chemistry, and magnetic properties [1]. IONPs can be used or released in terrestrial and aquatic environments, but little is known about the possible impacts and the risks that these materials may pose to human health and ecosystems [2]. Here, an overview is provided of some of the main IONPs that have been synthesized to date, discussing their possible toxic effects in cell cultures and terrestrial and aquatic organisms. This field is still relatively unexplored, and understanding the effects and impacts of magnetic nanoparticles (MNPs) in the environment can open new perspectives for use of these nanomaterials.

7.1.1 Magnetic Nanoparticles and Their Properties

MNPs is the term used for the class of engineered particulate materials with sizes <100 nm that can be manipulated under the influence of an external magnetic field [3]. Types of MNPs include nanoparticles of metals (e.g. nickel, cobalt, and iron), metal alloy nanoparticles (e.g. $CoPt_3$, FePt, and FeCo), iron oxides (e.g. magnetite (Fe_3O_4), maghemite (γ-Fe_2O_3), hematite (α-Fe_2O_3), and cobalt ferrite (Fe_2CoO_4)), among others.

The development and use of MNPs has been widely reported in the literature, including studies of the preparation, characterization, and applications of these systems [1, 3–7]. Among the different MNPs, IONPs, mainly in the form of

Toxicology of Nanoparticles and Nanomaterials in Human, Terrestrial and Aquatic Systems, First Edition. Edited by Marc A. Williams.

magnetite (Fe_3O_4) or its oxidized form known as maghemite (γ-Fe_2O_3), are most widely used, because of the following reasons: (i) they can be biodegraded and are biocompatible for use with humans [5, 8] since they are metabolized and degraded by the endogenous iron metabolic pathways and are subsequently used in the formation of red blood cells or are excreted via the kidneys [9, 10]; (ii) they can be easily functionalized by a variety of molecules including surfactants, polymers, antibodies, drugs, small peptides, and inorganic material (e.g. silica) and are stable in both hydrophobic and hydrophilic solvents [1, 11, 12]; and (iii) they possess magnetic properties, enabling these nanomaterials to be manipulated by means of an external magnetic field and thereby be delivered to specific targets. These and other advantages make IONPs excellent tools in a wide range of technological applications, notably in cancer therapy and as contrast agents in clinical diagnosis [4, 6, 7].

A variety of techniques can be used to prepare IONPs, with co-precipitation and thermal decomposition being most popular, followed by other less common methods such as microemulsions, electrochemical deposition, pyrolysis, and sol–gel processes [13, 14]. It is essential to have strict control over the synthesis of these IONPs, especially in relation to their size distribution, shape, and crystallinity, since these parameters are associated with the magnetic properties of the particles. Depending on the size of the nanoparticle, there can be different numbers of magnetic domains (regions with the same magnetization), resulting in two main types of magnetic behavior: ferromagnetic (FM) and superparamagnetic (SPM) [4] (Figure 7.1).

FM behavior occurs in materials with sizes above the critical diameter ($\sim C_d > 100$ nm, with the value varying according to the material), and for this reason they can exhibit many different magnetic domains [15]. These magnetic domains result from the balance between different energies (for example, the exchange energy, magnetocrystalline anisotropy, and the magnetostatic or dipole energy), which are responsible for generating permanent magnetization of these materials, even in the absence of a magnetic field [16, 17]. It is then possible to calculate the amount of residual magnetization (M_r) and the point of coercivity (H_c) (the strength of the field that must be applied in the negative direction to bring the magnetization of the sample back to zero), plotting a graph of the magnetic field (H) against the magnetization (M) of the material, as shown in Figure 7.1. However, SPM materials have sizes smaller than 30 nm, and so are not found divided into various magnetic domains (unlike FM), but have a single domain and therefore do not show a hysteresis loop. In other words, there is no residual magnetization in the absence of a magnetic field ($M_r = 0$), since the temperature must be above a specific value, known as the blocking temperature (T_B) [16]. This feature increases the stability of MNPs, and together with the

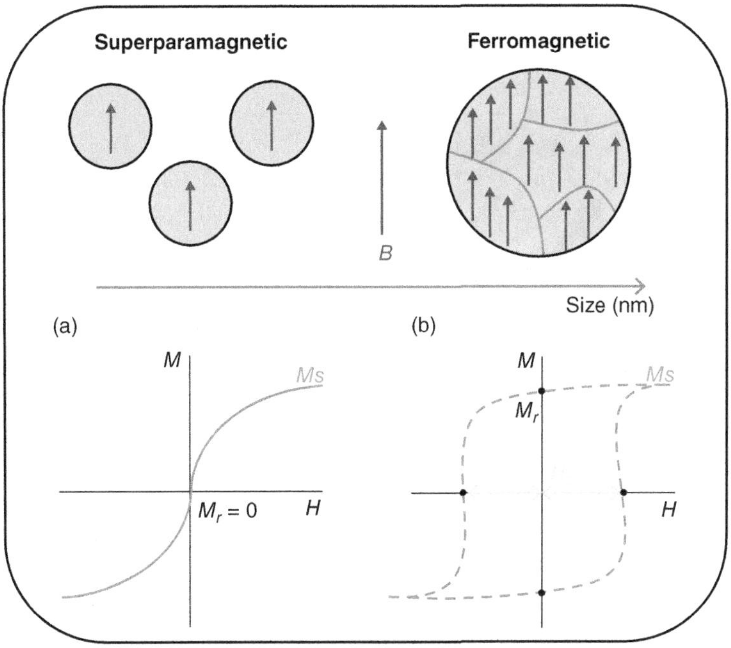

Figure 7.1 Schematic illustration of the main magnetic behaviors of IONPs: (a) a typical curve for a superparamagnetic material (SPM) and (b) a typical hysteresis loop for an array of single domain ferromagnetic (FM) nanoparticles, as a function of magnetic field (H) and magnetization (M). The presence of residual magnetization (M_r) and coercivity (H_c) can be seen in the ferromagnetic materials.

absence of residual magnetization makes SPM IONPs the main MNPs employed in biomedical applications.

7.1.2 Commercial Importance and Applications of IONPs

The large number of studies of IONPs in recent years, especially concerning SPM IONPs, reveals the promise shown by these nanomaterials, with new applications constantly emerging [5, 8, 12, 13, 18]. Some products, such as magnetic resonance imaging (MRI) contrast agents, are approved by the US Food and Drug Administration (FDA) and the European Commission (EC) and have been marketed for several years, examples being Feridex IV©, Sinerem©, and Resovist© [11, 19]. Recently, the FDA and subsequently the EC approved the use of another product containing SPM IONPs for clinical applications, called Ferumoxytol (Feraheme© or Rienso©), whose use is indicated for the treatment of iron deficiency in adults with chronic kidney disease [20]. Although the number of products containing SPM IONPs approved for use in humans is not very high, efforts to develop new

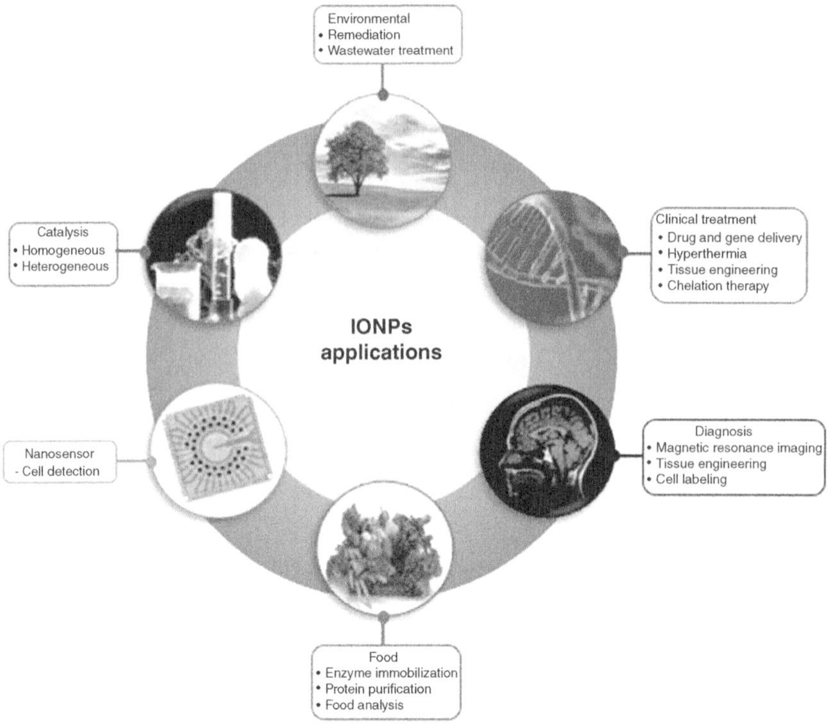

Figure 7.2 Different areas of application of IONPs: environmental, clinical treatment, diagnosis, food, nanosensors, and catalysis. Source: Refs. [25–30] Stock/Adobe Stock; vitstudio/Adobe Stock; Michal Ludwiczak/Adobe Stock.

compounds continue to increase, given the numerous clinical studies undertaken annually (https://clinicaltrials.gov) in anticipation of the introduction of new contrast agents for MRI and the development of drug delivery systems for purposes such as the treatment of anemia or for use in hyperthermia [11].

IONPs are not only used in the medical field for clinical diagnosis and treatment. They can also be used in food engineering [21], catalysis [22], the manufacture of sensors [23], and in the environmental area [24], as illustrated in Figure 7.2.

IONPs are increasing in importance in the environmental area, where they can be used for the removal of contaminants such as heavy metals [18, 25, 26, 31, 32], dyes [27, 28], drugs [33, 34], pesticides [35], and oil [36]. Other uses include acceleration of the coagulation of sewage [37] and the harvesting of microalgae by means of magnetophoresis [38].

One type of MNP, widely used to remove contaminants present in aquatic environments, is nanoscale zerovalent iron (nZVI) [32, 39]. However, although nZVI is very effective in removing contaminants, it is highly reactive and has poor stability, necessitating the use of polymers as protective agents in order to improve its applicability [40].

Despite the many applications of IONPs, few studies have been undertaken to understand the fate, behavior, and toxicity of these materials in organisms, or their interactions and long-term impacts on the environment and human health [18]. Although some studies have indicated that at low concentrations, IONPs do not cause toxicity in a range of cells and organisms, it is necessary to consider the nature of the applications of these systems. For example, when MNPs are concentrated in specific regions of the body, they may lead to toxic effects [10]. For this reason, a more detailed evaluation of the effects of IONPs in terrestrial and aquatic environments is required, as discussed in the following text.

7.1.3 Potential Toxic Effects of Magnetic Iron Oxide Nanoparticles

As already pointed out, IONPs have a range of applications in environmental decontamination (involving the removal and desorption of contaminants). However, given the potential widespread use of these nanoparticles for remediation purposes, the possible impacts of the systems should be evaluated in order to ensure the safe use of the technology. As an example, IONPs could be injected into the soil, where their small size and reactivity could cause the degradation of pollutants, but this practice might consequently lead to the contamination of soil and water bodies (surface and subterranean water) with nanoparticles or their by-products.

The use of nanoparticles in various scientific sectors (health, biotechnology, environment, electronics, etc.) means that studies to evaluate their toxicological potentials should be performed, considering their presence in different compartments (air, water, and soil). Factors that can affect the fate and behavior of nanoparticles in environmental compartments include the type of coating of the nanoparticle or its degradation products, its size, surface charge, concentration, and iron oxidation state [10]. Prior to use of these systems, detailed evaluation is required of the potential impacts on organisms (terrestrial and aquatic), including soil microbes and plants. It is notable that in recent years there have been increasing numbers of studies to assess the toxicological potential of MNPs. A useful review paper containing additional citations is the work of Tang and Lo [32], entitled “Magnetic nanoparticles: essential factors for sustainable environmental applications.” In this paper, the authors highlight an item concerning the toxicological aspects of the class of nanoparticles.

Figure 7.3 indicates some of the possible fates of MNPs and the environmental compartments in which they could present potential toxicity.

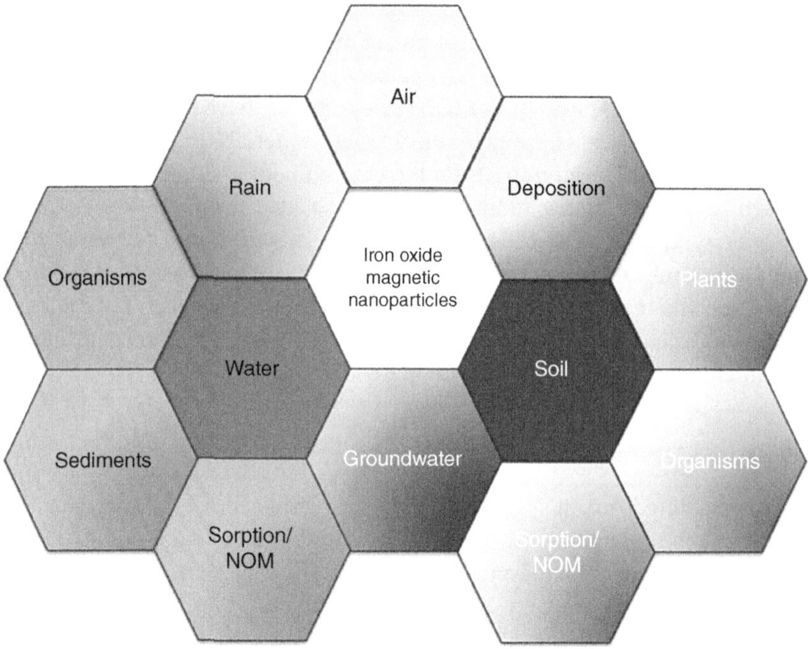

Figure 7.3 Possible fates of magnetic nanoparticles in the environment. In this scheme, it can be seen that magnetic nanoparticles can interact with the soil and with surface and subterranean waters, affecting a wide range of organisms.

A number of research groups have investigated the impacts and possible toxic effects of superparamagnetic iron oxide nanoparticles (SPIONs) on organisms, showing that the main impact is usually due to the excessive production of reactive oxygen species (ROS), including free radicals such as the superoxide anion and hydroxyl radicals, as well as hydrogen peroxide [18]. In their paper "Potential toxicity of SPM iron oxide nanoparticles," Singh et al. [10] present the main toxicity aspects of MNPs, including (i) generation of ROS, (ii) increased formation of micronuclei, (iii) induction of chromosomal condensation processes, (iv) altered mitochondrial function, (v) rupture of the cellular membrane, and (vi) inflammation pathways. Table 7.1 summarizes some of the studies conducted to date, together with the main results concerning effects on cells and organisms.

Additional to toxicity evaluated at the cellular level, studies have been conducted to determine the toxic potential toward aquatic organisms (Table 7.2) and terrestrial organisms (Table 7.3). In the case of aquatic organisms, the main organisms used to evaluate the toxic potential of metallic MNPs include the following: zebrafish (*Danio rerio*), Indian major carp (*Labeo rohita*), *Picochlorum* sp.

Table 7.1 Studies using assays with cell cultures or organisms to assess the impacts of IONPs: characteristics of the particles (functionalization and size distribution), concentration range evaluated, cell types and organisms, and main results.

Type of particle	Size (nm)/concentration range	Type of cell/organism	Main results	References
Fe_3O_4 at tetramethylammonium 11-aminoundecanoate	~5 nm/100 μg/mL	Human glia, human breast cancer, and normal cell lines	Affects cell function, phenotype, and viability	[41]
Fe_3O_4 at polyaspartic acid	8.5 nm/0.6×10^{16} or 1.6×10^{16} particles/mL	Swiss mice	Time and dose-dependent increase in micronucleus frequency	[42]
Fe_3O_4 at dextran	600 mg/kg Fe	Mouse cells (line: C57BL/10ScSn)	Increased number of DNA breaks in rats with dietary iron overload and oxidative damage to DNA (measured by the presence of 8-OH-dG adducts)	[43]
Fe_3O_4 and Fe_3O_4 at dextran	~10 nm/0.05 mg/mL	Human dermal fibroblasts (hTERT-BJ1)	Particles coated with dextran were endocytosed and caused cell death and reduced proliferation	[44]
Fe_3O_4 and Fe_3O_4 at polyvinyl alcohol (PVA)	20–30 nm/14.5–58 mg/mL	Mouse fibroblasts (line: L929)	Uncoated NPs induced greater toxicity than PVA-coated particles, probably due to the presence of a biodegradable polymer and fewer adsorption sites for proteins, ions, and other components in the medium	[45]
Fe_3O_4 at citrate	5 nm/1.5 and 3.0 mM	Rat macrophages (line: RAW 264.7)	Increases in the levels of malonyldialdehyde and protein carbonyls, without causing any cytotoxic effects	[46]
Fe_3O_4	1–3 μm/70 μg/mL	Human lung epithelial cells (line: A549)	Magnetite could cause genotoxicity via mitochondrial depolarization and induction of oxidative stress	[47]
Fe_3O_4 and Fe_3O_4 at pullulan	40–50 nm/0–2.0 mg/mL	Human dermal fibroblasts (line: hTERT-BJ1)	SPIONs induced changes in cytoskeleton organization and uncoated NPs were internalized in the cells, resulting in disruption of cytoskeleton organization of the cells	[48]

Table 7.2 Studies conducted to evaluate possible toxic effects on aquatic biota of magnetic nanoparticles containing iron oxide and cobalt.

Type of particle	Size (nm)/concentration range	Organism	Main results	References
Alpha-Fe_2O_3	30 nm/0.1–100 mg/L	Zebrafish (*D. rerio*)	≥10 mg/L and alpha-Fe_2O_3 aggregates caused developmental toxicity in embryos, with mortality, hatching delay, and malformation	[49]
Fe_3O_4	20 and 40 nm/200 mg/L	*Picochlorum* sp. (*Trebouxiophyceae*, Chlorophyta)	Affected growth and chlorophyll-a content during the exponential growth phase, but acquired adaptive mechanisms reversed these effects	[50]
Fe_3O_4	6.0 nm/ 0.064–0.64 mg/mL	Microcrustaceans (*D. Magna*) and bacteria (*V. Fischeri*)	Extreme sensitivity, indicating high toxicity	[51]
IONPs, IONPs@chitosan, and IONPs@poly (diallyldimethylammonium chloride)	20–30 nm/ 0–20 mg/mL	Microalgae (*Chlorella* sp.)	IONPs without and with surface functionalization were toxic to *Chlorella* sp., especially at high particle concentrations, exacerbated by indirect light shading effects of the suspended nanoparticles	[52]
Fe_3O_4, $Co_{0.2}Zn_{0.8}Fe_2O_4$, and $Co_{0.5}Zn_{0.5}Fe_2O_4$	Fe_3O_4 = 195.9 nm, $Co_{0.2}Zn_{0.8}Fe_2O_4$ = 176.5 nm, $Co_{0.5}Zn_{0.5}Fe_2O_4$ = 347.2 nm/ 12.5–400 μg/mL	Microalgae (*C. vulgaris*)	The NPs induced oxidative stress, adversely affected photosynthesis, and inhibited cell division	[53]
Fe_2O_3	100 nm/2–1000 mg/L	Indian major carp (*L. rohita*)	Affected hematological, biochemical, ionoregulatory, and enzymological parameters	[54]

(*Trebouxiophyceae*, Chlorophyta), microcrustaceans (*Daphnia magna*), bacteria (*Vibrio fischeri*), and microalgae (*Chlorella* sp., *Chlorella vulgaris*).

Similarly, studies have been conducted to evaluate the effects of SPIONs on terrestrial organisms including soil microorganisms (fungi and bacteria), plants, and animals (Table 7.3).

It can be seen from the previous discussion that the literature contains studies concerning evaluation of the toxicological potentials of nanoparticles in different environmental compartments. However, individual SPIONs possess different characteristics (charge, size, reactivity, etc.), which means that individual toxicity evaluation protocols are unlikely to be universally applicable. Another factor is that many studies do not consider aspects such as the aggregation of nanoparticles and their interaction with other species present in the environment, such as natural organic matter [66]. It should be stressed that such difficulties are not limited to evaluation of the toxicity potential of IONPs, but apply to the toxicity assessment of nanomaterials as a whole [67, 68].

7.2 Gaps and Obstacles

A current challenge in the study of magnetic IONPs is related to the lack of information concerning the speciation of these materials, as well as the interaction of SPIONs with species present in the environment under different environmental conditions of pH, level of natural organic matter, ionic strength, and temperature. A few studies have addressed these issues [69–72], but the number remains small, given the large quantity of functionalized and modified IONPs.

Another important aspect is related to the difficulty in quantifying nanomaterials at the concentration ranges in which they may be found in the environment. New and/or improved technologies are required in order to be able to characterize MNPs in complex matrices such as soils and surface and subterranean waters. Several techniques have emerged as ways of studying the fates of nanoparticles and their interactions with complex systems [73–76]. This issue is well described in a review article published by Sadik et al. [76], entitled "Current and emerging technologies for characterization of nanomaterials." The emerging techniques described include single particle inductively coupled plasma mass spectrometry (SP-ICP-MS), fluid flow fractionation inductively coupled plasma mass spectrometry (FFF-ICP-MS), and electrochemical tangential fluid flow (EC-TFF), among others. In the specific case of SPIONs, classical techniques that have been used to characterize these materials in biological matrices include superconducting quantum interference device (SQUID) magnetometry, scanning SQUID biosusceptometry (SSB), and magnetic force microscopy (MFM) [77]. It is expected that further development of these techniques will enable improved monitoring of these nanoparticles in the environment.

Table 7.3 Studies conducted to evaluate possible toxic effects of magnetic iron oxide nanoparticles in terrestrial organisms.

Type of particle	Size (nm)/concentration range	Organism	Main results	References
$CoFe_2O_4$	40 nm/0–25.5 μg/mL of Fe	Fungal strain (*Phanerochaete chrysosporium*)	Impeded cellulolytic fungi metabolism and their ecological role in wood waste decomposition	[55]
Fe_2O_3	20–40 nm/0.001–1000 mg/L	*Xenopus laevis*	NPs decreased the snout-vent length (SVL) of *Xenopus laevis* tadpoles at concentrations of 0.001, 10, 100, and 1000 mg/L	[56]
Fe_3O_4	<50 nm/0.1% and 1% w/w	Bacterial communities from sandy loam (Bet-Dagan) and sandy clay loam (Yatir) soils	NPs changed the hydrolytic activity and bacterial community composition in Bet-Dagan soil	[57]
Fe_3O_4 and γ-Fe_2O_3	10 nm/0.42, 0.84, and 1.26 mg/g	Oxalobacteraceae, Intrasporangium, Terrabacter, and Bacillus (Anthrosol soil)	NPs could potentially stimulate the growth of some bacteria and change the soil bacterial community structure	[58]
Fe_3O_4	20–30 nm	Soil microorganisms	NPs interacted with microbial communities inhabiting soil, affecting the microbial C/N ratio and increasing the metabolic quotient (qCO_2)	[59]
Fe_3O_4	20 nm/0.5 g/L (particles)	Pumpkin plants (*Cucurbita maxima*)	Particles could be absorbed, translocated, and accumulated in the plant tissues	[60]
Fe_3O_4 @carbon	10 nm/0–100 mg/kg	Sunflower (*Helianthus annuus*), tomato (*Lycopersicon esculentum*), pea (*Pisum sativum*), and wheat (*Triticum aestivum*)	NPs were absorbed by the root system and spread at different rates, depending on species, using the transpiration stream in the xylem vessels, reaching the aerial parts of the plants	[61]

Table 7.3 (Continued)

Type of particle	Size (nm)/concentration range	Organism	Main results	References
Fe_3O_4	30–50 nm/0–5000 μg/mL	Cucumber plants (*Cucumis sativus*)	Inhibitory effects were observed and the reduction in root growth was more obvious than the effect on seed germination percentage	[62]
Fe_3O_4	100 nm + aggregates/30, 100, and 500 mg/L	Perennial ryegrass (*Lolium perenne* L.) and pumpkin (*Cucurbita mixta*) plants	NPs induced oxidative stress in the ryegrass and pumpkin roots and shoots, indicating that prolonged exposure to the Fe_3O_4 NPs could pose toxicity risks to both plants	[63]
nZVI	300–400 nm/0–1000 mg/L	Cattail (*Typha latifolia*) and hybrid poplars (*Populus deltoides* × *Populus nigra*)	nZVI could exert phytotoxic effects on plants at concentrations >200 mg/L, and the extent of toxicity was dependent on plant species	[64]
nZVI	1–20 nm/0–5000 mg/L	Flax (*Linum usitatissimum* L.), ryegrass (*Lolium perenne* L.), and two-rowed barley (*Hordeum vulgare* L.)	At low concentrations, nZVI could be used on plants without detrimental effects, while concentrations in the range 1000–2000 mg/L caused inhibition of germination	[65]

In addition to the methodological aspects described earlier, evaluation of SPIONs should include assessments of the effects of exposure to these materials by means of molecular studies employing organisms used for environmental monitoring, in order to obtain specific information about genes that may be affected by the exposure. As an example, Qureshi et al. [78] developed a whole cell-based capacitive biochip (WCB) using bacteria (*Escherichia coli* cells) as biological sensors for detection of the biological nanotoxicity of Fe_3O_4 NPs. In other work, Pandey et al. [79] produced WCB using yeast (*Saccharomyces cerevisiae*) to detect the toxicity of carbon nanotubes.

Further studies are needed of the possible cumulative effects of these nanomaterials on the trophic transfer chain [80–82], as well as aspects related to bioaccumulation, bioconcentration, and biomagnification [83].

7.3 Conclusions

Magnetic IONPs s have indisputable potential applications in many areas. However, it is clear from the preceding discussion that many of the approaches described to date to evaluate the impacts of nanomaterials have not been adopted for these types of particles. As a result, there remains a lack of information concerning the safe use of these materials, especially in relation to their effects in the environment. There have been few studies to elucidate the mechanisms of possible toxicity of such particles, so there is a need for more detailed investigations and the use of emerging technologies to obtain information about the interactions of the particles with complex matrices, as well as to determine the toxic potentials of these nanomaterials at the molecular level.

This chapter provides an overview of current knowledge about the toxic potential of MNPs. However, it is clear that much still remains to be done in terms of characterizing the safety of these materials in contact with living organisms and the environment.

Acknowledgments

The authors are grateful for the financial support provided by the São Paulo State Science Foundation (FAPESP, grants #2015/26189-8 and #2015/15617-9), CAPES, and CNPq.

References

1 Colombo, M., Carregal-Romero, S., Casula, M.F. et al. (2012). Biological applications of magnetic nanoparticles. *Chem. Soc. Rev.* 41: 4306–4334.

2 Servin, A.D. and White, J.C. (2016). Nanotechnology in agriculture: next steps for understanding engineered nanoparticle exposure and risk. *Nanoimpact* 1: 9–12.

3 Indira, T.K. and Lakshmi, P.K. (2010). Magnetic nanoparticles – a review. *Int. J. Pharm. Sci. Nanotechnol.* 3: 1035–1042.

4 Frey, N.A., Peng, S., Cheng, K., and Sun, S.H. (2009). Magnetic nanoparticles: synthesis, functionalization, and applications in bioimaging and magnetic energy storage. *Chem. Soc. Rev.* 38: 2532–2542.

5 Arami, H., Khandhar, A., Liggitt, D., and Krishnan, K.M. (2015). In vivo delivery, pharmacokinetics, biodistribution and toxicity of iron oxide nanoparticles. *Chem. Soc. Rev.* 44: 8576–8607.

6 Hao, R., Xing, R.J., Xu, Z.C. et al. (2010). Synthesis, functionalization, and biomedical applications of multifunctional magnetic nanoparticles. *Adv. Mater.* 22: 2729–2742.

7 Reddy, L.H., Arias, J.L., Nicolas, J., and Couvreur, P. (2012). Magnetic nanoparticles: design and characterization, toxicity and biocompatibility, pharmaceutical and biomedical applications. *Chem. Rev.* 112: 5818–5878.

8 Hachani, R., Lowdell, M., Birchall, M. et al. (2016). Polyol synthesis, functionalisation, and biocompatibility studies of superparamagnetic iron oxide nanoparticles as potential MRI contrast agents. *Nanoscale* 8: 3278–3287.

9 Gupta, A.K. and Gupta, M. (2005). Synthesis and surface engineering of iron oxide nanoparticles for biomedical applications. *Biomaterials* 26: 3995–4021.

10 Singh, N., Jenkins, G.J., Asadi, R., and Doak, S.H. (2010). Potential toxicity of superparamagnetic iron oxide nanoparticles (SPION). *Nano Rev.* 1: 5358.

11 Cortajarena, A.L., Ortega, D., Ocampo, S.M. et al. (2014). Engineering iron oxide nanoparticles for clinical settings. *Nanobiomedicine* 1 (2): 1–20.

12 Gudovan, D., Balaure, P.C., Mihaiescu, D.E. et al. (2015). Functionalized magnetic nanoparticles for biomedical applications. *Curr. Pharm. Des.* 21: 6038–6054.

13 Barrow, M., Taylor, A., Murray, P. et al. (2015). Design considerations for the synthesis of polymer coated iron oxide nanoparticles for stem cell labelling and tracking using MRI. *Chem. Soc. Rev.* 44: 6733–6748.

14 Mulens, V., Morales, M.P., and Barber, D.F. (2013). Development of magnetic nanoparticles for cancer gene therapy: a comprehensive review. *Int. Shortly Res. Not.* 2013: 1–14.

15 Estelrich, J., Escribano, E., Queralt, J., and Busquets, M.A. (2015). Iron oxide nanoparticles for magnetically-guided and magnetically-responsive drug delivery. *Int. J. Mol. Sci.* 16: 8070–8101.

16 Issa, B., Obaidat, I.M., Albiss, B.A., and Haik, Y. (2013). Magnetic nanoparticles: surface effects and properties related to biomedicine applications. *Int. J. Mol. Sci.* 14: 21266–21305.

17 Trindade, T. and Daniel-Da-Silva, A.L. (2011). Advances in nanocomposite technology. In: *Biofunctional Composites of Polysaccharides Containing Inorganic Nanoparticles* (ed. A. Hashim). Intech.
18 Liu, G., Gao, J.H., Ai, H., and Chen, X.Y. (2013). Applications and potential toxicity of magnetic iron oxide nanoparticles. *Small* 9: 1533–1545.
19 Wang, Y.X. (2011). Superparamagnetic iron oxide based MRI contrast agents: current status of clinical application. *Quant. Imaging. Med. Surg.* 1: 35–40.
20 Lu, M., Cohen, M.H., Rieves, D., and Pazdur, R. (2010). FDA report: ferumoxytol for intravenous iron therapy in adult patientswith chronic kidney disease. *Am. J. Hematol.* 85: 315–319.
21 Cao, M., Li, Z.H., Wang, J.L. et al. (2012). Food related applications of magnetic iron oxide nanoparticles: enzyme immobilization, protein purification, and food analysis. *Trends Food Sci. Tech.* 27: 47–56.
22 Govan, J. and Gun'ko, Y.K. (2014). Recent advances in the application of magnetic nanoparticles as a support for homogeneous catalysts. *Nanomaterials (Basel)* 4: 222–241.
23 Koh, I. and Josephson, L. (2009). Magnetic nanoparticle sensors. *Sensors (Basel)* 9: 8130–8145.
24 Mueller, N.C. and Nowack, B. (2010). Nanoparticles for remediation: solving big problems with little particles. *Elements* 6: 395–400.
25 Yavuz, C.T., Mayo, J.T., Yu, W.W. et al. (2006). Low-field magnetic separation of monodisperse Fe_3O_4 nanocrystals. *Science* 314: 964–967.
26 Zhang, S.X., Zhang, Y.Y., Bi, G.M. et al. (2014). Mussel-inspired polydopamine biopolymer decorated with magnetic nanoparticles for multiple pollutants removal. *J. Hazard Mater.* 270: 27–34.
27 Wang, H. and Huang, Y.M. (2011). Prussian-blue-modified iron oxide magnetic nanoparticles as effective peroxidase-like catalysts to degrade methylene blue with H_2O_2. *J. Hazard Mater.* 191: 163–169.
28 Zhang, Y.R., Wang, S.Q., Shen, S.L., and Zhao, B.X. (2013). A novel water treatment magnetic nanomaterial for removal of anionic and cationic dyes under severe condition. *Chem. Eng. J.* 233: 258–264.
29 Zhu, H.Y., Fu, Y.Q., Jiang, R. et al. (2012). Novel magnetic chitosan/poly(vinyl alcohol) hydrogel beads: preparation, characterization and application for adsorption of dye from aqueous solution. *Bioresour. Technol.* 105: 24–30.
30 Zhao, M., Xu, Y., Zhang, C., Rong, H., Zeng, G. (2016). New trends in removing heavy metals from wastewater. *Appl. Microbiol. Biotechnol.* 100 (15): 6509–6518. https://doi.org/10.1007/s00253-016-7646-x. PMID: 27318819.
31 Ge, F., Li, M.M., Ye, H., and Zhao, B.X. (2012). Effective removal of heavy metal ions Cd^{2+}, Zn^{2+}, Pb^{2+}, Cu^{2+} from aqueous solution by polymer-modified magnetic nanoparticles. *J. Hazard Mater.* 211: 366–372.

32 Tang, S.C.N. and Lo, I.M.C. (2013). Magnetic nanoparticles: essential factors for sustainable environmental applications. *Water Res.* 47: 2613–2632.

33 Niu, H.Y., Dizhang, Meng, Z.F., and Cai, Y.Q. (2012). Fast defluorination and removal of norfloxacin by alginate/Fe@Fe_3O_4 core/shell structured nanoparticles. *J. Hazard Mater.* 227: 195–203.

34 Zhang, D., Niu, H.Y., Zhang, X.L. et al. (2011). Strong adsorption of chlorotetracycline on magnetite nanoparticles. *J. Hazard Mater.* 192: 1088–1093.

35 Xu, L.J. and Wang, J.L. (2012). Fenton-like degradation of 2,4-dichlorophenol using Fe_3O_4 magnetic nanoparticles. *Appl. Catal. B: Environ.* 123: 117–126.

36 Orbell, J.D., Godhino, L., Bigger, S.W. et al. (1997). Oil spill remediation using magnetic particles - An experiment in environmental technology. *J. Chem. Educ.* 74: 1446–1448.

37 Booker, N.A., Keir, D., Priestley, A.J. et al. (1991). Sewage clarification with magnetite particles. *Water Sci. Technol.* 23: 1703–1712.

38 Cerff, M., Morweiser, M., Dillschneider, R. et al. (2012). Harvesting fresh water and marine algae by magnetic separation: screening of separation parameters and high gradient magnetic filtration. *Bioresour. Technol.* 118: 289–295.

39 Stefaniuk, M., Oleszczuk, P., and Ok, Y.S. (2016). Review on nano zerovalent iron (nZVI): from synthesis to environmental applications. *Chem. Eng. J.* 287: 618–632.

40 Rao, J.P., Gruenberg, P., and Geckeler, K.E. (2015). Magnetic zero-valent metal polymer nanoparticles: current trends, scope, and perspectives. *Prog. Polym. Sci.* 40: 138–147.

41 Ankamwar, B., Lai, T.C., Huang, J.H. et al. (2010). Biocompatibility of Fe_3O_4 nanoparticles evaluated by in vitro cytotoxicity assays using normal, glia and breast cancer cells. *Nanotechnology* 21 (7): 75102. https://doi.org/10.1088/0957-4484/21/7/075102.

42 Sadeghiani, N., Barbosa, L.S., Silva, L.P. et al. (2005). Genotoxicity and inflammatory investigation in mice treated with magnetite nanoparticles surface coated with polyaspartic acid. *J. Magn. Magn. Mater.* 289: 466–468.

43 Faux, S.P., Francis, J.E., Smith, A.G., and Chipman, J.K. (1992). Induction of 8-hydroxydeoxyguanosine in Ah-responsive mouse-liver by iron and Aroclor 1254. *Carcinogenesis* 13: 247–250.

44 Berry, C.C., Wells, S., Charles, S., and Curtis, A.S.G. (2003). Dextran and albumin derivatised iron oxide nanoparticles: influence on fibroblasts in vitro. *Biomaterials* 24: 4551–4557.

45 Mahmoudi, M., Simchi, A., Imani, M. et al. (2010). A new approach for the in vitro identification of the cytotoxicity of superparamagnetic iron oxide nanoparticles. *Colloid Surf. B.* 75: 300–309.

46 Stroh, A., Zimmer, C., Gutzeit, C. et al. (2004). Iron oxide particles for molecular magnetic resonance imaging cause transient oxidative stress in rat macrophages. *Free Radic. Biol. Med.* 36: 976–984.

47 Karlsson, H.L., Holgersson, A., and Moller, L. (2008). Mechanisms related to the genotoxicity of particles in the subway and from other sources. *Chem. Res. Toxicol.* 21: 726–731.

48 Gupta, A.K. and Gupta, M. (2005). Cytotoxicity suppression and cellular uptake enhancement of surface modified magnetic nanoparticles. *Biomaterials* 26: 1565–1573.

49 Zhu, X.S., Tian, S.Y., and Cai, Z.H. (2012). Toxicity assessment of iron oxide nanoparticles in Zebrafish (*Danio rerio*) early life stages. *PLoS One* 7 (9): e46286. https://doi.org/10.1371/journal.pone.0046286.

50 Hazeem, L.J., Waheed, F.A., Rashdan, S. et al. (2015). Effect of magnetic iron oxide (Fe_3O_4) nanoparticles on the growth and photosynthetic pigment content of *Picochlorum* sp. *Environ. Sci. Pollut. Res.* 22: 11728–11739.

51 Garcia, A., Espinosa, R., Delgado, L. et al. (2011). Acute toxicity of cerium oxide, titanium oxide and iron oxide nanoparticles using standardized tests. *Desalination* 269: 136–141.

52 Toh, P.Y., Tai, W.Y., Ahmad, A.L. et al. (2016). Toxicity of bare and surfaced functionalized iron oxide nanoparticles towards microalgae. *Int. J. Phytoremediation* 18 (6): 643–650. https://doi.org/10.1080/15226514.2015.1086300.

53 Barhoumi, L. and Dewez, D. (2013). Toxicity of superparamagnetic iron oxide nanoparticles on green alga *Chlorella vulgaris*. *Biomed Res Int.* 2013: 647974.

54 Remyaa, A.S., Ramesha, M., Saravananb, M. et al. (2015). Iron oxide nanoparticles to an Indian major carp, *Labeo rohita*: impacts on hematology, iono regulation and gill Na^+/K^+ ATPase activity. *J. King Saud Univ.-Sci.* 27: 151–160.

55 Oprica, L., Nadejde, C., Andries, M. et al. (2015). Magnetic contamination of environment - laboratory simulation of mixed iron oxides impact on microorganism cells. *Environ. Eng. Manag. J.* 14: 581–586.

56 Nations, S., Wages, M., Canas, J.E. et al. (2011). Acute effects of Fe_2O_3, TiO_2, ZnO and CuO nanomaterials on *Xenopus laevis*. *Chemosphere* 83: 1053–1061.

57 Frenk, S., Ben-Moshe, T., Dror, I. et al. (2013). Effect of metal oxide nanoparticles on microbial community structure and function in two different soil types. *PLoS One* 8 (12): e84441. https://doi.org/10.1371/journal.pone.0084441.

58 He, S.Y., Feng, Y.Z., Ren, H.X. et al. (2011). The impact of iron oxide magnetic nanoparticles on the soil bacterial community. *J. Soil Sediments* 11: 1408–1417.

59 Antisari, L.V., Carbone, S., Gatti, A. et al. (2013). Toxicity of metal oxide (CeO_2, Fe_3O_4, SnO_2) engineered nanoparticles on soil microbial biomass and their distribution in soil. *Soil Biol. Biochem.* 60: 87–94.

60 Zhu, H., Han, J., Xiao, J.Q., and Jin, Y. (2008). Uptake, translocation, and accumulation of manufactured iron oxide nanoparticles by pumpkin plants. *J. Environ. Monitor.* 10: 713–717.

61 Cifuentes, Z., Custardoy, L., de la Fuente, J.M. et al. (2010). Absorption and translocation to the aerial part of magnetic carbon-coated nanoparticles through the root of different crop plants. *J. Nanobiotechnol.* 8 (8): 26. https://doi.org/10.1186/1477-3155-8-26.

62 Mushtaq, Y.K. (2011). Effect of nanoscale Fe_3O_4, TiO_2 and carbon particles on cucumber seed germination. *J. Environ. Sci. Health A* 46: 1732–1735.

63 Wang, H.H., Kou, X.M., Pei, Z.G. et al. (2011). Physiological effects of magnetite (Fe_3O_4) nanoparticles on perennial ryegrass (*Lolium perenne* L.) and pumpkin (*Cucurbita mixta*) plants. *Nanotoxicology* 5: 30–42.

64 Ma, X.M., Gurung, A., and Deng, Y. (2013). Phytotoxicity and uptake of nanoscale zero-valent iron (nZVI) by two plant species. *Sci. Total Environ.* 443: 844–849.

65 El-Temsah, Y.S. and Joner, E.J. (2012). Impact of Fe and Ag nanoparticles on seed germination and differences in bioavailability during exposure in aqueous suspension and soil. *Environ. Toxicol.* 27: 42–49.

66 Grillo, R., Rosa, A.H., and Fraceto, L.F. (2015). Engineered nanoparticles and organic matter: a review of the state-of-the-art. *Chemosphere* 119: 608–619.

67 Oostingh, G.J., Casals, E., Italiani, P. et al. (2011). Problems and challenges in the development and validation of human cell-based assays to determine nanoparticle-induced immunomodulatory effects. *Part. Fibre Toxicol.* 8 (1): 8. https://doi.org/10.1186/1743-8977-8-8.

68 Xue, H.Y., Liu, S.M., and Wong, H.L. (2014). Nanotoxicity: a key obstacle to clinical translation of siRNA-based nanomedicine. *Nanomedicine (UK)* 9: 295–312.

69 Baalousha, M. (2009). Aggregation and disaggregation of iron oxide nanoparticles: influence of particle concentration, pH and natural organic matter. *Sci. Total. Environ.* 407: 2093–2101.

70 Baalousha, M., Manciulea, A., Cumberland, S. et al. (2008). Aggregation and surface properties of iron oxide nanoparticles: influence of pH and natural organic matter. *Environ. Toxicol. Chem.* 27: 1875–1882.

71 Ghosh, S., Jiang, W., McClements, D.J., and Xing, B.S. (2011). Colloidal stability of magnetic iron oxide nanoparticles: influence of natural organic matter and synthetic polyelectrolytes. *Langmuir* 27: 8036–8043.

72 Hu, J.D., Zevi, Y., Kou, X.M. et al. (2010). Effect of dissolved organic matter on the stability of magnetite nanoparticles under different pH and ionic strength conditions. *Sci. Total. Environ.* 408: 3477–3489.

73 Hawkins, A.D., Bednar, A.J., Cizdziel, J.V. et al. (2014). Identification of silver nanoparticles in Pimephales promelas gastrointestinal tract and gill tissues using flow field flow fractionation ICP-MS. *RSC Adv.* 4: 41277–41280.

74 Meermann, B. (2015). Field-flow fractionation coupled to ICP-MS: separation at the nanoscale, previous and recent application trends. *Anal. Bioanal. Chem.* 407: 2665–2674.

75 Meisterjahn, B., Neubauer, E., Von der Kammer, F. et al. (2014). Asymmetrical flow-field-flow fractionation coupled with inductively coupled plasma mass spectrometry for the analysis of gold nanoparticles in the presence of natural nanoparticles. *J. Chromatogr. A* 1372: 204–211.

76 Sadik, O.A., Du, N., Kariuki, V. et al. (2014). Current and emerging technologies for the characterization of nanomaterials. *ACS Sustain. Chem. Eng.* 2: 1707–1716.

77 Hurley, K.R., Ring, H.L., Kang, H. et al. (2015). Characterization of magnetic nanoparticles in biological matrices. *Anal. Chem.* 87: 11611–11619.

78 Qureshi, A., Pandey, A., Chouhan, R.S. et al. (2015). Whole-cell based label-free capacitive biosensor for rapid nanosize-dependent toxicity detection. *Biosens. Bioelectron.* 67: 100–106.

79 Pandey, A., Chouhan, R.S., Gurbuz, Y. et al. (2015). *S. cerevisiae* whole-cell based capacitive biochip for the detection of toxicity of different forms of carbon nanotubes. *Sens. Actuator B-Chem.* 218: 253–260.

80 Chen, J.Y., Li, H.R., Han, X.Q., and Wei, X.Z. (2015). Transmission and accumulation of Nano-TiO_2 in a 2-step food chain (*Scenedesmus obliquus* to *Daphnia magna*). *Bull Environ. Contam. Toxicol.* 95: 145–149.

81 Lee, W.M., Yoon, S.J., Shin, Y.J., and An, Y.J. (2015). Trophic transfer of gold nanoparticles from *Euglena gracilis* or *Chlamydomonas reinhardtii* to *Daphnia magna*. *Environ. Pollut.* 201: 10–16.

82 Roche, R.D., Servin, A., Hawthorne, J. et al. (2015). Terrestrial trophic transfer of bulk and nanoparticle La_2O_3 does not depend on particle size. *Environ. Sci. Technol.* 49: 11866–11874.

83 Bour, A., Mouchet, F., Silvestre, J. et al. (2015). Environmentally relevant approaches to assess nanoparticles ecotoxicity: a review. *J. Hazard Mater.* 283: 764–777.

8

Carbon Nanotubes: Sublethal Effects and Unique Mechanisms of Toxicity in Aquatic Species

Tara Sabo-Attwood[1], Christine Ngan[2], Candice Lavelle[1], Jaime Plazas-Tuttle[2], and Navid B. Saleh[2]

[1] *University of Florida, Department of Environmental & Global Health, Center for Environmental and Human Toxicology, Gainesville, FL, USA*
[2] *University of Texas at Austin, Department of Civil, Architectural and Environmental Engineering, Austin, TX, USA*

8.1 Carbon Nanotubes in Aquatic Environments

Carbon nanotubes (CNTs) are allotropes of carbon comprised of a single tubule or multiple π-stacked concentric tubes [1–4]. These are considered to be one of the most commercially important and technologically relevant nanomaterials because of the unique physicochemical properties and are currently being used in microelectronics, energy storage, sports equipment, sensors, and vehicles, as well as in drug delivery [5]. Since there is a limited regulation regarding the use and handling of nanomaterials, it is highly likely that these will end up in aquatic ecosystems in waste streams through several routes as discussed in literature reports [6, 7]. One estimate puts yearly releases of CNTs directly into the aquatic environment at 33 metric tons [8]. Therefore, the increased use of CNTs and potential for environmental release calls for reliable data on the fate, transport, and biotoxicity to better evaluate the environmental impacts and potential health effects on relevant species.

CNTs have been shown to elicit toxic responses in biological systems including mammalian, invertebrate, and bacterial cells [9–11]. Several studies have shown that CNTs induce significant toxic effects in rodent models and human cell lines, with a particular emphasis on the pulmonary system, producing cytotoxic, inflammatory, pre-carcinogenic, and fibrogenic responses [12–14]. CNTs have also been shown to display antimicrobial effects by penetrating bacterial

Toxicology of Nanoparticles and Nanomaterials in Human, Terrestrial and Aquatic Systems, First Edition. Edited by Marc A. Williams.

membranes [15], whereas other reports have shown a lack of antimicrobial activity [16]. In many cases, cellular and molecular mechanisms have been studied that include oxidative stress (including reactive oxygen species [ROS] production), initiation of pro-fibrotic signaling pathways (i.e. TGF-β), genotoxicity, and immune modulation [17–19]. Compared to the aforementioned body of literature, investigations on specific physiological effects of CNTs in aquatic species are not as well defined. This chapter describes the current state of understanding of toxicity associated with CNT exposures to aquatic organisms. The focus will be on data collected from several model species including microinvertebrates that include small crustaceans (e.g. daphnia), macroinvertebrates (shrimp, snails, bivalves), amphibians, and fish, as these represent highly studied organisms in this area. A steady increase in the number of articles assessing CNT toxicity in these species has been observed since 2007 (Figure 8.1a) where nearly half of such work is represented by studies on fish and amphibians and micro- and macroinvertebrates represent the other half (Figure 8.1b). The chapter will be organized into several sections that include a summary of measured end points focusing primarily on traditional toxicity assays, a description of unique mechanisms of toxicity that have been more recently discovered, an introduction to horizon materials (nanohybrids [NHs]), and a future perspective.

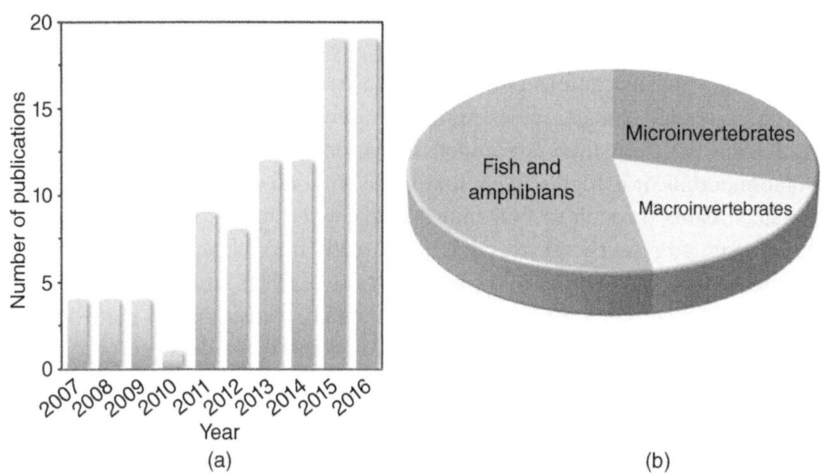

Figure 8.1 (a) Total number of publications that have investigated the effects of CNTs on micro- and macroinvertebrates and fish and amphibians for each year from 2007 to 2016. (b) The percentage of articles in total that studied effects of CNTs on the different species. Microinvertebrates are primarily represented by *Daphnia Magna* and macroinvertebrates by bivalves and shrimp.

8.2 Classical Toxicity: What We Have Learned

Most studies conducted to date have assessed toxicity from aqueous exposure to CNTs in a concentration range of 0.04–1000 mg/L. Most of this range is above what is expected to be present in the environment as determined by modeled concentrations of CNTs in surface water (0.0006–0.025 ng/L). Higher concentrations are likely to occur in sediments (40–1557 ng/kg). Traditional end points measured included mortality, growth, reproduction, and in some cases genotoxicity and measurements of oxidative stress [20–23]. Results of these studies, in general, do not show significant impacts of CNTs on mortality and growth in exposed organisms although there are several reports that show decreases in these measurements at high doses >10 mg/L [24–29]. A few studies have additionally probed effects on behavioral end points, such as swimming and feeding abilities. For example, *Daphnia* swimming velocity showed a non-monotonic decreased response at doses of 0.9, 6.9, and 22.5 mg/L [30], whereas newly hatched rare minnows (*Gobiocypris rarus*) exposed during development to 100 mg/L CNTs presented with malformations and reduced swimming speed [31]. Conversely, locomotor and social behaviors of rainbow trout (*Oncorhynchus mykiss*), including mean velocity when active, total distance moved, and the distribution of swimming speeds were not affected following exposure to 0.25 mg/L CNTs over 10 days [32]. Differences in behavioral end points are likely due to the CNT composition, i.e. single-walled carbon nanotube (SWNT) vs. multi-walled carbon nanotube (MWNT), variation in surface functionality, and the influence of aquatic exposure conditions (i.e. water quality, natural organic matter [NOM]). In general, no extreme overt toxicity has been observed despite that aquatic organisms, including vertebrates and invertebrates, were exposed to fairly high doses of CNTs. In addition, the majority of the exposures were performed via an aqueous route even though CNTs, in their natural state, are not likely to remain stable over significant periods of time and therefore will not remain suspended in the water column. This notion has been met with controversy as some studies report that CNT aggregates can last weeks to months in freshwater systems [33]. These data highlight the need for assessments of subtle physiological responses in response to more environmentally relevant doses through dietary means.

Virtually nothing is known about the molecular mechanisms that are responsible for sublethal toxicity imparted by CNTs in aquatic organisms and very few studies have begun to address this knowledge gap. Oxidative stress, possibly initiated by metal-containing CNTs, has been proposed to be critical; however, published data sets are contradictory [34]. Fraser et al. [35] showed that although trout fed with CNT-enriched diets exhibited no overt toxicity, lipid peroxidation was noted in the brain and intestine and suggested to be a result of ROS production.

Studies primarily performed in fish have indirectly shown significant increases in oxidative stress through measurements of lipid peroxidation (TBARS) [35, 36], antioxidant enzyme activity and expression (superoxide dismutase [SOD], catalase [CAT], glutathione peroxidase [GPx]) [37], and the levels of related transcription factors (Nrf2) [38]. Few studies have directly connected a role for oxidative stress in observed physiological end points such as growth and reproductive capacity. Handy et al. [39] suggested a link with oxidative stress and respiratory toxicity and Lee et al. [40] reported reduced growth. However, other reports show increased ROS production with decreased antioxidant capacity that is thought to drive cellular responses such as apoptosis [41], but the longer-term effects on organism outcomes have not been determined. Interestingly, exposure of the copepod *Paracyclopina nana* to CNTs caused activation of kinase signaling pathways and antioxidant responses that were determined to be independent of ROS production [42], and an observation was also documented in *Daphnia magna* [43] and mammalian systems [44]. While ROS production leading to oxidative damage has been one of the most widely studied mechanisms, these studies suggest that alternate mechanisms that contribute to toxicity should be considered.

8.3 Unique Mechanisms and Effects

There are several less traditional mechanisms of CNT toxicity that have been the focus of more recent studies that include nutrient depletion, immune modulation, and altered toxicity of co-contaminants. Evidences that have brought attention to these mechanisms are discussed in the following subsections.

8.3.1 Nutrient Depletion

Due to the ability to adsorb organic molecules, CNTs have been shown to induce indirect cytotoxicity, by limiting vital nutrients. This notion has been primarily supported by studies that utilize mammalian cell culture models, which show adsorption of serum components to CNTs [45]. These initial discoveries have led to a "nutrient depletion" hypothesis, which essentially states that CNTs that are taken up by organisms can bind to components of food, depleting nutrient supplies and eventually leading to inhibition of growth, reproduction, and energy storage in aquatic organisms. Several groups have shown that aquatic organisms, primarily microinvertebrates and fish, can take up CNTs via ingestion; however there is little evidence to support that these particles get absorbed across the gut epithelium [46]. In fact, tracking of CNTs after a single gavage in fathead minnows (FHM, *Pimephales promelas*) showed no evidence of intestinal absorption and total elimination of the CNTs after 7 days [47, 48]. The presence of CNTs in the gut tract of microinvertebrates has also been observed, yet the CNTs are not always eliminated and in many cases cause gut "clogging" in certain species that may be related to reduction in population growth rate, growth, and reduced reproductive capacity [26, 27, 29]. In general, there is no overwhelming evidence to support that CNTs are absorbed across the epithelium, regardless of species, although a few reports suggest CNTs penetrate cells through the cuticle layer [40]. It should be noted that these observations rely on TEM, which can have limitations in identifying CNTs in tissues and organisms. This is partly due to the challenges associated with carbon-based nanoparticle detection that are reviewed in Petersen et al. [49]. In brief, CNTs elude common methods of detection used routinely to identify small particles or fibers in organisms such as scanning electron microscopy/transmission electron microscopy (SEM/TEM) with elemental analysis and immunofluorescence that have significant limitations. The development of contemporary detection methods such as near-IR fluorescence (NIRF) has been shown to be a sensitive technique to detect and quantify trace levels of CNTs in biological systems [47], but there are drawbacks with this technology as well. While the tracking and quantitative abilities are excellent, the technique is limited to select types of pristine CNTs.

It is not clear whether limited absorption in the gut has to do with the size of the CNT aggregates or how they interact with other molecules, including food particles, within organisms. One interesting idea is that the presence of CNTs can perturb the normal gut environment by altering the expression of nutrient transporters and enzymes. Bisesi et al. [48] showed that FHMs that ingested CNT-containing feed modulated the gene expression of *pept2* in a manner that was specific to the section of intestinal tissue (proximal, Figure 8.2). The expression of such genes has been observed in fish under starvation conditions [50, 51] but interestingly did not exactly mimic the changes in fish that were starved as a comparison group. It is tempting to speculate that sorption of food particles and

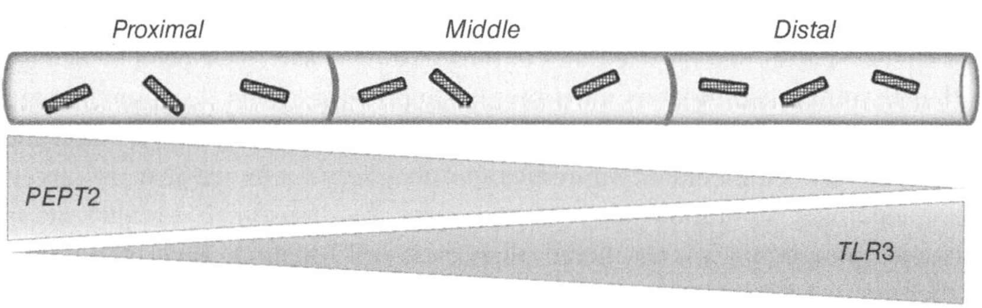

Figure 8.2 Diagram of a fish intestine exposed to CNTs showing the proximal, middle, and distal sections and the level of expression of pept2 and TLR3. The wider the triangle, the greater the expression; therefore the highest expression for pept2 occurred in the proximal intestine and the largest increase of TLR3 was observed in the distal intestine.

gastrointestinal (GI) proteins by CNTs is a mechanism whereby nutritional status is impacted. While these studies were performed over short duration (7 days) to assess growth, a recent study by our group showed a significant decrease in growth of bass fed CNTs, but the effect was not significant until after 12 weeks of continued feeding (manuscript in preparation). This is in contrast to other feeding studies where no change in growth was observed, but these experiments were only carried out for 6 weeks [35]. As we did not start to observe changes in growth until after 12 weeks, it suggests that chronic feeding studies may shed light on long-term effects on major health end points related to growth including reproductive capacity and population effects. One study performed on Daphnia showed that acute mortality was not a good predictor of chronic effects. No effects on mortality during acute exposures were noted; however chronic assays indicated significant differences in mortality, reproduction, and growth [52]. Furthermore, this same group assessed multigenerational effects showing negative impacts on Daphnid reproduction out to the F2 generation for select CNTs [53]. Only one other study has investigated multigenerational effects on CNT-injected zebrafish embryos that had lower survival rates compared with unexposed embryos [54]. It is not clear whether the adverse effect on swimming behavior, growth, reproductive, and generational effects have any relationship to reduced nutritional capacity and energy production, but this concept has not been explored (Figure 8.3).

8.3.2 Immune Modulation

The effects of CNTs on the immune system has been an intense area of research focus in mammalian models [55–57], but few studies have investigated the ability of CNTs to modulate the immune system in aquatic organisms. Of the reports published to date, most have focused on immune modulation by other types of nanomaterials such as TiO_2 [58], iron oxide [59], zinc oxide, copper [60], fullerenes [61], polystyrene, and polycarbonate [62] nanoparticles. Exposures of several fish species and invertebrates to these nanoparticles have been associated with both immunostimulatory and immunosuppressive effects. A comprehensive review by Jovanovic et al. [63] summarizes the current body of literature and points out that cell-mediated immunity and phagocytic cells are primary targets of nanoparticle immunotoxicity as measured by lysosomal destabilization, frustrated phagocytosis, and altered phagocytic cell function. Such modulation of the immune system can lead to impaired health and increased susceptibility to pathogens. This notion was shown in a study published in 2015 by the same group [58] where pre-exposure of FHM to TiO_2 nanoparticles significantly increased fish mortality during a subsequent bacterial challenge. Other studies using fish models have linked exposure of zinc oxide and copper polystyrene and

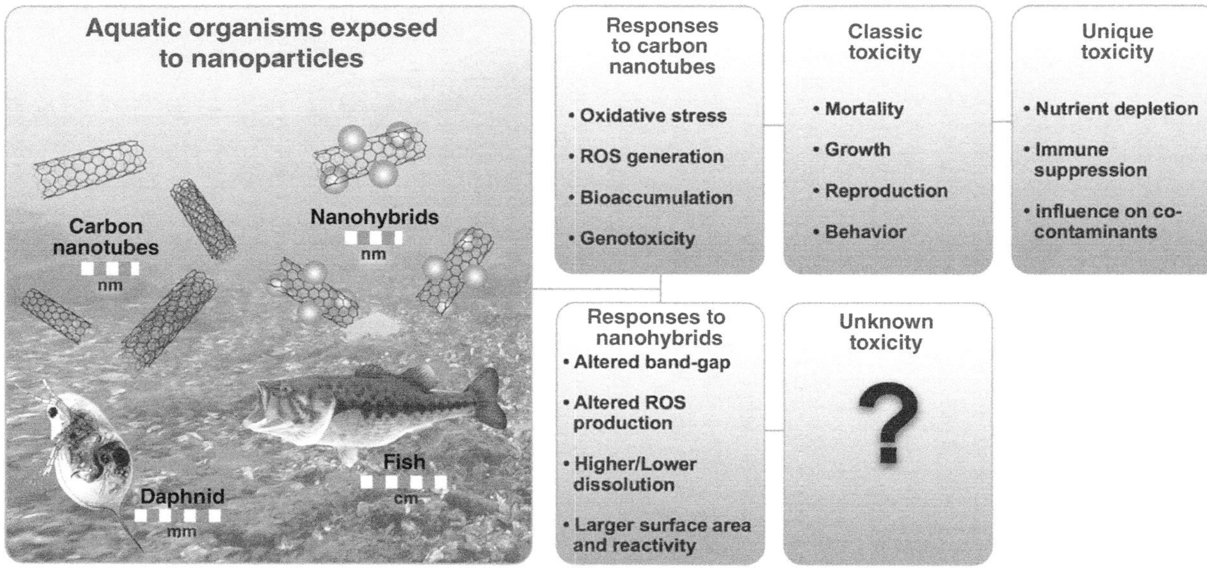

Figure 8.3 Diagram showing the known responses of aquatic organisms to CNTs, classic toxicity end points, and unique mechanisms of toxicity. Responses to nanohybrids are suggested but associated toxicity has not yet been determined.

polycarbonate nanoparticles to repress lysozyme activity, increase degranulation of primary granules and neutrophil extracellular trap release, result in varied changes in respiratory burst, and increase the expression of interleukin-1beta and tumor necrosis factor-alpha [60, 62]. A few investigations on macroinvertebrates have also reported the effects of nanomaterials on immune parameters such as lysosomal destabilization that occurred in oysters in response to fullerenes [61].

Very few studies have probed the immune response specifically to CNTs in aquatic species, and in fact, there are far more articles related to their use as delivery vehicles for vaccines of high importance to the aquaculture industry [64] than those focused on toxicity. Most investigations of CNT immunomodulatory effects have focused on cell responses *in vitro*. Klaper et al. [65] was the first to report the modulation of immune marker expression, including IL-1β, in a non-mammalian trout macrophage cell line. While the study did reveal stimulatory effects of some CNTs, the results were difficult to delineate from the vehicle control (cholate). Investigations on the immune reposes of freshwater invertebrates, such as *Hirudo medicinalis*, show that CNTs increased monocyte–macrophages (CD45+, CD68+) and the secretion of cytokines, specifically IL-8 [66]. While not exactly typical CNTs, the effects of surface-modified self-assembled rosette nanotubes on channel catfish B- and T-cell lines were evaluated and showed a reduced ability of cells to phagocytose [67].

Due to the substantial lack of studies that have investigated CNT-driven immune responses in aquatic organisms, the mechanisms that contribute to such alterations have not been elucidated. It has been suggested that interference or activation of immune receptors, such as leukocyte immune-type receptors (IpLITR) [67] or pathogen recognition receptors (toll-like receptor [TLRs]) [63] are a target. In fact, we have observed that CNTs can alter TLR activity and other innate immune signaling pathways in mammalian models systems [17] and have recently shown that TLR3, which interacts with viral nucleic acids, is upregulated in the distal intestine of FHM exposed to CNTs for 96 h via the diet (Figure 8.2) [68]. It has been speculated that due to their sorptive nature as described earlier, CNTs may also form complexes with pathogens in the water or sediments, essentially acting as a "carrier" of pathogens (e.g. LPS) and/or immune stimulating agents [63]. These results highlight a substantial gap in our knowledge of immune perturbations in aquatic species by CNTs that could have significant impact on overall health in contaminated environments.

8.3.3 Influence on Co-contaminants

Studies that have gained traction over the past decade related to CNT toxicity in aquatic organisms include those focused on their ability to alter the toxicity of co-contaminants [69–71]. These studies are primarily based on the sorptive nature of the CNTs and their ability to interact directly with co-contaminants that are known to be present in aqueous environments. Numerous studies have developed sorption isotherms for CNT binding to chemicals, but the effects of such interactions on aquatic species have only been studied for a subset of these. Co-contaminants studied in this capacity include arsenic [72], zinc [73, 74], fluoranthane [75], cadmium [74, 76, 77], lead [78], carbofuran [79], copper [40, 80], triclocarban [81], phenanthrene [82], ethinylestradiol [83], pentachlorophenol [84], carbofuran [79], and polychlorinated biphenyls (PCBs) [85], and results vary depending on the type of CNT, organism, co-contaminant, and end points measured (Table 8.1). For many of the co-contaminants, their toxicity increased in the presence of CNTs, likely due to the ability of the CNTs to enhance bioavailability; however, in some cases, CNTs decreased toxicity as in the case of fluoranthene [75] and triclocarban [81]. Interestingly, two independent studies performed by Lee et al. [40] and Kim et al. [80] showed the opposite effect of CNTs on copper toxicity, which was decreased in the former and increased in the latter. In both studies, bioavailability of copper was suspected of driving the toxicity observed, and the differences between studies may be due to the organism (Rotifer vs. Daphnia), type of CNT (MWNT vs. SWNT), and exposure conditions used. Unexpectedly, exposed monogonont rotifers (*Brachionus koreanus*) showed toxic effects such as impaired growth and antioxidant depletion despite a decrease in copper bioavailability. One question that remains in many of the studies performed is the ability of bound contaminants to be released from the CNTs and remain biologically active once they enter organisms. A report by Bisesi et al. [83] recently showed that CNT-sorbed ethinylestradiol could be released and absorbed

Table 8.1 Co-contaminants.

CNT + co-contaminant	Bioavailability/toxicity
Zinc	Increased [73, 74]
Fluoranthane	Decreased [75]
Cadmium	Increased [74, 76, 77]
Lead	Increased [78]
Carbofuran	Increased [79]
Copper	Decreased/increased [40, 80]
Arsenic	Increased [72]
Pentachlorophenol	Increased [86]
Triclocarban	Decreased [81]
Phenanthrene	Increased [82]

in the GI tract of largemouth bass (LMB, *micropterus salmoides*) and remain bioactive, retaining the ability to highly induce the hepatic expression of a target gene (vitellogenin). This is one of the few studies to assess downstream-specific biological molecular targets (estrogen receptor) that go beyond mass balance measures of co-contaminants and physiological end points (mortality, growth). A few other reports have also quantified more general cellular responses to suspected CNT released and absorbed metals such as reduced SOD and CAT activity [77]. These studies highlight the importance of understanding how CNTs impact co-contaminant behavior *in vivo*. The investigations with ethinylestradiol further suggest that the endocrine system is a possible target of SWNT-induced toxicity. It also points to the ability of CNTs to adsorb natural hormones and other endogenous like molecules that are critical for optimal health – an area that has not been explored in this capacity.

8.4 Next-Generation Nanomaterials: Nanohybrids

Advancement in nanotechnology has directed materials engineering toward achieving multifunctionality with hybridization of multiple nanoscale entities of unique chemical origin, known as NHs [87–91]. These hierarchical ensembles, for example, have shown to achieve desired band structure of metal/metal oxides by hybridization with CNTs and are now utilized as improved photocatalysts under ultraviolet and visible range irradiation [92, 93]. Another example includes controlling dissolution of nanosilver by over-coating with a chemically inert gold layer to attain long-lasting antimicrobial effects [94, 95]. New combinations of multicomponent NHs are being synthesized in an ever-expanding space of materials synthesis and engineering [91]. These "horizon" materials are improving products and devices while leaving behind increased uncertainty on their environmental behavior and aquatic toxicity.

8.4.1 Variation in Nanohybrid Composition and Environmentally Relevant Properties

From an extensive literature review, Aich et al. [91] identified four classes of NHs: carbon–carbon NH (CCNH), carbon–metal NH (CMNH), metal–metal NH (MMNH), and organic molecule-coated NH (OMCNH). Here, we will focus on those classes containing CNTs. The CCNH class consists of a conjugation of carbonaceous NMs such as CNTs, fullerenes, and graphene sheets, which are hybridized in multiple ways, usually consisting of one NM, endohedrally or exohedrally bound to the other. One such structure is a nano-peapod where the fullerene is endohedrally conjugated to a CNT [96]. How the NHs are combined plays a large role on the emergent properties of the NHs despite the use of the same individual NMs. For example, the dispersibility of nano-grapevines varies depending on whether the fullerenes are covalently or non-covalently bonded to the CNT [97]. Moreover, if the fullerenes were inserted inside the CNT, this formation has been shown to alter physical properties such as stiffness, morphology, and surface potential [98]. These CCNHs are used in a wide range of applications in electronics/energy devices such as field effect transistors and fuel cells, environmental monitoring devices such as chemical sensors, and biomedical applications such as drug delivery [99–102].

The CMNH combines metal or metal oxide NMs with carbonaceous nanostructures; e.g. quantum dot encapsulated inside a CNT. Because of their improved charge separation and electron density, CMNHs are popular in electronics and energy devices such as transistors (e.g. graphene–ZnO), transparent conductive films (e.g. SWNT–Au), optical limiting devices (e.g. ZnO–graphene quantum dots), and fuel cell applications (e.g. CNT/TiO_2-Pt) [103–107]. CMNHs are also used as MRI contrast agents (e.g. gadofullerene) and as many types of chemical sensors for environmental monitoring (e.g. Pt–graphene, MWNT–Fe^0, graphene–Fe) [108–111].

8.4.2 Toxic Responses Demonstrated by NHs

Such novel combinations of singular NMs in the formation of NHs are expected to lead to the emergence of novel properties such as stiffness, band gap energetics, and ROS generation potential [89, 90]. These properties may contribute to unanticipated toxicity to aquatic species. Similar to the discussion on co-contaminants where CNTs can influence metal toxicity, covalent linkage of the two could have a significant effect on metal dissolution characteristics and ROS production, the ability of CNTs to sorb biological molecules, and stimulation or repression of the immune system. Conjugation may also alter the fate and transport of the NHs, which will affect their distribution in environmental media (water vs. sediments) and biodistribution within exposed organisms. As NHs have not been well characterized with respect to toxicity, the question remains as to whether a CNT-metal hybrid will produce biological responses in a manner similar to the co-contaminant effects noted earlier. If nanohybrids bring emergent properties, significant uncertainty in predicted toxic responses are likely, which go beyond synergistic, additive, or antagonistic effects (i.e. CNT and metal). Increased synthesis of NHs with new compositions and emergent properties is imminent; therefore the evaluation of their biological behavior is needed with considerations for development of standard protocols for accurate measurements of new properties that advance our understanding of underlying cellular mechanisms of action.

8.5 Future Perspectives

Our understanding of toxicity of CNTs to aquatic species has primarily focused on mortality, growth, reproduction, and in some cases behavior with limited focus on mechanisms of action. Of the mechanisms most well studied, oxidative stress and genotoxicity have shown to be of primary interest; however alternate mechanisms have recently surfaced that include nutrient depletion, immune modulation, and effects on co-contaminants. Further, the development of novel materials – nanohybrids – has not been investigated yet with respect to toxicity. Emergent properties of such horizon materials may present new mechanistic challenges in elucidating toxicity pathways in aquatic species.

Acknowledgments

This work was supported by NSF EAGER 00126963 (to T. S-A and N. B. S) and USDA 00114317 (to T. S-A and C. L).

References

1 Khan, I.A., Flora, J.R.V., Afrooz, A.R.M.N. et al. (2015). Change in chirality of semiconducting single-walled carbon nanotubes can overcome anionic surfactant stabilisation: a systematic study of aggregation kinetics. *Environ. Chem.* 12 (6): 652–661.

2 Khan, I.A., Aich, N., Afrooz, A.R.M.N. et al. (2013). Fractal structures of single-walled carbon nanotubes in biologically relevant conditions: role of chirality vs. media conditions. *Chemosphere* 93 (9): 1997–2003.

3 Khan, I.A., Afrooz, A.R.M.N., Flora, J.R.V. et al. (2013). Chirality affects aggregation kinetics of single-walled carbon nanotubes. *Environ. Sci. Technol.* 47 (4): 1844–1852.

4 Khan, I.A., Ferguson, P.L., Sabo-Attwood, T., and Saleh, N.B. (2010). Systematic change in chirality affects aggregation kinetics of single-walled carbon nanotubes. *Abstr. Pap. Am. Chem. Soc.* 240: 185.

5 Bianco, A., Kostarelos, K., Partidos, C.D., and Prato, M. (2005). Biomedical applications of functionalised carbon nanotubes. *Chem. Commun. (Camb.)* 5: 571–577.

6 Wise, K. and Brasuel, M. (2011). The current state of engineered nanomaterials in consumer goods and waste streams: the need to develop nanoproperty-quantifiable sensors for monitoring engineered nanomaterials. *Nanotechnol. Sci. Appl.* 4: 73–86.

7 Nowack, B., Brouwer, C., Geertsma, R.E. et al. (2013). Analysis of the occupational, consumer and environmental exposure to engineered nanomaterials used in 10 technology sectors. *Nanotoxicology* 7 (6): 1152–1156.
8 Keller, A.A., McFerran, S., Lazareva, A., and Suh, S. (2013). Global life cycle releases of engineered nanomaterials. *J. Nanopart. Res.* 15 (6): 1–17.
9 Pietroiusti, A., Somma, G., Torriero, A. et al. (2016). Adverse health effects following exposure to engineered nanomaterials. *G. Ital. Med. Lav. Ergon.* 38 (3): 140–143.
10 Kwak, J.I. and An, Y.J. (2015). Ecotoxicological effects of nanomaterials on earthworms: a review. *Hum. Ecol. Risk. Assess.* 21 (6): 1566–1575.
11 Berdjeb, L., Pelletier, E., Pellerin, J. et al. (2013). Contrasting responses of marine bacterial strains exposed to carboxylated single-walled carbon nanotubes. *Aquat. Toxicol.* 144: 230–241.
12 Dong, J. and Ma, Q. (2016). Myofibroblasts and lung fibrosis induced by carbon nanotube exposure. *Part. Fibre Toxicol.* 13: 1–22.
13 Ema, M., Gamo, M., and Honda, K. (2016). A review of toxicity studies of single-walled carbon nanotubes in laboratory animals. *Regul. Toxicol. Pharmacol.* 74: 42–63.
14 Landsiedel, R., Sauer, U.G., Ma-Hock, L. et al. (2014). Pulmonary toxicity of nanomaterials: a critical comparison of published in vitro assays and in vivo inhalation or instillation studies. *Nanomedicine* 9 (16): 2557–2585.
15 Kang, S., Pinault, M., Pfefferle, L.D., and Elimelech, M. (2007). Single-walled carbon nanotubes exhibit strong antimicrobial activity. *Langmuir* 23 (17): 8670–8673.
16 Wang, X., Mansukhani, N.D., Guiney, L.M. et al. (2016). Toxicological profiling of highly purified metallic and semiconducting single-walled carbon nanotubes in the rodent lung and E. coli. *ACS Nano* 10 (6): 6008–6019.
17 Sanpui, P., Zheng, X., Loeb, J.C. et al. (2014). Single-walled carbon nanotubes increase pandemic influenza a H1N1 virus infectivity of lung epithelial cells. *Part. Fibre Toxicol.* 11: 1–15.
18 Kuempel, E.D., Jaurand, M.C., Moller, P. et al. (2017). Evaluating the mechanistic evidence and key data gaps in assessing the potential carcinogenicity of carbon nanotubes and nanofibers in humans. *Crit. Rev. Toxicol.* 47 (1): 1–58.
19 Vietti, G., Lison, D., and van den Brule, S. (2016). Mechanisms of lung fibrosis induced by carbon nanotubes: towards an adverse outcome pathway (AOP). *Part. Fibre Toxicol.* 13: 11.
20 Vale, G., Mehennaoui, K., Cambier, S. et al. (2016). Manufactured nanoparticles in the aquatic environment-biochemical responses on freshwater organisms: a critical overview. *Aquat. Toxicol.* 170: 162–174.

21 Rocha, T.L., Gomes, T., Sousa, V.S. et al. (2015). Ecotoxicological impact of engineered nanomaterials in bivalve molluscs: an overview. *Mar. Environ. Res.* 111: 74–88.

22 Baun, A., Hartmann, N.B., Grieger, K., and Kusk, K.O. (2008). Ecotoxicity of engineered nanoparticles to aquatic invertebrates: a brief review and recommendations for future toxicity testing. *Ecotoxicology* 17 (5): 387–395.

23 Scown, T.M., van Aerle, R., and Tyler, C.R. (2010). Review: do engineered nanoparticles pose a significant threat to the aquatic environment? *Crit. Rev. Toxicol.* 40 (7): 653–670.

24 Mouchet, F., Landois, P., Datsyuk, V. et al. (2011). International amphibian micronucleus standardized procedure (ISO 21427-1) for in vivo evaluation of double-walled carbon nanotubes toxicity and genotoxicity in water. *Environ. Toxicol.* 26 (2): 136–145.

25 Mottier, A., Mouchet, F., Laplanche, C. et al. (2016). Surface area of carbon nanoparticles: a dose metric for a more realistic ecotoxicological assessment. *Nano Lett.* 16 (6): 3514–3518.

26 Li, H., Zhang, T., Liang, G.Y. et al. (2011). In vivo evaluation of acute toxicity of water-soluble carbon nanotubes. *Toxicol. Environ. Chem.* 93 (3): 603–615.

27 Lukhele, L.P., Mamba, B.B., Musee, N., and Wepener, V. (2015). Acute toxicity of double-walled carbon nanotubes to three aquatic organisms. *J. Nanomater.* 1–19.

28 Alloy, M.M. and Roberts, A.P. (2011). Effects of suspended multi-walled carbon nanotubes on daphnid growth and reproduction. *Ecotoxicol. Environ. Saf.* 74 (7): 1839–1843.

29 Roberts, A.P., Mount, A.S., Seda, B. et al. (2007). In vivo biomodification of lipid-coated carbon nanotubes by *Daphnia magna*. *Environ. Sci. Technol.* 41 (8): 3025–3029.

30 Stanley, J.K., Laird, J.G., Kennedy, A.J., and Steevens, J.A. (2016). Sublethal effects of multiwalled carbon nanotube exposure in the invertebrate *Daphnia magna*. *Environ. Toxicol. Chem.* 35 (1): 200–204.

31 Zhu, B., Liu, G.L., Ling, F. et al. (2015). Development toxicity of functionalized single-walled carbon nanotubes on rare minnow embryos and larvae. *Nanotoxicology* 9 (5): 579–590.

32 Boyle, D., Fox, J.E., Akerman, J.M. et al. (2014). Minimal effects of waterborne exposure to single-walled carbon nanotubes on behaviour and physiology of juvenile rainbow trout (*Oncorhynchus mykiss*). *Aquat. Toxicol.* 146: 154–164.

33 Garner, K.L. and Keller, A.A. (2014). Emerging patterns for engineered nanomaterials in the environment: a review of fate and toxicity studies. *J. Nanopart. Res.* 16 (8): 1–28.

34 Skjolding, L.M., Sorensen, S.N., Hartmann, N.B. et al. (2016). Aquatic ecotoxicity testing of nanoparticles-the quest to disclose nanoparticle effects. *Angew. Chem.* 55 (49): 15224–15239.

35 Fraser, T.W.K., Reinardy, H.C., Shaw, B.J. et al. (2011). Dietary toxicity of single-walled carbon nanotubes and fullerenes (C-60) in rainbow trout (*Oncorhynchus mykiss*). *Nanotoxicology* 5 (1): 98–108.

36 Carrillo, Y., Torres-Duarte, C., Oviedo, M.J. et al. (2015). Lipid peroxidation and protein oxidation induced by different nanoparticles in zebrafish organs. *Appl. Ecol. Environ. Res.* 13 (3): 709–723.

37 Wang, X.H., Qu, R.J., Huang, Q.G. et al. (2015). Hepatic oxidative stress and catalyst metals accumulation in goldfish exposed to carbon nanotubes under different pH levels. *Aquat. Toxicol.* 160: 142–150.

38 da Rocha, A.M., Ferreira, J.R., Barros, D.M. et al. (2013). Gene expression and biochemical responses in brain of zebrafish *Danio rerio* exposed to organic nanomaterials: carbon nanotubes (SWCNT) and fullerenol (C-60(OH)(18-22)(OK4)). *Comp Biochem Phys A* 165 (4): 460–467.

39 Handy, R.D., Al-Bairuty, G., Al-Jubory, A. et al. (2011). Effects of manufactured nanomaterials on fishes: a target organ and body systems physiology approach. *J. Fish Biol.* 79 (4): 821–853.

40 Lee, J.W., Kang, H.M., Won, E.J. et al. (2016). Multi-walled carbon nanotubes (MWCNTs) lead to growth retardation, antioxidant depletion, and activation of the ERK signaling pathway but decrease copper bioavailability in the monogonont rotifer (*Brachionus koreanus*). *Aquat. Toxicol.* 172: 67–79.

41 Lee, J.W., Won, E.J., Kang, H.M. et al. (2016). Effects of multi-walled carbon nanotube (MWCNT) on antioxidant depletion, the ERK signaling pathway, and copper bioavailability in the copepod (*Tigriopus japonicus*). *Aquat. Toxicol.* 171: 9–19.

42 Kim, D.H., Puthumana, J., Kang, H.M. et al. (2016). Adverse effects of MWCNTs on life parameters, antioxidant systems, and activation of MAPK signaling pathways in the copepod *Paracyclopina nana*. *Aquat. Toxicol.* 179: 115–124.

43 Kim, K.T., Edgington, A.J., Klaine, S.J. et al. (2009). Influence of multiwalled carbon nanotubes dispersed in natural organic matter on speciation and bioavailability of copper. *Environ. Sci. Technol.* 43 (23): 8979–8984.

44 Shvedova, A.A., Pietroiusti, A., Fadeel, B., and Kagan, V.E. (2012). Mechanisms of carbon nanotube-induced toxicity: focus on oxidative stress. *Toxicol. Appl. Pharmacol.* 261 (2): 121–133.

45 Casey, A., Herzog, E., Lyng, F.M. et al. (2008). Single walled carbon nanotubes induce indirect cytotoxicity by medium depletion in A549 lung cells. *Toxicol. Lett.* 179 (2): 78–84.

46 Selck, H., Handy, R.D., Fernandes, T.F. et al. (2016). Nanomaterials in the aquatic environment: a European Union-United States perspective on the status of ecotoxicity testing, research priorities, and challenges ahead. *Environ. Toxicol. Chem.* 35 (5): 1055–1067.

47 Bisesi, J.H., Merten, J., Liu, K. et al. (2014). Tracking and quantification of single-walled carbon nanotubes in fish using near infrared fluorescence. *Environ. Sci. Technol.* 48 (3): 1973–1983.

48 Bisesi, J.H., Ngo, T., Ponnavolu, S. et al. (2015). Examination of single-walled carbon nanotubes uptake and toxicity from dietary exposure: tracking movement and impacts in the gastrointestinal system. *Nano* 5 (2): 1066–1086.

49 Petersen, E.J., Flores-Cervantes, D.X., Bucheli, T.D. et al. (2016). Quantification of carbon nanotubes in environmental matrices: current capabilities, case studies, and future prospects. *Environ. Sci. Technol.* 50 (9): 4587–4605.

50 Orozco, Z.G.A., Soma, S., Kaneko, T., and Watanabe, S. (2017). Effects of fasting and refeeding on gene expression of slc15a1a, a gene encoding an oligopeptide transporter (PepT1), in the intestine of *Mozambique tilapia*. *Comp. Biochem. Physiol. B* 203: 76–83.

51 Rimoldi, S., Bossi, E., Harpaz, S. et al. (2015). Intestinal B(0)AT1 (SLC6A19) and PEPT1 (SLC15A1) mRNA levels in European sea bass (*Dicentrarchus labrax*) reared in fresh water and fed fish and plant protein sources. *J Nutr Sci* 4: 1–13.

52 Arndt, D.A., Moua, M., Chen, J., and Klaper, R.D. (2013). Core structure and surface functionalization of carbon nanomaterials alter impacts to daphnid mortality, reproduction, and growth: acute assays do not predict chronic exposure impacts. *Environ. Sci. Technol.* 47 (16): 9444–9452.

53 Arndt, D.A., Chen, J., Moua, M., and Klaper, R.D. (2014). Multigeneration impacts on Daphnia magna of carbon nanomaterials with differing core structures and functionalizations. *Environ. Toxicol. Chem.* 33 (3): 541–547.

54 Cheng, J.P., Chan, C.M., Veca, L.M. et al. (2009). Acute and long-term effects after single loading of functionalized multi-walled carbon nanotubes into zebrafish (*Danio rerio*). *Toxicol. Appl. Pharmacol.* 235 (2): 216–225.

55 Ali, A., Suhail, M., Mathew, S. et al. (2016). Nanomaterial induced immune responses and cytotoxicity. *J. Nanosci. Nanotechnol.* 16 (1): 40–57.

56 Orecchioni, M., Bedognetti, D., Sgarrella, F. et al. (2014). Impact of carbon nanotubes and graphene on immune cells. *J. Transl. Med.* 12: 1–11.

57 Dumortier, H. (2013). When carbon nanotubes encounter the immune system: desirable and undesirable effects. *Adv. Drug Deliv. Rev.* 65 (15): 2120–2126.

58 Jovanovic, B., Whitley, E.M., Kimura, K. et al. (2015). Titanium dioxide nanoparticles enhance mortality of fish exposed to bacterial pathogens. *Environ. Pollut.* 203: 153–164.

59 Ates, M., Demir, V., Arslan, Z. et al. (2016). Chronic exposure of tilapia (*Oreochromis niloticus*) to iron oxide nanoparticles: effects of particle morphology on accumulation, elimination, hematology and immune responses. *Aquat. Toxicol.* 177: 22–32.

60 Wang, T., Long, X.H., Liu, Z.P. et al. (2015). Effect of copper nanoparticles and copper sulphate on oxidation stress, cell apoptosis and immune responses in the intestines of juvenile *Epinephelus coioides*. *Fish Shellfish Immunol.* 44 (2): 674–682.

61 Ringwood, A.H., Levi-Polyachenko, N., and Carroll, D.L. (2009). Fullerene exposures with oysters: embryonic, adult, and cellular responses. *Environ. Sci. Technol.* 43 (18): 7136–7141.

62 Greven, A.C., Merk, T., Karagoz, F. et al. (2016). Polycarbonate and polystyrene nanoplastic particles act as stressors to the innate immune system of fathead minnow (*Pimephales promelas*). *Environ. Toxicol. Chem.* 35 (12): 3093–3100.

63 Jovanovic, B. and Palic, D. (2012). Immunotoxicology of non-functionalized engineered nanoparticles in aquatic organisms with special emphasis on fish-review of current knowledge, gap identification, and call for further research. *Aquat. Toxicol.* 118: 141–151.

64 Ji, J., Torrealba, D., Ruyra, A., and Roher, N. (2015). Nanodelivery systems as new tools for immunostimulant or vaccine administration: targeting the fish immune system. *Biology* 4 (4): 664–696.

65 Klaper, R., Arndt, D., Setyowati, K. et al. (2010). Functionalization impacts the effects of carbon nanotubes on the immune system of rainbow trout, *Oncorhynchus mykiss*. *Aquat. Toxicol.* 100 (2): 211–217.

66 Girardello, R., Tasselli, S., Baranzini, N. et al. (2015). Effects of carbon nanotube environmental dispersion on an aquatic invertebrate, *Hirudo medicinalis*. *PLoS One* 10 (12): e0144361.

67 Ede, J.D., Ortega, V.A., Boyle, D. et al. (2016). The effects of rosette nanotubes with different functionalizations on channel catfish (*Ictalurus punctatus*) lymphocyte viability and receptor function. *Environ Sci-Nano* 3 (3): 578–592.

68 Johnson, H.; Robinson, S.; Simons, A.; Bikkasani, A.; Kane, A. B.; Bisesi, J. H.; Sabo-Attwood, T. (2016) Novel mechanisms of nanomaterial toxicity through direct exposure or indirect interactions with environmental components. *SETAC Annual Conference Abstract Proceedings.*

69 Sanchis, J., Olmos, M., Vincent, P. et al. (2016). New insights on the influence of organic co-contaminants on the aquatic toxicology of carbon nanomaterials. *Environ. Sci. Technol.* 50 (2): 961–969.

70 Boncel, S., Kyziol-Komosinska, J., Krzyzewska, I., and Czupiol, J. (2015). Interactions of carbon nanotubes with aqueous/aquatic media containing organic/inorganic contaminants and selected organisms of aquatic ecosystems – a review. *Chemosphere* 136: 211–221.

71 Canesi, L., Ciacci, C., and Balbi, T. (2015). Interactive effects of nanoparticles with other contaminants in aquatic organisms: friend or foe? *Mar. Environ. Res.* 111: 128–134.

72 Wang, X.H., Qu, R.J., Allam, A.A. et al. (2016). Impact of carbon nanotubes on the toxicity of inorganic arsenic [AS(III) and AS(V)] to *Daphnia magna*: the role of certain arsenic species. *Environ. Toxicol. Chem.* 35 (7): 1852–1859.

73 Yan, L.Q., Feng, M.B., Liu, J.Q. et al. (2016). Antioxidant defenses and histological changes in *Carassius auratus* after combined exposure to zinc and three multi-walled carbon nanotubes. *Ecotoxicol. Environ. Saf.* 125: 61–71.

74 Yu, Z.G. and Wang, W.X. (2013). Influences of ambient carbon nanotubes on toxic metals accumulation in *Daphnia magna*. *Water Res.* 47 (12): 4179–4187.

75 Linard, E.N., van den Hurk, P., Karanfil, T. et al. (2015). Influence of carbon nanotubes on the bioavailability of fluoranthene. *Environ. Toxicol. Chem.* 34 (3): 658–666.

76 Wang, X.H., Qu, R.J., Liu, J.Q. et al. (2016). Effect of different carbon nanotubes on cadmium toxicity to *Daphnia magna*: the role of catalyst impurities and adsorption capacity. *Environ. Pollut.* 208: 732–738.

77 Qu, R.J., Wang, X.H., Wang, Z.Y. et al. (2014). Metal accumulation and antioxidant defenses in the freshwater fish *Carassius auratus* in response to single and combined exposure to cadmium and hydroxylated multi-walled carbon nanotubes. *J. Hazard. Mater.* 275: 89–98.

78 Barbieri, E., Campos-Garcia, J., Martinez, D.S. et al. (2016). Histopathological Effects on Gills of Nile Tilapia (*Oreochromis niloticus*, Linnaeus, 1758) Exposed to Pb and carbon nanotubes. *Microscopy Microanal.* 22 (6): 1162–1169.

79 Campos-Garcia, J., Martinez, D.S.T., Alves, O.L. et al. (2015). Ecotoxicological effects of carbofuran and oxidised multiwalled carbon nanotubes on the freshwater fish Nile tilapia: nanotubes enhance pesticide ecotoxicity. *Ecotoxicol. Environ. Saf.* 111: 131–137.

80 Kim, K.T., Klaine, S.J., Lin, S.J. et al. (2010). Acute toxicity of a mixture of copper and single-walled carbon nanotubes to *Daphnia magna*. *Environ. Toxicol. Chem.* 29 (1): 122–126.

81 Simon, A., Preuss, T.G., Schaffer, A. et al. (2015). Population level effects of multiwalled carbon nanotubes in *Daphnia magna* exposed to pulses of triclocarban. *Ecotoxicology* 24 (6): 1199–1212.

82 Su, Y., Yan, X.M., Pu, Y.B. et al. (2013). Risks of single-walled carbon nanotubes acting as contaminants-carriers: potential release of phenanthrene in Japanese medaka (*Oryzias latipes*). *Environ. Sci. Technol.* 47 (9): 4704–4710.

83 Bisesi, J.H. Jr., Robinson, S.E., Lavelle, C.M. et al. (2017). Influence of the gastrointestinal environment on the bioavailability of ethinyl estradiol sorbed to single-walled carbon nanotubes. *Environ. Sci. Technol.* 51 (2): 948–957.

84 Sun, H.W., Ruan, Y.F., Zhu, H.K. et al. (2014). Enhanced bioaccumulation of pentachlorophenol in carp in the presence of multi-walled carbon nanotubes. *Environ. Sci. Pollut. R.* 21 (4): 2865–2875.

85 Parks, A.N., Chandler, G.T., Portis, L.M. et al. (2014). Effects of single-walled carbon nanotubes on the bioavailability of PCBs in field-contaminated sediments. *Nanotoxicology* 8 (Suppl 1): 111–117.

86 Sun, G.Q., Wang, P.P., Ge, S.G. et al. (2014). Photoelectrochemical sensor for pentachlorophenol on microfluidic paper-based analytical device based on the molecular imprinting technique. *Biosens. Bioelectron.* 56: 97–103.

87 Plazas-Tuttle, J., Rowles, L.S., Chen, H. et al. (2015). Dynamism of stimuli-responsive nanohybrids: environmental implications. *Nano* 5 (2): 1102–1123.

88 Saleh, N.B., Aich, N., Plazas-Tuttle, J. et al. (2015). Research strategy to determine when novel nanohybrids pose unique environmental risks. *Environ. Sci. Nano* 2 (1): 11–18.

89 Saleh, N.B., Sabo-Attwood, T., and Huang, X.Y. (2014). Emergent properties of nanohybrids and their potential environmental implications. *Abstr. Pap. Am. Chem. Soc.* 248.

90 Saleh, N.B., Afrooz, A.R.M.N., Bisesi, J.H. et al. (2014). Emergent properties and toxicological considerations for nanohybrid materials in aquatic systems. *Nanomaterials* 4 (2): 372–407.

91 Aich, N., Plazas-Tuttle, J., Lead, J.R., and Saleh, N.B. (2014). A critical review of nanohybrids: synthesis, applications and environmental implications. *Environ. Chem.* 11 (6): 609–623.

92 Wu, C.H., Kuo, C.Y., and Chen, S.T. (2013). Synergistic effects between TiO_2 and carbon nanotubes (CNTs) in a TiO_2/CNTs system under visible light irradiation. *Environ. Technol.* 34 (17): 2513–2519.

93 Lee, W.J., Lee, J.M., Kochuveedu, S.T. et al. (2012). Biomineralized N-doped CNT/TiO_2 core/shell nanowires for visible light photocatalysis. *ACS Nano* 6 (1): 935–943.

94 Li, T., Albee, B., Alemayehu, M. et al. (2010). Comparative toxicity study of ag, au, and ag-au bimetallic nanoparticles on *Daphnia magna*. *Anal. Bioanal. Chem.* 398 (2): 689–700.

95 Ding, S., Chen, J.S., Luan, D. et al. (2011). Graphene-supported anatase TiO_2 nanosheets for fast lithium storage. *Chem. Commun. (Camb.)* 47 (20): 5780–5782.

96 Smith, B.W. and Luzzi, D.E. (2000). Formation mechanism of fullerene peapods and coaxial tubes: a path to large scale synthesis. *Chem. Phys. Lett.* 321 (1–2): 169–174.

97 Nasibulin, A.G., Pikhitsa, P.V., Jiang, H. et al. (2007). A novel hybrid carbon material. *Nat. Nanotechnol.* 2 (3): 156–161.

98 Diez-Pascual, A.M., Naffakh, M., and Gomez-Fatou, M.A. (2011). Mechanical and electrical properties of novel poly(ether ether ketone)/carbon nanotube/inorganic fullerene-like WS2 hybrid nanocomposites: experimental measurements and theoretical predictions. *Mater. Chem. Phys.* 130 (1–2): 126–133.

99 Shimada, T., Ohno, Y., Okazaki, T. et al. (2004). Transport properties of C-78, C-90 and Dy@C-82 fullerenes-nanopeapods by field effect transistors. *Physica E.* 21 (2–4): 1089–1092.

100 Jha, N., Jafri, R.I., Rajalakshmi, N., and Ramaprabhu, S. (2011). Graphene-multi walled carbon nanotube hybrid electrocatalyst support material for direct methanol fuel cell. *Int. J. Hydrogen Energy* 36 (12): 7284–7290.

101 Chen, X., Zhu, J.N., Xi, Q., and Yang, W.S. (2012). A high performance electrochemical sensor for acetaminophen based on single-walled carbon nanotube-graphene nanosheet hybrid films. *Sensor Actuators B. Chem.* 161 (1): 648–654.

102 Simon, F., Peterlik, H., Pfeiffer, R. et al. (2007). Fullerene release from the inside of carbon nanotubes: a possible route toward drug delivery. *Chem. Phys. Lett.* 445 (4–6): 288–292.

103 Kong, B.S., Jung, D.H., Oh, S.K. et al. (2007). Single-walled carbon nanotube gold nanohybrids: application in highly effective transparent and conductive films. *J. Phys. Chem. C* 111 (23): 8377–8382.

104 Rigdon, W.A., Larrabee, D.D., and Huang, X.Y. (2013). Composite carbon nanotube and titania supports for resilent oxidation electrocatalysts in polymer electrolyte fuel cells. *Abstr. Pap. Am. Chem. Soc.* 246.

105 Son, D.I., Kwon, B.W., Park, D.H. et al. (2012). Emissive ZnO-graphene quantum dots for white-light-emitting diodes. *Nat. Nanotechnol.* 7 (7): 465–471.

106 Song, W., Kwon, S.Y., Myung, S. et al. (2014). High-mobility ambipolar ZnO-graphene hybrid thin film transistors. *Sci. Rep.* 4: 1–6.

107 Zhang, J.F., Ge, J.C., Shultz, M.D. et al. (2010). In vitro and in vivo studies of single-walled carbon nanohorns with encapsulated metallofullerenes and exohedrally functionalized quantum dots. *Nano Lett.* 10 (8): 2843–2848.

108 Bolskar, R.D. (2008). Gadofullerene MRI contrast agents. *Nanomedicine* 3 (2): 201–213.

109 Guo, S.J., Wen, D., Zhai, Y.M. et al. (2010). Platinum nanoparticle ensemble-on-graphene hybrid nanosheet: one-pot, rapid synthesis, and used as new electrode material for electrochemical sensing. *ACS Nano* 4 (7): 3959–3968.

110 Zhang, S., Wang, D., Zhou, L. et al. (2013). Intensified internal electrolysis for degradation of methylene blue as model compound induced by a novel hybrid material: multi-walled carbon nanotubes immobilized on zero-valent iron plates (Fe-0-CNTs). *Chem. Eng. J.* 217: 99–107.

111 Jabeen, H., Chandra, V., Jung, S. et al. (2011). Enhanced Cr(VI) removal using iron nanoparticle decorated graphene. *Nanoscale* 3 (9): 3583–3585.

9

Surface Reactivity of Anatase and Rutile Samples: Relationship with Toxicity on Aquatic Organisms

Charlotte Hurel[1], Norbert Jordan[2], Ulrike Gerber[2], Stephan Weiss[2], Bernd Kubier[3], and Reinhard Kleeberg[4]

[1] *CNRS, Institut de Physique de Nice (INPHYNI), Université Côte d'Azur, 28 Avenue Valrose, 06108 Nice Cedex 2, France*
[2] *Helmholtz-Zentrum Dresden-Rossendorf e.V., Institute of Resource Ecology, Bautzner Landstraße 400, 01328 Dresden, Germany*
[3] *Institute of Mechanical Process Engineering and Minerals Processing, TU Bergakademie Freiberg, Agricolastraße 1, 09599 Freiberg, Germany*
[4] *Institute of Mineralogy, TU Bergakademie Freiberg, Brennhausgasse 14, 09596 Freiberg, Germany*

9.1 Introduction

The increasing use of engineered nanosized particles resulted in their release into the environment. Among engineered nanoparticles, titanium dioxide (TiO_2) is extensively used in industrial applications such as cosmetics, paints, or building materials [1] and agricultural and environmental remediation [2]. Therefore, the study of their potential impact to the environment is of major concern. Indeed, once in a biological or environmental context, nanoparticles never consist of "bare" particles [2]. Environmental factors such as pH, ionic strength, presence of organic matter, and ions influence the stability of nanoparticles in aquatic media. Dispersibility of TiO_2 nanoparticles depends on their surface area, primary particle size, allotropic form [3], sample preparation (sonication prior use, continuous shaking, etc.), surface modification (functionalization, coating, etc.), and the medium they are dispersed in. Among TiO_2 intrinsic properties, surface area is one of the parameters that determines the degree of agglomeration since high surface area favors agglomeration. This degree of agglomeration corresponds to the sum of the attractive and repulsive forces between particles. The attraction is due to the van der Waals force, while the electrostatic repulsion results in interactions of the electrical double layer (EDL) surrounding each particle. The surface charge of nanoparticles is controlled by several mechanisms,

Toxicology of Nanoparticles and Nanomaterials in Human, Terrestrial and Aquatic Systems, First Edition. Edited by Marc A. Williams.

including surface ionization, ion adsorption, and lattice ion dissolution, while the thickness of EDL is a function of solution ionic strength (an increase in ionic strength leads to a decrease in double layer thickness). In the absence of other ions, surface ionization controls the surface charge of $nTiO_2$. Generally, at low pH, particles have a positively charged surface, while it is negatively charged at high pH. The isoelectric point (pH_{IEP}) corresponds to the pH at which particles have zero net surface charge. When pH is far from the isoelectric point, zeta potential increases (positively or negatively). At these conditions, the electrostatic repulsive force overcome the van der Waals force, suppressing agglomeration. Near to pH_{IEP}, the repulsive force is weakened due to low surface charge and van der Waals forces are dominating. When cations are present in the media, particles agglomeration is enhanced significantly [4], probably due to a screening effect of repulsive charges during adsorption of cations onto the surface. It has been demonstrated that during sorption of divalent cations (Ba, Ni, Pb, Cd, Ca, and Mg) onto mineral surface (anatase, rutile, Al_2O_3–SiO_2, and SnO_2), an increase of the pH_{IEP} toward higher pH values as well as an increase of the zeta potential at $pH > pH_{IEP}$ occurred. In addition, a charge reversal point (CRP) has been noticed, at which the sign of the zeta potential was reversed again. These changes have been explained by the overcharging of the compact part of the EDL at the inner Helmholtz plane of the Stern layer from specific adsorption of divalent cations onto surface hydroxyl group and introducing positive charge units into the compact layer of EDL, leading to an increase of the surface potential [5–8]. Kosmulski et al. [9] observed that adsorption of Ca^{2+} onto rutile was higher than the one of Mg^{2+}, since the latter gave zeta potential of higher magnitude at alkaline pH value. The influence of ionic strength on particle agglomeration can be evaluated through the measurement of the hydrodynamic diameter of nanoparticles. Increasing ionic strength results in the compression of the EDL, leading to unstable particles tending to agglomerate and resulting in a substantial increase of hydrodynamic diameter.

To evaluate the ecotoxicological impact of NPs on organisms, protocols have been developed, including several invertebrates' taxa for toxicity assessment. These protocols were adapted to be suitable for assessment of NPs toxicity. Among the "model organisms," cladocerans are the most used, with protocols for acute (lethal) and (sub)chronic (sublethal) toxicity assays. The genus *Daphnia* is the most studied taxon within cladocerans but is suitable only for freshwater media. Results on *Daphnia magna* have shown a higher toxicity of TiO_2 than their bulk counterparts [10]. These results highlighted the significant effect of the primary particle size on toxicity [11]. Micrometer-sized TiO_2 agglomerates are concentrated in *D. magna* gut by active ingestion, while well-dispersed nanoparticles are taken up through endocytosis and passive drinking [12]. By this way, bioaccumulation of nanoparticles may interfere with food intake and ultimately affect growth and reproduction in long-term exposition [13]. Toxicity of TiO_2 nanoparticles was enhanced when exposure time was extended from 48 to 72 h [13]. Under UVA irradiation, toxicity of anatase and rutile toward *Daphnia similis* was enhanced, due to the generation of reactive oxygen species [14].

Rotifers tests involve similar conditions to the cladocerans but are suitable both in freshwater and in marine water media [15]. Rotifers are major components of zooplankton in freshwater and coastal marine ecosystems. They serve as food resource for young fishes and some species contribute to the decomposition of soil organic matter. They can be useful indicator species, providing valuable insight into the effects of NPs on microinvertebrate. NPs of small size (37 nm) have been shown to depress the population growth rate [16, 17]. On the other hand, larger particles have been demonstrated to be confined in the gut, implicating nanoparticle size as a critical factor in the ability to penetrate the gut wall and enter tissues [16].

Diatoms are important planktons ubiquitous in marine and freshwater ecosystems and contribute to approximately 20% of the world primary photosynthesis production [18, 19]. Diatoms are currently generating the major part of the organic matter used as food for life in the sea. Algae growth inhibition tests are widely distributed in the risk assessment of waters and are very sensitive to heavy metal pollution. Ecotoxicity of nanosized TiO_2 on algae has been recently reported, but high variabilities and no clear relationship were found between the concentrations, primary particle size, and toxic effect [20].

All these aquatic species are part of a complex community belonging to the beginning of the food chain. Thus, any alteration in their population might impact endemism and aquatic ecosystem function, global climate, and atmospheric CO_2 concentration [18]. All of them are model organisms used as test species to study the depletion of aquatic systems [21–29]. Furthermore, these populations are well suited for experimental studies at the laboratory scale with a high reproducibility.

In this book chapter, the surface and bulk parameters factors driving the toxicity of TiO_2 nanoparticles on aquatic organisms like zooplanktons (crustacean freshwater water flea and rotifers) and phytoplankton (diatom) were assessed. To compare the results, three organisms were chosen: marine rotifers *Brachionus plicatilis*, marine microalgae *Phaeodactylum tricornutum*, and freshwater crustacean *D. magna*. The aim of this work was to investigate how the environmental fate of aquatic organisms is influenced by the particle size, surface site concentration, surface charge, agglomeration state, and allotropic form of TiO_2 commercially available substances. An extended and accurate characterization of four TiO_2 particles with different primary particle size and allotropic form was performed (by means of X-ray diffraction [XRD], transmission electron microscopy [TEM], scanning electron microscopy [SEM], inductively coupled plasma-mass spectrometry [ICP-MS], and N_2-BET). The surface site concentration was determined by means of potentiometric titrations. The zeta potential, giving tendencies of the stability and the potential agglomeration of nanoparticles suspensions, was determined by electrophoretic mobility measurements in two different background electrolytes at identical ionic strength, namely, $NaNO_3$ (used for comparison) and synthetic freshwater (SFW). The agglomeration state of particles suspended in these two electrolytes was determined by laser diffraction, in order to check the potential influence on the TiO_2 response. Finally, the three monocultures were exposed to increasing concentrations of TiO_2 to assess their ecotoxicity.

9.2 TiO_2 Solid Phase Characterization

Three anatase substances of different primary particle size, i.e. 15 nm (A1), 32 nm (A2), and 325 MESH (A3), and one rutile substance of 1 μm size (R1), were purchased from Alfa Aesar. The characteristics of these substances, as given by the provider, are detailed in Table 9.1.

The TiO_2 substances were characterized by X-ray powder diffraction (XRD-3000TT Seifert). Qualitative phase analysis and phase quantification were performed on the basis of the Rietveld method. The shape, morphology, and primary particle size of A1 and A2 commercial anatase substances were obtained by TEM (Titan 80-300 FEI) and those of A3 and R1 by SEM (S-4800 Hitachi). The specific surface area (SSA) of each sample was determined by applying the Brunauer–Emmett–Teller (BET) equation based on nitrogen adsorption isotherms at 77 K (Multi-point Beckman Coulter surface analyzer SA 3100). The presence of metal impurities in each substance was determined by ICP-MS (ELAN 9000 Perkin Elmer) after digestion with a mixture of concentrated HNO_3, HCl, HF, and H_3BO_3. The total surface site concentration was determined by potentiometric titrations of TiO_2 suspensions, and surface charge and isoelectric point were determined by electrophoretic mobility measurements (Zetasizer Nano ZS, Malvern Instruments). The bulk and surface properties as well as the results of the qualitative phase analysis of the four solids are given in Table 9.2.

Table 9.1 Data from the provider for anatase and rutile samples.

Sample	Size	Purity (%)	Surface area (m^2/g)
A1	15 nm	99.7	240
A2	32 nm	99.9	45
A3	325 MESH	99.6	n.d
R1	0.9–1.6 μm	99.8	3.3–6.1

Note: n.d: not determined.

Table 9.2 Surface and bulk properties of TiO_2 materials.

Material	**A1**	**A2**	**A3**	**R1**
Composition (wt% ± 3σ):	Anatase (95.4 ± 0.4) Rutile (4.6 ± 0.4)	100% anatase	Anatase (98.3 ± 0.2) Rutile (1.7 ± 0.2)	Anatase (0.36 ± 0.12) Rutile (98.45 ± 0.13) $BaTiO_3$ (1.19 ± 0.05)
Crystallite size (nm ± 3σ)[1]	a, b: 15.7 ± 0.3 c: 9.8 ± 0.3	a, b: 16.8 ± 0.3 c: 11.2 ± 0.3	a, b: 128.0 ± 3.0 c: 106.0 ± 3.0	515 ± 105
Average primary particle size (nm)[2]	10–20	10–20	70–200	100–500
Specific surface area (m^2/g)	124.0	101.7	10.7	4.8
Particle surface area (m^2/L)[3]	31.0	25.4	2.7	1.2
pH_{PZC} (in $NaNO_3$)	6.1	6.0	6.0	6.0
pH_{IEP} (in $NaNO_3$)	6.1	6.3	3.5	4.2

1 Calculated from XRD line broadening, with respect to the crystallographical axes (a, b, c) when specified.
2 Estimated from SEM and TEM observations.
3 For a particle concentration of 0.25 g/L.

By comparing the XRD patterns to the ICDD card 00-021-1272 (anatase), ICDD card 00-021-1276 (rutile), and ICDD card 00-034-0129 ($BaTiO_3$), the samples A1 and A3 can be identified as a polycrystalline anatase phase, with indications for the presence of rutile (Figure 9.1).

Sample A2 is a pure polycrystalline anatase phase. The main crystal phase identified in substance R1 is rutile, with additional contributions from anatase and a tetragonal $BaTiO_3$ phase. Based on the obtained X-ray diffractograms, the presence of brookite can be excluded for all substances. According to data from Table 9.2, the abundance of other phases than those expected was estimated to be always less than 4.6 wt%. The amount of Ba determined by ICP-MS after digestion was found to be of minor importance (<0.1 wt%) in all substances except R1, where a higher amount was revealed (~0.8 wt%), in agreement with Rietveld refinement analysis. The amount of Mg, Al, K, and Ca was found to be negligible in all substances (<0.1 wt%). The obtained XRD patterns show well-defined peaks and clearly indicates that the substances are crystalline. The crystallite size, defined as the average size of a coherent scattering domain, was also determined

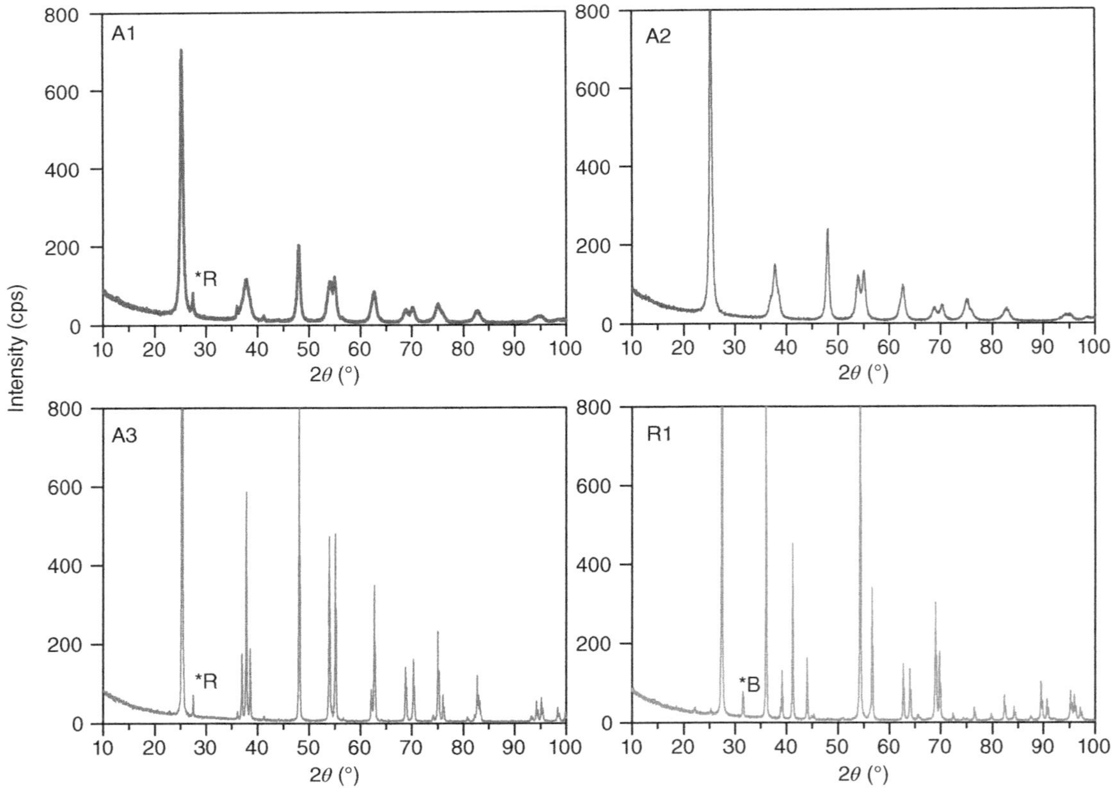

Figure 9.1 X-ray diffraction patterns of A1, A2, A3 and R1 nanoparticles (*R and *B indicate contamination by rutile and $BaTiO_3$, respectively).

by XRD. Substance-related peak broadening was described by a classical model assuming isotropic size-related broadening from a lognormal distribution of crystallite size and isotropic microstrain [30]. This model can be slightly modified for some of the nanosized powders. For materials grown in anisotropic crystallite shape, ellipsoidal models have been introduced [31], allowing a calculation of average dimensions in the direction of the crystallographic axes, based on the *hkl* dependency of the line broadening. Such model can be chosen for substances with obvious anisotropic broadening effects or when indications for ellipsoidal shapes are derived from TEM observations, as it was the case for substances A1 and A2. Line broadening arising from complicated crystallite size distributions is hard to deduce from XRD data, but of course some simple models are available. Assuming a bimodal distribution of crystallites with different broadening effects, the peaks could be fitted much better by two profiles, sharing the same lattice parameters and structural information but assuming two groups of crystallites. Another reliable interpretation could be that the calculated "less abundant but more strained sub-phase" (profile) is nothing else than the contribution of the strained parts of the crystallite at the surface of the crystallites. However, as such model could not be verified by other investigations, the results (especially average crystallite sizes) should be interpreted in a conservative manner. Substances A1 and A2 showed significant line broadening, indicating a truly nanosize of the slightly ellipsoidal crystallites of the anatase. In substances A3 and R1, the main phases, namely, anatase and rutile, respectively, display less line broadening, indicating a significantly bigger crystallite size. The profile shape of the rutile substance R1 is extremely sharp, and, consequently, this minor broadening can hardly be recalculated in serious crystallite size values. The assumption of a bimodal size distribution and contribution of minor microstrain results in an average size of about 500 nm but with a significant uncertainty (Table 9.2).

TEM images (Figure 9.2) show that A1 and A2 powders have similar slightly ellipsoidal shape, and their primary particle size was estimated to be around 10–20 nm. According to SEM, the A3 particles have similar shape than A1 and A2, with a bigger primary particle size, ranging from 70 to 200 nm. Particles of R1 have a different morphology than the anatase materials, with irregular shape, approximately 100–500 nm long and 10–50 nm wide and sharper edges (Table 9.2).

The solids vary in their SSA with values ranging from 4.8 m^2/g for R1 to 124 m^2/g for A1. As expected, an increase of primary particle size led to a decrease of SSA among the anatase substances (Table 9.2). The apparent discrepancy in the SSA trend between A1 and A2 substances (i.e. different values for identical primary particle size) can be due to the uncertainty on their TEM-based primary particle size estimation as well as a different porosity. In the following, SEM/TEM results will be used for the discussion concerning the influence of the primary particle size on the pH_{IEP} and agglomeration of TiO_2 particles.

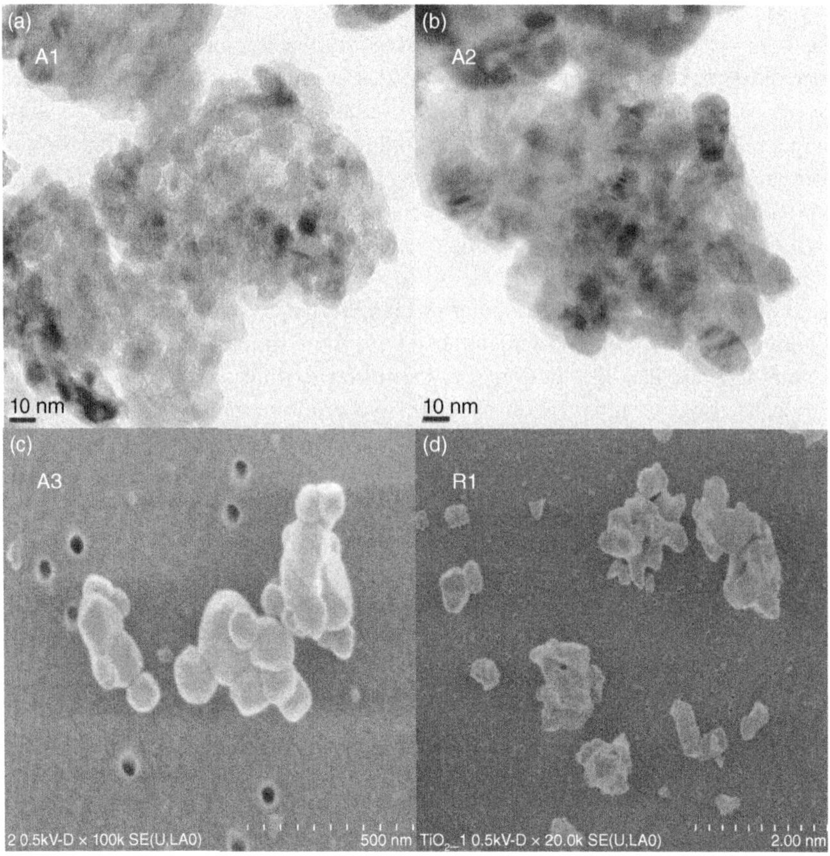

Figure 9.2 TEM images of A1 and A2 samples. SEM images of A3 and R1 samples.

9.3 Potentiometric Titrations

The acid and base titration curves of titanium dioxide particles (20 g/L) suspended in $NaNO_3$ (0.1 M) electrolytic media are shown in Figure 9.3a,b, respectively.

The values for $[H^+]_{sat}$, $[OH^-]_{sat}$, and $[SOH]_{tot}$ are summarized in Table 9.3. The surface sites concentration of A1 and A2 substances are nearly similar, and a difference of one order of magnitude between the surface site concentration of A3 and R1 substances is observed. According to the surface properties of each substance from Table 9.2, as the primary particle size is increased, the surface reactivity ($[SOH]_{tot}$) is decreased. The pH_{PZC}, at which the surface charge is zero, was found to be at ~6 for all substances (Table 9.2).

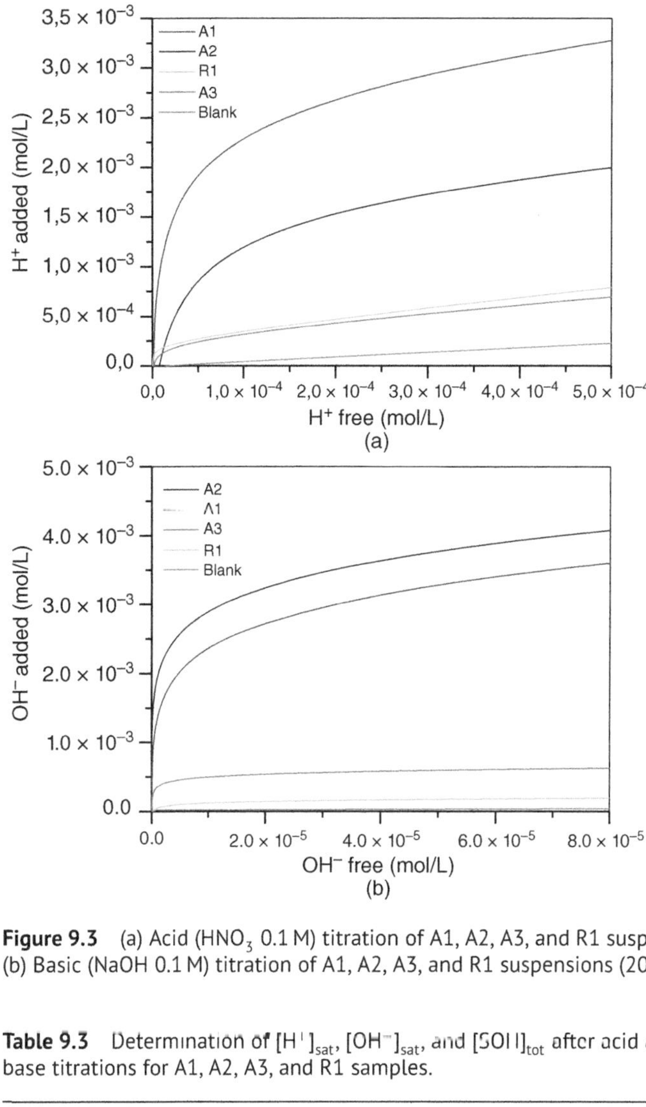

Figure 9.3 (a) Acid (HNO_3 0.1 M) titration of A1, A2, A3, and R1 suspensions (20 g/L). (b) Basic (NaOH 0.1 M) titration of A1, A2, A3, and R1 suspensions (20 g/L).

Table 9.3 Determination of $[H^+]_{sat}$, $[OH^-]_{sat}$, and $[SOH]_{tot}$ after acid and base titrations for A1, A2, A3, and R1 samples.

Sample	$[H^+]_{sat}$ (mol/L)	$[OH^-]_{sat}$ (mol/L)	$[SOH]_{tot}$ (mol/L)
A1	8.8×10^{-3}	5.1×10^{-3}	6.95×10^{-3}
A2	5.3×10^{-3}	8.4×10^{-3}	6.85×10^{-3}
A3	2.5×10^{-3}	4.5×10^{-3}	3.50×10^{-3}
R1	3.2×10^{-4}	3.0×10^{-4}	3.10×10^{-4}

9.4 Electrophoresis Measurements

The isoelectric point (pH_{IEP}) is the pH at which the zeta potential is zero, i.e. at which particles do not move under electrical field. At this pH_{IEP}, the positive and negative charges determining the electrophoretic mobility of the particles are equal. When the pH is shifted from the pH_{IEP}, the net surface charge of the particles increases, and therefore the dispersibility of the particles increases due to higher electrostatic repulsion.

Electrophoresis measurements were performed on each TiO_2 substance, in $NaNO_3$ and in SFW, at $I = 0.00085$ M, with a 0.25 g/L mass-to-volume ratio (see Table 9.4).

9.4.1 In $NaNO_3$

The pH_{IEP} was found to be located at pH 6.1, pH 6.3, pH 3.5, and pH 4.2 for A1, A2, A3, and R1 substances (Table 9.2), respectively, in agreement with former reported values [3, 32, 33]. pH_{IEP} and pH_{PZC} are very close for A1 and A2 samples (both values are approx. 6). In contrast, pH_{IEP} was significantly lower than pH_{PZC} for A3 and R1 samples. In an extensive literature survey, Kosmulski [32, 34, 35] demonstrated that fluctuations of pH_{IEP} and pH_{PZC} values of rutile and anatase were independent of the allotropic form. Low IEP could be explained in terms of phosphate or other anionic impurities in commercial titanias. On the other hand, the low pH of titania dispersions could be explained in terms of traces of acid occluded in pores or in intragrain spaces in titania powders. From Kosmulski's [32, 34, 35] work, we can admit that A1 and A2 samples were synthesized with similar protocols, while A3 and R1 were obtained from a different synthesis protocol, bringing various impurities or traces of acid. According to Figure 9.4, for anatase substances (A1, A2 ,and A3), the pH_{IEP} depends on the primary particle size. Indeed, the pH_{IEP} decreased from 6.1 and 6.3 for A1 and A2 to 3.5 for A3 with increasing primary particle size from ~10–20 nm (A1 and A2) to ~70–200 nm (A3).

Suttiponparnit et al. [3] reported similar observations; however, a proper mechanistic explanation of this phenomenon cannot be given at the actual state of knowledge.

Table 9.4 Composition of the synthetic freshwater (SFW).

Substance	KCl	$CaCl_2 \cdot 2H_2O$	$MgSO_4 \cdot 7H_2O$	$NaHCO_3$
SFW (g/L)	0.006	0.294	0.123	0.065

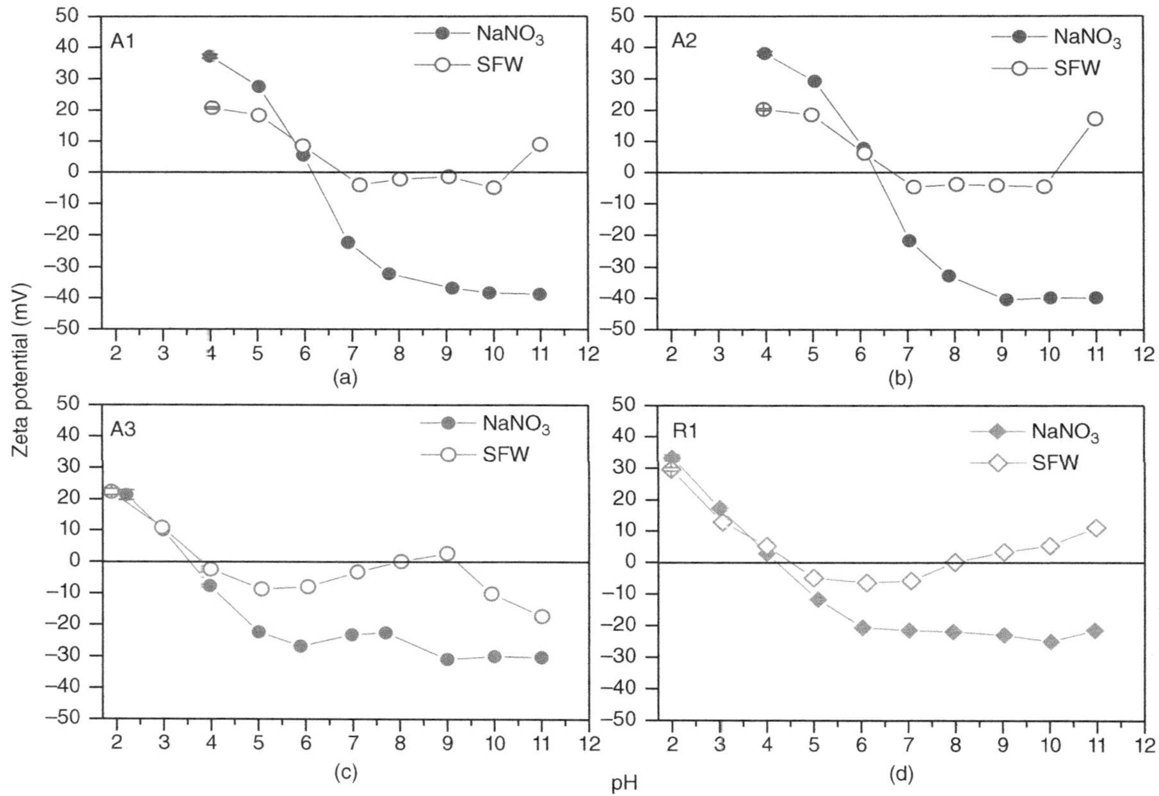

Figure 9.4 Zeta potential of TiO_2 substances, $m/v = 0.25$ g/L, 4 days of shaking. $I = 0.00885$ M $NaNO_3$, A1 (●), A2 (●), A3 (●), and R1 (◆); $I = 0.00885$ M SFW, A1 (○), A2 (○), A3 (○), and R1 (◇).

9.4.2 In Synthetic Freshwater (SFW)

According to Figure 9.4, for all studied substances suspended in SFW, an increase of the pH_{IEP} toward higher pH values as well as an increase of the zeta potential at $pH > pH_{IEP}$ took place. In addition, we could also notice the occurrence of a CRP, at which the sign of the zeta potential reversed again. This is coherent with observations made by Janusz et al. [5], Janusz and Matysek [6], Janusz and Skwarek [7], Skwarek et al. [8], and Kosmulski et al. (1999). These changes can be explained by the overcharging of the compact part of the EDL at the inner Helmholtz plane of the Stern layer by specific adsorption of Ca^{2+} and Mg^{2+} onto surface hydroxyl group and introducing positive charge units into the compact layer of EDL, leading to an increase of the surface potential. Kosmulski et al. (1999) observed that the sorption of Ca^{2+} onto rutile was higher than the one of Mg^{2+}, since the latter gave zeta potential of higher magnitude at alkaline pH value. Thus, the surface reactivity of TiO_2 substances suspended in SFW is modified at $pH > pH_{IEP}$, due to the sorption of Ca^{2+} and Mg^{2+} ions.

9.5 Size Measurements of the Agglomerates

The extent of agglomeration of TiO_2 nanoparticles dispersed in aqueous media can significantly impact their reactivity toward organisms. Therefore, agglomerate size analysis of TiO_2 suspensions was performed by laser diffraction. Measurements were carried out using a Sympatec Helos instrument with a 100 mm focal length lens, giving particle size ranges between 0.9 and 175 µm. Agglomerate size distribution data were processed by the Fraunhofer diffraction theory. Since it was not possible to determine the agglomeration size distribution at the particle mass concentration used for zeta potential measurements (0.25 g/L), suspensions with a mass-to-volume ratio of 0.01 g/L were analyzed.

The laser diffraction equivalent diameter D_L, corresponding to the diameter of spherical particles with equivalent diffraction characteristics, was determined as well as the mass median diameter d_{50}, which corresponds to the diameter at which half of all particles (by mass) are larger and the other half smaller. Obtained values were averaged over three measurements.

In order to determine the effect of primary particle size, surface area, and electrolyte on the agglomeration of TiO_2 particles in 0.00885 M $NaNO_3$, samples A1, A2, and A3 suspensions at pH 4 and samples A1, A2, A3, and R1 suspensions at pH 8.1 were measured. Other measurements were done for all samples dispersed in synthetic freshwater at pH 8.1, at which the ecotoxicity tests were performed. All samples were equilibrated four days before the analysis. The zeta potential of these samples prepared at 0.01 g/L was also measured, in order to check whether potential difference with results obtained at 0.25 g/L could be found.

9.5.1 In $NaNO_3$

Figure 9.5 represents the particle frequency distribution, i.e. the percentage of particles found in each size range divided by the width of size range in each case. According to Figure 9.5, the most frequent particle size is similar for substances from A1 to A3 (0.7 μm), while no significant frequent particle size is obtained for R1. The maximum frequency distribution increased from substances A1 and A2 to substance A3. The mass median diameter d_{50}, obtained from the cumulative distribution (data not shown), increased from 1.44 μm for A1 and 1.18 μm for A2 to 5.67 μm for A3 with increasing primary particle size, indicating a higher agglomerate size. Measurements in $NaNO_3$ were performed at pH 4, at this pH value, surface charge of A1 and A2 is positive and pH value is far from IEP. In these conditions electrostatic repulsive forces are predominant, and particle agglomeration is not enhanced. This is not the case for A3 and R1, indeed pH 4 is very close to IEP, and in these conditions repulsive forces are not predominant, particles tend to agglomerate [36].

9.5.2 In Synthetic Freshwater (SFW)

In SFW at pH 8.1 (Figure 9.5), the profile of particle frequency distribution is different from $NaNO_3$ at pH 4, showing a higher heterogeneity of the particle sizes. The mass median diameter d_{50}, obtained from the cumulative distribution (data not shown), was increased from 3.79 μm (A1) and 3.59 μm (A2) to 8.14 μm (A3) with increasing primary particle size from ~10–20 to ~70–200 nm, among the anatase substances, confirming the expected higher agglomerate size in comparison with $NaNO_3$ media, due to a reduction of electrostatic repulsive forces and the predominance of van der Waals forces. The mass median diameter of A3 substance (8.14 μm) was lower than the one observed for R1 substance (19.51 μm), due to its lower primary particle size. For all TiO_2 suspensions in SFW at pH ~8, the mass median diameter and the agglomerate size was higher in comparison to suspensions in $NaNO_3$. From Figure 9.4, one can notice that electrophoretic mobility of all tested substances is close to zero in SFW media at pH 8.1. In these conditions electrostatic repulsion is not favored; this can explain the higher agglomerate size. The elevated agglomerate size of rutile compared with anatase substances could also be linked to its higher hydrophobia. In conclusion, in SFW and at pH ~8, the increase of primary particle size leads to a higher mass median diameter and a higher agglomerate size.

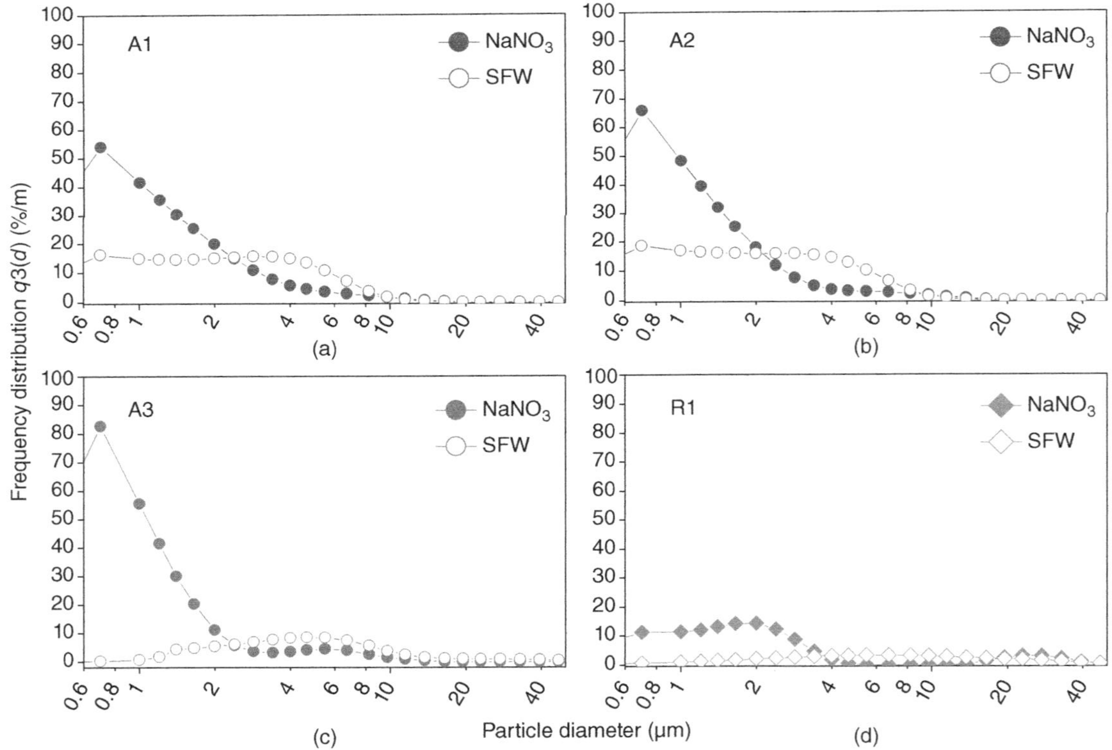

Figure 9.5 Particle frequency distribution of TiO_2 substances, $m/v = 0.01$ g/L, 4 days of equilibration. $I = 0.00885$ M $NaNO_3$ pH 4, A1 (●), A2 (●), A3 (●), and R1 (◆); $I = 0.00885$ M SFW pH 8.1, A1 (○), A2 (○), A3 (○), and R1 (◇).

9.6 Ecotoxicity Tests

The toxicity of TiO_2 particles was evaluated through acute toxicity tests considering rotifer, microcrustacean, and diatoms. These species are encountered in moderate to high salinity media, freshwaters, and marine waters, respectively. For all organisms, the median effective concentration EC_{50}, corresponding to the concentration of TiO_2 NP (mg/L), which induces 50 % of the maximal response, was estimated by graphical interpolation after plotting the calculated percent mortalities on a log concentration vs. % mortality, immobilization, and growth inhibition sheet. Normality and homogeneity of variance were checked using the Shapiro-Wilk and the Levene test, respectively. Statistical differences between groups to assess for concentration effects were analyzed using one-way analysis of variance (ANOVA). Afterward, Tukey's test was applied to compare the differences between groups and controls. For nonparametric data (no variance equality), the Kruskal–Wallis ANOVA test was applied to explore significant differences between groups. The two sample Student's *t*-test assuming unequal variances was applied to check differences between pairs of data sets. Statistically significant values were reported in the results when the probability of the result assuming the null hypothesis (p) is inferior to 0.05.

All statistical analyses were carried out using Origin 9.1 (OriginLab, Northampton, MA, USA) and the Microsoft Excel 2013 data analysis program.

9.6.1 Rotifer Toxicity Test

Brachionus plicatilis test is given to be a short-term (24 h) acute toxicity test [37]. Cysts of *B. plicatilis* were commercially obtained (Rotoxkit, Microbiotests Inc., Belgium). They were incubated for 28 h in synthetic marine water (SMW) of reduced salinity (20 ppt) (Table 9.5) at 25 °C under continuous illumination of 3000–4000 lux. After hatching, the cells were transferred to multiwell test plates containing the TiO_2 suspensions with increasing concentrations (0, 1, 10, 50, 100, and 150 mg/L for substances A1 and A2; 0, 200, 400, 600, 800, and 1000 mg/L for substance A3; and 0, 100, 200, 300, 400, and 500 mg/L for substance R1), prepared in SMW (5 rotifers per replicate, hence 30 rotifers per concentration). The test plates were incubated at 25 ± 2 °C in darkness, and after 24 and 48 h, the number of dead rotifers in each well was counted. Rotifers are considered dead

Table 9.5 Composition of the synthetic marine water (SMW) pH 8.1.

Substance	NaCl	KCl	$CaCl_2 \cdot 2H_2O$	$MgCl_2 \cdot 6H_2O$	$MgSO_4 \cdot 7H_2O$	$NaHCO_3$	H_3BO_3
SMW (g/L)	26.4	0.84	1.67	4.60	5.58	0.17	0.03

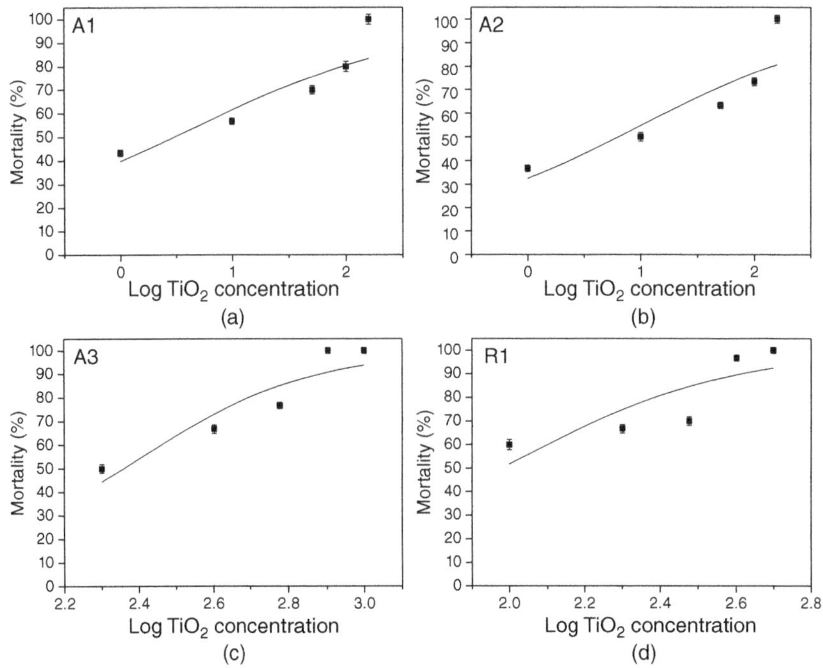

Figure 9.6 Dose response curve for *B. plicatilis* (mortality after 48 h) of four different TiO_2 substances: A1, A2, A3, and R1 in synthetic marine water (SMW).

if they do not exhibit any internal or external movement within 5 s of observation (Figure 9.6). For the calculation of the EC_{50} values, the results after 48 h were used. The data were obtained from five independent samples investigated simultaneously for error analysis.

For all substances the mortality of nearly 100% is reached at a concentration over 100 mg/L. EC_{50} values at 48 h equal to 5.1 mg/L for A1, 10.2 mg/L for A2, 256.7 mg/L for A3, and 109.4 mg/L for R1. The results of the acute toxicity test with the different TiO_2 substances after 48 h incubation showed that the mortality of rotifer increased with decreasing size of TiO_2 substances. In SFW, size distribution of TiO_2 is heterogeneous and different from $NaNO_3$ media, ranging from 0.6 to 4 μm for A1 and A2 anatase samples and from 1 to 10 μm for A3 and R1 samples (Figure 9.5). Despite the particle aggregation, A1 and A2 anatase samples displayed a 20–50-fold higher toxicity compared with the A3 and R1 samples. Though the laser diffraction measurements were carried out in SFW, the substances should exhibit similar trends in SMW in terms of agglomeration behavior. This result is in good accordance with experiments carried out on *Brachionus manjavacas* [16]. Indeed, it has been shown that both the size and the concentration of the nanoparticles are responsible for the decrease of the population growth rate. Small particles are able to penetrate the gut wall of rotifers, while large particles do not provoke any toxic effect on the growth of *B. manjavacas*, since they are not ingested by the organism [16, 38]. At pH 8 in SFW, TiO_2 surface charge is neutral for all the samples, hence toxicity cannot be associated to electrostatic interactions.

9.6.2 Microcrustacean Toxicity Test

Daphnia magna is a long-term (72 h) acute toxicity test. A commercial test kit Daphtoxkit F™ (MicroBioTests Inc., Gent, Belgium) was used. Test was carried out according to the Organisation for Economic Co-operation and Development (OECD) Guideline 202 [39]. Test suspensions of TiO_2 nanoparticles (0, 0.001, 0.01, 0.1, 1, 10, and 100 mg/L for samples A1 and A2; 0, 200, 400, 600, 800, and 1000 mg/L for sample A3; and 0, 100, 200, 300, 400, and 500 mg/L for sample R1) were prepared in SFW. The hatching of ephippia of *D. magna* was performed at 22 °C for 72 h under continuous illumination (6000 lux). Then, neonates were fed with cyanobacteria (*Spirulina* powder). The nanoparticles suspensions were placed in test plates containing test wells (5 neonates per well). After incubation in the dark at 20 ± 2 °C for 24 hours, the immobility was determined every 24 h during 72 h and compared to the control. The test was valid if the number of immobile organisms in the control vial did not exceed 10% after 72 h of incubation.

Immobilization (%) of *D. magna* was measured after 24, 48, and 72 h of contact with TiO_2 concentrations ranging from 0 to 100 mg/L for A1 and A2 samples, 0–1000 mg/L for A3 sample, and 0–500 mg/L for R1 sample. Results after 72 h are shown in Figure 9.7.

The traditional 48 h acute toxicity test may not be sufficient (data not shown) for the toxicity assessment of TiO_2 because no or low toxicity was found within 48 h of exposure (contrary to observations with *B. plicatilis*), but high toxicity was observed when the exposure time was extended to 72 h, in agreement with former observations [13, 40]. The low uptake kinetic of nanoparticles by *D. magna* can explain that it is necessary to extend the duration of the test in order to get toxicity data [17]

After 72 h of contact, the percentage of immobilization ranged from 0% to 100% for all the TiO_2 samples. Thus, EC_{50} values were calculated for 72 h of contact time. EC_{50} values equal to 1.1 mg/L for A1, 2.5 mg/L for A2, 239.2 mg/L for A3, and 86.7 mg/L for R1. In the test conditions A1 and A2 samples exhibited a much higher immobilization of cells compared with A3 and R1 samples, even if particles tend to agglomerate in SFW media. This indicates that the influx rate on nanoparticles is size dependent, which is in good agreement with data from the literature [27]. Toxicity results are in good accordance with the ones obtained for *B. plicatilis* test; i.e. A1 and A2 samples have comparable toxicity, while A3 and R1 samples

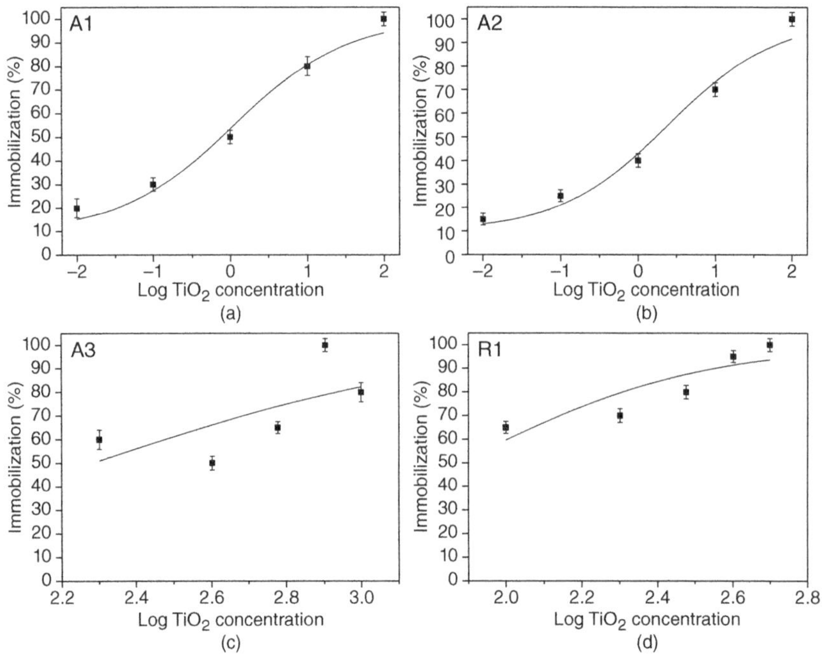

Figure 9.7 Dose response curve for *D. Magna* (% immobilization after 72 h) of four different TiO_2 substances: A1, A2, A3, and R1 in synthetic freshwater.

are less toxic. Due to the very low solubility of TiO_2, the toxic effect observed for *D. magna* cannot be linked to the presence of free titanium(IV) ions. Thus, toxicity can be correlated to the size of the particles, allowing their ingestion by the microcrustaceans.

9.6.3 Diatoms Toxicity Test

The 72 h diatom acute growth inhibition test was performed following the prescription of the ISO guideline "Water Quality – Marine Algal Growth Inhibition Tests with *Skeletonema costatum* and *Phaeodactylum tricornutum*" Standardization [41]. *P. tricornutum* diatom inoculum was obtained commercially from the test system Marine Algaltoxkit (MicroBioTests Inc., Belgium). Diatom culture in exponential growth phase was exposed to different concentrations of TiO_2 suspensions prepared with the SMW. The cell suspensions were transferred to spectrophotometric cells and incubated for 72 h in an incubator under controlled temperature (20 ± 2 °C) and under continuous illumination (10 000 lux). Daily

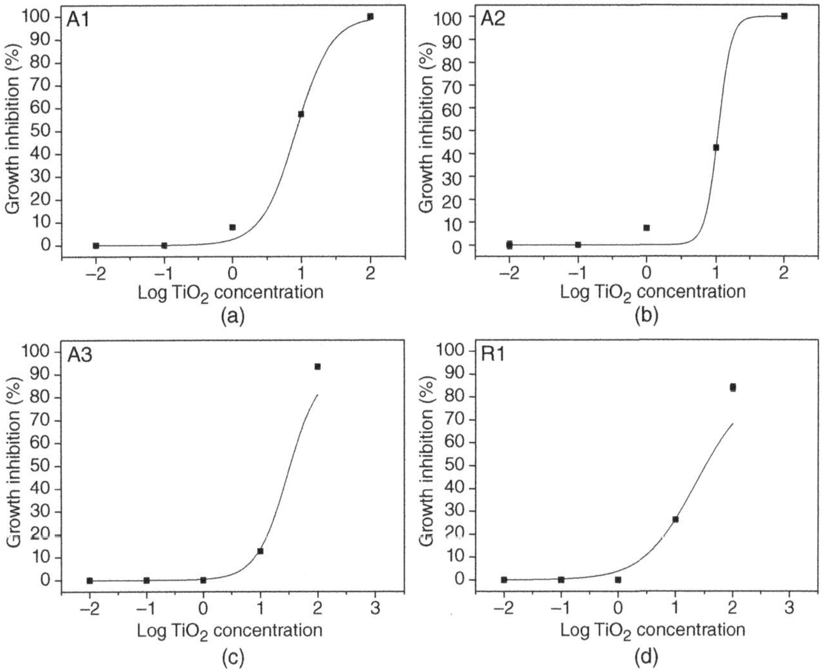

Figure 9.8 Dose response curve for *P. tricornutum* (growth inhibition after 72 h) of four different TiO_2 substances: A1, A2, A3, and R1 in synthetic marine water.

measurements for 72 h of growth in each spectrophotometric cell were carried out by determining the optical density of the diatom suspension at $\lambda = 670$ nm (Metertech, Spectrophotometer UV/Visible, SP-8001). The data were obtained from three independent samples that were carried out simultaneously for error analysis.

The growth inhibition (%) of *P. tricornutum* was measured after 72 h of contact with TiO_2 concentrations ranging from 0 to 100 mg/L for A1, A2, A3, and R1 substances (Figure 9.8). TiO_2 substances induce a toxicity response face to *P. tricornutum* from 24 h of contact. According to the standard procedure, the most significant growth inhibition is obtained after 72 h of contact. After this time, the estimated EC_{50} values equal to 8.4, 10.7, 30.2, and 22.2 mg/L for A1, A2, A3, and R1 samples, respectively. These values are in the same order of magnitude for the four TiO_2 samples. TiO_2 nanoparticles were found to be located inside cell wall and membrane of the green algae *Chlamydomonas reinhardtii* and that cell surface was surrounded by TiO_2 particles, which could present an obstacle to the exchange of substances between the cell and its surrounding environment [42].

9.7 Discussion

TEM and SEM pictures of TiO_2 substances have shown that A1, A2, and A3 substances were of slightly ellipsoidal shape, while R1 exhibited a more irregular shape. XRD analysis evidenced that A1 and A3 substances contained some rutile, and R1 contained some anatase and $BaTiO_3$ impurities. A2 was therefore the only pure substance. According to BET measurements, the SSAs, and the particle surface areas decrease as follows: A1 > A2 > A3 > R1, corresponding to the increase of the primary particle size. The pH_{IEP} of A1, A2, and A3 decreases as the primary particle size increases. This phenomenon was highlighted by Suttiponparnit et al. [3] who demonstrated that a decrease of the primary particle size provokes an increase of the pH_{IEP} of anatase particles. Again, at this stage of knowledge, a proper mechanistic explanation cannot be given in this study. It has been shown that the isoelectric point was rather insensitive to the crystal structure (anatase or rutile) and to the experimental determination method [32, 34, 35], but the synthesis routes of TiO_2 substances can influence the isoelectric point [3, 33]. Contrary to what could be expected according to the primary particle size, the pH_{IEP} of R1 substance was found to be superior to the pH_{IEP} of A3 substance. The presence of $BaTiO_3$ impurities in R1 substance could be one of the explanations.

The surface reactivity deduced from potentiometric titrations has shown that A1 and A2 substances exhibited the same total surface site concentration and the same pH_{IEP} (i.e. the same surface reactivity). A3 and R1 substances have a lower total surface site concentration, which is in good agreement with their primary particle sizes. The electrolytic media can influence the surface reactivity of TiO_2 substances. Indeed, electrophoresis measurements have shown that the adsorption of calcium and magnesium on TiO_2 surface influences the zeta potential value for pH value superior to 8 (for Ca adsorption) and 10 (for Mg adsorption) (data not shown). For acidic pH values (pH inferior to 6) adsorption of sulfate ions is possible [43]. The decrease in zeta potential value for A1 and A2 substances can be explained by the adsorption of sulfate ions from the SFW. Even if adsorption of sulfate measured by ion chromatography was not significant, a very low amount (inferior to the detection limit of ion chromatography) of adsorbed sulfates could explain the decrease of zeta potential at acidic pH values (data not shown). The decrease of the surface charge at $pH < pH_{IEP}$ has been explained by the adsorption of oxyanions creating negatively charged surface complexes possibly at the vicinity of the shear plane and, thus, indicating a close association to the surface [44–46]. This phenomenon was not observed for A3 and R1 substances, but considering the lower surface reactivity of these substances, no adsorption of sulfate ions was expected. Toxicity tests were performed at pH 8.1 in synthetic natural waters. In these conditions, zeta potential is close to zero for all TiO_2 substances, indicating that surface is near neutral. This leads to a reduction of electrostatic repulsive forces and to an increase of the extent of agglomeration as well as the mass median diameter, due to the predominance of van der Waals forces. At pH 8.1, adsorption of calcium is expected to influence the surface reactivity of TiO_2 substances. Toxicity tests consist in determining the EC_{50} value when the mass-to-volume ratio of TiO_2 substances is increased. In $NaNO_3$ at pH 4, a decrease of the magnitude of the zeta potential was observed with decreasing particle mass concentration from 0.25 to 0.01 g/L, due to the decrease in particle surface area. The subsequent decrease of the magnitude of the zeta potential reduces the electrostatic repulsive forces, leading to an enhanced agglomeration and therefore an increased mass median diameter d_{50}. Nevertheless, even if the zeta potential value decreased with decreasing mass-to-volume ratio, the same tendency was observed; i.e. A1 and A2 substances had still a higher zeta potential than A3 (Figure 9.4). In SFW at pH 8.1, no significant decrease of the magnitude of the zeta potential was observed with decreasing particle mass concentration. Then, whatever the electrolytic media ($NaNO_3$ or SFW) and the pH value (4 or 8), even if the change in the mass-to-volume ratio can change the surface reactivity, the relative comparison of our substances remains valid.

Toxicity tests were carried out on the three aquatic species *D. magna, P. tricornutum,* and *B. plicatilis*. The results of the toxicity test show a dependency on the size of the nanoparticles and also on the TiO_2 allotropic form. In general, the tests showed that small TiO_2 nanoparticles have a stronger toxic effect. Smaller particles could easily pass the cell membrane and lead intracellular to toxic effects. Bigger TiO_2 nanoparticles may not be able to pass this barrier system of the cells and show only extracellular toxic effects, but the organisms itself show differences in toleration. A1 and A2 substances show the strongest toxic effects against the aquatic organism. Even if EC_{50} values are of same order of magnitude, A1 substance exhibits a twice higher toxicity than A2 substance. According to surface characterization, shape, SSA, pH_{IEP}, primary particle size, mass median diameter d_{50}, and the total site concentration are of same order of magnitude for both substances. The difference between A1 and A2 substances resides in the presence of rutile impurities in A1. The presence of rutile could explain the increased toxicity of A1 substance. For comparable primary particle size (100–500 and 70–200 nm, for R1 and A3, respectively), the rutile substance displays a two times higher toxicity than anatase substance (based on EC_{50} values). This result could be due to the shape of rutile particles in R1 substance or to the allotropic form itself or the presence of $BaTiO_3$ impurities. Indeed, none of the parameters from surface characterization of R1 substance can be correlated to toxicity results.

Considering toxicity data obtained for anatase substances, it appears that toxicity results increase when the primary particle size (and mass median diameter d_{50}) is decreased, and it depends on the surface reactivity (i.e. when the surface area increases, the total surface site decreases, and the toxicity is decreased). Considering that the surface reactivity is linked to the particle size, toxicity for a same allotropic form depends on the primary particle size. Finally, the most sensitive test to evaluate toxicity of TiO_2 substances in aquatic media is *D. Magna* test, and the least sensitive is *B. plicatilis* test.

9.8 Conclusions

Based on the extensive surface and bulk characterization as well as toxicity test on aquatic organisms in fresh and marine synthetic waters, the followings can be concluded:

- For anatase, toxicity depends on the primary particle size and the agglomerate size (mass median diameter d_{50}) and consequently depends on surface reactivity (total surface site concentration, SSA, pH_{IEP}).
- For anatase substances of comparable nanometric size, agglomerate size, and surface reactivity, toxicity could be increased by the presence of rutile impurities in the substance.
- For the two allotropic forms of TiO_2 under investigation and substances of comparable primary particle size, rutile has a higher toxicity than anatase. This can be attributed either to the shape of particles, to the allotropic form itself, or to the presence of $BaTiO_3$ impurities.
- Among *D. magna, B. plicatilis*, and *P. tricornutum*, the most toxic effects of TiO_2 nanoparticles at low concentrations was shown for *D. magna*. In contrast *B. plicatilis* cells show small toxic effects also at high concentrations of TiO_2 (*D. magna* is widely used in aquatic risk assessment [11]).

The comparison of TiO_2 nanoparticles toxicity of different allotropic forms cannot be based only on their primary particle size, and their acute toxicity in the natural aquatic environment depends on several bulk and surface parameters that have to be deeply investigated.

Acknowledgments

The authors would like to thank Christa Müller and Heidrun Neubert for their technical assistance for electrophoretic mobility measurements, Aline Ritter for ICP-MS measurements, and Carola Eckardt for BET determination and Ionic Chromatography measurements. We are also grateful to Dr. Arndt Mücklich for TEM, Elfi Christalle for SEM, and Annett Kästner for particle size distribution measurements.

References

1 Aitken, R.J., Chaudhry, M.Q., Boxall, A.B.A., and Hull, M. (2006). Manufacture and use of nanomaterials: current status in the UK and global trends. *Occup. Med.* 56 (5): 300–306.

2 Navarro, E., Baun, A., Behra, R. et al. (2008). Environmental behavior and ecotoxicity of engineered nanoparticles to algae, plants, and fungi. *Ecotoxicology* 17 (5): 372–386.

3 Suttiponparnit, K., Jiang, J.K., Sahu, M. et al. (2011). Role of surface area, primary particle size, and crystal phase on titanium dioxide nanoparticle dispersion properties. *Nanoscale Res. Lett.* 6: 8.

4 Miao, L., Wang, C., Hou, J. et al. (2016). Effect of alginate on the aggregation kinetics of copper oxide nanoparticles (CuO NPs): bridging interaction and hetero-aggregation induced by Ca^{2+}. *Environ. Sci. Pollut. Res.* 23 (12): 11611–11619.

5 Janusz, W., Galgan, A., and Skwarek, E. (2004). Effects of the adsorption of Ba^{2+} ions on the electrical double layer at the 4th group metal oxide/electrolyte interface. *Adsorpt. Sci. Technol.* 22 (10): 795–806.

6 Janusz, W. and Matysek, M. (2006). Coadsorption of Cd(II) and oxalate ions at the TiO_2/electrolyte solution interface. *J. Colloid Interface Sci.* 296 (1): 22–29.

7 Janusz, W. and Skwarek, E. (2011). Adsorption of Ca(II) and Fe(III) ions at the SnO_2/electrolyte solution interface. *Physicochem. Probl. Miner. Process.* 46: 73–82.

8 Skwarek, E., Matysek-Nawrocka, M., Janusz, W. et al. (2008). Adsorption of heavy metals at the Al_2O_3–SiO_2/$NaClO_4$ electrolyte interface. *Physicochem. Probl. Miner. Process.* 42: 153–163.

9 Kosmulski, M., Gustafsson, J., and Rosenholm, J.B. (1999). Ion specificity and viscosity of rutile dispersions. *Colloid Polym. Sci.* 277 (6): 550–556.

10 Zhu, X.S., Zhu, L., Chen, Y.S., and Tian, S.Y. (2009). Acute toxicities of six manufactured nanomaterial suspensions to *Daphnia magna*. *J. Nanopart. Res.* 11 (1): 67–75.

11 Bang, S.H., Le, T.-H., Lee, S.K. et al. (2011). Toxicity assessment of titanium (IV) oxide nanoparticles using *Daphnia magna* (water flea). *Environ. Health Toxicol.* 26.

12 Tan, L.-Y., Huang, B., Xu, S. et al. (2016). TiO_2 nanoparticle uptake by the water flea *Daphnia magna* via different routes is calcium-dependent. *Environ. Sci. Technol.* 50 (14): 7799–7807.

13 Zhu, X.S., Chang, Y., and Chen, Y.S. (2010). Toxicity and bioaccumulation of TiO_2 nanoparticle aggregates in *Daphnia magna*. *Chemosphere* 78 (3): 209–215.

14 Marcone, G.P.S., Oliveira, A.C., Almeida, G. et al. (2012). Ecotoxicity of TiO_2 to *Daphnia similis* under irradiation. *J. Hazard. Mater.* 211: 436–442.

15 Rico-Martinez, R., Arzate-Cardenas, M.A., Robles-Vargas, D. et al. (2016). Rotifers as models in toxicity screening of chemicals and environmental samples. *Invert. Exp. Models Toxicity Screen.* 57.

16 Snell, T.W. and Hicks, D.G. (2011). Assessing toxicity of nanoparticles using *Brachionus manjavacas* (Rotifera). *Environ. Toxicol.* 26 (2): 146–152.

17 Clément, L., Hurel, C., and Marmier, N. (2013). Toxicity of TiO_2 nanoparticles to cladocerans, algae, rotifers and plants - effects of size and crystalline structure. *Chemosphere* 90 (3): 1083–1090.

18 Armbrust, E.V. (2009). The life of diatoms in the world's oceans. *Nature* 459 (7244): 185–192.

19 Huang, A.Y., He, L.W., and Wang, G.C. (2011). Identification and characterization of microRNAs from *Phaeodactylum tricornutum* by high-throughput sequencing and bioinformatics analysis. *BMC Genomics* 12.

20 Fekete-Kertész, I., Maros, G., Gruiz, K., and Molnar, M. (2016). The effect of TiO_2 nanoparticles on the aquatic ecosystem: a comparative ecotoxicity study with test organisms of different trophic levels. *Period. Polytech., Chem. Eng.* 60 (4): 231–243.

21 Domingues, N., Matos, A.R., da Silva, J.M., and Cartaxana, P. (2012). Response of the diatom *Phaeodactylum tricornutum* to photooxidative stress resulting from high light exposure. *PLoS One* 7 (6).

22 Ghazy, M.M.E.-D., Habashy, M.M., Kossa, F.I., and Mohammady, E.Y. (2009). Effects of salinity on survival, growth and reproduction of the water flea, *Daphnia magna*. *Nat. Sci.* 7 (11): 29–42.

23 Goix, S., Leveque, T., Xiong, T.T. et al. (2014). Environmental and health impacts of fine and ultrafine metallic particles: assessment of threat scores. *Environ. Res.* 133: 185–194.

24 Heinlaan, M., Ivask, A., Blinova, I. et al. (2008). Toxicity of nanosized and bulk ZnO, CuO and TiO_2 to bacteria *Vibrio fischeri* and crustaceans *Daphnia magna* and *Thamnocephalus platyurus*. *Chemosphere* 71 (7): 1308–1316.

25 Kostopoulou, V., Carmona, M.J., and Divanach, P. (2012). The rotifer *Brachionus plicatilis*: an emerging bio-tool for numerous applications. *J. Biol. Res.-Thessaloniki* 17: 97–112.

26 Robinson, J.D., Wares, J.P., and Drake, J.M. (2013). Extinction hazards in experimental *Daphnia magna* populations: effects of genotype diversity and environmental variation. *Ecol. Evol.* 3 (2): 233–243.

27 Zhao, C.M. and Wang, W.X. (2012). Size-dependent uptake of silver nanoparticles in *Daphnia magna*. *Environ. Sci. Technol.* 46 (20): 11345–11351.

28 Mortimer, M., Kasemets, K., Heinlaan, M. et al. (2008). High throughput kinetic *Vibrio fischeri* bioluminescence inhibition assay for study of toxic effects of nanoparticles. *Toxicol. in Vitro* 22 (5): 1412–1417.

29 Parvez, S., Venkataraman, C., and Mukherji, S. (2006). A review on advantages of implementing luminescence inhibition test (*Vibrio fischeri*) for acute toxicity prediction of chemicals. *Environ. Int.* 32 (2): 265–268.

30 Bertaut, E.F. (1950). Raies de Debye–Scherrer et repartition des dimensions des domaines de Bragg dans les poudres polycristallines. *Acta Cryst., Sect. A: Found. Crystallogr.* 3 (1): 14–18. https://scripts.iucr.org/cgi-bin/paper?a00207.

31 Zunic, T.B. and Dohrup, J. (1999). Use of an ellipsoid model for the determination of average crystallite shape and size in polycrystalline samples. *Powder Diffr.* 14 (3): 203–207.

32 Kosmulski, M. (2009a). Compilation of PZC and IEP of sparingly soluble metal oxides and hydroxides from literature. *Adv. Colloid Interface Sci.* 152 (1–2): 14–25.

33 Kosmulski, M. (2009b). *Surface Charging and Points of Zero Charge*. CRC Press.

34 Kosmulski, M. (2002). The significance of the difference in the point of zero charge between rutile and anatase. *Adv. Colloid Interface Sci.* 99 (3): 255–264.

35 Kosmulski, M. (2003). A literature survey of the differences between the reported isoelectric points and their discussion. *Colloids Surf., A* 222 (1-3): 113–118.

36 Berg, J.M., Romoser, A., Banerjee, N. et al. (2009). The relationship between pH and zeta potential of approx. 30 nm metal oxide nanoparticle suspensions relevant to in vitro toxicological evaluations. *Nanotoxicology* 3 (4): 276–283.

37 Snell, T.W. and Persoone, G. (1989). Acute toxicity bioassays using rotifers 1. A test for brackish and marine environments with *Brachionus plicatilis*. *Aquat. Toxicol.* 14 (1): 65–80.

38 Snell, T.W., Kubanek, J., Carter, W. et al. (2006). A protein signal triggers sexual reproduction in *Brachionus plicatilis* (Rotifera). *Mar. Biol.* 149 (4): 763–773.

39 OECD (2004), *Test No. 202: Daphnia sp. Acute Immobilisation Test*, OECD Guidelines for the Testing of Chemicals, Section 2, Éditions OCDE, Paris, https://doi.org/10.1787/9789264069947-en.

40 Strigul, N., Vaccari, L., Galdun, C. et al. (2009). Acute toxicity of boron, titanium dioxide, and aluminum nanoparticles to *Daphnia magna* and *Vibrio fischeri*. *Desalination* 248 (1-3): 771–782.

41 International Organization for Standardization ISO 10253 (2006). *Water quality-Marine Algal Growth Inhibition Test with Skeletonema COSTATUM and Phaeodactylum tricornutum*, 2e, stage 90, 20, ISO/TC 147/SC 5. International Organization for Standardization.

42 Chen, L.Z., Zhou, L.N., Liu, Y.D. et al. (2012). Toxicological effects of nanometer titanium dioxide (nano-TiO_2) on *Chlamydomonas reinhardtii*. *Ecotoxicol. Environ. Saf.* 84: 155–162.

43 Horanyi, G. (2003). Investigation of the specific adsorption of sulfate ions on powdered TiO_2. *J. Colloid Interface Sci.* 261 (2): 580–583.
44 Elzinga, E.J., Tang, Y.Z., McDonald, J. et al. (2009). Macroscopic and spectroscopic characterization of selenate, selenite, and chromate adsorption at the solid–water interface of γ-Al_2O_3. *J. Colloid Interface Sci.* 340 (2): 153–159.
45 Jordan, N., Foerstendorf, H., Weiss, S. et al. (2011). Sorption of selenium(VI) onto anatase: macroscopic and microscopic characterization. *Geochim. Cosmochim. Acta* 75 (6): 1519–1530.
46 Jordan, N., Ritter, A., Foerstendorf, H. et al. (2013). Adsorption mechanism of selenium(VI) onto maghemite. *Geochim. Cosmochim. Acta* 103: 63–75.

10

Cardiorespiratory Toxicity of Nanoparticles in Aquatic Environments

Christopher A. Dieni and Tyson J. MacCormack

Department of Chemistry and Biochemistry, Mount Allison University, Sackville, New Brunswick, Canada

10.1 Introduction

Nanotechnology has facilitated the development of new technologies and contributed markedly to refinements in existing technologies. Broadly defined, engineered nanomaterials (ENMs) are synthetic structures with dimensions between ~1 and 100 nm but with surface coatings and adsorbed biomolecules (the biomolecular corona); actual sizes encountered in aquatic environments may reach into the high hundreds of nanometers [1, 2]. The high reactivity and often unique physicochemical properties of nanoscale structures make them desirable for use in a variety of consumer and industrial applications. Toxicity is unlikely to be a major issue for targeted, small volume applications, but agricultural and environmental remediation technologies involving the intentional release of large quantities of ENMs into the environment are a concern. There is great potential for such applications to improve the targeting of traditional pesticides [3] or break down existing contaminants [4], so understanding the risks associated with these products will facilitate the introduction of greener alternatives. The aquatic environment will be a sink for ENMs through terrestrial runoff, wastewater input, and atmospheric deposition. We now have a considerable knowledge base on the aquatic toxicity of many ENM formulations, but there are still significant gaps in our understanding. Numerous studies have demonstrated minimal acute ENM toxicity at what could be considered environmentally relevant exposure doses, but far fewer have addressed chronic toxicity under realistic exposure conditions.

In terrestrial vertebrates, the inhalation of airborne nanomaterials, either incident or engineered, is frequently linked to increases in cardiorespiratory risk factors. Nanomaterials are able to penetrate deep into the alveoli

Toxicology of Nanoparticles and Nanomaterials in Human, Terrestrial and Aquatic Systems, First Edition. Edited by Marc A. Williams.

of the lungs where they can trigger immunological and oxidative stress responses, subsequently leading to vascular inflammation and cardiorespiratory dysfunction [5]. Direct or indirect indicators of oxidative stress are frequently observed in aquatic organisms exposed to ENMs, and a growing body of evidence suggests suspended ENMs may also trigger cardiorespiratory dysfunction in these animals. The cardiorespiratory physiology of fish has been well characterized [6], but relatively few studies have addressed the impact of environmental contaminants on the heart and gill function in this diverse group of animals. The role of the gills in gas and ion exchange necessitates that they present a large surface area to the environment and a small diffusion distance to the blood. These anatomical features render them susceptible to exposure to suspended particulates and make the cardiorespiratory system particularly vulnerable to any associated toxicity. All aquatic organisms have evolved in the presence of a variety of natural nanomaterials, but it is uncertain how fish will cope with the unique combinations of physicochemical properties found in ENMs. A number of reviews have addressed the acute toxicity of metallic and nonmetallic ENMs to fish [7, 8], which tends to be modest relative to common dissolved chemical toxins. This chapter will provide an overview of the potential mechanisms of ENM toxicity specifically as they relate to impacts on cardiorespiratory function. Considerable attention will be focused on the role of the immune system in contributing to the oxidative stress response associated with ENM exposure, as this topic has yet to be fully addressed in the literature. We conclude by providing perspectives on how the cardiorespiratory toxicity of ENMs may impact fish under ecologically relevant conditions and highlight current gaps in knowledge.

10.2 Cellular and Molecular Mechanisms of Engineered Nanomaterial Toxicity

While many studies demonstrate that ENMs can induce toxicity in aquatic organisms, the exact mechanisms by which they are proposed to be toxic vary significantly. These may be due in part to the study conditions, such as exposure duration, dose, the characteristics of the ENM, or the model organism in question. However, mechanisms of ENM toxicity may also vary due to the multitude of interactions that they can form; the very physicochemical characteristics that are manipulated to allow nanostructures to remain dispersed in aqueous solutions, and to target them to specific cellular components, are the same characteristics that allow them to interact nonspecifically with a plethora of biomolecules (Figure 10.1). In this section, we explore the mechanisms by which ENMs induce toxicity on a molecular and cellular scale.

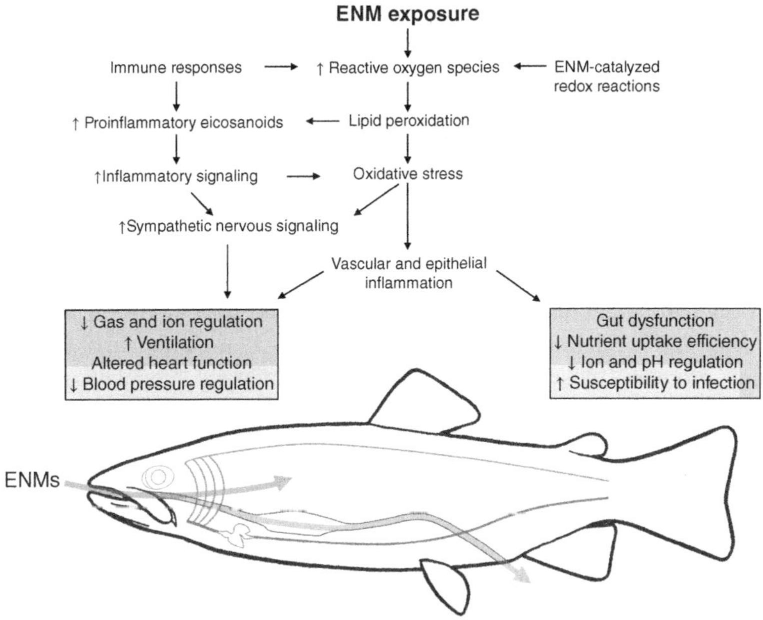

Figure 10.1 Potential mechanisms of engineered nanomaterial (ENM) toxicity in fish. Exposure is most likely to occur at the gill or gut epithelia where ENMs can directly or indirectly increase reactive oxygen species production, leading to oxidative stress and inflammation. These responses can disrupt membrane-associated processes and alter neural and humoral regulation of cardiorespiratory function.

10.2.1 Uptake-Independent Mechanisms

Toxins must often be internalized to exert their most damaging effects, but there are numerous extracellular or cell-surface interactions (e.g. in respiratory tract or digestive lumen) capable of causing mild to moderate toxicity, even in the absence of significant uptake. In general terms, toxins classified as irritants or corrosives may not result in significant systemic poisoning *per se*, but may cause localized cellular and tissue dysfunction or damage, which if serious can trigger widespread inflammatory or other physiological responses. This is especially relevant to the toxicity of ENMs.

ENMs rely predominantly on endocytic mechanisms for cellular internalization, so uptake may therefore be a slow and inefficient process, highly dependent on the physicochemical properties of the particle [2, 9–11]. It is therefore important to consider uptake-independent mechanisms of nanotoxicity in addition to those that depend on internalization, and we present some examples of these uptake-independent mechanisms here.

10.2.1.1 Accumulation on Cell Surfaces and Interference with Membrane and Transport Functions

In fish, the gill and gut represent the most sensitive epithelial barriers, and their function is crucial for ionoregulation, acid–base homeostasis, waste excretion, and nutrient uptake. These tissues are arguably the most vulnerable to interactions with ENMs on their apical membrane surfaces. One mechanism by which ENMs disrupt apical epithelial function is by direct or indirect dysregulation of membrane proteins. A noteworthy example is the adenosine triphosphate (ATP)-dependent sodium–potassium pump (NKA), but since this is typically found in the basolateral membrane, it will be discussed with uptake-dependent mechanisms. ENMs can impair epithelial function by accumulation on or around membrane proteins. In synthetic phospholipid membrane monolayers, gold ENMs preferentially colocalize with membrane proteins, rather than dispersing throughout the lipid phase (Figure 10.2; MacCormack et al. [12]). Protein function is significantly influenced by the presence of ENMs in this *in vitro* system, but it is still unclear whether this phenomenon occurs *in vivo* or if this behavior can be generalized to other ENM formulations. Epithelial function can also be impaired by excessive mucous secretion that impedes diffusion of substrates to and from transporters, and there are many examples where this is a significant toxicological mechanism in fish. In rainbow trout exposed to single-walled carbon nanotubes (SWCNTs), there was a dose-dependent increase in gill mucus secretion [13], and similar responses were noted in trout exposed to TiO_2 ENMs [14]. While substantial ENM uptake was unlikely in these instances, significant localized and systemic toxicity was observed. For ZnO ENMs, there is some evidence of uptake by fish, but accumulation on the apical surface and in mucus appears to account for at least a portion of their toxicity in tilapia [11] and carp [15]. Excessive mucous accumulation or changes to gill morphology will likely impair epithelial ion and gas exchange [14] and may explain why fish are more sensitive to environmental hypoxia following ENM exposure [16].

10.2.1.2 Activation of Cell Surface Inflammatory Receptors

Even without uptake into systemic circulation, ENMs are capable of interactions leading to the activation or suppression of both branches of the immune system: innate (or inherent, nonspecific) and adaptive (or specific, acquired). The innate immune system is particularly relevant here, since adaptive immunity is generally considered incomplete and inadequate in fish when compared with mammals [17–20]. Interactions between ENMs and the immune system in aquatic organisms will therefore predominantly involve the innate immune system and inflammation, a complex mechanism of innate immunity. In mammals, the cardiorespiratory dysfunction associated with inhalation of airborne nanoparticles is

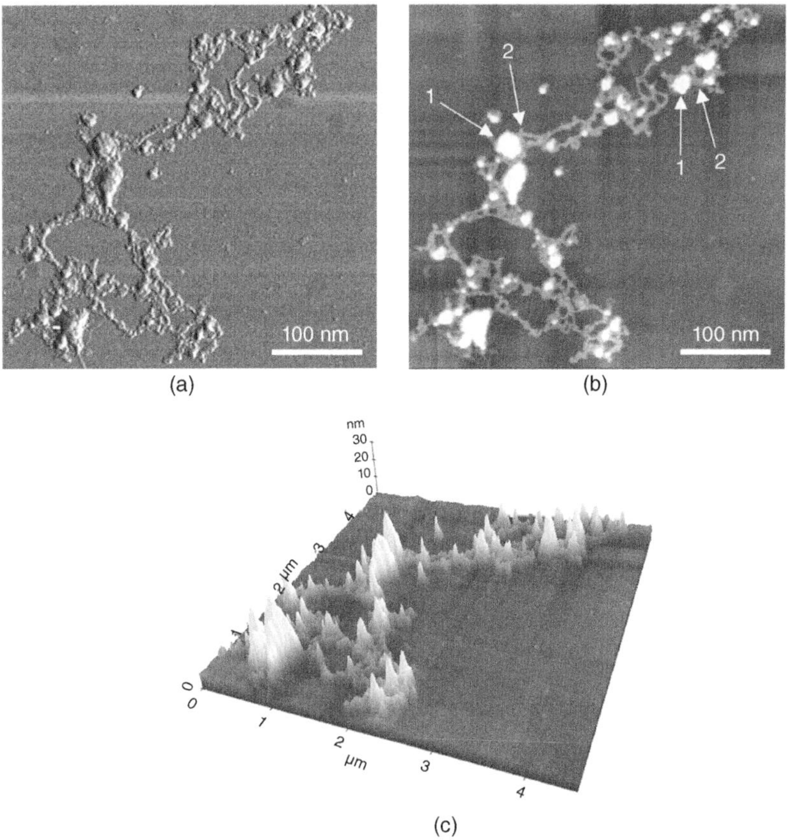

Figure 10.2 Atomic force microscopy phase (a), two-dimensional topography (b), and three-dimensional topography images of phospholipid membrane monolayers containing surface-functionalized gold ENMs. Two-dimensional topography images show peripheral membrane proteins approximately 18 nm in height (white, indicated by arrow 1) surrounded by gold ENMs approximately 5 nm in height (darker yellow, indicated by arrow 2). Three-dimensional topography images clearly show protruding peaks corresponding to membrane proteins, surrounded by smaller clusters of gold ENMs. The phospholipid phase is represented by the orange background. Source: MacCormack et al. [12].

tightly linked to the disruption of immune function [21] and the propagation of inflammatory responses, and it is likely that similar mechanisms exist in aquatic animals.

Uptake-independent activation of innate immunity is most commonly a result of direct contact between ENMs and cell surface receptors that sense and protect the host organism against infectious pathogens, nonbiological contaminants, and host pathophysiologies. These receptors are divided into two broad categories based on molecular patterns present in their activating ligands, hence their alias "pattern recognition receptors" or PRRs. Pathogen-associated molecular pattern (PAMP) receptors traditionally bind biological macromolecules such as nucleic acid structures, protein features, and carbohydrate–lipid motifs unique to microbes. Damage-associated molecular pattern (DAMP) receptors bind host ligands indicative of cellular damage, such as heat shock proteins (HSPs), nuclear proteins, or purine metabolites released from injured cells. ZnO ENM exposure upregulates gill HSP expression in fish in the absence of ENM uptake [22], so the necessary signals for DAMP receptor activation are clearly triggered by aquatic ENM exposure. ENMs that have been coated with any of these PAMPs or DAMPs, either deliberately (e.g. for targeting or sensing applications) or through incidental adsorption, can efficiently activate cell surface PRRs [23–25]. Certain types of ENMs may also be capable of activating cell surface receptors independently of traditional PAMPs or DAMPs. Divalent metal cations bind to and activate TLR4, a member of the family of toll-like receptors [26, 27], and a similar process is thought to occur with metal ENMs [28–30]. In some cases, ENMs trigger the same inflammatory activation that metal ions do, but in others toxicity can be manifested through a noninflammatory competitive inhibition of PRRs.

Immune Cell Receptors Uptake-independent activation of inflammatory pathways often stems from contact with innate immune cells such as macrophages, which are present in barrier tissues such as the gut epithelium. In air-breathing vertebrates, alveolar macrophages are a primary defense against respiratory infections and irritations, but in fish, the gill epithelium is directly exposed to the aquatic environment, so there is no analogous process. Primary immune function at the apical surface of the gill is carried out by mucous cells interspersed between respiratory and ionoregulatory cells. Although mucous cells are part of the immune system, they do not typically display the phagocytic capabilities of macrophages or dendritic cells (DCs), nor do they have a broad assortment of PRRs. Apart from mucous cells, the gill epithelium is not rich in immune cells capable of sensing or phagocytosing potential threats [17, 19, 31], so uptake-independent induction of innate immunity is likely most prevalent in the gut.

Secretory goblet (i.e. mucous cells), Paneth, and microfold cells are innate immune cells located on the luminal surface of the gut epithelium, interspersed among the intestinal epithelial cells. Goblet and Paneth cells function by secreting mucus and antibacterial peptides, respectively, whereas microfold cells transcytose antigens and contaminants to antigen presenting cells in the gut-associated lymphoid tissues on the basal side of the basement membrane. In aquatic and nonaquatic organisms, DCs are the predominant type of innate immune cell capable of triggering an immune response to ENMs without a requirement for uptake. In mammals, DCs extend branched projections, known as dendrites, through the gaps between epithelial cells to sample luminal contents for potential antigens. DCs, along with other phagocytic cells, express the widest variety and highest levels of PRRs, which when activated can produce an inflammatory cascade. In the gut this inflammation response is tightly controlled by spatiotemporal regulation of PRR expression and/or higher activation thresholds of immune responses, likely to minimize the provocation of inflammatory responses by commensal bacteria. It is therefore possible that uptake-independent activation of immune responses to ENMs in the gut will be will be somewhat dulled compared to other tissues.

Non-immune Cell Receptors In fish, immune cells capable of sensing threats are seemingly not present at the apical surface of gill epithelia; rather, they are typically present in basolateral spaces immediately adjacent to the gill, such as the gill-associated lymphoid tissue and interbrachial lymphoid tissue [32]. As such, uptake-independent triggering of inflammation by ENMs at the gill is less clear than other mechanisms of nanotoxicity such as disruption of cell surface integrity and transport functionality. Nonetheless, it is very likely that even nonimmune cells have mechanisms to sense and sample potential threats as they come into contact with the epithelium, and this is commonplace in many organisms [33–35]. It remains unclear what types of cells would be most directly involved in sensing and responding to ENMs or whether uptake-independent clustering of PRRs is a significant mechanism of gill nanotoxicity [36–40].

Regardless of the specific cell type or receptor that may be activated by ENMs, the innate immune signaling mechanisms in any nonimmune cell can provoke responses in proximal immune cells. For instance, the expression of several cytokines, including interleukins, has been established in the gills of multiple species of fish. Several of these interleukins, and other signals generated by stressed or hyperplasic epithelial cells, are capable of stimulating secretion from gill mucous cells [17, 18, 41, 42], which can be an effective short-term defense against ENMs. With high doses or chronic exposures, ENMs are eventually able to inflict cell damage, and these inflammatory cascades have been implicated in the systemic pathophysiologies observed in response to SWCNT [13] and TiO_2 [14], even where ENM uptake is not significant.

10.2.1.3 Uptake-Independent Generation of Reactive Oxygen Species

Much of the oxidative stress associated with ENMs is attributed to uptake-dependent mechanisms. After phagocytic cells internalize ENMs, they attempt to degrade them via high levels of protons and reactive oxygen species (ROS), including superoxide ($O_2^{\cdot -}$) and hydroxyl radical ($OH^{\cdot}$), as well as reactive nitrogen and carbon species. However, ENMs are also able to generate ROS without internalization. The high surface area and active electronic configurations of ENMs allow them to function as highly reactive semiconductors capable of accepting or donating electrons [43, 44]. In cell-free systems, redox reactions can be catalyzed by ENMs, including the depletion of antioxidants and the generation of oxidized species [45–48], and oxygen concentration in the system influences the rate of these reactions. Gill tissue is clearly vulnerable to this type of toxicity as it has a large surface area for ENM adsorption and a continuous supply of oxygen to accept electrons and generate $O_2^{\cdot -}$ [45, 47, 48]. $O_2^{\cdot -}$ can then react with other biological radicals or transition metals to perpetuate the production of reactive species, with membrane lipids being the most immediate targets [11, 14, 22]. As cell membranes are compromised, secondary gill injuries can occur, including the induction of apoptosis and osmoregulatory/ionoregulatory stress [22, 49], which can trigger significant tissue remodeling (Figure 10.3; [50]). Such responses to ENM exposure may impair oxygen uptake [16], cause cardiorespiratory stress [22], and ultimately reduce the animal's aerobic scope [50].

The activation of PRRs is also associated with ROS generation, particularly in phagocytic cells that produce reactive intermediates to destroy pathogens or larger macromolecules [26, 31]. Toll-like receptors and cytokine receptors will initiate a signaling cascade, leading to the increased expression of phagocyte oxidase and nitric oxide synthase, which catalyze the production of superoxide and nitric oxide, respectively. Together, these can react to form peroxynitrite, a very powerful oxidant capable of inducing cell death through necrosis and apoptosis [51, 52] and is toxic to both pathogens and host cells alike. While ENMs are most likely to activate DC toll-like receptors through uptake-independent interactions in the gut, it is possible that they may also activate ROS production in non-phagocytic epithelial cells [53]. ROS generation in non-phagocytes is scarcely elucidated in fish, but if viable, it would be yet another mechanism by which reactive intermediates might be produced by uptake-independent PRR activation in the gill epithelium.

Once ROS are present at sufficient levels in cells, propagation of inflammation to systemic levels of oxidative stress becomes more plausible. Intracellular nucleotide binding and oligomerization domain (NOD)-like receptors, the best-characterized of which is NLRP3 (for which fish orthologues exist), are found to be sensitive to intracellular DAMPs, including ROS [26, 54, 55]. The presence of ROS triggers the assembly of the inflammasome, a protein complex containing NLRP3 (or its orthologues in other species), a caspase-recruiting adaptor protein, and one or more caspases, which together are capable of processing the IL-1β precursor (pro-IL-1β) into the mature, secretable form of IL-1β. This major cytokine is then capable of systemic circulation and activation of inflammatory responses in all tissues. In fish acutely exposed to ZnO ENMs, markers of oxidative stress were observed in liver

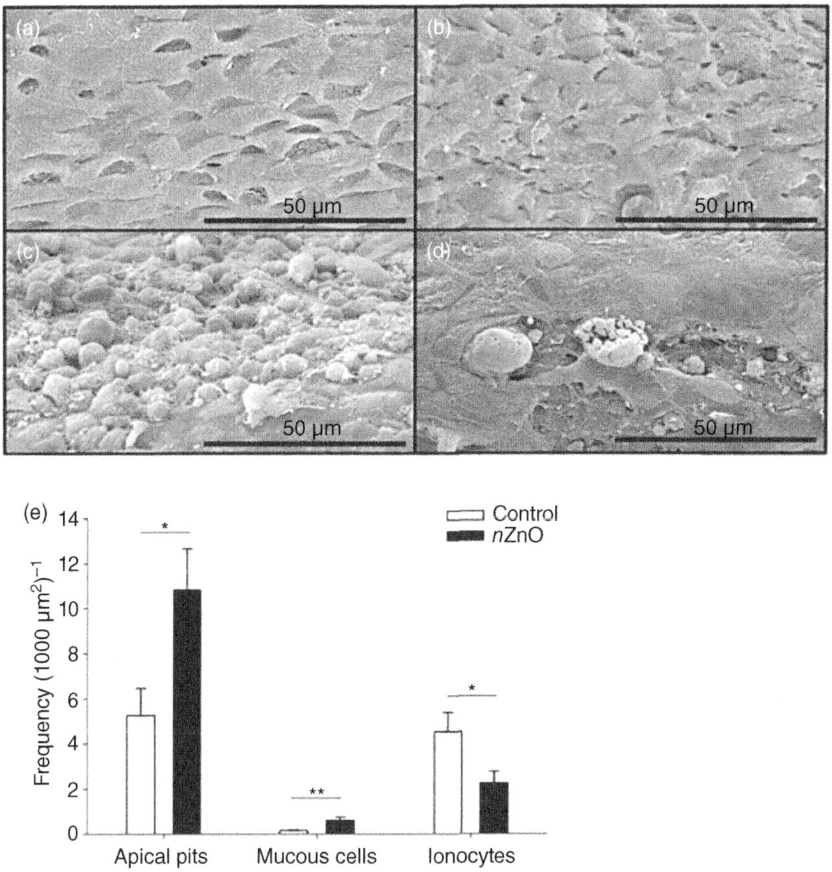

Figure 10.3 Scanning electron micrographs of gill filaments from white suckers (*Catostomus commersonii*) exposed to a sham (a) or 1 mg/L *n*ZnO suspension (b)–(d). Images show examples of ionocytes between pavement cells (a), apical pits (b), mucous globules (c), and mucous cells (d). Frequency of occurrence of apical pits, mucous cells, and ionocytes in the epithelium of the primary gill arch filament (e). Asterisks denote a significant difference between treatments (*$p < 0.05$, **$p < 0.01$). Source: Callaghan et al. [50]/with permission Elsevier.

tissue [56] in the absence of ENM uptake [22], suggesting that such mechanisms are likely relevant to ENM toxicity *in vivo*.

10.2.2 Uptake-Dependent Mechanisms

As ENMs accumulate at thin epithelial barriers, both natural and pathophysiological mechanisms may allow them to penetrate into the intracellular environment.

Once they have infiltrated cells at the epithelial layer, ENMs can cause intracellular damage within the epithelium and/or penetrate into the circulation, at which point they become bioavailable to other organs and tissues.

Endocytosis is one mechanism of ENM uptake that occurs predominantly at the gut epithelia but also at the gill. Gill endocytosis is poorly addressed in the literature, and it remains largely unclear why it occurs in pavement epithelial cells. Gills are not a site where significant uptake of micronutrients occurs, and most gases, ions, or other small molecules either diffuse passively across membranes or have access to pores, channels, or pumps that would preclude the need for vesicular processes. The notion of gill endocytosis is analogous to endocytosis by alveolar epithelial cells. Although of minor importance compared to the extensive phagocytosis carried out by macrophages patrolling the airspace of air-breathing vertebrates [57], alveolar epithelial cells do exhibit endocytosis to facilitate clearance of protein and mucus [58, 59] and assist in antigen presentation [60]. Without macrophages or immune cell access to the gill-water interface in fish, endocytosis may be more important in clearing mucus, protein, debris, and sampling the external environment for pathogenic threats than in air-breathing organisms. While the role of gill endocytosis may not be conclusively defined, many studies have pointed to vesicular processes as a major mechanism for ENM uptake at the gills [7, 61–64]. Similar vesicular mechanisms may also allow for ENMs to affect the gut epithelium and ancillary tissues. DCs, which extend dendrites into the gut lumen, can sample luminal contents and phagocytose antigens and microbes to present them to T cells within the gut associated lymphoid tissue and potentially activate adaptive immunity. M cells, discussed earlier because of their minimal or absent cell surface PRRs, also transport antigens in a similar manner; in contrast to DCs, M cells do not process their endocytosed contents extensively, but seemingly transcytose intact antigens or microbes to the gut associated lymphoid tissue for eventual capture and processing. Enterocytes may also endocytose ENMs as part of their natural uptake of gut contents. In mammals, micronutrients such as iron and vitamin B_{12} are internalized by receptor-mediated endocytosis [65]. Enterocytes, while not necessarily specialized immune cells, are also capable of transporting and presenting opsonized antigens across the luminal barrier via receptor-dependent uptake [66]. Although much of fish immunity is poorly defined compared with mammals, fish appear to have cells and/or functions analogous to those of M cells, gut DCs (though this is hotly debated at present), and intestinal epithelial cells capable of receptor-mediated and non-receptor-mediated uptake [31, 32, 67–69].

ENMs may not only penetrate cells via endocytic mechanisms but also "punch" their way into cells via damaging mechanisms. If ENMs locally generate sufficient levels of ROS, the integrity of cell membranes may be compromised, leading to formation of pores or holes through which ENMs can penetrate into cells [70, 71].

Similar mechanisms may also compromise basolateral membranes, allowing ENMs to access to the interstitium and ultimately the vasculature. Once ENMs have passed these barriers, there are multiple mechanisms by which they can induce toxicity.

10.2.2.1 Disruption of Ion Transporters by Intact Nanostructures and Ion Products and Physiological Regulation

The NKA is an ion antiporter ubiquitously expressed in the plasma membranes of all animal cells. This protein is largely responsible for establishing resting membrane potential in cells and energizing secondarily active membrane transport processes. In fish gill and gut tissue, NKA is critical for, among other things, ionoregulation and the maintenance of acid–base homeostasis, which are processes intimately linked with cardiorespiratory function. While the biophysical and topological basis of ENM-NKA interactions have yet to be elucidated, numerous studies in fish have found NKA activity to be either depressed or increased with exposure (Table 10.1). In acute exposures, gill NKA activity is decreased following exposure to TiO_2 ENMs in rainbow trout [14], Cu and Al ENMs in zebra fish [73, 77], and Ag ENMs in juvenile rainbow trout [49]. By contrast, gill NKA activity was increased in white suckers exposed to ZnO ENMs [22]. Goldfish exposed for periods of 4 days to metal oxide ENMs (or mixtures thereof) showed concentration-dependent increases of NKA activity; independent of the metal composition or mixture used, NKA activity increased with concentrations up to 80 mg/L, and then decreased thereafter [74].

The proposed mechanisms by which ENMs may affect NKA are not well-defined, and changes in NKA activity are not always associated with changes in plasma osmolality or ion concentrations. As previously discussed, some ENMs preferentially associate with proteins and alter their activity in synthetic membrane systems (Figure 10.2, AM Rundle, TJ MacCormack, MV Meli, unpublished), so it is possible that NKA activity is altered through a direct interaction with ENMs. ENM-induced lipid peroxidation may also alter gill permeability, affecting the demand for ion pumping and NKA activity [22]. A primary confounding factor is that the dissolution products of metallic ENMs (e.g. Ag^+, Al^{3+}, $Cu^{+/2+/3+/4+}$, $Fe^{2+/3+}$, Ti^{4+}, Zn^{2+}, etc.) may also impact NKA function, although a number of studies have provided evidence suggesting this may not be the only mechanism involved [76, 78]. It is also possible that respiratory stress linked to excessive mucous accumulation on the gills may secondarily influence NKA activity. Ventilation rate often increases in fish exposed to ENMs, and there is some evidence that ENM exposure may trigger a hypoxia response in fish [22]. Such responses could lead to disruptions in water and ion flux, as they are often associated with increased epithelial permeability in what is termed the osmorespiratory compromise [79].

Table 10.1 The effects of aquatic engineered nanomaterial (ENM) exposure on fish gill Na^+/K^+-ATP-dependent sodium–potassium pump (NKA) activity.

Species	ENM	Dose/exposure	Effect	References
Oncorhynchus mykiss	SWCNT ~1–2 nm diameter × 5–30 μm long	0.1–0.5 mg/L for 10 d	Dose-dependent changes in NKA activity	[13]
Oncorhynchus mykiss	Copper 87 nm	20 or 100 μg/L$^-$	Decreased NKA activity	[72]
Danio rerio	Copper 80 nm	0.25–1.5 mg/L$^-$	Decreased NKA activity	[73]
Carassius auratus	Copper 40 nm	40–160 mg/L for 4 d	Increased NKA activity	[74]
Catostomus commersonii	ZnO ENMs 25 nm	1.0 mg/L for 25 h	Increased NKA activity	[22]
Carassius auratus	ZnO 40 nm	40–160 mg/L for 4 d	Increased NKA activity	[74]
Oreochromis niloticus	ZnO 10–30 or 100 nm	1–10 mg/L for 7 or 14 d	Dose-, size-, and time-dependent changes in NKA activity	[75]
Danio rerio	Ag 20 or 110 nm, citrate stabilized	1 mg/L for 96 h	20 and 110 nm ENMs decreased NKA activity, but inhibition was greater with 20 nm ENMs	[76]
Oncorhynchus mykiss	Silver 18 ± 9 nm, citrate stabilized	1.0 mg/L for 48 h	Decreased NKA activity, Na^+ uptake	[49]
Carassius auratus	CeO_2 20 nm	80–160 mg/L for 4 d	Increased NKA activity	[74]
Oncorhynchus mykiss	TiO_2 21 nm	1 mg/L for 14 d	Decreased NKA activity	[14]

It is interesting to note that NKA is becoming increasingly linked to immunity and inflammation in its purported role as a signal transducer, serving to modulate immune cells and responses in nonimmune cells [80–82]. Since the immune system is central to animal responses to ENMs, it is conceivable that this system may play a role in the regulation of NKA activity following exposure. Alternatively, the regulation of NKA for immune prioritization may be what leads, in turn, to the ionoregulatory and osmoregulatory stress frequently observed in fish exposed to ENMs.

10.2.2.2 Activation of Systemic Immunity

As ENMs progress from a localized accumulation and deposition to systemic levels of infiltration, they can interact with additional components of the innate and adaptive immune systems, both of which are present in jawed fish and higher vertebrates [20]. Uptake-independent interactions with innate immune components (e.g. epithelial PRRs) can trigger clear immune reactions, but systemic responses are capable of inducing significant physiological stresses in their attempts to protect the host organism.

While ENMs are presumed capable of activating the immune system (i.e. antibody-secreting B cells and helper or cytotoxic T cells), ENM-adaptive immune interactions have received only preliminary attention and are poorly understood [23]. ENMs will adsorb immunoglobulins into their biomolecular coronae, but these interactions are presumed to be largely nonspecific, so the complementarity determining region loops would likely not bind to epitopes on the ENM structure itself. More specific interactions may arise with conjugated biomolecules or with other biomolecules adsorbed into the corona. Interactions between ENMs and the innate immune system are more well-studied and appreciated because the broad, nonspecific nature of innate immune components makes them more prone to interacting with ENMs.

10.3 Complement

While antibodies are the humoral components of the adaptive immune system capable of recognizing specific epitopes, complement and affiliated proteins are the humoral components of the innate immune system, instead recognizing PAMPs, DAMPs, and other motifs [17, 18]. As an effector mechanism of innate immunity, complement is present in all aquatic organisms. Complement can exert its effects through three widely known pathways: (i) the classical pathway, in which the C1 complement protein binds F_c region of an immunoglobulin already bound to an antigen or a microbial cell surface, (ii) the lectin pathway, in which the C2 and C4 complement proteins are processed by lectins bound to nonhost glycosidic motifs (e.g. mannosidic) on microbial cell surfaces and antigens, and (iii) the alternative pathway, in which basal levels of autoproteolyzed C3a complement protein binds directly to microbial cell surfaces. All three pathways are possible in jawed fish or higher aquatic organisms, as they have adaptive immune systems with immunoglobulins; C2 and C4 may bind to glycosidic motifs functionalized to the ENM surface or in the biomolecular corona, and C3a may bind to the ENM surface. C1 may bind to immunoglobulins that have in turn bound specifically to epitopes functionalized to the ENM surface, bound specifically to epitopes adsorbed within the biomolecular corona, or immunoglobulins themselves adsorbed into the biomolecular corona. Alternatively, all these complement proteins may themselves have been adsorbed to the biomolecular corona, initiating their cascades entirely by nonspecific protein-ENM interactions [23, 83–85].

Following these initial binding events, the three pathways of the complement cascade converge at the binding of C3b. In the earlier steps of some of the pathways and subsequent steps of the later pathways, proteolytic fragments of some of the complement proteins are released into circulation; the complement cascade is essentially a proteolytic cascade. C3a and C5a are proteolytic fragments, which function as chemoattractants and vasodilators to promote the migration of lymphocytes and delivery of acute phase proteins to the site of infection or injury [17, 18, 42]. Clearly these pathways may have important implications for cardiovascular function in fish exposed to ENMs, but data is not available to assess their toxicological relevance. In short, complement can be self-proliferating and sustaining, particularly with ENMs where there is a high degree of possible nonspecific protein interactions at the ENM surface or within the biomolecular corona [86]. The continued assembly of the complement complex leads to two broad outcomes: (i) the formation of the membrane attack complex, which bores holes into microbial cells, and (ii) the formation of ligands for phagocytic receptors (this will be further elaborated upon in the following sections).

10.4 Phagocytosis

One of the main mechanisms by which ENMs are able to penetrate epithelial barriers is by endocytosis. DCs and M cells, should they be present in fish guts, are more specialized in the uptake and processing of foreign invaders. Beyond the epithelial barriers of fish are additional cells focused on the uptake, destruction, processing, and presentation of microbes, antigens, and contaminants. These include circulating neutrophils and resident macrophages in tissues. These phagocytes internalize foreign bodies and traffic them to specialized vesicles known as phagolysosomes via receptor- and non-receptor-mediated mechanisms. By combining acidification via ATP-dependent proton pumps and the generation of reactive intermediates via phagocyte oxidase and nitric oxide synthase (the oxidative burst), phagolysosomal contents may be destroyed and degraded to small, presentable molecules used to activate the adaptive immune system [17, 18]. All fish have the necessary innate immune cells to phagocytose ENMs [64, 86], and in goldfish, ENMs have been shown to alter neutrophil viability and function *in vitro* [87]. *In vivo*, ENM exposure has been shown to induce phagocyte activation, including the upregulation of the oxidative burst machinery, secretion of pro-inflammatory cytokines, and proliferation of phagocytes in fathead minnow [88].

While some ENM compositions are susceptible to phagolysosomal degradation, other inert or nonreactive materials are more resistant [70, 89, 90]. As phagocytes pump protons and reactive species into their phagolysosomes, frustrated phagocytosis may develop, whereby ENMs act as a sponge, and harmful species accumulate within the phagolysosome to the point of phagocytic necrosis, allowing for escape of the unperturbed ENM, protons and ROS, and DAMPs. This can stimulate additional inflammation and oxidative stress, so it is conceivable that ENMs labeled as "unreactive" may in fact be more harmful to aquatic organisms than those that are readily degradable in biological environments (e.g. metal oxide ENMs; [64]).

10.5 Conclusions and Ecological Perspectives

The previous discussion outlined potential mechanisms of ENM toxicity, each with some relevance to cardiorespiratory function in fish. Cardiorespiratory capacity is a key determinant of scope for survival in fish, and in some situations cardiac failure is considered the limiting factor in environmental stress tolerance [91]. Through multiple pathways, the oxidative stress often linked to ENM exposure in fish may impact vascular tone and alter blood pressure, necessitating cardiorespiratory adjustments to maintain homeostasis in convection and exchange processes. Gas and ion exchange activities at the gill both rely on and regulate convective transport in the circulatory system, so any toxicant affecting the gill will necessarily impact the heart to some degree. The increase in energy production to fuel immune and osmoregulatory processes will also increase basal nutrient and oxygen uptake requirements, both of which can be negatively impacted by ENM exposure. In the absence of an increase in maximum metabolic rate, the concomitant reduction in aerobic scope could decrease the animal's chances of surviving common environmental stressors such as hypoxia or high temperatures.

ENMs may also impact cardiorespiratory physiology via a disparate mechanism. In studies on adult white sucker, we have shown that acute exposure to ZnO

ENMs temporarily reduces heart rate [22]. This was not seen with CeO_2 ENM exposure [92], but similar responses have been observed in other fish species exposed to Ag [93, 94] and SiO_2 ENMs [95]. Oxidative tissue damage and the stimulation of gill neuroepithelial cells or a reduction in acetylcholinesterase activity may be increasing acetylcholine levels at the cardiac pacemaker, slowing the heart [50]. White suckers exposed to ZnO ENMs exhibit a mild but significant reduction in aerobic scope, suggesting that the cardiorespiratory response to ENMs was large enough to impact the animal's scope for survival.

The toxicity of common ENM formulations is often considered modest relative to chemical toxins (e.g. pesticides), but realistic exposure scenarios are still not well characterized, so conclusions regarding long-term impacts on aquatic organisms should be made cautiously. We know that ENM exposure can sensitize fish to hypoxia [16] and reduce aerobic scope, but few studies have examined the physiological or ecological consequences of ENM toxicity, particularly in chronic exposure situations. The cardiorespiratory system will likely act as the nexus between the cellular toxicity of ENMs and their ecological consequences. Studies incorporating direct measurements of cardiac function or proxies of cardiorespiratory capacity (e.g. aerobic scope or swimming performance) will provide valuable context for cellular and biochemical nanotoxicology studies, as they have for traditional chemical toxicants [96]. As climate change increases the frequency and severity of environmental stressors and ENM use continues to expand, the true significance of aquatic nanotoxicity will undoubtedly become clear.

References

1 Petosa, A.R., Jaisi, D.P., Quevedo, I.R. et al. (2010). Aggregation and deposition of engineered nanomaterials in aquatic environments: role of physicochemical interactions. *Environ. Sci. Technol.* 44: 6532–6549.

2 Monopoli, M.P., Åberg, C., Salvati, A., and Dawson, K.A. (2012). Biomolecular coronas provide the biological identity of nanosized materials. *Nat. Nanotechnol.* 7: 779–786.

3 Khot, L.R., Sankaran, S., Maja, J.M. et al. (2012). Applications of nanomaterials in agricultural production and crop protection: a review. *Crop Prot.* 35: 64–70.

4 Fu, F., Dionysios, D.D., and Liu, H. (2014). The use of zero-valent iron for groundwater remediation and wastewater treatment: a review. *J. Hazard Mat.* 267: 194–205.

5 Duffin, R., Mills, N.L., and Donalson, K. (2007). Nanoparticles – a thoracic toxicology perspective. *Yonsei Med. J.* 48: 561–572.

6 Gamperl, A.K. and Driedzic, W.R. (2009). Cardiovascular function and cardiac metabolism during environmental hypoxia. In: *Fish Physiology*, vol. 27 (ed. J.G. Richards, A.P. Farrell and C.J. Brauner), 301–360. London: Elsevier.
7 Shaw, B.J. and Handy, R.D. (2011). Physiological effects of nanoparticles on fish: a comparison of nanometals versus metal ions. *Environ. Int.* 37: 1083–1097.
8 MacCormack, T.J., Goss, G.G., and Handy, R.D. (2014). Emerging threats: engineered organic nanomaterials. In: *Fish Physiology: Organic Chemical Toxicology of Fishes*, vol. 33 (ed. K. Tierney, A.P. Farrell and C.J. Brauner), 436–479. London: Elsevier.
9 Bannunah, A.M., Vllasaliu, D., Lord, J., and Stolnik, S. (2014). Mechanisms of nanoparticle internalization and transport across an intestinal epithelial cell model: effect of size and surface charge. *Mol. Pharm.* 11: 4363–4373.
10 Yameen, B., Choi, W.I., Vilos, C. et al. (2014). Insight into nanoparticle cellular uptake and intracellular targeting. *J. Control. Release* 190: 485–499.
11 Kaya, H., Aydın, F., Gürkan, M. et al. (2015). Effects of zinc oxide nanoparticles on bioaccumulation and oxidative stress in different organs of tilapia (*Oreochromis niloticus*). *Environ. Toxicol. Pharmacol.* 40: 936–947.
12 MacCormack, T.J., Rundle, A.M., Malek, M. et al. (2017). Gold nanoparticles partition to and increase the activity of glucose-6-phosphatase in a synthetic phospholipid membrane system. *PLoS One* 12 (8): e0183274.
13 Smith, C.J., Shaw, B.J., and Handy, R.D. (2007). Toxicity of single walled carbon nanotubes to rainbow trout, (*Oncorhynchus mykiss*): respiratory toxicity, organ pathologies, and other physiological effects. *Aquat. Toxicol.* 82: 94–109.
14 Federici, G., Shaw, B.J., and Handy, R.D. (2007). Toxicity of titanium dioxide nanoparticles to rainbow trout (*Oncorhynchus mykiss*): gill injury, oxidative stress, and other physiological effects. *Aquat. Toxicol.* 84: 415–430.
15 Lee, J.W., Kim, J.E., Shin, Y.J. et al. (2014). Serum and ultrastructure responses of common carp (*Cyprinus carpio* L.) during long-term exposure to zinc oxide nanoparticles. *Ecotoxicol. Environ. Saf.* 104: 9–17.
16 Bilberg, K., Malte, H., Wang, T., and Baatrup, E. (2010). Silver nanoparticles and silver nitrate cause respiratory stress in Eurasian perch (*Perca fluviatilis*). *Aquat. Toxicol.* 96: 159–165.
17 Tort, L., Balasch, J.C., and Mackenzie, S. (2003). Fish immune system. A crossroads between innate and adaptive responses. *Immunologia* 22: 277–286.
18 Magnadóttir, B. (2006). Innate immunity of fish (overview). *Fish Shellfish Immunol.* 20: 137–151.
19 Uribe, C., Folch, H., Enriquez, R., and Morgan, G. (2011). Innate and adaptive immunity in teleost fish: a review. *Vet. Med.-Czech* 56: 486–503.
20 Eddy, F.B. and Handy, R.D. (2012). *Ecological and Environmental Physiology of Fishes*. Oxford: Oxford University Press.

21 Morimoto, Y., Izumi, H., and Kuroda, E. (2014). Significance of persistent inflammation in respiratory disorders induced by nanoparticles. *J. Immunol. Res.* 2014: 962871.

22 Bessemer, R.A., Butler, K.M.A., Tunnah, L. et al. (2015). Cardiorespiratory toxicity of environmentally relevant zinc oxide nanoparticles in the freshwater fish *Catostomus commersonii*. *Nanotoxicology* 9: 861–870.

23 Hussain, S., Vanoirbeek, J.A.J., and Hoet, P.H.M. (2012). Interactions of nanomaterials with the immune system. *WIREs Nanomed. Nanobiotechnol.* 4: 169–183.

24 Mathieu, C., Rioux, G., Dumas, M.C., and Leclerc, D. (2013). Induction of innate immunity in lungs with virus-like nanoparticles leads to protection against influenza and *Streptococcus pneumoniae* challenge. *Nanomedicine* 9: 839–848.

25 Smith, D.M., Simon, J.K., and Baker, J.R. Jr., (2013). Applications of nanotechnology for immunology. *Nat. Rev. Immunol.* 13: 592–605.

26 Bryant, C.E. and Monie, T.P. (2012). Mice, men and the relatives: cross-species studies underpin innate immunity. *Open Biol.* 2: 120015.

27 Rachmawati, D., Bontkes, H.J., Verstege, M.I. et al. (2013). Transition metal sensing by toll-like receptor-4: next to nickel, cobalt and palladium are potent human dendritic cell stimulators. *Contact Derm.* 68: 331–338.

28 Luo, Y.H., Chang, L.W., and Pinpin, L. (2015). Metal-based nanoparticles and the immune system: activation, inflammation, and potential applications. *Biomed Res. Int.* 2015: 143720.

29 Petrarca, C., Clemente, E., Amato, V. et al. (2015). Engineered metal based nanoparticles and innate immunity. *Clin. Mol. Allergy.* 13: 13.

30 Pinsino, A., Russo, R., Bonaventura, R. et al. (2015). Titanium dioxide nanoparticles stimulate sea urchin immune cell phagocytic activity involving TLR/p38 MAPK-mediated signalling pathway. *Sci. Rep.* 5: 14492.

31 Bassity, E. and Clark, T.G. (2012). Functional identification of dendritic cells in the teleost model, rainbow trout (*Oncorhynchus mykiss*). *PLoS One* 7: e33196.

32 Parra, D., Reyes-Lopez, F.E., and Tort, L. (2015). Mucosal immunity and B cells in teleosts: effect of vaccination and stress. *Front. Immunol.* 6: 354.

33 Bals, R. and Hiemstra, P.S. (2004). Innate immunity in the lung: how epithelial cells fight against respiratory pathogens. *Eur. Respir. J.* 23: 327–333.

34 Marques, R. and Boneca, I.G. (2011). Expression and functional importance of innate immune receptors by intestinal epithelial cells. *Cell. Mol. Life Sci.* 68: 3661–3673.

35 Aoki, M.P., Carrera-Silva, E.A., Cuervo, H. et al. (2012). Nonimmune cells contribute to crosstalk between immune cells and inflammatory mediators in the innate response to *Trypanosoma cruzi* infection. *J. Parasitol. Res.* 2012: 737324.

36 Ishii, A., Matsuo, A., Sawa, H. et al. (2007). Lamprey TLRs with properties distinct from those of the variable lymphocyte receptors. *J. Immunol.* 178: 397–406.

37 Haugarvoll, E., Bjerkås, I., Nowak, B.F. et al. (2008). Identification and characterization of a novel intraepithelial lymphoid tissue in the gills of *Atlantic salmon*. *J. Anat.* 213: 202–209.

38 Zhao, F., Li, Y.W., Pan, H.J. et al. (2013). Expression profiles of toll-like receptors in channel catfish (*Ictalurus punctatus*) after infection with *Ichthyophthirius multifiliis*. *Fish Shellfish Immunol.* 35: 993–997.

39 Basu, M., Paichha, M., Swain, B. et al. (2015). Modulation of TLR2, TLR4, TLR5, NOD1 and NOD2 receptor gene expressions and their downstream signaling molecules following thermal stress in the Indian major carp catla (*Catla catla*). *3 Biotech* 5: 1021–1030.

40 Poynter, S., Lisser, G., Monjo, A., and DeWitte-Orr, S. (2015). Sensors of infection: viral nucleic acid PRRs in fish. *Biology (Basel)* 4: 460–493.

41 Esteban, M.A. (2012). An overview of the immunological defenses in fish skin. *ISRN Immunol.* 2012: 853470.

42 Reyes-Cerpa, S., Maisey, K., Reyes-López, F. et al. (2012). Fish cytokines and immune response. In: *New Advances and Contributions to Fish Biology* (ed. H. Turker), 3–58. Rijeka: InTech.

43 Nel, A.E., Xia, T., Mädler, L., and Li, N. (2006). Toxic potential of materials at the nanolevel. *Science* 311: 622–627.

44 Fu, P.P., Xia, Q., Hwang, H.M. et al. (2014). Mechanisms of nanotoxicity: generation of reactive oxygen species. *J. Food Drug Anal.* 22: 64–75.

45 Foucaud, L., Wilson, M.R., Brown, D.M., and Stone, V. (2007). Measurement of reactive species production by nanoparticles prepared in biologically relevant media. *Toxicol. Lett.* 174: 1–9.

46 Rogers, E.J., Hsieh, S.F., Organti, N. et al. (2008). A high throughput in vitro analytical approach to screen for oxidative stress potential exerted by nanomaterials using a biologically relevant matrix: human blood serum. *Toxicol. In Vitro* 22: 1639–1647.

47 Luna-Velasco, A., Field, J.A., Cobo-Curiel, A., and Sierra-Alvarez, R. (2011). Inorganic nanoparticles enhance the production of reactive oxygen species (ROS) during the autoxidation of L-3,4-dihydroxyphenylalanine (L-dopa). *Chemosphere* 85: 19–25.

48 Sun, W., Luna-Velasco, A., Sierrra-Alvarez, R., and Field, J.A. (2013). Assessing protein oxidation by inorganic nanoparticles with enzyme-linked immunosorbent assay (ELISA). *Biotechnol. Bioeng.* 110: 694–701.

49 Schultz, A.G., Ong, K.J., MacCormack, T.J. et al. (2012). Silver nanoparticles inhibit sodium uptake in juvenile rainbow trout (*Oncorhynchus mykiss*). *Environ. Sci. Technol.* 46: 10295–10301.

50 Callaghan, N.I., Allen, G.J., Robart, T.E. et al. (2016). Zinc oxide nanoparticles trigger cardiorespiratory stress and reduce aerobic scope in the white sucker, *Catostomus commersonii. NanoImpact* 2: 29–37.
51 Cookson, M.R., Ince, P.G., and Shaw, P.J. (1998). Peroxynitrite and hydrogen peroxide induced cell death in the NSC34 neuroblastoma x spinal cord cell line: role of poly (ADP-ribose) polymerase. *J. Neurochem.* 70: 501–508.
52 He, X., Aker, W.G., Leszczynski, J., and Hwang, H.M. (2014). Using a holistic approach to assess the impact of engineered nanomaterials inducing toxicity in aquatic systems. *J. Food Drug Anal.* 22: 128–146.
53 Boncompain, G., Schneider, B., Delevoye, C. et al. (2010). Production of reactive oxygen species is turned on and rapidly shut down in epithelial cells infected with *Chlamydia trachomatis. Infect Immun.* 78: 80–87.
54 Angosto, D. and Mulero, V. (2014). The zebrafish as a model to study the inflammasome. *Inflammasome* 1: 27–29.
55 Abais, J.M., Xia, M., Zhang, Y. et al. (2015). Redox regulation of NLRP3 inflammasomes: ROS as trigger or effector? *Antioxid. Redox Signal.* 22: 1111–1129.
56 Dieni, C.A., Callaghan, N.I., Gormley, P.T. et al. (2014). Physiological hepatic response to zinc oxide nanoparticle exposure in the white sucker, *Catostomus commersonii. Comp. Biochem. Physiol. Part - C: Toxicol. Pharmacol.* 162: 51–61.
57 Hussell, T. and Bell, T.J. (2014). Alveolar macrophages: plasticity in a tissue-specific context. *Nat. Rev. Immunol.* 14: 81–93.
58 Hastings, R.H., Wright, J.R., Albertine, K.H. et al. (1994). Effect of endocytosis inhibitors on alveolar clearance of albumin, immunoglobulin G, and SP-A in rabbits. *Am. J. Physiol.* 266: L544–L552.
59 Hastings, R.H., Folkesson, H.G., and Matthay, M.A. (2004). Mechanisms of alveolar protein clearance in the intact lung. *Am. J. Physiol.* 286: L679–L789.
60 Harriff, M.J., Cansler, M.E., Toren, K.G. et al. (2014). Human lung epithelial cells contain *Mycobacterium tuberculosis* in a late endosomal vacuole and are efficiently recognized by $CD8^+$ T cells. *PLoS One* 9: e97515.
61 Handy, R.D., von der Kammer, F., Lead, J.R. et al. (2008). The ecotoxicology and chemistry of manufactured nanoparticles. *Ecotoxicology* 17: 287–314.
62 Scown, T.M., Santos, E.M., Johnston, B.D. et al. (2010). Effects of aqueous exposure to silver nanoparticles of different sizes in rainbow trout. *Toxicol. Sci.* 115: 521–534.
63 Farkas, J., Christian, P., Gallego-Urrea, J.A. et al. (2011). Uptake and effects of manufactured silver nanoparticles in rainbow trout (*Oncorhynchus mykiss*) gill cells. *Aquat. Toxicol.* 101: 117–125.
64 Handy, R.D., Cornelis, G., Fernandes, T. et al. (2012). Ecotoxicity test methods for engineered nanomaterials: practical experiences and recommendations from the bench. *Environ. Toxicol. Chem.* 31: 15–31.

65 Tuma, P. and Hubbard, A.L. (2003). Transcytosis: crossing cellular barriers. *Physiol. Rev.* 83: 871–932.

66 Miron, N. and Cristea, V. (2012). Enterocytes: active cells in tolerance to food and microbial antigens in the gut. *Clin. Exp. Immunol.* 167: 405–412.

67 Lugo-Villarino, G., Balla, K.M., Stachura, D.L. et al. (2010). Identification of dendritic antigen-presenting cells in the zebrafish. *Proc. Natl. Acad. Sci. U.S.A.* 107: 15850–15855.

68 Gomez, D., Sunyer, J.O., and Salinas, I. (2013). The mucosal immune system of fish: the evolution of tolerating commensals while fighting pathogens. *Fish Shellfish Immunol.* 35: 1729–1739.

69 Rombout, J.H., Yang, G., and Kiron, V. (2014). Adaptive immune responses at mucosal surfaces of teleost fish. *Fish Shellfish Immunol.* 40: 633–643.

70 Nel, A.E., Mädler, L., Velegol, D. et al. (2009). Understanding biophysicochemical interactions at the nano-bio interface. *Nat. Mater.* 8: 543–557.

71 Chou, L.Y., Ming, K., and Chan, W.C. (2011). Strategies for the intracellular delivery of nanoparticles. *Chem. Rev. Soc.* 40: 233–245.

72 Shaw, B.J., Al-Bairuty, G.A., and Handy, R.D. (2012). Effects of waterborne copper nanoparticles and copper sulphate on rainbow trout, (*Oncorhynchus mykiss*): physiology and accumulation. *Aquatic. Toxicol.* 116–117: 90–101.

73 Griffitt, R.J., Weil, R., Hyndman, K.A. et al. (2007). Exposure to copper nanoparticles causes gill injury and acute lethality in zebrafish (*Danio rerio*). *Environ. Sci. Technol.* 41: 8178–8186.

74 Xia, J., Zhao, H.Z., and Lu, G.H. (2013). Effects of selected metal oxide nanoparticles on multiple biomarkers in *Carassius auratus. Biomed. Environ. Sci.* 9: 742–749.

75 Kaya, H., Aydın, F., Gürkan, M. et al. (2016). A comparative toxicity study between small and large size zinc oxide nanoparticles in tilapia (*Oreochromis niloticus*): organ pathologies, osmoregulatory responses and immunological parameters. *Chemosphere* 144: 571–582.

76 Osborne, O.J., Lin, S., Chang, C.H. et al. (2015). Organ-specific and size-dependent Ag nanoparticle toxicity in gills and intestines of adult zebrafish. *ACS Nano* 9: 9573–9584.

77 Griffitt, R.J., Feswick, A., Weil, R. et al. (2011). Investigation of acute nanoparticulate aluminum toxicity in zebrafish. *Environ. Toxicol.* 26: 541–551.

78 Griffitt, R.J., Hyndman, K., Denslow, N.D., and Barber, D.S. (2009). Comparison of molecular and histological changes in zebrafish gills exposed to metallic nanoparticles. *Toxicol. Sci.* 107: 404–415.

79 Robertson, L.M., Val, A.L., Almeida-Val, V.M.F., and Wood, C.M. (2015). Ionoregulatory aspects of the osmorespiratory compromise during acute environmental hypoxia in 12 tropical and temperate teleosts. *Physiol. Biochem. Zool* 88: 357–370.

80 Aizman, O. and Aperia, A. (2003). Na,K-ATPase as a signal transducer. *Ann. N. Y. Acad. Sci.* 986: 489–496.
81 Rodrigues-Mascarenhas, S., Da Silva de Oliveira, A., Amoedo, N.D. et al. (2009). Modulation of the immune system by ouabain. *Ann. N. Y. Acad. Sci.* 1153: 153–163.
82 Kinoshita, P.F., Yshii, L.M., Vasconcelos, A.R. et al. (2014). Signaling function of Na,K-ATPase induced by ouabain against LPS as an inflammation model in hippocampus. *J. Neuroinflammation* 11: 218.
83 Tenzer, S., Docter, D., Rosfa, S. et al. (2011). Nanoparticle size is a critical physicochemical determinant of the human blood plasma corona: a comprehensive quantitative proteomic analysis. *ACS Nano* 5: 7155–7167.
84 Docter, D., Distler, U., Storck, W. et al. (2014). Quantitative profiling of the protein coronas that form around nanoparticles. *Nat. Protoc.* 9: 2030–2044.
85 Kononenko, V., Narat, M., and Drobne, D. (2015). Nanoparticle interaction with the immune system. *Arh Hig Rada Toksikol* 66: 97–108.
86 Ji, J., Torrealba, D., Ruyra, À., and Roher, N. (2015). Nanodelivery systems as new tools for immunostimulant or vaccine administration: targeting the fish immune system. *Biology (Basel)* 4: 664–696.
87 Ortega, V.A., Katzenback, B.A., Stafford, J.L. et al. (2015). Effects of polymer-coated metal oxide nanoparticles on goldfish (*Carassius auratus* L.) neutrophil viability and function. *Nanotoxicology* 9: 23–33.
88 Jovanović, B., Anastasova, L., Rowe, E.W. et al. (2011). Effects of nanosized titanium dioxide on innate immune system of fathead minnow (*Pimephales promelas Rafinesque*, 1820). *Ecotoxicol. Environ. Saf.* 74: 675–683.
89 Stone, V., Johnston, H., and Schins, R.P. (2009). Development of *in vitro* systems for nanotoxicology: methodological considerations. *Crit. Rev. Toxicol.* 39: 613–626.
90 Manke, A., Wang, L., and Rojanasakul, Y. (2013). Mechanisms of nanoparticle-induced oxidative stress and toxicity. *Biomed. Res. Int.* 2013: 942916.
91 Pörtner, H.O. and Farrell, A.P. (2008). Physiology and climate change. *Science* 322: 690–692.
92 Rundle, A., Robertson, A.B., Blay, A.M. et al. (2015). Cerium oxide nanoparticles exhibit minimal cardiac and cytotoxicity in the freshwater fish *Catostomus commersonii. Comp. Biochem. Physiol. Part - C: Toxicol. Pharmacol.* 181–182: 19–26.
93 Asharani, P.V., Wu, Y.L., Gong, Z., and Valiyaveettil, S. (2008). Toxicity of silver nanoparticles in zebrafish model systems. *Nanomedicine* 19: 1–8.
94 Cho, J., Kim, K., Ryo, T. et al. (2013). Stepwise embryonic toxicity of silver nanoparticles on *Oryzias latipes*. *BioMed Res. Int.* 2013: 494671.

95 Duan, J., Yu, Y., Li, Y. et al. (2013). Cardiovascular toxicity evaluation of silica nanoparticles in endothelial cells and zebrafish model. *Biomaterials* 34: 5853–5862.

96 Gerger, C.J. and Weber, L.P. (2015). Comparison of the acute effects of benzo-a-pyrene on adult zebrafish (*Danio rerio*) cardiorespiratory function following intraperitoneal injection versus aqueous exposure. *Aquat. Toxicol.* 165: 19–30.

Section III

Human Systems

11

Air Pollution and Neurodevelopmental Disorders

Joshua L. Allen[1], Carolyn Klocke[2], Keith Morris-Schaffer[3], Katherine Conrad[4], Marissa Sobolewski[4], and Deborah A. Cory-Slechta[4,5]

[1] *Sarepta, Columbus, OH, USA*
[2] *Bausch Health, Rochester, NY, USA*
[3] *Exponent, Sacramento, CA, USA*
[4] *Department of Environmental Medicine, University of Rochester Medical Center, Rochester, NY 14642, USA*
[5] *University of Rochester Medical Center, Box EHSC, Department of Environmental Medicine, Rochester NY 14642, USA*

11.1 Air Pollution and the Brain

Air pollution is a complex mixture of particles, gases, trace metals, and adsorbed organic contaminants. Particulate sizes range from coarse (<10 μm) to fine (<2.5 μm) to ultrafine (<0.1 μm). While not a significant component of air pollution by mass, ultrafine particles (UFPs) achieve orders of magnitude higher particle count concentrations and surface area to volume ratios compared with larger particle sizes. The high surface area to volume ratios of UFPs allows for great adsorption of other toxic air pollutants such as oxidant gases (e.g. O_3, NO_x), organic compounds, and transition metals per unit mass. Levels of up to 50% of inhaled UFP are deposited in pulmonary alveolar regions of the lung from where they can traverse the alveolocapillary barrier to access pulmonary interstitium. From there, particles can cross endothelial cells into blood circulation to subsequently impact other organs, including the heart and the brain, and thereby lead to more serious health consequences. These properties of UFPs, coupled with data indicating deposition that can also occur in the nasal cavity allowing translocation to the brain directly across the olfactory epithelium [1–3], are why they are generally considered among the most reactive elements of air pollution [4, 5].

Air pollution is a worldwide environmental health problem, to which exposures begin in gestation and are cumulative over the life span. Considered in a global context, new studies ascribe 3.3 million premature deaths per year to outdoor air

Toxicology of Nanoparticles and Nanomaterials in Human, Terrestrial and Aquatic Systems, First Edition. Edited by Marc A. Williams.

pollution [6], and as of 2013, air pollution was listed as the 12th leading global risk factor for disability adjusted life year reductions [7]. While air quality standards have clearly improved in many countries, even as of 2012, the U.S. Census Bureau estimated that approximately 142 million U.S. residents live in counties where regulated levels of various air pollutants exceeded standards set by the US Environmental Protection Agency (EPA). Further, ongoing urbanization trends and expanding road traffic in many areas of the world are predicted to further increase population exposures to UFP air pollution [12]. Given these trends, a clear potential exists for ambient particulate matter-induced neurotoxicity. Currently, the US EPA sets regulatory standards for levels of PM_{10} and $PM_{2.5}$. In addition to the fact that it represents a significant technological challenge, regulation of UFPs is said to await additional evidence of their toxicity.

11.1.1 The Brain as a Target of Air Pollution

The focus of air pollution research has long centered on cardiovascular and pulmonary systems, yet increasing evidence now indicates that air pollution also adversely impacts the brain. Mechanisms thought to induce air pollution toxicity in the peripheral organs, particularly inflammation and oxidative stress, are also likely operative in the central nervous system (CNS) [9]. Indeed, human studies have reported associations of various components of air pollution with neurodevelopmental disorders and with neurodegenerative diseases [8–11].

In children, for example, various components of air pollution have been associated with multiple different neurodevelopmental conditions. These include autism spectrum disorder (ASD) [13–23], schizophrenia [24, 25], impaired cognitive abilities [26–29], attention deficit hyperactivity disorder (ADHD) and deficits in attention-related behaviors and behavioral regulation [28, 30–32], reduced mental development index and IQ scores [33], symptoms of anxiety/depression [34, 35], decreased nonverbal reasoning ability [36], delayed psychomotor development [37], and even altered brain connectivity [38].

Numerous studies also point to the vulnerability of adult brain to air pollution exposures. Several epidemiological studies have reported associations between traffic-related air pollution exposures with Alzheimer's disease [10, 39, 40] and with various impairments in cognitive function [41–43]. Some, but not all [44], studies have associated exposures to air pollution with Parkinson's disease [45]. In addition, loss of white matter in older women was reported in association with exposure to $PM_{2.5}$ [46].

Animal models confirm behavioral and CNS toxicity in response to gestational, postnatal, and adult exposures to air pollution and/or its components [47], including findings of depressive-like behaviors, impaired spatial learning and memory, reduced apical dendritic spine density and dendritic branching in hippocampus [48], reduced numbers of hippocampal neuronal GluA1 glutamate receptor subunits, altered neuronal differentiation of cortical neurons [49], altered locomotor activity and levels of brain catecholamines [50], and deficits in spatial learning and memory and mitochondrial dysfunction [51]. Animal studies also reveal the development in the brain of oxidative stress and inflammatory responses to inhaled nanoparticulate matter [52] or to diesel exhaust [8, 53, 54] as well as altered blood–brain barrier permeability [55].

11.2 Air Pollution and Neurodevelopmental Disorders

As noted earlier, air pollution exposures in children have been associated with neurodevelopmental disorders, including ASD [13–23], schizophrenia [24, 25], and ADHD and related behaviors [28, 30–32].

ASD has now reached a worldwide prevalence as high as 1–2% [56, 57], with a clear bias toward males being more affected [57–60]. Diagnosis of ASD, a highly heterogeneous disorder, relies on the identification of impairments of its core behavioral features, social deficits, impaired communication, and repetitive/perseverative behaviors early in life. While the etiology of this highly intractable and extremely heterogeneous neurodevelopmental disorder remains unknown, numerous epidemiological studies of populations from different states within the United States and from different countries now report associations of ASD diagnosis with various air pollution components or surrogates, including residential distance from the freeway [20], ambient levels of metals [19, 22], aromatic solvents [16, 19, 22], styrene and chromium [61], gases [13, 15, 21], diesel exhaust [19, 22], and particulate matter ($PM_{2.5}$ and/or PM_{10}) [13, 15, 17, 18, 21]. Some report increased odds ratios for ASD diagnoses in relation to levels of $PM_{2.5}$ [15, 18, 21], which includes UFPs and/or NO_2, a surrogate for UFP [14]. One exception to such findings was a recent study that combined data across four cohorts, but given that each cohort used different measures of autistic traits, power and comparability may have been limited [37].

Schizophrenia, another male predominant neurodevelopmental disorder [62], has been estimated to have a prevalence of up to 1% in the population [63]. Like ASD, it is highly heterogenous and etiologically is thought to arise from genetic and/or environmental factors [62]. Its core features include delusions, hallucinations, impaired motivation, and social withdrawal as well as evidence of cognitive impairment. Although far less investigated than ASD, two studies have linked markers of air pollution to schizophrenia. Yackerson et al. [25] reported an effect of solid air-suspended particles with numbers of suicide attempts and exacerbation of psychotic episodes, with effects dependent upon prevailing wind directions, suggesting interaction effects. Pedersen et al. [24] found after adjustment for degree of urbanization that children living with the highest level of traffic had a risk for schizophrenia of 4.40 (95% CI: 1.25–15.50) as compared with children living with the lowest level of traffic.

ADHD is a neurodevelopmental disorder estimated to affect anywhere from 2% to 10% of school-aged children worldwide [64]. As recently revised in terms of diagnostic criteria, ADHD is a chronic condition characterized by a persistent pattern of inattention and/or hyperactivity and impulsivity that results in functional impairments in multiple settings. Two categories for ADHD symptoms are recognized, i.e. inattention and hyperactivity–impulsive behaviors. Although considered highly heritable, prenatal and perinatal risk factors, including environmental exposures, have been recognized [65]. Suggestive evidence of an involvement of air pollution with ADHD has also been reported. For example, Siddique et al. [32] found that levels of PM_{10} were associated with a significantly increased odds ratio for ADHD (OR = 2.07) in school-aged children in India. Newman et al. [31] found that exposure to air pollution during infancy, measured using a surrogate of traffic-related air pollution (elemental carbon attributed to traffic), was associated with higher hyperactivity scores in children at 7 years of age. Using black carbon as a marker of traffic particles, Chiu et al. [30] found associations with commission errors and slower reaction times that were more apparent in boys, consistent with the higher prevalence of ADHD in males. In contrast, measures of traffic-related air pollution in a population of Swedish twins born during 1992–2000 were not associated with ADHD, or with ASD [66].

11.2.1 Shared Co-morbidities of Neurodevelopmental Disorders

Of particular note is that although such conditions as ASD, schizophrenia, and ADHD have unique features, these neurodevelopmental disorders are often co-morbid and also share many features [67–72]. Schizophrenia, for example, particularly early-onset childhood schizophrenia, was for many years considered to share the same spectrum with autism, based on overlaps in features [73–76], with ASD considered an early manifestation of schizophrenia. ASD and schizophrenia can occur as co-morbidities, and behaviorally, both include deficits in social interactions, cognition, executive functions, emotional processing, and sensorimotor gating. Both can involve auditory and visual hallucinations [75, 77], abnormalities of serotonergic, dopaminergic and glutamatergic systems [78–80], and shared genetic causes [81, 82]. Schizophrenia and ASD also share neuropathological features, including ventriculomegaly (VM: enlarged lateral ventricles) [83–85], a clinical indicator of white matter alterations that are crucial to brain interhemispheric connectivity.

ADHD is reported at high rates in children with ASD [86]. Both disorders exhibit high heritability and some shared genetic factors [87]. ASD and ADHD can also share symptoms and features. Both involve alterations in sensory processing and in motor function, as well as impulsivity; both have also been considered disorders of hemispheric dysconnectivity [68]. Like ASD, ADHD involves alterations in white matter microstructure and myelination [88, 89].

While schizophrenia and ADHD are also clearly unique disorders, they too share features. For example, studies have reported in one case that 40% of children with childhood-onset schizophrenia had histories of attention deficits and hyperactivity [90] and that 24% of such children displayed ADHD-like behaviors in another case [91]. In another study [92], nearly half of a group of children with childhood-onset schizophrenia had a prior diagnosis of ADHD. Impaired social interactions characterize both disorders [93]. Dopamine dysfunction occurs in both disorders [94–98]. Recent evidence also suggests shared genetic susceptibility between the two disorders [67].

Even with their clear unique features, the co-morbidities and overlapping presentations of features and underlying genetic susceptibilities and biochemical features have significant implications. Specifically, if air pollution exposures can lead to such features, then such exposures may serve as a risk factor for multiple different neurodevelopmental disorders.

11.2.2 Potential Mechanisms of Air Pollution Associations with Neurodevelopmental Disorders

11.2.2.1 Microglial Activation and Inflammation

Underlying reasons for the numerous reports of associations of air pollution with ASD are not known, but several lines of evidence suggest that air pollution-induced inflammatory mechanisms during development could play a key role [99–103]. Mechanistically, it has been proposed that air pollution-associated CNS pathology reflects interactive pathways that include systemic inflammation, as mediated e.g. by vagal nerve afferents [104, 105], in conjunction with direct effects of particulate matter on brain tissue [47].

For example, markers of inflammation are elevated in amniotic fluid [106] and through life [107–109] in ASD, consistent with a chronic inflammatory state in both the periphery [102, 108, 110, 111] and the brain [109, 112–114]. Increased cytokine levels, a consequence of inflammation, are associated with the severity of ASD diagnostic features [115] and enhanced microglial activation seen in multiple regions in brains from individuals with ASD [116, 117]. Perhaps not surprisingly, levels of asthma, a disease associated with air pollution-induced inflammation [118–121], are substantially higher in children with ASD, consistent with common inflammatory mechanisms in the lung and brain. In fact, in a nationally representative sample of 77 951 children, asthma was 35% more common in those with ASD after controlling for age, gender, body mass index, race, brain injury, secondhand smoke, and socioeconomic status [122]; further the prevalence of both ASD and asthma has been increasing [123, 124].

Peripheral and brain inflammation has also long been considered as a pathophysiological mechanism of schizophrenia [125, 126] with involvement of CNS microglial activation [126]. As with ASD, elevated levels of oxidative stress markers have been reported in both periphery [127] and in nervous tissue, including the brain of individuals with schizophrenia [128].

Inflammation is also implicated in ADHD [68, 69]. For example, levels of soluble intercellular adhesion molecule (siCAM), considered important to inflammation, were found to be higher in individuals diagnosed with ADHD than in healthy controls [129]. A genome-wide analysis for ADHD found genes involved in regulation of inflammation through pathway analysis [130]. In children born prematurely, elevated levels of inflammation-related proteins in blood were associated with attention problems at 2 years of age [131].

11.2.2.2 Ventriculomegaly, White Matter Damage, and Consequent Interhemispheric Dysconnectivity

Schizophrenia and autism also share neuropathological features, including VM [83–85, 132] and white matter injury [133, 134]. Causes of VM are not entirely understood, but impaired circulation and absorption of cerebral spinal fluid (CSF) leading to the most severe form, hydrocephalus, defects in brain development, and damage or loss of brain tissue, have all been suggested. VM occurs in 1% of the general fetal population. While in 50% of cases it does not lead to CNS abnormalities, the other 50% can be associated with multiple developmental CNS disorders, including autism and schizophrenia [84, 132, 135–139], idiopathic mental retardation, periventricular leukomalacia [140–142], fragile X syndrome, and attention deficit disorder, and with developmental delays [143, 144]. At a tissue level, its consequences can include progressive hydrocephalus, gray matter migration abnormalities, loss of parenchymal brain tissue, agenesis of the corpus callosum (CC), and delayed or abnormal maturation of white matter, i.e. reduced myelin basic protein (MBP) expression, diminished total axon volume, and microcephaly [136, 143–150]. VM associated with such deficits persists after birth [143] and is more prevalent in males [145]. The reasons for this sex bias are not understood. Mild VM has also been linked to later cases of ADHD [143, 151].

White matter damage represents another potential mechanism that could contribute to such disorders. White matter is critical to brain connectivity. The largest white matter tract in the brain, the CC, is critical to connectivity across the cerebral ventricles and is thus highly implicated in interhemispheric communication/integration for sensory, motor, and higher cognitive functions [152]. Interhemispheric dysconnectivity, or more specifically, long-range underconnectivity, and short-range overconnectivity have been reported in ASD [153, 154], schizophrenia [155–157], and ADHD [158].

Several studies have reported reductions in the size and the myelination of the CC in children with ASD [99, 159, 160]. The consequences of decreased myelination and/or size of the CC include underconnectivity of the two cerebral hemispheres and increased localized connectivity, changes that correlate with interhemispherically based functions including social and language deficits and altered hand preference/usage in children with ASD [133, 161–164]. Thinning of a callosal subregion, the genu, in ASD is notable given its importance to frontal cortical connectivity and its relationship to increased brain size and gyrification [165]. In schizophrenics, reductions in functional connectivity increased with duration of disease and correlated with psychotic symptoms [156]. Meta-analysis has also demonstrated reductions in CC size in ADHD, particularly the splenium [166]. This does not appear to be due to medication, which instead tended to normalize splenium size [167].

11.2.2.3 Altered Glutamate and Dopamine

Hyperglutamate function has been hypothesized to underlie ASD [168], and air pollution impacts brain glutamate function [49]. Both developing and adult microglia have functional glutamate receptors, and microglial activation can engender glutamate release [169, 170]. Furthermore, glutamate release can also activate microglia, leading to a vicious cycle [171] and a potential mechanism of chronic inflammation. Single nucleotide polymorphisms in glutamate transporter genes [168], and astrocyte specific dysfunctions that affect glutamate metabolism [172], have been found in genetic variants in ASD. These variants affect genes in the neurexin–neuroligin gene superfamily that are important to the regulation of glutamatergic synapse formation [173]. Collectively these finding contribute to the "hyperglutamate theory" of ASD [168].

Moreover, glutamatergic changes can also produce dysregulation of dopamine systems [174, 175]. Dopaminergic dysfunction has been suggested to underlie various behavioral features of ASD, such as executive function [176] and repetitive behaviors [177], and has long been associated with manifestations of both schizophrenia [178, 179] and ADHD [180], with drugs in these classes serving as the clinical bases for some pharmacological treatments for these disorders.

11.3 An Animal Model of UFP-Induced Developmental Neuropathology and Behavioral Disorders

Human studies relating air pollution exposures to various neurodevelopmental disorders can be significantly bolstered by animal models. Such models are beginning to emerge, and this chapter summarizes the results of several studies from our laboratory, describing the consequences of developmental exposure of mice to concentrated ambient ultrafine particle matter (CAPS; <0.1 μm diameter) that collectively appear to provide biological plausibility for reported epidemiological associations of air pollution with neurodevelopmental disorders. The focus on UFPs arises from the consideration that it represents the most reactive of air pollution particulates, including its ability to adsorb other toxic moieties and to bypass epithelial and other cell barriers and enter the bloodstream and CNS.

In these studies, mice were exposed to real time, road inlet ambient concentrated (10–20×) UFP using the Harvard University Concentrated Ambient Particle System (HUCAPS) fitted with a size-selective inlet and high-volume UFPs (≤100 nm) concentrator (10–20×) that takes in outdoor air at 5000 L/min and concentrates ambient UFP, as previously described [181–185]. Exposures lasted for 4 h/d from 0700 to 1200 for 4 days per week from PND4-7 and PND10-13, with exposure timing corresponding to peak vehicular traffic outside the intake valve of the HUCAPS instrumentation. A condensation particle counter (TSI 3022A) provided particle counts. Mass concentration was calculated using idealized particle density (1.5 g/cm^3). A scanning mobility particle sizer (SMPS) was used to determine particle size distribution and median particle diameter + geometric standard deviation (GSD). Due to the presence of the particle impactor in the HUCAPS system, mice in the CAPS exposure chamber may be held at a slightly higher negative pressure compared with the air-exposed mice in an adjacent chamber. Flows of CAPS-enriched and filtered air were both maintained at 35–40% relative humidity and 77–79 °F.

Timing of this CAPS exposure is equivalent to human third trimester [186–188] and encompasses a period of marked neuro- and gliogenesis in mice [189]. These exposures unexpectedly produced a pattern of developmental neurotoxicity notably similar to multiple hypothesized mechanistic underpinnings of neurodevelopmental disabilities characteristic of ASD, schizophrenia, and ADHD. Moreover, clear sex-dependent differences, as seen in many neurodevelopmental disorders, were found in the model. This exposure paradigm is highly relevant to human exposures as UFPs occurred at levels consistent with high traffic areas of major US cities [190–196].

11.3.1 Developmental CAPS Exposures of Mice Produce Male-Specific Microglial Activation

As previously noted, microglial activation has been reported in the context of ASD and of schizophrenia [116, 117, 126] and implicated in ADHD [68]. Studies from the CAPS model demonstrate persistent increases in microglial presence, as visualized by ionized calcium-binding adapter molecule 1 (IBA-1) immunostaining [182, 183, 185]. Further, these effects were found in male, but not in female, CAPS-treated mice. CAPS-induced microglial activation was observed in the

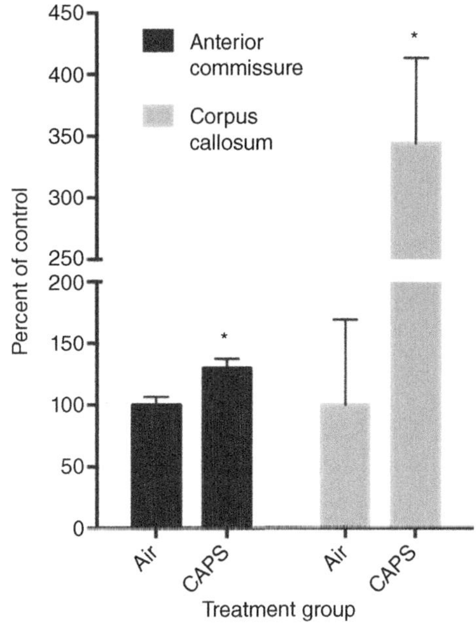

Figure 11.1 Developmental exposures to CAPS produced male-specific and time-dependent increases in microglial presence as measured via immunoreactivity of ionized calcium binding adapter molecule 1 (IBA-1). Male IBA-1 levels (group mean ± S.E.) as percent of time point control for air and CAPS-treated groups at PND14 in anterior commissure and at PND270 in corpus callosum. CAPS = concentrated ambient ultrafine particles; PND = postnatal day; * = significantly different from corresponding time point control. $N = 4-5$/group. Source: Modified from Allen et al. [182, 183].

major white matter tracts of the brain including the anterior commissure as seen at postnatal day (PND) 14 and in the CC at PND270 (Figure 11.1), where an increase of 340% relative to control was observed, despite the fact that exposures had ended at PND13. The latter indicates a persistent, lifelong, microglial activation.

11.3.2 Developmental CAPS Exposures of Mice Produce Male-Specific Ventriculomegaly

Disruption to the normal development of white matter and concomitant VM are sometimes observed in the brains of children with ASD and schizophrenia and have been linked to later ADHD [83–85, 132, 143, 151]. Initial observations from our studies included an unexpected, male-specific VM of variable severity in CAPS-treated males that was not observed in CAPS-treated females [182, 185]. Quantitation of ventricle size was determined at several time points, being statistically higher in CAPS males out to at least PND270 [182, 185] (Figure 11.2). These findings suggest persistent VM and corresponding white matter damage.

11.3.3 Developmental CAPS Exposures of Mice Produce Male-Specific White Matter Tract Disruption

Appearance of VM suggested that the neighboring CC may also be impacted by CAPS exposure, thus prompting quantification of CC size and assessment of

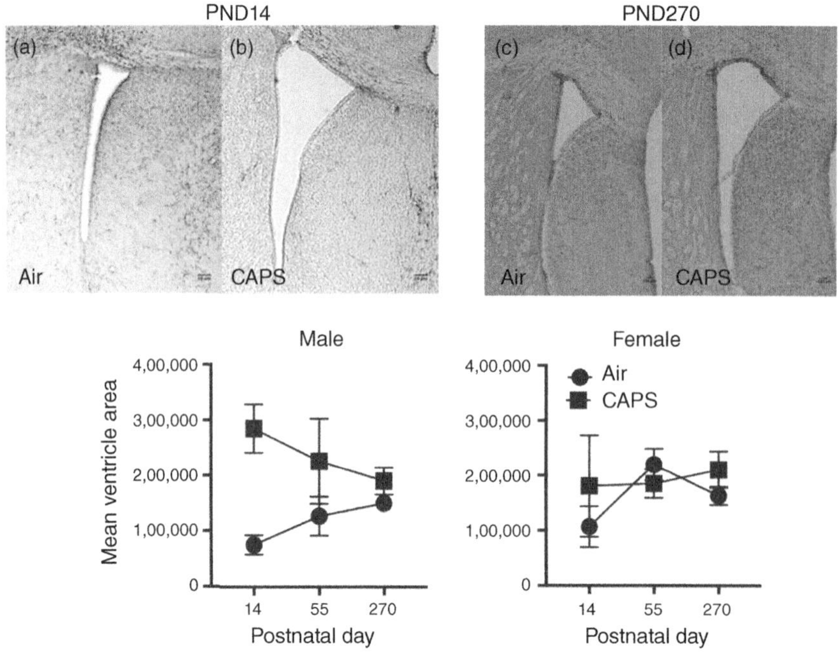

Figure 11.2 Developmental exposures to CAPS lead to increased lateral ventricle size in the brains of male mice. Panels (a)–(d): Images of lateral ventricle from air-exposed (a), (c) and CAPS-exposed (b), (d) male mice at PND14 (a), (b) and approximate PND270 (c), (d). Bottom: Group mean ± S.E. ventricle area for air vs. CAPS-exposed male mice at the indicated postnatal day of age. Effects were graded with this example demonstrating the most severe of effects. * = significantly different from corresponding time point control. $N = 4–5$/group. Source: Allen et al. [182] / with permission from Environmental Health Perspectives.

its developmental trajectory through determination of MBP expression in the region [182, 185].

11.3.3.1 Corpus Callosum Size

Reductions in size of CC, the largest white matter tract in the brain, have been reported in ASD [160] and ADHD [197]. In first-episode schizophrenics, enlarged lateral ventricle sizes were seen and inversely correlated with CC volume [198].

Prior studies have indicated rapid growth in the CC of mice, with adult-like patterns of connections by PND15 [199]. In the CAPS studies, the CC was slightly smaller in females compared with males at PND14, but larger at PND270 [185]. These findings are consistent with previous reports of greater increases in CC size in female rats between adulthood and middle age [152] that is considered to reflect the initially larger number and greater size of myelinated axons in males [152] and more delayed maturation to the adult pattern in females [200].

At PND14, both males and females exhibited statistically significant CAPS-induced reductions in CC size (Figure 11.3a,c), whereas no significant differences in size were found at PND270 (Figure 11.3b,d) [185]. While CAPS-treated

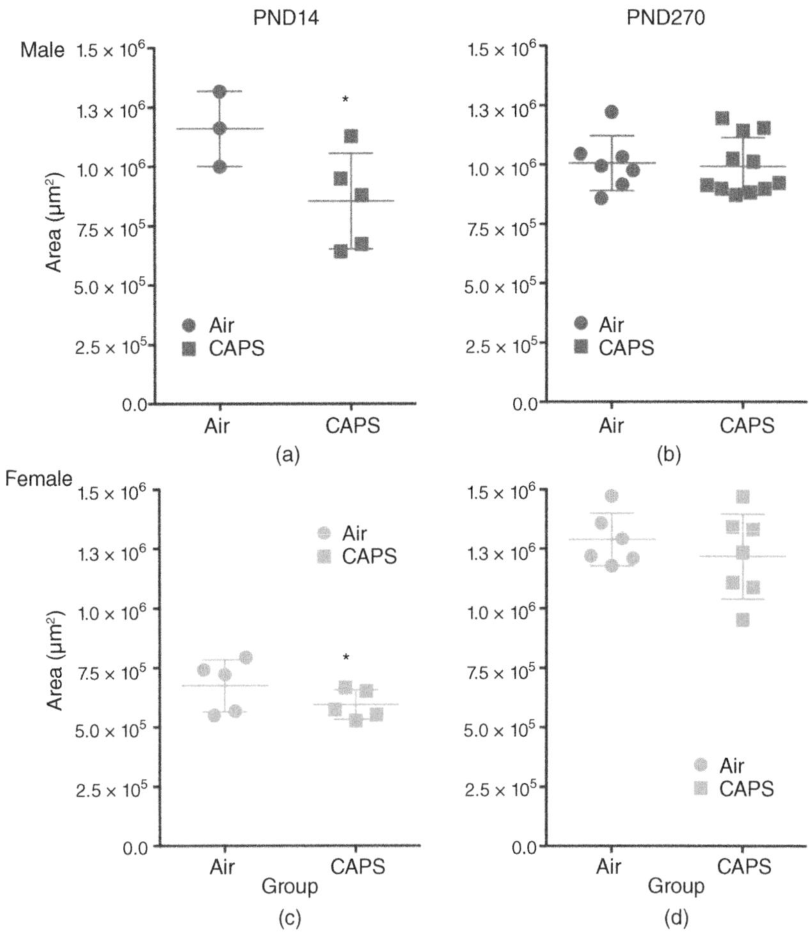

Figure 11.3 Developmental exposures to CAPS reduces the size of corpus callosum in both males and females at PND14 with slightly enhanced effects in males. Panels (a)–(d): Callosal size (μm^2) of individual male and female mice following air or postnatal CAPS treatment as measured at PND14 and PND270. Each data point represents an individual value. Sample sizes of 3–5 for PND14 and 7–11 for PND270. CAPS = concentrated ambient ultrafine particles; PND = postnatal day. * = significantly different from corresponding time point control. N = 4–5/group. Source: From Allen et al. [185].

males showed a 30% size reduction compared with 10% in CAPS-treated females at PND14 relative to air controls (Figure 11.3a,c), statistical analysis of PND14 data confirmed only an effect of treatment, but not a treatment by sex interaction that would statistically confirm greater magnitude effects in males. However, as can be seen from the plotted individual data, differences among male at PND14 included a broader decrement with some overlap with air controls, while evidence for female effects based on individual data were less biologically compelling.

11.3.3.2 Corpus Callosum Myelination

Alterations in myelination patterns were examined using MBP staining at three different sites along the CC as follows: truncus, medial external capsule (MEC), and lateral external capsule (LEC) [185]. At PND14 (Figure 11.4a,b), marked reductions of approximately 35% in MBP staining within the CC were seen in males at the truncus that also extended to the medial external capsule (approximately 15%), whereas no significant differences in MBP staining area were seen in females at any point along the CC. At PND270 (Figure 11.4c,d), no significant treatment-related effects were found in either sex despite a visual trend toward hypomyelination in CAPS-treated females. The data suggests a male-specific hypomyelination in the truncus and MEC without impacting the LEC. In contrast to the CC, striatal and cortical myelination were not impacted by CAPS (not shown in the figure).

11.3.4 Developmental CAPS Exposures of Mice Elevate Glutamate Levels and Result in Male-Specific Excitatory–Inhibitory Imbalance

Changes in glutamate and/or dopamine are observed in ASD [168, 201], schizophrenia [179], and ADHD [202]. Reports indicate that children and adults with ASD show altered glutamatergic balance (excitatory–inhibitory imbalance) in both the periphery and in the CNS [203–208]. Developmental exposures to CAPS may, in part, recapitulate these findings of ASD. CAPS increased total levels of frontal cortex glutamate, glutamine, and gabba aminobutyric acid (GABA) in both sexes following postnatal exposures [182, 183]. Hippocampal glutamate was selectively elevated in males, but not females, suggesting greater potential for CAPS-associated excitotoxicity in males. Indeed, an imbalance of excitatory/inhibitory neurotransmitters, measured as the ratio of glutamate/GABA, was observed only in the hippocampus of males treated with CAPS at PND14 and PND60 (Figure 11.5), as GABA increases were less pronounced than those of glutamate in males. These findings are consistent with the male specificity of microglial activation (Figure 11.1) and its potential to produce glutamate release, as seen in ASD [169, 170].

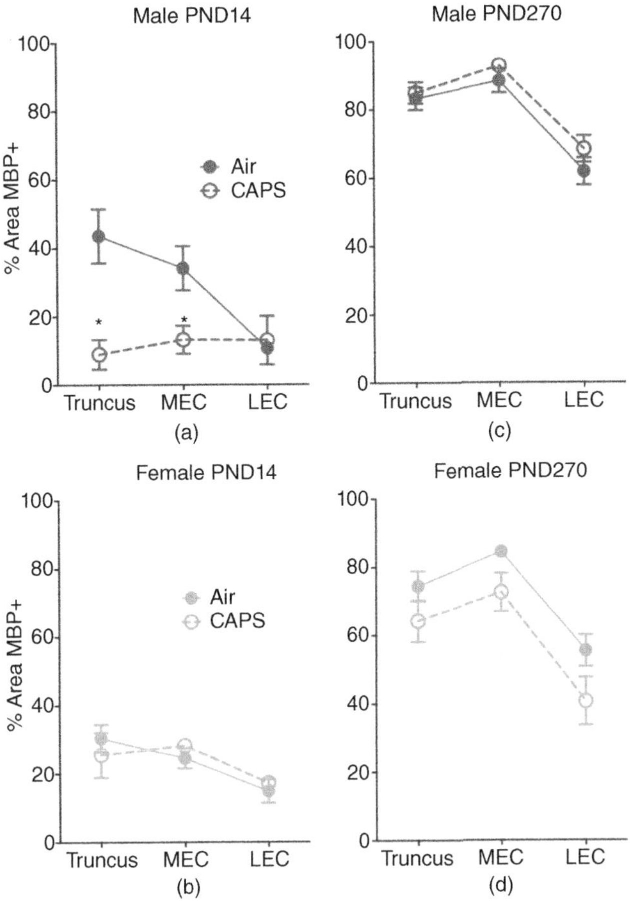

Figure 11.4 Developmental exposures to CAPS produced a male-specific hypomyelination of the corpus callosum at PND14. Intensity of myelin basic protein labeling expressed as percent of total callosal area at PND14 (a), (b) and PND270 (c), (d) for male and female air vs. postnatal CAPS-treated mice as measured at three different sites along the corpus callosum: truncus, medial external capsule, and lateral external capsule for treatment groups as indicated. Shown is group mean ± S.E. CAPS = concentrated ambient ultrafine particles; PND = postnatal day. * = significantly different from corresponding time point control. $N = 4-5$/group. Source: From Allen et al. [185].

11.3.5 Developmental CAPS Exposures of Mice Are Associated with Impulsive-Like Behavior

ASD diagnosis is predicated on primary behavioral characteristics that include social deficits, impaired communication, and repetition/perseveration [56], yet

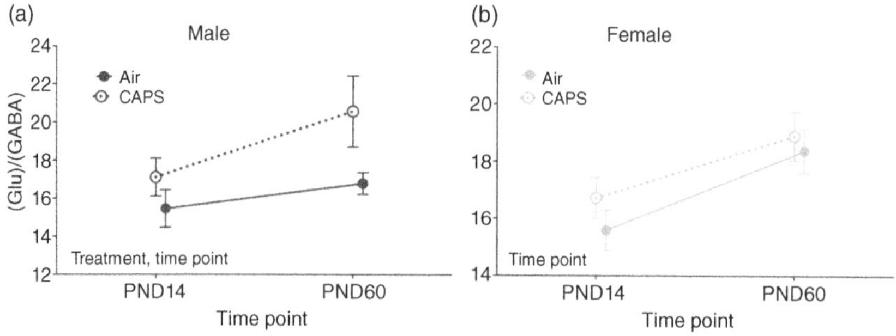

Figure 11.5 Developmental exposures to CAPS produced a male-specific frontal cortex excitatory/inhibitory imbalance. Glutamate/GABA ratios based on levels of frontal cortex at PND14 and PND60 as based on extracellular glutamate, glutamine, and GABA in air vs. postnatal CAPS-treated males (a) and female (b) mice. Shown are group means ± S.E. CAPS = concentrated ambient ultrafine particles; PND = postnatal day; treatment = main effect of CAPS in statistical analysis; time point = main effect of time point in statistical analysis, $N = 9-12$/group. Source: Data modified from Allen et al. [182] and [183].

executive functions including set shifting can also be impaired in ASD and are frequent co-morbid features of the disorder [165, 209–214]. Impulsive-like behavior can also characterize some forms of ADHD [215] and of schizophrenia [216]. The unexpected observations of disrupted CC white matter development, VM, and excitatory/inhibitory imbalance in response to CAPS treatment were found only after our completion of behavioral testing in our studies, and therefore explicit testing for the primary behavioral domains of ASD, ADHD, or schizophrenia have not yet been completed, nor were all behavioral assays that have been implemented carried out in both sexes. Consequently, the full extent to which postnatal CAPS reproduces behavioral features of these disorders, and its sex-dependence, is not yet fully delineated.

However, developmental treatment with CAPS appears to produce perseverative/impulsive behaviors in male mice as measured using a fixed ratio waiting-for-reward paradigm. This behavioral task required 25-level press responses in an operant chamber to initiate a waiting component during which free reward deliveries occurred [217]. The time between each successive free reward delivery doubled during the waiting component, thus requiring increasingly long wait times, and any lever press response during the waiting component reset the fixed ratio 25 response requirement to restart the wait component. Developmental exposures to CAPS markedly decreased the length of time male mice waited and increased the number of fixed ratio resets despite the added cost in terms of number of responses required for each reward delivery [181]. Figure 11.6 presents a cumulative record of responding of an individual subject from each treatment group depicting this perseverative/impulsive behavior, behaviors consistent with ASD-like repetition and impulsivity.

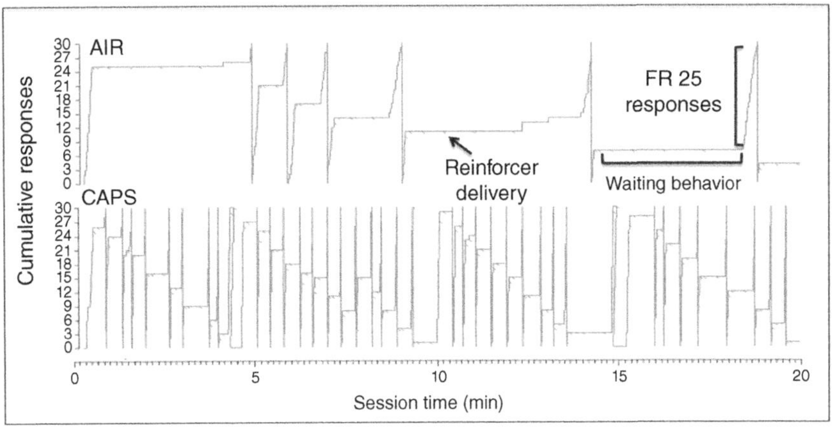

Figure 11.6 Neonatal CAPS exposure results in impulsive behavior in males. Graph depicts cumulative records of representative individual males from indicated treatment group performing on a fixed ratio (FR) waiting-for-reward schedule. *Y*-axis response counter resets to baseline after each 30 responses; pips (backslashes) = reward delivery; horizontal lines = periods of waiting; vertical lines = FR responding. CAPS animals characterized as having more FR responses (resets) and shorter periods of waiting. Source: From Allen et al. [181].

11.4 Summary and Conclusions

Postnatal exposures of mice to ambient UFP appear to reproduce many features consistent with characteristics of ASD, schizophrenia, and ADHD, disorders known to share many features. Effects following developmental exposure to CAPS have included inflammation/microglial activation (Figure 11.1), VM (Figure 11.2), reduced CC size (Figure 11.3), reduction in CC myelination (Figure 11.4), an altered ratio of glutamate and GABA, consistent with excitatory/inhibitory imbalance (Figure 11.5), and repetitive/impulsive-like behaviors (Figure 11.6), many of which were either male-specific or of greater magnitude in males.

Data discussed herein indicate UFP-induced white matter alterations at PND14 selectively in commissural tracts (CC), with effects that were of significantly greater magnitude in males, consistent with our previous observation of male-specific and persistent microglial activation and VM [182, 183]. These changes included both a significantly reduced CC size, as well as male-specific hypomyelination of CC at PND14, with marked reductions particularly in the truncus and the medial, as opposed to the lateral external capsule. CC size ultimately reached control levels later in life following postnatal CAPS in males, whereas small but nonsignificant reductions in myelination were suggestive in females at PND270, findings that could indicate delayed neurotoxicity in this sex and perhaps repair in males. In contrast, no changes were seen in white matter in either striatum or in frontal cortex in either sex.

Adverse pregnancy outcomes, specifically preterm birth and reduced birth weight are often associated with brain white matter alterations [141, 218], as seen in conditions such as periventricular leukomalacia [219], that have been linked to inflammatory mechanisms [220–222] and can lead to cognitive impairments. In up to 50% of extremely and very preterm infants, alterations in brain white matter constitute the most important basis of axonal/neuronal damage associated with adverse neurocognitive and motor development [223]. This is notable given that links between prematurity and reduced birth weight with air pollution are increasingly reported [224, 225]. Alterations in brain white matter are also observed in ADHD [226]. Dysconnectivity, i.e. functional and/or anatomical damage to white matter axons in communicating pathways found in ASD and schizophrenia [133, 163, 164, 227], is critical to language, social behavior, and some cognitive functions. VM associated with such deficits persists after birth [143] and is more prevalent in males [145]. Prenatal VM is predictive of time to first-episode psychosis and schizophrenia diagnosis [84, 137, 144, 145, 148].

Animal models yielding a similar neuropathological profile to that produced by CAPS treatment can result from exposures to hypoxia, ischemia, maternal–fetal infection, and lipopolysaccharide (endotoxin) administration [152]. Frequently observed in such pathology is an early necrosis in the subventricular zone adjacent to the lateral ventricle with depletion of stem cells, followed by later proliferation, as also observed in conditions such as periventricular leukomalacia [219]. Numerous studies now demonstrate the role of reactive microglia, with a particular vulnerability of premyelinating oligodendroglia, to such insults [223, 228].

Collectively, the findings have generated several important questions. For example, what underlies the preferential male vulnerability to the early reduction in CC size and hypomyelination? One possibility relates to the reports of greater numbers of colonizing activated microglia in male brain at the time during which these CAPS exposures occurred [229]. Neuro- and gliogenesis, which derive from germinal ventricular and subventricular zones next to the ventricle walls, occur at peak levels during the postnatal period when CAPS exposures occurred [189] and continue at a constant level into young adulthood [230, 231], with activated microglia actually *required* for neuro- and oligodendrogenesis through the release of cytokines (potentially in combinations), including IL-1β and IL-6 [232]. This state, however, despite its requirement for neuro- and gliogenesis to proceed, could also render male brain microglia more vulnerable to dysfunction or even cell death in response to an ongoing inflammatory insult. Females exhibit increased numbers and more activated microglia than males only at a later postnatal age (PND30) [229]. Thus CAPS exposures in these studies may have preceded the period of female vulnerability; correspondingly, it is possible that later exposures would have enhanced vulnerability preferentially in females.

The reduction of CC size and myelination coupled with microglial activation and VM [182] is markedly similar to diffuse periventricular leukomalacia or white matter injury of prematurity/neonatal white matter damage [228, 233] produced by fetal inflammatory or hypoxic–ischemic injury. Based on its well-known inflammatory properties [234], UFP exposures would be expected to activate brain microglia, resulting in the release of proinflammatory cytokines [235–237] that could lead to hypomyelination via toxicity to oligodendrocytes, the myelinating cells of the brain [238]. Indeed, particular vulnerability of premyelinating oligodendrocytes to such consequences of microglial activation has been described [239, 240].

Reductions in myelination were seen in CC but not in either striatum or frontal cortex. This could reflect regional differences in timing of myelination in early development relative to the CAPS exposure period. Interestingly, as in humans [241], myelination in mouse brain occurs earliest in the caudal part of the CC (PND14) and only later in the anterior segments, i.e. truncus and genu, e.g. between PND21 and 28 [242]. Myelination of regions such as striatum and cortex occurs even later, with studies showing myelination levels of only 1% in frontal cortex at PND21, up to about 15% at PND90, and up to approximately 70% at PND180. At PND21, myelination of striatum is only about 5% of the area and increases to about 28% by PND90 [243]. Thus, it may be that the absence of white matter myelination changes in striatum and frontal cortex reflects the fact that these processes are occurring at a sufficiently later time relative to CAPS exposures and thus are spared and/or that any ongoing inflammation levels have subsided in that period.

It is critical to consider that different components of the observed profile of CAPS-induced neuropathology and behavioral deficits are actually consistent with multiple neurodevelopmental disorders, including ASD [99, 159, 244, 245], schizophrenia [246], and ADHD [70, 167, 226, 247–251] among others, all of which also show a higher prevalence in males, consistent with the male dominance of outcomes observed here. The implications of this are that air pollution could serve as a risk factor for a wider range of neurodevelopmental disorders than is currently appreciated.

One cited limitation of rodent models of inhaled air pollution has been that rodents, unlike humans, are obligate nose breathers and thus dosing of mice results in higher exposures relative to humans. However, this may be a minimal or negligent concern according to a recent study based on computational modeling using an anatomically accurate model of the human nasal cavity. It was found that the olfactory dose per unit surface area is actually estimated to be higher in humans depending on the size of the particle [252, 253]. The latter is of particular note as emphasized in a recent editorial [254], given the presence in fly ash and car exhaust of a significant fraction of particles below 10 nm, often overlooked because of measurement difficulty [255].

In addition, given the vast variability in phenotypic expression and severity of neurodevelopmental disorders, a risk factor such as air pollution may likewise show significant heterogeneity. Air pollution levels and composition vary greatly across season, climate, geography, etc. and could contribute to the variability seen in the phenotypes, a phenomenon that could be further compounded by the developmental timing of exposure. The multifactorial genetic and environmental risks for neurodevelopmental disorders, superimposed on different periods of brain development, could underlie phenotypic variability. Air pollution, as one such risk factor, could pose a greater threat in more highly polluted regions and/or in areas where it includes more toxic elements. Correspondingly, multiple components of air pollution including diesel exhaust, fine as well as UFP, and toxicants adsorbed to particles, such as endotoxin (lipopolysaccharide; LPS), also produce inflammatory changes in the brain [256–262] and their individual contributions warrant investigation.

References

1 Elder, A., Gelein, R., Silva, V. et al. (2006). Translocation of inhaled ultrafine manganese oxide particles to the central nervous system. *Environ. Health Perspect.* 114 (8): 1172–1178.

2 Hunter, D.D. and Dey, R.D. (1998). Identification and neuropeptide content of trigeminal neurons innervating the rat nasal epithelium. *Neuroscience* 83 (2): 591–599.

3 Lewis, J., Bench, G., Myers, O. et al. (2005). Trigeminal uptake and clearance of inhaled manganese chloride in rats and mice. *Neurotoxicology* 26 (1): 113–123.

4 Oberdorster, G., Ferin, J., and Lehnert, B.E. (1994). Correlation between particle size, in vivo particle persistence, and lung injury. *Environ. Health Perspect.* 102 (Suppl 5): 173–179.

5 Wilson, M.R., MacNee, W., Stone, V. et al. (2001). Size-dependent proinflammatory effects of ultrafine polystyrene particles: a role for surface area and oxidative stress in the enhanced activity of ultrafines. *Toxicol. Appl. Pharmacol.* 175 (3): 191–199.

6 Lelieveld, J., Evans, J.S., Fnais, M. et al. (2015). The contribution of outdoor air pollution sources to premature mortality on a global scale. *Nature* 525 (7569): 367–371.

7 Forouzanfar, M.H., Alexander, L., Anderson, H.R. et al. (2016). Global, regional, and national comparative risk assessment of 79 behavioural, environmental and occupational, and metabolic risks or clusters of risks in 188 countries, 1990–2013: a systematic analysis for the global burden of disease study 2013. *Lancet* 388 (10053): 1659–1724.

8 Costa, L.G., Cole, T.B., Coburn, J. et al. (2017). Neurotoxicity of traffic-related air pollution. *Neurotoxicology* 59: 133–139.

9 Heusinkveld, H.J., Wahle, T., Campbell, A. et al. (2016). Neurodegenerative and neurological disorders by small inhaled particles. *Neurotoxicology* 56: 94–106.

10 Power, M.C., Adar, S.D., Yanosky, J.D. et al. (2016). Exposure to air pollution as a potential contributor to cognitive function, cognitive decline, brain imaging, and dementia: a systematic review of epidemiologic research. *Neurotoxicology* 56: 235–253.

11 Xu, X., Ha, S.U., and Basnet, R. (2016). A review of epidemiological research on adverse neurological effects of exposure to ambient air pollution. *Front. Public Health* 4: 157.

12 Kumar, P., Morawska, L., Birmili, W. et al. (2014). Ultrafine particles in cities. *Environ. Int.* 66: 1–10.

13 Becerra, T.A., Wilhelm, M., Olsen, J. et al. (2013). Ambient air pollution and autism in Los Angeles county, California. *Environ. Health Perspect.* 121 (3): 380–386.

14 Beckerman, B., Jerrett, M., Brook, J.R. et al. (2008). Correlation of nitrogen dioxide with other traffic pollutants near a major expressway. *Atmos. Environ.* 42: 275–290.

15 Jung, C.R., Lin, Y.T., and Hwang, B.F. (2013). Air pollution and newly diagnostic autism spectrum disorders: a population-based cohort study in Taiwan. *PLoS One* 8 (9): e75510.

16 Kalkbrenner, A.E., Daniels, J.L., Chen, J.C. et al. (2010). Perinatal exposure to hazardous air pollutants and autism spectrum disorders at age 8. *Epidemiology* 21 (5): 631–641.

17 Kalkbrenner, A.E., Windham, G.C., Serre, M.L. et al. (2015). Particulate matter exposure, prenatal and postnatal windows of susceptibility, and autism spectrum disorders. *Epidemiology* 26 (1): 30–42.

18 Raz, R., Roberts, A.L., Lyall, K. et al. (2015). Autism spectrum disorder and particulate matter air pollution before, during, and after pregnancy: a nested case-control analysis within the Nurses' health study II cohort. *Environ. Health Perspect.* 123 (3): 264–270.

19 Roberts, A.L., Lyall, K., Hart, J.E. et al. (2013). Perinatal air pollutant exposures and autism spectrum disorder in the children of Nurses' health study II participants. *Environ. Health Perspect.* 121 (8): 978–984.

20 Volk, H.E., Hertz-Picciotto, I., Delwiche, L. et al. (2011). Residential proximity to freeways and autism in the CHARGE study. *Environ. Health Perspect.* 119 (6): 873–877.
21 Volk, H.E., Lurmann, F., Penfold, B. et al. (2013). Traffic-related air pollution, particulate matter, and autism. *JAMA Psychiat.* 70 (1): 71–77.
22 Windham, G.C., Zhang, L., Gunier, R. et al. (2006). Autism spectrum disorders in relation to distribution of hazardous air pollutants in the San Francisco bay area. *Environ. Health Perspect.* 114 (9): 1438–1444.
23 Talbott, E.O., Arena, V.C., Rager, J.R. et al. (2015). Fine particulate matter and the risk of autism spectrum disorder. *Environ. Res.* 140: 414–420.
24 Pedersen, C.B., Raaschou-Nielsen, O., Hertel, O. et al. (2004). Air pollution from traffic and schizophrenia risk. *Schizophr. Res.* 66 (1): 83–85.
25 Yackerson, N.S., Zilberman, A., Todder, D. et al. (2014). The influence of air-suspended particulate concentration on the incidence of suicide attempts and exacerbation of schizophrenia. *Int. J. Biometeorol.* 58 (1): 61–67.
26 Harris, M.H., Gold, D.R., Rifas-Shiman, S.L. et al. (2015). Prenatal and childhood traffic-related pollution exposure and childhood cognition in the project viva cohort (Massachusetts, USA). *Environ. Health Perspect.* 123 (10): 1072–1078.
27 Suglia, S.F., Gryparis, A., Wright, R.O. et al. (2008). Association of black carbon with cognition among children in a prospective birth cohort study. *Am. J. Epidemiol.* 167 (3): 280–286.
28 Chiu, Y.H., Hsu, H.H., Coull, B.A. et al. (2016). Prenatal particulate air pollution and neurodevelopment in urban children: examining sensitive windows and sex-specific associations. *Environ. Int.* 87: 56–65.
29 Sunyer, J., Esnaola, M., Alvarez-Pedrerol, M. et al. (2015). Association between traffic-related air pollution in schools and cognitive development in primary school children: a prospective cohort study. *PLoS Med.* 12 (3): e1001792.
30 Chiu, Y.H., Bellinger, D.C., Coull, B.A. et al. (2013). Associations between traffic-related black carbon exposure and attention in a prospective birth cohort of urban children. *Environ. Health Perspect.* 121 (7): 859–864.
31 Newman, N.C., Ryan, P., Lemasters, G. et al. (2013). Traffic-related air pollution exposure in the first year of life and behavioral scores at 7 years of age. *Environ. Health Perspect.* 121 (6): 731–736.
32 Siddique, S., Banerjee, M., Ray, M.R. et al. (2011). Attention-deficit hyperactivity disorder in children chronically exposed to high level of vehicular pollution. *Eur. J. Pediatr.* 170 (7): 923–929.
33 Perera, F.P., Li, Z., Whyatt, R. et al. (2009). Prenatal airborne polycyclic aromatic hydrocarbon exposure and child IQ at age 5 years. *Pediatrics* 124 (2): e195–e202.

34 Perera, F.P., Rauh, V., Whyatt, R.M. et al. (2006). Effect of prenatal exposure to airborne polycyclic aromatic hydrocarbons on neurodevelopment in the first 3 years of life among inner-city children. *Environ. Health Perspect.* 114 (8): 1287–1292.

35 Perera, F.P., Tang, D., Wang, S. et al. (2012). Prenatal polycyclic aromatic hydrocarbon (PAH) exposure and child behavior at age 6–7 years. *Environ. Health Perspect.* 120 (6): 921–926.

36 Edwards, S.C., Jedrychowski, W., Butscher, M. et al. (2010). Prenatal exposure to airborne polycyclic aromatic hydrocarbons and children's intelligence at 5 years of age in a prospective cohort study in Poland. *Environ. Health Perspect.* 118 (9): 1326–1331.

37 Guxens, M., Ghassabian, A., Gong, T. et al. (2016). Air pollution exposure during pregnancy and childhood autistic traits in four European population-based cohort studies: the ESCAPE project. *Environ. Health Perspect.* 124 (1): 133–140.

38 Pujol, J., Martinez-Vilavella, G., Macia, D. et al. (2016). Traffic pollution exposure is associated with altered brain connectivity in school children. *NeuroImage* 129: 175–184.

39 Oudin, A., Forsberg, B., Adolfsson, A.N. et al. (2016). Traffic-related air pollution and dementia incidence in northern Sweden: a longitudinal study. *Environ. Health Perspect.* 124 (3): 306–312.

40 Ranft, U., Schikowski, T., Sugiri, D. et al. (2009). Long-term exposure to traffic-related particulate matter impairs cognitive function in the elderly. *Environ. Res.* 109 (8): 1004–1011.

41 Clifford, A., Lang, L., Chen, R. et al. (2016). Exposure to air pollution and cognitive functioning across the life course – a systematic literature review. *Environ. Res.* 147: 383–398.

42 Schikowski, T., Vossoughi, M., Vierkotter, A. et al. (2015). Association of air pollution with cognitive functions and its modification by APOE gene variants in elderly women. *Environ. Res.* 142: 10–16.

43 Weuve, J., Puett, R.C., Schwartz, J. et al. (2012). Exposure to particulate air pollution and cognitive decline in older women. *Arch. Intern. Med.* 172 (3): 219–227.

44 Palacios, N., Fitzgerald, K.C., Hart, J.E. et al. (2014). Particulate matter and risk of Parkinson disease in a large prospective study of women. *Environ. Health* 13: 80.

45 Ritz, B., Lee, P.C., Hansen, J. et al. (2016). Traffic-related air pollution and Parkinson's disease in Denmark: a case-control study. *Environ. Health Perspect.* 124 (3): 351–356.

46 Chen, J.C., Wang, X., Wellenius, G.A. et al. (2015). Ambient air pollution and neurotoxicity on brain structure: evidence from Women's health initiative memory study. *Ann. Neurol.* 78 (3): 466–476.
47 Block, M.L. and Calderon-Garciduenas, L. (2009). Air pollution: mechanisms of neuroinflammation and CNS disease. *Trends Neurosci.* 32 (9): 506–516.
48 Fonken, L.K., Xu, X., Weil, Z.M. et al. (2011). Air pollution impairs cognition, provokes depressive-like behaviors and alters hippocampal cytokine expression and morphology. *Mol. Psychiatry* 16 (10): 987–995, 973.
49 Davis, D.A., Bortolato, M., Godar, S.C. et al. (2013). Prenatal exposure to urban air nanoparticles in mice causes altered neuronal differentiation and depression-like responses. *PLoS One* 8 (5): e64128.
50 Suzuki, T., Oshio, S., Iwata, M. et al. (2010). In utero exposure to a low concentration of diesel exhaust affects spontaneous locomotor activity and monoaminergic system in male mice. *Part. Fibre Toxicol.* 7: 7.
51 Ku, T., Ji, X., Zhang, Y. et al. (2016). PM2.5, SO_2 and NO_2 co-exposure impairs neurobehavior and induces mitochondrial injuries in the mouse brain. *Chemosphere* 163: 27–34.
52 Cheng, H., Saffari, A., Sioutas, C. et al. (2016). Nano-scale particulate matter from urban traffic rapidly induces oxidative stress and inflammation in olfactory epithelium with concomitant effects on brain. *Environ. Health Perspect.* 124 (10): 1537–1546.
53 Bolton, J.L., Auten, R.L., and Bilbo, S.D. (2014). Prenatal air pollution exposure induces sexually dimorphic fetal programming of metabolic and neuroinflammatory outcomes in adult offspring. *Brain Behav. Immun.* 37: 30–44.
54 Roque, P.J., Dao, K., and Costa, L.G. (2016). Microglia mediate diesel exhaust particle-induced cerebellar neuronal toxicity through neuroinflammatory mechanisms. *Neurotoxicology* 56: 204–214.
55 Oppenheim, H.A., Lucero, J., Guyot, A.C. et al. (2013). Exposure to vehicle emissions results in altered blood brain barrier permeability and expression of matrix metalloproteinases and tight junction proteins in mice. *Part. Fibre Toxicol.* 10: 62.
56 Lai, M.C., Lombardo, M.V., and Baron-Cohen, S. (2014). Autism. *Lancet* 383 (9920): 896–910.
57 Elsabbagh, M., Divan, G., Koh, Y.J. et al. (2012). Global prevalence of autism and other pervasive developmental disorders. *Autism Res.* 5 (3): 160–179.
58 Frazier, T.W., Georgiades, S., Bishop, S.L. et al. (2014). Behavioral and cognitive characteristics of females and males with autism in the Simons simplex collection. *J. Am. Acad. Child Adolesc. Psychiatry* 53 (3): 329–40 e1-3.
59 Van Wijngaarden-Cremers, P.J., van Eeten, E., Groen, W.B. et al. (2014). Gender and age differences in the core triad of impairments in autism

spectrum disorders: a systematic review and meta-analysis. *J. Autism Dev. Disord.* 44 (3): 627–635.

60 Werling, D.M. and Geschwind, D.H. (2013). Sex differences in autism spectrum disorders. *Curr. Opin. Neurol.* 26 (2): 146–153.

61 Talbott, E.O., Marshall, L.P., Rager, J.R. et al. (2015). Air toxics and the risk of autism spectrum disorder: the results of a population based case-control study in southwestern Pennsylvania. *Environ. Health* 14: 80.

62 Owen, M.J., Sawa, A., and Mortensen, P.B. (2016). Schizophrenia. *Lancet* 388 (10039): 86–97.

63 Chen, J., Cao, F., Liu, L. et al. (2015). Genetic studies of schizophrenia: an update. *Neurosci. Bull.* 31 (1): 87–98.

64 Hawi, Z., Cummins, T.D., Tong, J. et al. (2015). The molecular genetic architecture of attention deficit hyperactivity disorder. *Mol. Psychiatry* 20 (3): 289–297.

65 Thapar, A. and Cooper, M. (2016). Attention deficit hyperactivity disorder. *Lancet* 387 (10024): 1240–1250.

66 Gong, T., Almqvist, C., Bolte, S. et al. (2014). Exposure to air pollution from traffic and neurodevelopmental disorders in Swedish twins. *Twin Res. Hum. Genet.* 17 (6): 553–562.

67 Hamshere, M.L., Stergiakouli, E., Langley, K. et al. (2013). Shared polygenic contribution between childhood attention-deficit hyperactivity disorder and adult schizophrenia. *Br. J. Psychiatry* 203 (2): 107–111.

68 Kern, J.K., Geier, D.A., King, P.G. et al. (2015). Shared brain connectivity issues, symptoms, and comorbidities in autism spectrum disorder, attention deficit/hyperactivity disorder, and Tourette syndrome. *Brain Connect.* 5 (6): 321–335.

69 Mitchell, R.H. and Goldstein, B.I. (2014). Inflammation in children and adolescents with neuropsychiatric disorders: a systematic review. *J. Am. Acad. Child Adolesc. Psychiatry* 53 (3): 274–296.

70 Paul, L.K. (2011). Developmental malformation of the corpus callosum: a review of typical callosal development and examples of developmental disorders with callosal involvement. *J. Neurodev. Disord.* 3 (1): 3–27.

71 Rommelse, N.N., Geurts, H.M., Franke, B. et al. (2011). A review on cognitive and brain endophenotypes that may be common in autism spectrum disorder and attention-deficit/hyperactivity disorder and facilitate the search for pleiotropic genes. *Neurosci. Biobehav. Rev.* 35 (6): 1363–1396.

72 Takahashi, Y., Yu, Z., Sakai, M., and Tomita, H. (2016). Linking activation of microglia and peripheral monocytic cells to the pathophysiology of psychiatric disorders. *Front. Cell. Neurosci.* 10: 144.

73 Cheung, C., Yu, K., Fung, G. et al. (2010). Autistic disorders and schizophrenia: related or remote? An anatomical likelihood estimation. *PLoS One* 5 (8): e12233.

74 King, B.H. and Lord, C. (2011). Is schizophrenia on the autism spectrum? *Brain Res.* 1380: 34–41.

75 Meyer, U., Feldon, J., and Dammann, O. (2011). Schizophrenia and autism: both shared and disorder-specific pathogenesis via perinatal inflammation? *Pediatr. Res.* 69 (5 Pt 2): 26R–33R.

76 Rapoport, J., Chavez, A., Greenstein, D. et al. (2009). Autism spectrum disorders and childhood-onset schizophrenia: clinical and biological contributions to a relation revisited. *J. Am. Acad. Child Adolesc. Psychiatry* 48 (1): 10–18.

77 Barneveld, P.S., Pieterse, J., de Sonneville, L. et al. (2011). Overlap of autistic and schizotypal traits in adolescents with autism Spectrum disorders. *Schizophr. Res.* 126 (1–3): 231–236.

78 Brown, M.S., Singel, D., Hepburn, S., and Rojas, D.C. (2013). Increased glutamate concentration in the auditory cortex of persons with autism and first-degree relatives: a (1)H-MRS study. *Autism Res.* 6 (1): 1–10.

79 Shimmura, C., Suzuki, K., Iwata, Y. et al. (2013). Enzymes in the glutamate-glutamine cycle in the anterior cingulate cortex in postmortem brain of subjects with autism. *Mol. Autism.* 4 (1): 6.

80 Sodhi, M.S. and Sanders-Bush, E. (2004). Serotonin and brain development. *Int. Rev. Neurobiol.* 59: 111–174.

81 de Lacy, N. and King, B.H. (2013). Revisiting the relationship between autism and schizophrenia: toward an integrated neurobiology. *Annu. Rev. Clin. Psychol.* 9: 555–587.

82 Smoller, J.W., Craddock, N., Kendler, K. et al. (2013). Identification of risk loci with shared effects on five major psychiatric disorders: a genome-wide analysis. *Lancet* 381 (9875): 1371–1379.

83 Palmen, S.J., Hulshoff Pol, H.E., Kemner, C. et al. (2005). Increased gray-matter volume in medication-naive high-functioning children with autism spectrum disorder. *Psychol. Med.* 35 (4): 561–570.

84 Sanfilipo, M., Lafargue, T., Arena, L. et al. (2000). Fine volumetric analysis of the cerebral ventricular system in schizophrenia: further evidence for multifocal mild to moderate enlargement. *Schizophr. Bull.* 26 (1): 201–216.

85 Shen, M.D., Nordahl, C.W., Young, G.S. et al. (2013). Early brain enlargement and elevated extra-axial fluid in infants who develop autism spectrum disorder. *Brain* 136 (Pt 9): 2825–2835.

86 Taurines, R., Schwenck, C., Westerwald, E. et al. (2012). ADHD and autism: differential diagnosis or overlapping traits? A selective review. *Atten Defic Hyperact Disord* 4 (3): 115–139.

87 Rommelse, N.N., Franke, B., Geurts, H.M. et al. (2010). Shared heritability of attention-deficit/hyperactivity disorder and autism spectrum disorder. *Eur. Child Adolesc. Psychiatry* 19 (3): 281–295.

148 Kyriakopoulou, V., Vatansever, D., Elkommos, S. et al. (2014). Cortical overgrowth in fetuses with isolated ventriculomegaly. *Cereb. Cortex* 24 (8): 2141–2150.
149 Li, Y., Estroff, J.A., Khwaja, O. et al. (2012). Callosal dysgenesis in fetuses with ventriculomegaly: levels of agreement between imaging modalities and postnatal outcome. *Ultrasound Obstet. Gynecol.* 40 (5): 522–529.
150 Manfredi, R., Tognolini, A., Bruno, C. et al. (2010). Agenesis of the corpus callosum in fetuses with mild ventriculomegaly: role of MR imaging. *Radiol. Med.* 115 (2): 301–312.
151 Ball, J.D., Abuhamad, A.Z., Mason, J.L. et al. (2013). Clinical outcomes of mild isolated cerebral ventriculomegaly in the presence of other neurodevelopmental risk factors. *J. Ultrasound Med.* 32 (11): 1933–1938.
152 Yates, M.A. and Juraska, J.M. (2007). Increases in size and myelination of the rat corpus callosum during adulthood are maintained into old age. *Brain Res.* 1142: 13–18.
153 Peterson, D., Mahajan, R., Crocetti, D. et al. (2015). Left-hemispheric microstructural abnormalities in children with high-functioning autism spectrum disorder. *Autism Res.* 8 (1): 61–72.
154 Rane, P., Cochran, D., Hodge, S.M. et al. (2015). Connectivity in autism: a review of MRI connectivity studies. *Harv. Rev. Psychiatry* 23 (4): 223–244.
155 Chen, J., Yao, Z., Qin, J. et al. (2015). Abnormal inter- and intra-hemispheric integration in male paranoid schizophrenia: a graph-theoretical analysis. *Shanghai Arch. Psychiatry* 27 (3): 158–166.
156 Ribolsi, M., Daskalakis, Z.J., Siracusano, A., and Koch, G. (2014). Abnormal asymmetry of brain connectivity in schizophrenia. *Front. Hum. Neurosci.* 8: 1010.
157 Sun, Y., Chen, Y., Collinson, S.L. et al. (2017). Reduced hemispheric asymmetry of brain anatomical networks is linked to schizophrenia: a connectome study. *Cereb. Cortex* 27 (1): 602–615.
158 Ercan, E.S., Suren, S., Bacanli, A. et al. (2016). Altered structural connectivity is related to attention deficit/hyperactivity subtypes: a DTI study. *Psychiatry Res.* 256: 57–64.
159 Barnea-Goraly, N., Kwon, H., Menon, V. et al. (2004). White matter structure in autism: preliminary evidence from diffusion tensor imaging. *Biol. Psychiatry* 55 (3): 323–326.
160 Casanova, M.F. and Pickett, J. (2013). The neuropathology of autism. In: *Imaging the Brain in Autism* (ed. M.F. Casanova, A. El-Baz and J.S. Suri), 27–43. New York: Springer.
161 Floris, D.L., Chura, L.R., Holt, R.J. et al. (2013). Psychological correlates of handedness and corpus callosum asymmetry in autism: the left hemisphere dysfunction theory revisited. *J. Autism Dev. Disord.* 43 (8): 1758–1772.

162 Forrester, G.S., Pegler, R., Thomas, M.S., and Mareschal, D. (2014). Handedness as a marker of cerebral lateralization in children with and without autism. *Behav. Brain Res.* 268: 14–21.

163 Kana, R.K., Libero, L.E., and Moore, M.S. (2011). Disrupted cortical connectivity theory as an explanatory model for autism spectrum disorders. *Phys Life Rev* 8 (4): 410–437.

164 Wass, S. (2011). Distortions and disconnections: disrupted brain connectivity in autism. *Brain Cogn.* 75 (1): 18–28.

165 Vidal, C.N., Nicolson, R., DeVito, T.J. et al. (2006). Mapping corpus callosum deficits in autism: an index of aberrant cortical connectivity. *Biol. Psychiatry* 60 (3): 218–225.

166 Hutchinson, A.D., Mathias, J.L., and Banich, M.T. (2008). Corpus callosum morphology in children and adolescents with attention deficit hyperactivity disorder: a meta-analytic review. *Neuropsychology* 22 (3): 341–349.

167 Schnoebelen, S., Semrud-Clikeman, M., and Pliszka, S.R. (2010). Corpus callosum anatomy in chronically treated and stimulant naive ADHD. *J. Atten. Disord.* 14 (3): 256–266.

168 Jacob, S., Brune, C.W., Badner, J.A. et al. (2011). Family-based association testing of glutamate transporter genes in autism. *Psychiatr. Genet.* 21 (4): 212–213.

169 Murugan, M., Ling, E.A., and Kaur, C. (2013). Glutamate receptors in microglia. *CNS Neurol. Disord. Drug Targets* 12 (6): 773–784.

170 Takeuchi, H., Mizuno, T., Zhang, G. et al. (2005). Neuritic beading induced by activated microglia is an early feature of neuronal dysfunction toward neuronal death by inhibition of mitochondrial respiration and axonal transport. *J. Biol. Chem.* 280 (11): 10444–10454.

171 Biber, K., Neumann, H., Inoue, K., and Boddeke, H.W. (2007). Neuronal 'On' and 'Off' signals control microglia. *Trends Neurosci.* 30 (11): 596–602.

172 Kondapalli, K.C., Hack, A., Schushan, M. et al. (2013). Functional evaluation of autism-associated mutations in NHE9. *Nat. Commun.* 4: 2510.

173 Chih, B., Engelman, H., and Scheiffele, P. (2005). Control of excitatory and inhibitory synapse formation by neuroligins. *Science* 307 (5713): 1324–1328.

174 de Bartolomeis, A., Fiore, G., and Iasevoli, F. (2005). Dopamine-glutamate interaction and antipsychotics mechanism of action: implication for new pharmacological strategies in psychosis. *Curr. Pharm. Des.* 11 (27): 3561–3594.

175 Reiderer, P., Lange, K.W., Kornhuber, J., and Danielczyk, W. (1992). Glutamatergic-dopaminergic balance in the brain. Its importance in motor disorders and schizophrenia. *Arzneimittelforschung* 42: 265–268.

176 Kriete, T. and Noelle, D.C. (2015). Dopamine and the development of executive dysfunction in autism spectrum disorders. *PLoS One* 10 (3): e0121605.

177 Staal, W.G. (2015). Autism, DRD3 and repetitive and stereotyped behavior, an overview of the current knowledge. *Eur. Neuropsychopharmacol.* 25 (9): 1421–1426.

178 Abi-Dargham, A. (2014). Schizophrenia: overview and dopamine dysfunction. *J. Clin. Psychiatry* 75 (11): e31.

179 Howes, O., McCutcheon, R., and Stone, J. (2015). Glutamate and dopamine in schizophrenia: an update for the 21st century. *J. Psychopharmacol.* 29 (2): 97–115.

180 Sharma, A. and Couture, J. (2014). A review of the pathophysiology, etiology, and treatment of attention-deficit hyperactivity disorder (ADHD). *Ann. Pharmacother.* 48 (2): 209–225.

181 Allen, J.L., Conrad, K., Oberdorster, G. et al. (2013). Developmental exposure to concentrated ambient particles and preference for immediate reward in mice. *Environ. Health Perspect.* 121 (1): 32–38.

182 Allen, J.L., Liu, X., Pelkowski, S. et al. (2014). Early postnatal exposure to ultrafine particulate matter air pollution: persistent ventriculomegaly, neurochemical disruption, and glial activation preferentially in male mice. *Environ. Health Perspect.* 122 (9): 939–945.

183 Allen, J.L., Liu, X., Weston, D. et al. (2014). Developmental exposure to concentrated ambient ultrafine particulate matter air pollution in mice results in persistent and sex-dependent behavioral neurotoxicity and glial activation. *Toxicol. Sci.* 140 (1): 160–178.

184 Allen, J.L., Liu, X., Weston, D. et al. (2014). Consequences of developmental exposure to concentrated ambient ultrafine particle air pollution combined with the adult paraquat and maneb model of the Parkinson's disease phenotype in male mice. *Neurotoxicology* 41: 80–88.

185 Allen, J.L., Oberdorster, G., Morris-Schaffer, K. et al. (2017). Developmental neurotoxicity of inhaled ambient ultrafine particle air pollution: parallels with neuropathological and behavioral features of autism and other neurodevelopmental disorders. *Neurotoxicology* 59: 140–154.

186 Clancy, B., Finlay, B.L., Darlington, R.B., and Anand, K.J. (2007). Extrapolating brain development from experimental species to humans. *Neurotoxicology* 28 (5): 931–937.

187 Clancy, B., Kersh, B., Hyde, J. et al. (2007). Web-based method for translating neurodevelopment from laboratory species to humans. *Neuroinformatics* 5 (1): 79–94.

188 Rice, D. and Barone, S. Jr. (2000). Critical periods of vulnerability for the developing nervous system: evidence from humans and animal models. *Environ. Health Perspect.* 108 (Suppl 3): 511–533.

189 Bandeira, F., Lent, R., and Herculano-Houzel, S. (2009). Changing numbers of neuronal and non-neuronal cells underlie postnatal brain growth in the rat. *Proc. Natl. Acad. Sci. U. S. A.* 106 (33): 14108–14113.

190 Bukowiecki, N., Dommen, J., Prevot, A.S.H. et al. (2002). A mobile pollutant measurement laboratory-measuring gas phase and aerosol ambient concentrations with high spatial and temporal resolution. *Atmos. Environ.* 36: 36–37.

191 Bukowiecki, N., Kittelson, D.B., Watts, W.F. et al. (2002). Real-time characterization of ultrafine and accumulation mode particles in ambient combustion aerosols. *J. Aerosol Sci.* 33: 1139–1154.

192 Kittelson, D.B., Watts, W.F., Johnson, J.P. et al. (2004). On-road exposure to highway aerosols. 1. Aerosol and gas measurements. *Inhal. Toxicol.* 16 (Suppl 1): 31–39.

193 Kittelson, D.B. (2004). Nanoparticle emissions on Minnesota highways. *Atmos. Environ.* 38: 9–19.

194 Pirjola, L., Parviainen, H., Hussein, T. et al. (2004). "Sniffer" – a novel tool for chasing vehicles and measuring traffic pollutants. *Atmos. Environ.* 38: 3625–3635.

195 Weijers, E.P., Khylstov, A.Y., Kos, G.P.A., and Erisman, J.W. (2004). Variability of particulate matter concentrations along roads and motorways determined by a moving measurement unit. *Atmos. Environ.* 38: 2993–3002.

196 Westerdahl, D., Fruin, S., Sax, T. et al. (2005). Mobile platform mesurements of ultrafine particles and associated pollutant concentrations on freeways and residential streets in Los Angeles. *Atmos. Environ.* 39: 3597–3610.

197 Chen, L., Hu, X., Ouyang, L. et al. (2016). A systematic review and meta-analysis of tract-based spatial statistics studies regarding attention-deficit/hyperactivity disorder. *Neurosci. Biobehav. Rev.* 68: 838–847.

198 Del Re, E.C., Konishi, J., Bouix, S. et al. (2016). Enlarged lateral ventricles inversely correlate with reduced corpus callosum central volume in first episode schizophrenia: association with functional measures. *Brain Imaging Behav.* 10 (4): 1264–1273.

199 Wahlsten, D. (1984). Growth of the mouse corpus callosum. *Brain Res.* 317 (1): 59–67.

200 Kim, J.H. and Juraska, J.M. (1997). Sex differences in the development of axon number in the splenium of the rat corpus callosum from postnatal day 15 through 60. *Brain Res. Dev. Brain Res.* 102 (1): 77–85.

201 Quaak, I., Brouns, M.R., and Van de Bor, M. (2013). The dynamics of autism spectrum disorders: how neurotoxic compounds and neurotransmitters interact. *Int. J. Environ. Res. Public Health* 10 (8): 3384–3408.

202 Pattij, T. and Vanderschuren, L.J. (2008). The neuropharmacology of impulsive behaviour. *Trends Pharmacol. Sci.* 29 (4): 192–199.

203 Naushad, S.M., Jain, J.M., Prasad, C.K. et al. (2013). Autistic children exhibit distinct plasma amino acid profile. *Indian J. Biochem. Biophys.* 50 (5): 474–478.

204 Kuwabara, H., Yamasue, H., Koike, S. et al. (2013). Altered metabolites in the plasma of autism spectrum disorder: a capillary electrophoresis time-of-flight mass spectroscopy study. *PLoS One* 8 (9): e73814.

205 Horder, J., Lavender, T., Mendez, M.A. et al. (2013). Reduced subcortical glutamate/glutamine in adults with autism spectrum disorders: a [(1)H]MRS study. *Transl. Psychiatry* 3: e279.

206 Bejjani, A., O'Neill, J., Kim, J.A. et al. (2012). Elevated glutamatergic compounds in pregenual anterior cingulate in pediatric autism spectrum disorder demonstrated by 1H MRS and 1H MRSI. *PLoS One* 7 (7): e38786.

207 Shimmura, C., Suda, S., Tsuchiya, K.J. et al. (2011). Alteration of plasma glutamate and glutamine levels in children with high-functioning autism. *PLoS One* 6 (10): e25340.

208 Choudhury, P.R., Lahiri, S., and Rajamma, U. (2012). Glutamate mediated signaling in the pathophysiology of autism spectrum disorders. *Pharmacol. Biochem. Behav.* 100 (4): 841–849.

209 Troyb, E., Rosenthal, M., Eigsti, I.M. et al. (2014). Executive functioning in individuals with a history of ASDs who have achieved optimal outcomes. *Child Neuropsychol.* 20 (4): 378–397.

210 Ravizza, S.M., Solomon, M., Ivry, R.B., and Carter, C.S. (2013). Restricted and repetitive behaviors in autism spectrum disorders: the relationship of attention and motor deficits. *Dev. Psychopathol.* 25 (3): 773–784.

211 Rosenthal, M., Wallace, G.L., Lawson, R. et al. (2013). Impairments in real-world executive function increase from childhood to adolescence in autism spectrum disorders. *Neuropsychology* 27 (1): 13–18.

212 Christ, S.E., Kester, L.E., Bodner, K.E. et al. (2011). Evidence for selective inhibitory impairment in individuals with autism spectrum disorder. *Neuropsychology* 25 (6): 690–701.

213 Lemon, J.M., Gargaro, B., Enticott, P.G., and Rinehart, N.J. (2011). Executive functioning in autism spectrum disorders: a gender comparison of response inhibition. *J. Autism Dev. Disord.* 41 (3): 352–356.

214 Scott-Van Zeeland, A.A., Dapretto, M., Ghahremani, D.G. et al. (2010). Reward processing in autism. *Autism Res.* 3 (2): 53–67.

215 Nigg, J.T. (2013). Attention deficits and hyperactivity-impulsivity: what have we learned, what next? *Dev. Psychopathol.* 25 (4 Pt 2): 1489–1503.

216 Hoptman, M.J. (2015). Impulsivity and aggression in schizophrenia: a neural circuitry perspective with implications for treatment. *CNS Spectr.* 20 (3): 280–286.

217 Brockel, B.J. and Cory-Slechta, D.A. (1998). Lead, attention, and impulsive behavior: changes in a fixed-ratio waiting-for-reward paradigm. *Pharmacol. Biochem. Behav.* 60: 545–552.

218 Elitt, C.M. and Rosenberg, P.A. (2014). The challenge of understanding cerebral white matter injury in the premature infant. *Neuroscience* 276: 216–238.

219 Back, S.A. (2006). Perinatal white matter injury: the changing spectrum of pathology and emerging insights into pathogenetic mechanisms. *Ment. Retard. Dev. Disabil. Res. Rev.* 12 (2): 129–140.

220 Melhem, E.R. et al. (2000). Periventricular leukomalacia: relationship between lateral ventricular volume on brain MR images and severity of cognitive and motor impairment. *Radiology* 214 (1): 199–204.

221 Perlman, J.M. (1998). White matter injury in the preterm infant: an important determination of abnormal neurodevelopment outcome. *Early Hum. Dev.* 53 (2): 99–120.

222 Wang, S., Fan, G., Xu, K., and Wang, C. (2013). Potential of diffusion tensor MR imaging in the assessment of cognitive impairments in children with periventricular leukomalacia born preterm. *Eur. J. Radiol.* 82 (1): 158–164.

223 Back, S.A. and Rosenberg, P.A. (2014). Pathophysiology of glia in perinatal white matter injury. *Glia* 62 (11): 1790–1815.

224 Shah, P.S. and Balkhair, T. (2011). Air pollution and birth outcomes: a systematic review. *Environ. Int.* 37 (2): 498–516.

225 Stieb, D.M., Chen, L., Eshoul, M., and Judek, S. (2012). Ambient air pollution, birth weight and preterm birth: a systematic review and meta-analysis. *Environ. Res.* 117: 100–111.

226 Gilliam, M., Stockman, M., Malek, M. et al. (2011). Developmental trajectories of the corpus callosum in attention-deficit/hyperactivity disorder. *Biol. Psychiatry* 69 (9): 839–846.

227 Guo, S., Kendrick, K.M., Zhang, J. et al. (2013). Brain-wide functional inter-hemispheric disconnection is a potential biomarker for schizophrenia and distinguishes it from depression. *Neuroimage Clin* 2: 818–826.

228 Leviton, A. and Gilles, F. (1996). Ventriculomegaly, delayed myelination, white matter hypoplasia, and "periventricular" leukomalacia: how are they related? *Pediatr. Neurol.* 15 (2): 127–136.

229 Schwarz, J.M., Sholar, P.W., and Bilbo, S.D. (2012). Sex differences in microglial colonization of the developing rat brain. *J. Neurochem.* 120 (6): 948–963.

230 Tabata, H., Yoshinaga, S., and Nakajima, K. (2012). Cytoarchitecture of mouse and human subventricular zone in developing cerebral neocortex. *Exp. Brain Res.* 216 (2): 161–168.

231 Tramontin, A.D., Garcia-Verdugo, J.M., Lim, D.A. et al. (2003). Postnatal development of radial glia and the ventricular zone (VZ): a continuum of the neural stem cell compartment. *Cereb. Cortex* 13 (6): 580–587.

232 Shigemoto-Mogami, Y., Hoshikawa, K., Goldman, J.E. et al. (2014). Microglia enhance neurogenesis and oligodendrogenesis in the early postnatal subventricular zone. *J. Neurosci.* 34 (6): 2231–2243.

233 Rezaie, P. and Dean, A. (2002). Periventricular leukomalacia, inflammation and white matter lesions within the developing nervous system. *Neuropathology* 22 (3): 106–132.

234 Kelly, F.J. and Fussell, J.C. (2015). Linking ambient particulate matter pollution effects with oxidative biology and immune responses. *Ann. N. Y. Acad. Sci.* 1340: 84–94.

235 Baburamani, A.A., Supramaniam, V.G., Hagberg, H., and Mallard, C. (2014). Microglia toxicity in preterm brain injury. *Reprod. Toxicol.* 48: 106–112.

236 Harry, G.J. and Kraft, A.D. (2012). Microglia in the developing brain: a potential target with lifetime effects. *Neurotoxicology* 33 (2): 191–206.

237 Chew, L.J., Fusar-Poli, P., and Schmitz, T. (2013). Oligodendroglial alterations and the role of microglia in white matter injury: relevance to schizophrenia. *Dev. Neurosci.* 35 (2–3): 102–129.

238 El Waly, B., Macchi, M., Cayre, M. et al. (2014). Oligodendrogenesis in the normal and pathological central nervous system. *Front. Neurosci.* 8: 145.

239 Liu, X.B., Shen, Y., Plane, J.M., and Deng, W. (2013). Vulnerability of premyelinating oligodendrocytes to white-matter damage in neonatal brain injury. *Neurosci. Bull.* 29 (2): 229–238.

240 Volpe, J.J., Kinney, H.C., Jensen, F.E., and Rosenberg, P.A. (2011). The developing oligodendrocyte: key cellular target in brain injury in the premature infant. *Int. J. Dev. Neurosci.* 29 (4): 423–440.

241 Bird, C.R., Hedberg, M., Drayer, B.P. et al. (1989). MR assessment of myelination in infants and children: usefulness of marker sites. *AJNR Am. J. Neuroradiol.* 10 (4): 731–740.

242 Vincze, A., Mazlo, M., Seress, L. et al. (2008). A correlative light and electron microscopic study of postnatal myelination in the murine corpus callosum. *Int. J. Dev. Neurosci.* 26 (6): 575–584.

243 Mengler, L., Khmelinskii, A., Diedenhofen, M. et al. (2014). Brain maturation of the adolescent rat cortex and striatum: changes in volume and myelination. *NeuroImage* 84: 35–44.

244 Aoki, Y., Abe, O., Nippashi, Y., and Yamasue, H. (2013). Comparison of white matter integrity between autism spectrum disorder subjects and typically developing individuals: a meta-analysis of diffusion tensor imaging tractography studies. *Mol. Autism.* 4 (1): 25.

245 Keller, T.A., Kana, R.K., and Just, M.A. (2007). A developmental study of the structural integrity of white matter in autism. *Neuroreport* 18 (1): 23–27.

246 Lener, M.S., Wong, E., Tang, C.Y. et al. (2015). White matter abnormalities in schizophrenia and schizotypal personality disorder. *Schizophr. Bull.* 41 (1): 300–310.

247 Langevin, L.M., Macmaster, F.P., Crawford, S. et al. (2014). Common white matter microstructure alterations in pediatric motor and attention disorders. *J. Pediatr.* 164 (5): 1157–1164 e1.

248 Chaim, T.M., Zhang, T., Zanetti, M.V. et al. (2014). Multimodal magnetic resonance imaging study of treatment-naive adults with attention-deficit/hyperactivity disorder. *PLoS One* 9 (10): e110199.

249 Newbury, A.J. and Rosen, G.D. (2012). Genetic, morphometric, and behavioral factors linked to the midsagittal area of the corpus callosum. *Front. Genet.* 3: 91.

250 Qiu, M.G., Ye, Z., Li, Q.Y. et al. (2011). Changes of brain structure and function in ADHD children. *Brain Topogr.* 24 (3–4): 243–252.

251 Cao, Q., Sun, L., Gong, G. et al. (2010). The macrostructural and microstructural abnormalities of corpus callosum in children with attention deficit/hyperactivity disorder: a combined morphometric and diffusion tensor MRI study. *Brain Res.* 1310: 172–180.

252 Dong, J., Shang, Y., Inthavong, K. et al. (2016). From the cover: comparative numerical modeling of inhaled nanoparticle deposition in human and rat nasal cavities. *Toxicol. Sci.* 152 (2): 284–296.

253 Garcia, G.J., Schroeter, J.D., and Kimbell, J.S. (2015). Olfactory deposition of inhaled nanoparticles in humans. *Inhal. Toxicol.* 27 (8): 394–403.

254 Pedata, P., Stoeger, T., Zimmermann, R. et al. (2015). Are we forgetting the smallest, sub 10 nm combustion generated particles? *Part. Fibre Toxicol.* 31 (12): 34.

255 Ronkko, T., Virtanen, A., Kannosto, J. et al. (2007). Nucleation mode particles with a nonvolatile core in the exhaust of a heavy duty diesel vehicle. *Environ. Sci. Technol.* 41: 6384–6389.

256 Bolton, J.L., Smith, S.H., Huff, N.C. et al. (2012). Prenatal air pollution exposure induces neuroinflammation and predisposes offspring to weight gain in adulthood in a sex-specific manner. *FASEB J.* 26 (11): 4743–4754.

257 Campbell, A., Araujo, J.A., Li, H. et al. (2009). Particulate matter induced enhancement of inflammatory markers in the brains of apolipoprotein E knockout mice. *J. Nanosci. Nanotechnol.* 9 (8): 5099–5104.

258 Hartz, A.M., Bauer, B., Block, M.L. et al. (2008). Diesel exhaust particles induce oxidative stress, proinflammatory signaling, and P-glycoprotein up-regulation at the blood-brain barrier. *FASEB J.* 22 (8): 2723–2733.

259 Levesque, S., Surace, M.J., McDonald, J. et al. (2011). Air pollution & the brain: subchronic diesel exhaust exposure causes neuroinflammation and elevates early markers of neurodegenerative disease. *J. Neuroinflammation* 8: 105.

260 Levesque, S., Taetzsch, T., Lull, M.E. et al. (2011). Diesel exhaust activates and primes microglia: air pollution, neuroinflammation, and regulation of dopaminergic neurotoxicity. *Environ. Health Perspect.* 119 (8): 1149–1155.

261 Morgan, C.P. and Bale, T.L. (2011). Early prenatal stress epigenetically programs dysmasculinization in second-generation offspring via the paternal lineage. *J. Neurosci.* 31 (33): 11748–11755.

262 van Berlo, D., Albrecht, C., Knaapen, A.M. et al. (2010). Comparative evaluation of the effects of short-term inhalation exposure to diesel engine exhaust on rat lung and brain. *Arch. Toxicol.* 84 (7): 553–562.

12

Toxicity of Nanomaterials to the Gastrointestinal Tract

Penelope A. Rice

Division of Food Contact Notifications, Office of Food Additive Safety, Center for Food Safety and Nutrition, Food and Drug Administration, College Park, MD 20740, USA

12.1 Introduction

New technologies that image, measure, model, and manipulate matter on the nanoscale (≤100 nm) have the potential to virtually affect every sector of the economy and our daily lives. According to the definition of nanomaterials put forward by the International Organization for Standardization (ISO) document series ISO TS 80004, the nanoscale is defined as being in the size range of ~1–100 nm, and nanomaterials are defined to be those materials with any external dimension in the nanoscale or having external or surface structure in the nanoscale [1]. These nanomaterials comprise several large families, which have diverse product applications, and the number of products containing, or claiming to contain, nanomaterials is growing, as evidenced by the fact that the U.S. Environmental Protection Agency (US EPA) has received over 160 new chemical notices under Toxic Substances Control Act (TSCA) for nanoscale materials since 2005 [2]. Many uses of these nanoscale materials may result in oral exposure to consumers. Oral exposure is an extremely important pathway to consider when assessing overall consumer exposure to nanomaterials. Direct oral exposure may occur via nano-enabled medicines or dietary supplements; toothpaste, mouthwash, or lip gloss that contain nanomaterials; migration of nanomaterials from food packaging into food; or direct addition of nanomaterials to food. For example, oral exposure to titanium dioxide and synthetic amorphous silica (SAS), a portion of which is nanosized, has been estimated to be 40 mg/person/day in the United Kingdom [3] from their uses as excipients in oral drugs, color additives, and ingredients in cosmetics, toothpaste, and sunscreen. Zinc oxide nanoparticles (ZnONPs) are

Toxicology of Nanoparticles and Nanomaterials in Human, Terrestrial and Aquatic Systems, First Edition. Edited by Marc A. Williams.

widely used in sunscreens and other cosmetics that may be associated with some adventitious oral exposure [4]. Internationally, silver nanoparticles (NPs) have a wide array of potential applications as antimicrobials in cosmetic products and food packaging [5]. Other materials that contain a particle size distribution extending into the nanoscale range that have been investigated for use in food or food contact applications include nanoclays; calcium, zinc, and iron nanoparticle supplements; nanogold, copper, palladium, and vanadium; nano-emulsions; and lipid nanoparticles [6]. Novel medical applications of nanoparticles include polymeric nanoparticles or nano-liposomes designed for increased systemic bioavailability of orally administered drugs [7].

Indirect oral exposure to nanomaterials may occur via environmental degradation of nano-enabled products and subsequent incorporation of these degradation products into drinking water and the food chain. For example, titanium dioxide, silicon dioxide, and silver nanoparticles may be incorporated into polymeric substances used directly in contact with food or as components of paints, siding, and other items. Degradation of these items in landfills may result in leaching of the nanoparticles into groundwater, which would then be used for drinking water or irrigation of crops [8].

Nanoparticulates that are naturally present in the external environment may also become incorporated in food or drinking water, resulting in oral exposure to these materials. For instance, amorphous nanoparticulate forms of carbon are produced in food during cooking [9]. Fullerenes or carbon nanotubes (CNTs), such as single-walled carbon nanotubes (SWCNTs) or multiwalled carbon nanotubes (MWCNTs), produced as combustion products from natural or man-made fires may become incorporated in food [10]. Silver or copper nanoparticles produced from photoreduction of Ag- or Cu-ions released from copper or silver objects may adventitiously contaminate food or water in contact with these materials [11]. The intestinal epithelium is also naturally exposed to calcium phosphate nanoparticles that form within the lumen of the small intestine [3]. Therefore, oral exposure to nanoparticles is neither rare nor new; the human body has always been exposed to nanoparticulates of various types, and the structure and function of the gastrointestinal tract (GIT) reflects, in part, the body's adaptation to such chronic oral exposure.

The GIT is the human body's barrier between all ingested substances and the systemic circulation and internal organs, such as the brain or the liver. Maintenance of an intact, healthy GIT is essential for limiting systemic exposure to ingested substances, and any substance that compromises this barrier, either through direct toxicity or decreased barrier function, has the potential to display systemic toxicity as well. Therefore, the issue of toxicity to the GIT should be considered when assessing the overall toxicity of oral exposure This chapter will summarize the available evidence concerning the direct effects of nanomaterials on the GIT, focusing on the nanomaterials with greatest potential for oral exposure: titanium dioxide, silica, nanosilver, ZnONP, fullerenes and CNTs, and other selected metal oxide nanomaterials. It should be emphasized that this review considers all potential sources of oral exposure to nanomaterials, except for nano-enabled drugs. Although nano-enabled drugs are also a source of oral exposure to nanomaterials, the toxicity of these materials is beyond the scope of this chapter due to space limitations and the fact that data for the nano-carrier alone (no drug) is either nonexistent or limited to uptake studies only. The chapter will also only briefly discuss the effects of nanomaterials on the immune system and will not address the effects of nanomaterials on intestinal microbiota, as these subjects are covered elsewhere in this book. Furthermore, as almost all of the available data concern effects of nanomaterials on the intestine and the majority of systemic exposure to orally administered nanomaterials occur via intestinal uptake, the effects of nanomaterial on intestinal physiology and barrier function will be the particular focus of this chapter. Before discussing the toxicity data for the selected nanomaterials, the chapter will briefly review the structure and function of the GIT and commonly used methods for assessing the toxicity of chemicals to the GIT. The chapter will also briefly discuss the challenges inherent in the toxicity testing of nanomaterials. An overall analysis of the quality and data-trends follows the discussions of nanomaterial-specific toxicity data.

12.2 GI Physiology and Toxicity Testing

The GIT is a four-layered tube comprising the pharynx and oral cavity, the esophagus, the stomach, the small and large intestines, and the rectum (as reviewed by [12]). Its purpose is to break down ingested substances, absorb components of the substance that are required for physiological function, and expel those components that are either potentially harmful or that have no nutritional value. The GIT is a barrier between pathogens, toxins, and abrasive particulates in the ingesta and the systemic circulation and participates in innate immune responses against ingested pathogens. Ingested substances are mechanically broken down by the teeth in the oral cavity followed by acidic and proteolytic hydrolysis in the stomach to form chyme. After passage into the duodenum through the gastric pyloric sphincter, the chyme is mixed with bile, comprised of bile salts cholesterol and phospholipids, and pancreatic secretions containing enzymes that hydrolyze proteins, carbohydrates, and lipids into their constituent amino acids, monosaccharides, monoacyl glycerides, and fatty acids [12]. These constituents are then absorbed through the wall of the small intestine. The non-absorbed portion is then passed to the large intestine (colon and cecum), which absorbs water, sodium, fatty acids, and minerals from the ingesta, forming it into a compact mass prior to passage out of the body as feces through the rectum and anus.

The four layers that comprise the GIT are, in order from lumen to outer wall, the mucosa, the submucosa, the smooth muscle layer, and the adventitia (as reviewed in [13]). The mucosa is lined on the luminal side by a layer of epithelial cells, beneath which is the lamina propria. The lamina propria contains large numbers of lymphatic vessels and capillaries through which substances absorbed by epithelial cells are transported to the rest of the body via the large blood vessels and lymphatics of the submucosa. The lamina propria also contains large numbers of immune cells that participate in immune surveillance of luminal environment. The enteric nervous system in the submucosa and smooth muscle layer regulates secretion, fluid transport, and blood flow in the submucosa and mucosal layers, coordinates peristaltic contractions in the smooth muscle layer, and modulates GIT immune function via synaptic contacts with immune cells in the lamina propria.

The type of epithelial cell lining the mucosal layer is specialized for each compartment of the GIT, and significant species differences exist with respect to each compartment's physiological parameters [12]. In areas of the body where nutrient absorption is minimal, such as the oral cavity, pharynx, esophagus, and anus, the function of the epithelium is primarily to protect deeper layers of the GIT from damage by abrasive or corrosive elements present in the ingesta; as such, these compartments are lined with a stratified squamous epithelium interspersed with mucus-secreting goblet cells.

The stomach is lined with long, close-packed tubular glands that secrete acid and pepsinogen (the secretory epithelium), containing neck cells that secrete large amounts of thick, alkaline mucus, pepsinogen-secreting chief cells, and acid-secreting parietal cells [12]. The pH of the gastric lumen in humans varies between 1.7 in the fasted state and 5 in the fed state; although similar in the fasted state to commonly used animal models, human fed-state gastric pH is significantly more alkaline than the value of 2 for pigs and dogs [12] and 2.98 and 3.2 for mice and rats, respectively [14]. Gastric emptying times vary by species also (99.8 minutes in dogs vs. 59.7 minutes in humans [15]]). Gastric anatomy and distribution of cell types within the gastric gland differs significantly between humans and the most commonly used laboratory animal models, rodents, non-human primates, and dogs. For instance, while dogs, monkeys, and humans possess a glandular-type stomach, rodents' stomachs are divided into a glandular portion similar to humans' stomachs and a non-glandular portion, the forestomach [12]. Unlike the thick mucus covering and protecting the glandular portion of the stomach, the forestomach is covered by a relatively thin layer of mucus and a thin, stratified squamous epithelium. As such, the rodent forestomach is uniquely vulnerable to lesions related to gavage administration of highly irritating substances.

Once the chyme enters the small intestine, it passes through the primary absorptive portion of the GIT: the duodenum, the jejunum, and the ileum, in order, with the jejunum and ileum having the largest absorptive surface areas within the GIT of 60 m^2 each in humans [12]. The small intestine mucosal surface is formed into villi, which increases the absorptive surface area of the mucosa. Small intestinal length, numbers of villi/unit area, and villi shape vary significantly by species [12]. These villi are lined with a single layer of enterocytes linked together by tight junctions interspersed with goblet cells, antibacterial-peptide-secreting Paneth cells, enteroendocrine cells, and tuft cells [16]. M-cells, specialized enterocytes covering the lymphoid Peyer's patches (PPs) in the small intestine, sample the intestinal lumen for antigenic stimuli; areas containing these M-cells are known as the follicle-associated epithelium (FAE). The numbers and distribution of PP also varies between species and organism life stage; human PPs are mainly located in the jejunum, with some ileal patches, and vary in number from 59 in a 30-week-old fetus to 239 at puberty [12]. In contrast, rats and mice have significant amounts of lymphoid tissue in the cecum and colon as well as in the small intestine, and the total numbers of follicles in the GIT range 49–160 in mice and 166–382 in rats [14]). The pH and water content of the intestinal fluid also differs significantly between humans and rodents; rodent intestinal contents are significantly more acidic (pH of 4–5 vs. 5.5–6.5 in humans) and have threefold to fivefold more fluid content [14, 15]. Species differences in total absorptive surface area, especially the total amount of absorptive surface represented by FAE covering PP, pH, bile content and flow rate, and fecal water content may result in significant differences in nanomaterial physicochemical and toxic properties in the gut, which may confound extrapolation of the results of animal studies to humans.

The mucosa of the colon and cecum is comprised of specialized water-absorbing enterocytes and goblet cells and is divided into transverse furrows and sacculated; the cecum region in humans is continuous with the colon, which is divided into four sections [12]. Here, too, there are significant species differences with respect to the relative colon and cecal length and size, the degree of differentiation of the cecum, the presence or absence of sacculation, and the presence or absence of lymphoid follicles in the sub-epithelium [12, 14].

Absorption of ingesta primarily occurs in small intestine enterocytes. Enterocytes are tall, columnar polarized cells, with the apical pole in contact with the intestinal lumen and the basolateral pole in contact with the systemic circulation and dendritic cells, macrophages, and intraepithelial lymphocytes in the sub-epithelium [16]. The apical pole has a brush border composed of microvilli, which increases the absorptive surface area of the cell. The brush border is covered with transmembrane mucins (MUC1, MUC3, MUC4, MUC12,

MUC13, MUC17), known as the glycocalyx in aggregate, which serve as a protective barrier and participate in signal transduction via their cytoplasmic tails [16]. Dipeptidases, oligo-1,6-glucosidase, glucoamylase, saccharase, lactase, maltase, enterokinase, and intestinal lipase are embedded in the glycocalyx and participate in the digestion of absorbed food to monosaccharides and amino acids, which are then endocytosed into the enterocyte via special transport systems and transferred to capillaries at the basolateral pole [17]. Fatty acids are either actively transported or, as part of micelles, passively diffuse into the enterocyte, where they are packaged into chylomicrons and transferred to the lymphatics [17, 18]. Lipophilic compounds may diffuse unhindered from the apical to the basolateral pole, whereas more hydrophilic compounds are either taken up via special transporters and transcytosed across the cell or, if the compound is ≤5 nm in diameter, it may enter the systemic circulation via paracellular transport [19]. Enterocytes also express several drug transport proteins within the brush border, including peptide transporters PepT1 and HPT1, P-glycoprotein (MDR1), and organic anion transporter proteins (OATPs) [20]. Enterocyte expression of toll receptors (TLR) and nucleotide-binding oligomerization domain (NOD-R)-like receptors coupled with the NF-kappa-B pathway and a B7-related protein and secretion of inflammatory cytokines allows modulation of T-cell function in the lamina propria [16]. A subpopulation of goblet cells can also take up antigen via the autophagy/secretory pathway and present that antigen to dendritic cells in the lamina propria, but this process has a size-exclusion limit of 10 kDa [16].

M-cells in the FAE can take up bacteria, viruses, protozoa, and particles of up to 10 μm in diameter via phagocytosis [21]. These cells are lower and flatter than the intestinal enterocyte, contain microfolds instead of villi on their apical side, and are in direct contact with T- and B-lymphocytes in the lamia propria [17]. Although the uptake process itself is rapid and facile due to the fact that the thinner mucus layer covering these cells, the mass rate of transport through M-cells is extremely low because M-cells cover <1% of the intestinal surface [22]. Uptake of particulates chiefly occurs through the FAE, with transport being the fastest and most extensive for particles with diameters between 20 and 250 nm [3]. Once taken up by M-cells, submicron-sized particulates are transported to PP in the underlying subepithelial layer, which are part of the gut-associated lymphoid tissue (GALT), and subsequently translocated to lymph nodes throughout the body, whereas 3–10 μm particulates appear to be sequestered in PP [21].

Mucus is secreted throughout the GIT by goblet cells embedded in the epithelium and is a vital component of the barrier function of the GIT. Mucus is formed from 0.5 to 40 MDa O-linked, branched glycoproteins (mucins) secreted by goblet cells, which is intermixed with lipids, bacteria, antibodies, carbohydrates, and cellular debris [22]. The mucins produced by goblet cells vary by region of the GIT, with MUC2, MUC5AC and MUC6, and MUC5B being the primary mucin proteins secreted by goblet cells in the intestine, stomach, and salivary gland, respectively [16]. The glycosylated side chains of mucins interact to form a hydrophilic mesh interspersed with hydrophobic regions formed from the lipid-coated, non-glycosylated portions of the mucin protein [22]. The mucus layer may only be visualized histopathologically via use of Carnoy's fixative during sample preparation; otherwise, dehydration of the sample causes the mucus layer to shrink into invisibility [16]. This mesh is resistant to proteolysis and retards the progress of particles or other substances through the thickness of the mucus layer based on particle size, charge, and lipophilicity, with small (<100 nm), neutrally charged particles being more mobile than large, highly charged particles [22]. In the small intestine, the mucus only forms a single layer, whereas mucus in the colon forms a bilayered structure: an unstirred, adhesive layer on top of the epithelium and a loosely adherent layer in contact with the ingesta [16]. The high turnover rate of the loosely adherent layer decreases the residence time of substances entrapped in the mucus. However, in some cases, mucoadhesive particles show prolonged intestinal retention times, and this property has been deliberately engineered into nanoparticle drug carriers for improved oral bioavailability [22]. The thickness and turnover rate of mucus varies throughout the gut, with the thickest mucus layer being present in the stomach and colon in humans [22]. Dietary components, such as fiber, water intake, and certain toxins, may also affect mucus barrier properties and turnover rate by affecting the viscoelastic properties of the hydrogel or increasing mucus flow. Due to its extremely high turnover and ability to entrap substances of highly varied size and composition, the mucus layer represents a formidable barrier to the gastrointestinal (GI) absorption of nanoparticles and is, no doubt, one of the reasons why animal studies have reported such low values for systemic bioavailability of ingested nanomaterials.

From the preceding mini-review of GI physiology, it should be clear that a nanomaterial entering the GIT will experience a wide range of varying environments that will, in turn, affect the physicochemical characteristics of the nanomaterial. Moreover, how this modified nanomaterial interacts with the mucus layer and with each of the specialized types of cells present in different regions of the GIT will determine what toxic effects may be expected upon ingestion and what types of test systems are most appropriate for use in characterizing the toxicity of the nanomaterial to the GIT.

Traditionally, toxicologists have used oral toxicity studies of varying duration in rodent or non-rodent animal models to assess the toxicological profile of a given substance. These studies usually include gross and microscopic assessment of organs of the GIT, along with observations of clinical signs related to adverse effects on the gut, such as alterations in motility (constipation, diarrhea), vomiting (in non-rodent species), and gross alterations in stools, such as the presence of blood [23]. Oral toxicity studies use a method of administration that closely models human exposures, such as exposure via drinking water or food; or, for rodents, gavage administration may be used [23]. Although the protocols for oral toxicity testing have been well established for decades, nanomaterials may present unique challenges when used in these assays due to the fact that their physicochemical properties are highly dependent upon the environment to which they are exposed, which includes the exposure milieu (gavage vs. drinking water) and environmental conditions within the GIT. Species differences in the location and numbers of PP, the relative length of intestinal segments, GI transit times and pH, digesta water content, numbers of villi/mm^2, differences in bile flow, and differences in mucus layer thickness may affect the interaction of nanomaterials with the GIT such that interspecies extrapolation becomes extremely complex [12]. The high volumes of fluid typically used in gavage studies (5–10 mL/kg) may disrupt the protective mucus layer of the stomach and small intestine, leading to greater exposure of the epithelium to the test article than would occur if other methods of test article administration are used [24]. This is particularly concerning for evaluating nanomaterials, given that mucus normally represents a formidable barrier to particulate uptake. Additional confounding factors related to gavage administration include lack of nanomaterial exposure to the buccal cavity, absence of chewing forces, potential gavage accidents (aspiration, esophageal tears), handling stress, abnormal GI transit times, and nonspecific binding of the nanomaterial to the gavage tubing [25]. Possibly the least confounded and yet most human-relevant method of administration is via capsule or tablet, which is only possible with non-rodents.

Animal models of inflammatory bowel disease (IBD) have also been used to assess the effects on nanomaterials on the GIT in a compromised state. These models involve pretreating mice with sulfonic acid derivatives, producing an acute, inflammatory injury to the intestinal epithelium that recapitulates some histopathological aspects of human IBD [26]. Disadvantages of this model include high mortality rate, species differences in small intestine length and extent of histopathological alteration along the length of the gut, lack of responsiveness to commonly used drugs for treating IBD, and the fact that IBD in humans is a chronic, immune-mediated disease [26].

Additional models of intestinal absorption in the rat, such as the everted and non-everted gut sac model, the intestinal perfusion and loop models, and the Ussing chamber method have been widely used to assess the intestinal absorption of medicines encapsulated in nanomaterial carriers [27–29]. These methods [25, 30] have the advantage of assessing nanomaterial interaction with the intestinal mucosa in a physiologically relevant setting, where all or most of the relevant cell types and anatomical structures are intact and the mucus barrier is present. The disadvantages to these methods include the fact that the nanomaterials bypass the other portions of the GIT, which themselves may alter the toxicological properties of the nanomaterial, the morphological changes in the tissue due to handling or degeneration, and the absence of the normal food and gut flora matrix. In addition, as these models are primarily meant to assess drug/nanoparticle transport, the models are unsuitable for use in assessing the toxicity of repeated doses of nanomaterials to the intestinal epithelium, due to the fact that the tissues degenerate relatively quickly *ex vivo* [30]. In particular, nanoparticle transport using the Using chamber method, in which isolated small intestinal mucosa is fixed between the receiving and diffusion pools of a perfusion device, or the rat gut sac method, in which a ligated segment of intestine is cultured *ex vivo* in a nutrient bath, can only be assessed for very short time periods because of degeneration of the mucosa due to isolation from its blood supply [30]. Although the blood supply and neural network of the intestine in intestinal perfusion models are intact, the studies are performed in an anesthetized animal, which limits the model to assessing single dose, acute effects only [30].

Given the cost, time, and confounding factors of tests in mammalian animal models, toxicologists have developed several cell-based model systems to assess the effects of various test materials on the GIT. The vast majority of these test systems are comprised of one or more cell lines intended to model the mucosal lining of the small intestine. The most frequently used model system is the differentiated Caco-2 cell cultured in transwell plates. The Caco-2 cell line was originally derived from human colon carcinoma cells, and there have since been several different cell lineages generated from the original clonal line [31]. When cultured for 14–21 days, Caco-2 cells form a confluent monolayer displaying many of the characteristics of small intestine epithelium, including polarization, microvilli on the apical cell surface, formation of tight junctions with neighboring cells, and expression of transporter proteins [17, 32]. The formation of the differentiated epithelial structure is assessed by monitoring brush border hydrolase alkaline phosphatase expression or transepithelial electrical resistance (TEER) of the differentiated monolayer [32]. Effects of test substances on cell viability may be assessed in culture using various methods, and effects on tight junction integrity may be assessed via the effects of the test substance on monolayer TEER, transport of hydrophilic markers such as [14C]-mannitol from the apical to the basolateral side of the culture, and expression of tight junction proteins [20]. Disruption of barrier function by the test article in this context would result in decreased TEER values and increased permeability to hydrophilic markers. There are several disadvantages to using this cell line, including absence of a mucus layer, increased TEER values and lower paracellular permeability compared with small intestine epithelium, and large interlaboratory variation in paracellular permeability values for Caco-2 isolates [17, 19, 31]. Culture conditions have not been standardized for this cell line; variations in passage number, presence or absence of serum in the culture media, cell density at initial plating, culture on solid vs. permeable surface, polymer type comprising the cell culture surface, and time in culture prior to use have significant effects on cell morphology and differentiation as well as epithelial barrier integrity [25, 33]. The lack of a mucus barrier and of a standardized culture protocol for Caco-2 cells may confound interpretation of nanotoxicity assays using this model system, as the system may overestimate the potential toxicity of the nanomaterial to the gut, and comparing results for a given test material between different studies may be impossible due to variable Caco-2 culture conditions, even if the test article is sufficiently characterized as to its physicochemical properties.

Recent publications have described co-culture model systems using two or more cell lines that attempt to more faithfully reproduce intestinal physiology. Co-cultures of differentiated Caco-2 cells and HT-29 cells [34], which differentiate into a mucus-secreting goblet cell phenotype, and triple co-cultures of differentiated Caco-2 cells, HT-29, and Raji-B cells that reproduce an M-cell/PP phenotype in culture [35] have been used to assess the toxicity of nanomaterials to the GIT. Additional models used with nanomaterials include cultures of various intestinal cell lines on semi-wet interfaces in the presence of mechanical stimulation [36] and culture of Caco-2 cells under everted or microgravity conditions [37]. A published triple culture system of differentiated Caco-2 cells, macrophages, and dendritic cells reportedly closely recapitulates the interaction between enterocytes and antigen-presenting cells in the lamina propria and has been used to model the interaction of nanoparticles with the intestinal epithelium in both the intact and a diseased state [26].

12.3 Nanomaterial Toxicity Assessment: Challenges

It should be emphasized that the cell-based test systems described earlier have traditionally been used to assess the toxicity and pharmacokinetics of test materials that satisfy four basic assumptions: (i) solubility in aqueous media, (ii) no interference with the assays used to assess toxicity, (iii) biological relevance of the observed toxicity, and (iv) equivalence of test article between the cell-based system and *in vivo*. Nanomaterials may violate one or all of these assumptions, as will be discussed in the following text.

Cell-based systems generally assume that distribution of the test material within the test medium is homogeneous and that each test cell receives an equal dose of the test article. However, nanomaterials may remain partially or completely in particulate form in media, and the media's high protein content and ionic strength may produce significant changes in nanomaterial hydrodynamic size, size distribution, and zeta potential, triggering aggregate formation [38]. Partially or completely insoluble nanomaterials exhibit gravitational settling, with larger agglomerates and aggregates settling faster than smaller single particles. Therefore, instead of each cell receiving the same, stable dose over the exposure period, cells receive a heterogeneous distribution of doses that steadily increases over the exposure period as greater numbers of particles settle to the bottom of the culture vessel [39]. Cells in suspension also receive a heterogeneous dose and see a steady diminution of dose as the test article settles out of solution and/or adheres to the culture vessel walls. The gravitational settling of particulates may be so extensive as to form a micron-thick layer over the cells, which prevents oxygen and nutrients from reaching them, effectively smothering the cells and preventing assessment of biologically relevant effects of the test article [39]. Interpolation from an effective dose value *in vitro* to *in vivo* toxicity would be difficult in this context, due to the highly complex nature of the dosimetry *in vitro* [38]. Several groups have published strategies for evaluating nanomaterial dosimetry in cell-based systems; however, these strategies seem to be mainly applicable to metal oxide nanoparticles or quantum dots of roughly spherical morphology [40, 41].

Potential interference with fluorescence- or absorbance-based assay systems is another issue. The fact that carbonaceous nanomaterials, like fullerenes and nanotubes, can interfere with the MTT assay is well-known [42]. However, potential interference with other assay systems is still not widely recognized or accounted for in the nanotoxicology literature. Metallic nanoparticles, such as nanosilver or nanogold, can either enhance or quench the spectral properties of the fluorophores used in assays (as reviewed in Ref. [43] to detect cytotoxicity (lactate dehydrogenase [LDH] release; WST-1 assay), reactive oxygen species (ROS) (oxidation of dichloro-dihydro-fluorescein diacetate [DCFH-DA]), glutathione content, and changes in intracellular Ca concentration, leading to potential artifacts [42, 44–46]. Carbonaceous materials may interfere with both fluorescent and colorimetric assay systems by absorbing and/or neutralizing the readout dye [42]. Small metallic and metal oxide nanoparticles (4–15 nm) are known to strongly absorb light at most of the wavelengths commonly used in colorimetric assays and can interfere with the readout of optical density or absorbance at certain wavelengths [47]. Large aggregates may also interfere with COMET or cytokinesis-block micronucleus assays, either by obstructing the view of the comet tail or cell [44] or by reacting with isolated DNA from exposed cells, producing post-hoc DNA damage. Nanomaterials may even interfere with the efficiency of RNA extraction from cells [47] and reverse transcription efficiency during PCR [48]. Because of these factors, published analyses have recommended that use of multiple orthogonal assay systems to measure the same end point or to include a control within each assay to account for any interference of the nanomaterial with the test system [49–51].

Several reviews have discussed the issue of non-biologically relevant cytotoxicity produced by contaminants in nanomaterials. For instance, cytotoxic impurities from the nanomaterial production process may account for the cytotoxicity seen with the nanomaterial itself [52, 53]. Nanomaterials may also produce cytotoxicity by absorbing nutrients or other necessary components of the medium, starving the cells [42, 54]. Endotoxin contamination may produce cytotoxicity and confound assays assessing immune function end points, such as cytokine production [55]. Unfortunately, some nanomaterials may also interfere with the *Limulus* amebocyte assay used to measure endotoxin levels [55]. Finally, the presence or absence of serum in the exposure medium may also significantly alter the nanomaterial's toxicological properties *in vitro*, with a reduction in toxicity noted in the presence of serum in many cases [56].

Differences in the test material environment between cell-based studies and animal studies may also complicate interpretation of oral toxicity data for nanomaterials, as the physicochemical properties of nanomaterials are heavily matrix dependent. For instance, during digestion, nanomaterials may undergo numerous processes that can affect their size distribution, such as flocculation, coalescence, and Ostwald ripening [57]; or they may be partially or completely dissolved in the acidic environment of the stomach or the lumen of the small intestine, a phenomenon particularly relevant for partially or completely soluble metallic nanomaterials. Ingested nanomaterials can conceivably interact with several components of the food matrix [58], producing a corona that will be divergent from that produced after incubation with cell culture medium [49]. The divergence in corona composition between an ingested nanomaterial and a nanomaterial suspended in cell culture media may produce differences in toxicological effects *in vivo* vs. in cell-based systems for the same nanomaterial [59]. The amount of protein present in the exposure matrix will also affect the hydrodynamic size and size distribution of the nanomaterial dependent upon the kinetics and extent of protein absorption on the nanomaterial surface and the ability of a given protein to bridge adjacent nanoparticles [60]. Studies have found that low protein concentrations can increase aggregate formation compared with protein-free aqueous solutions, whereas high protein concentrations decrease aggregate formation. This effect is mediated by unfolded proteins on the surface of the nanomaterial binding to adjacent nanomaterial surfaces [61]. The sections in this chapter discussing the available data on the toxicity of specific nanomaterials will also discuss the effects of digestion on the physicochemical characteristics of those nanomaterials in order to provide a perspective from which to analyze the biological relevance of *in vitro* data.

As should be clear from the preceding discussion, assessment of the toxicity of nanomaterials to the GIT involves consideration of a multitude of potentially confounding factors, particularly when using a cell-based assay system. The majority of the published toxicity studies discussed in the following sections do not appear to have considered these factors in their design, as will be discussed in more detail in Section 12.6. The following section for each specific nanomaterial will first discuss the sources of oral exposure to these materials, followed by the discussion of the effects of digestion on the physicochemical characteristics of the nanomaterial, the evidence of toxicity to the GIT from animal studies, the evidence of toxicity to the GIT from cell-based studies, and the uptake of each nanomaterial type from the GIT.

12.4 Toxicity of Specific Nanomaterial Types

12.4.1 Titanium Dioxide

TiO_2 is one of the most common nanomaterials to which consumers are exposed due to its widespread use as a food additive. Food-grade TiO_2 (E171 in Europe) is a mixture of 30–400 nm diameter spherical rutile or anatase particles, of which a fraction (reported for one sample to be c. 36%) is in the nanoscale range of 1–100 nm in diameter [62]. A nanosized TiO_2 particle product, P25, is also commercially used as a photocatalyst, a catalyst carrier, or a heat stabilizer in rubber; P25 is a 15/85 mixture of 30–40 nm rutile/anatase particles [62]. TiO_2 is also used in sunscreen and lip balm, in biocidal and UV-protectant coatings, and in paint for surface applications [8, 62]. Nano-TiO_2, a non-food-grade material also known as P25, is one of the most frequently occurring nanomaterials in waste due to its presence in building and demolition waste, plastics, textiles, electronics waste, and scrapped motor vehicles [8]; as such, wastewater contamination with TiO_2 is extremely prevalent. Oral exposure to TiO_2 for the average US adult was estimated to be 0.3 mg/person/day from drinking water and 0.2–0.7 mg/kg bw/day from food [62], of which some fraction will be nanosized particles.

Following oral ingestion, TiO_2 undergoes significant transformation during the process of digestion. One study reported that incubating a TiO_2 nanoparticle (TiO_2-NP) test article in artificial gastric fluid (AGF) increased the mean hydrodynamic diameter and zeta potential from 473.6 nm and −33.46 mV in water to 1702 nm and + 6.98 mV [63]; subsequent incubation in artificial intestinal fluid (AIF) decreased the nanoparticles' zeta potential to −2.47 mV, while the hydrodynamic size further increased to ~2 μm. Increased TiO_2-NP hydrodynamic size after incubation with artificial digestive fluids has been independently confirmed in other published studies [64, 65], although another published study reported no change in hydrodynamic size after incubation of TiO_2-NP in media with serum followed by AGF [66].

The majority of the *in vivo* and cell-based studies report very little uptake of orally administered TiO_2-NP into the systemic circulation of rats [63, 67–69] or translocation to the basolateral pole of Caco-2 cells [64–66, 69–74]. However, a small fraction of the administered dose of nanoparticles is taken up into PP and mesenteric lymph nodes (MLNs) in rodents [65, 67, 70, 75, 76]. This uptake into the lymphatic circulation is recapitulated in Caco-2/RajiB co-culture model systems, indicating that an M-cell phenotype is necessary for systemic uptake of nano-TiO_2 [65]. In contrast to the minimal uptake of nanoparticles in rats, studies conducted in mice report significantly higher TiO_2-NP uptake from the gut and translocation into non-GIT tissues, particularly the spleen and thymus [77–79]. These disparate results may be due to anatomical differences in the GIT between the two species or compromised barrier function in mice.

The extent of TiO_2 uptake from the GIT in humans is uncertain. A recently published physiologically based pharmacokinetic model for human systemic exposure to dietary TiO_2 estimated that only 0.06% of a given oral dose of TiO_2 (particle size range not specified) is absorbed into the systemic circulation [80], which would appear to rule out significant accumulation of TiO_2-NPs in secondary lymphoid organs like the spleen in humans. However, studies have shown that TiO_2 particulates are present in tissue macrophages, also known as pigment cells, in the terminal ileum [81, 82], although their presence in these macrophages does not seem to correlate with markers of activation in these cells. An additional study reported increased Ti concentrations in the serum of patients with IBD, as compared with healthy controls [83]. It should be noted that one study also reported the presence of TiO_2-NP in the blood of healthy volunteers after capsule administration of 50 mg food-grade 260 nm anatase TiO_2-NP; however, this study did not confirm via inductively coupled plasma mass spectrometry (ICP-MS) or another technique that the observed particulates were, in fact, TiO_2 [84]. An additional oral uptake study performed in healthy volunteers reported no changes in Ti concentrations in blood and urine samples after single administration of oral doses of nanosized or micron-sized TiO_2 [69]. Therefore, it would seem that uptake of TiO_2-NP in healthy humans is limited to accumulation in regional lymphoid tissue, although increased systemic exposure to ingested TiO_2-NP cannot be ruled out in disease states associated with compromised GIT barrier function.

Many of the oral toxicity studies conducted with TiO_2-NP did not assess the effects of the test substance on the GIT. A literature search located only nine such studies that assessed apical end points relevant to the assessment of GIT toxicity, such as histopathology of some or all GI organs, organ weights, transport function, or other markers of mucosal toxicity. In contrast, 15 studies conducted with TiO_2-NPs in mono- or co-culture systems using various intestinal cell lines were available (Table 12.1). The majority of studies did not identify adverse histopathological effects of oral exposure toTiO_2-NPs on the GIT of healthy rats [63, 76, 85, 87], and one study reported no effects of TiO_2-NPs on the GIT of healthy mice [83]. In contrast, several studies have noted adverse effects of TiO_2-NP administration on the murine GIT. Three studies conducted in mice with TiO_2-NPs reported adverse effects of the test article on the stomach, from increased bodyweight-relative stomach weights [78] to mucosal inflammation [77] and even ulceration and necrosis [88]. Although increased numbers of mast cells were noted in the stomach mucosa of 3-week-old rats gavaged with TiO_2-NP for 30 days, this finding was not present in adult rats in the same study, and the toxicological significance of this finding is uncertain [63]. Additional toxicological effects were noted in the jejunum and ileum of TiO_2-NP-treated mice. A study conducted with two sizes of uncoated anatase TiO_2-NPs, 260 and 60 nm, noted mucosal epithelial hypertrophy/hyperplasia in the small intestine, induction of enterocyte inflammatory cytokine secretion, and increased numbers of CD4+ T-cell in small intestine lamina propria after dosing mice by gavage for 10 days with 100 mg/kg of either test articles [86]. Rutile 20–50 nm TiO_2-NPs increased the severity of acute inflammation in a murine dextran sodium sulfate (DSS) colitis model and increased Caco-2 cell inflammasome activation in response to lipopolysaccharide (LPS) in a dose-dependent manner and to a threefold greater extent than 360 nm anatase particles [83], providing further evidence of a pro-inflammatory effect of TiO_2-NPs in the mouse small intestine. An additional study in mice reported increased small intestine mucosal permeability to dextran-fluorescein isothiocyanate (FITC) and accumulation of Ti in ileal villi and draining lymphoid nodules after a single gavage dose of 12.5 mg/kg 12 nm anatase TiO_2-NPs [65]. The test articles employed in these studies encompassed a wide range of particle diameters, included both anatase and rutile TiO_2-NPs, and were administered by either gavage or in the diet. There does not seem to be any specific trend in the data linking a specific set of physicochemical parameters or crystal form to increased or decreased potential for toxicity. Moreover, food additive-grade TiO_2 produced similar toxic effects in the small intestine to TiO_2-NPs when compared side by side in the same studies [76, 86], indicating a lack of effect of average particle diameter. However, these findings do imply a species-specific adverse effect of TiO_2-NPs on the murine GIT, which may or may not be relevant to assessing the toxicity of TiO_2-NPs in humans.

Studies in cell-based systems also show disparate effects of TiO_2-NPs. TiO_2-NPs were cytotoxic to undifferentiated Caco-2 cells or SW480 cells in some studies, cell death being associated with increased reactive oxygen species (ROS) generation

Table 12.1 Toxicity studies conducted with TiO_2.

References	Test article[a)]	Test system[b)]	Effects seen
In vivo studies			
[77]	NP: 80 and 25 nm, fine: 155 nm, >92% pure	CD1 mice, gavaged w/5 g/kg NP vs. fine in 0.5% HPMC, sacrifice 14 d after dose	No clinical toxicity in any group. Inflammatory cells in chorion layer of the stomach in 80 nm NP mice. Accumulation of Ti in the liver and the kidney (80 nm NP) or the kidney only (25 nm NP) 2 wk after dosing
[85]	Anatase: 20–1600 nm in H_2O; 13% < 100 nm (SEM), 30–1600 nm in DMEM, 33% < 100 nm	Sprague-Dawley rats ♂♀, 60 d old, gavage w/NP in water 0, 1, or 2 mg/kg/d, 5 d→histopathology duodenum and jejunum	No clinical signs or qualitative histopathology changes. ↑Villus height and width in high-dose males only
[86]	MP: 260 nm; NP: 60 nm	C57Bl/6 mice, 20–25 g, gavaged w/0 or 100 mg/kg MP or NP in water for 10 d→small intestine histopathology (duodenum, jejunum, ileum), cytokine content	MP: ↑IFNγ and IL-23 (jejunum); IL-12, TNF, IFNγ, IL-4, IL-23, and TGF-β (ileum); ↑#s CD4 T-cells all sections. Hypertrophy/hyperplasia mucosal epithelium in ileum only NP: ↑TNF, IFNγ, IL-4, IL-23, TGF-β (ileum), ↑#s CD4 T-cells and hypertrophy/hyperplasia mucosal epithelium in all sections
[63]	NP: 75 nm, 63.95 m^2/g, anatase w/hydroxyl surface groups; 473.6 nm and −33.46 mV in water, 1702 nm and 6.98 mV in AGF, 2081 nm, and −2.47 mV in AIF	Sprague-Dawley ♂rats, 3-wk.o. vs. 8-wk.o., gavaged w/0, 10, 50, or 200 mg/kg NP in water, 30 d→histopathology, small intestine permeability (plasma D-lactate and diamine oxidase activity)	Adults: No histopathology in the stomach or small intestine, ↓permeability small intestine Young rats: ↑#s mast cells in stomach at 200 mg/kg Some NPs adsorbed to surface of the stomach/intestine mucosa. No change in Ti content in the blood, liver, kidney, spleen

[87]	Anatase and rutile NP vs. E171 food additive	♂ Wistar rats exposed to NP or MP via water infusion into tied-off ileal loop at 50 mg/kg for 3 h or via gavage 0, 1, or 100 mg/kg 28 d→EM of small intestine mucosa (ileal loop study only)	3 h direct-ileal exposure nontoxic to enterocytes, small amount of dose penetrated into enterocyte, aggregates found on enterocyte surface. ↑Ti from basal levels in the liver of rats dosed w/100 mg/kg rutile NP
[65]	95% anatase, 12 nm; 132 nm, 20 mV, 0.188 in water; 218.4 nm, 30.6 mV, 0.33 PDI in AGF; >1000 nm, −21.3 mV, >0.5 PDI in AIF; 320.5 nm, −10.8 mV, 0.29 PDI in cDMEM; 325 nm, −13.2 mV, 0.298 PDI in Ringer's + FCS. Protein corona major proteins: albumin, apolipoproteins, complement (cDMEM and Ringer's + FCS). # corona proteins: cDMEM > Ringer's + FCS	Excised Peyer's patches and ileum in Ussing chambers exposed to 50 μg/mL NPs in Ringer's + 10% FCS 2 h→transport and imaging Mice gavaged once w/12.5 mg/kg NPs in water→sacrifice after 6 h→imaging of Peyer's patches and ileum and paracellular permeability (dextran-FITC)	Ti-rich regions in ileum villi and peripheral zone lymphoid nodules in epithelial layer and deeper tissues of FAE and regular ileum. ↑Dextran-FITC permeability, ileum and colon. ↓Expression zonula occludins proteins in ileum, not colon
[88]	77% rutile/22% anatase; 46.2 nm; −17.2 mV and 0.261 PDI in water	Swiss Webster ♂ mice gavage w/0, 5, 50, or 500 mg/kg/d NP in water→sacrifice 24 h, 7 or 14 d post-last dose→stomach histopathology, in gastric cells: COMET, DNA fragmentation and laddering assays, malondialdehyde (MDA) and NO levels, GSH and Ti content, catalase activity	↑[Ti] gastric cells, all time points. ↑DNA damage, all doses and times. ↑p53 mutation frequency, MDA and NO content. ↓GSH content and catalase activity. ↑Severity stomach histopathology, all doses: submucosal edema, ulceration and necrosis, inflammation, degenerated mucosal glands

Table 12.1 (Continued)

References	Test article[a)]	Test system[b)]	Effects seen
[76]	NP 1: pigment-grade, rutile, alumina coat, 10 g/cm^2, 145 nm (TEM), 152 nm (XSDC), 21% < 100 nm (TEM) NP 2: pigment, uncoated, rutile, 7.1 g/cm^2, 173 nm (TEM), 167 nm (XSDC), 11% < 100 nm (TEM) NP 3: surface-treated, ultrafine, rutile/anatase, 42.2 g/cm^2, 73 nm (TEM), 71 nm (XDSC), 73% < 100 nm (TEM)	8-wk.o. Crl CD (SD) rats 90-d (OECD 408) study: gavage w/0, 100, 300, 1000 mg/kg/d NP 1 in 0.5% aq methylcellulose for 92 or 93 d, 10 mL/kg volume 28-d study: gavage w/0 or 24 000 mg/kg/d NP2 in NANO-pure H_2O 28 d Acute (OECD TG 425 study), female rats, single gavage w/0, 175, 550, 1750, 5000 mg/kg NP 3 in unspecified vehicle→necropsy D14 after dose	No adverse GI histopathology. Presence of NP in gut and draining lymph nodes noted w/treatment
[78]	Fine NP: anatase, 119.66 nm; 414.61 nm, and −58.4 mV in PBS Nano-NP: anatase, 18.31 nm; 55.34 nm, and −24.1 mV in PBS	CD-1 mice males, 6-wk.o., gavaged w/0 or 64 mg/kg/d NP in PBS for 28 wk→stomach: body ratio and [Ti] in the blood, liver, pancreas (ICP-OES)	↑Stomach:BW ratio NP only. ↑[Ti] in the blood, pancreas, and liver NP only
[75]	Rutile NP: 500 nm (nominal), 475 nm in water	Adult ♀ Sprague-Dawley rats, 12–14-wk.o., gavage w/12.5 mg/kg/d NP in water 10 d→tissue Ti (ICP-AES) and SEM on tissues	NPs in Peyer's patches and MLN and liver. Total uptake of dose: 11.9%, colon (4%) > Peyer's patches (2.86%) ~ liver (2.4%) > lungs (1.2%). Trace amounts in the blood, small intestine, stomach, peritoneal tissue, spleen, heart, kidney
[70]	Spherical, 18.8 nm, 89.8 m^2/g, 12.9 nm pores; 117–202 nm in H_2O; 270–376 nm in DMEM	Sprague-Dawley ♂ rats, 180–230 g, gavage w/100 mg/kg NPs in water + 2 mM citrate→sacrifice 24 h after dose→ICP-MS for tissue [Ti] and EM on Peyer's patches and small intestine	No ↑ [Ti] in the spleen, liver, intestines, Peyer's patches, MLN. No NPs in small intestine epithelium. Some cells in Peyer's patches had several NP aggregates distributed in cytoplasm

[67]	NM100: anatase, 267 nm, 10 m^2/g; NM101: anatase, 38 nm, 320 m^2/g, 278.49 nm and 0.32 PDI in ethanol/rat serum albumin/PBS (vehicle) NM102: anatase, 132 nm, 90 m^2/g; 366.83 nm and 0.29 PDI in vehicle NM103: rutile, 186 nm, 60 m^2/g, 127.18 nm and 0.38 PDI in vehicle NM104: rutile, 67 nm, 60 m^2/g, 107.73 nm, and 0.27 PDI in vehicle	Wistar rats, 7-wk.o., gavaged w/2–3 mg/rat NPs in vehicle, single dose vs. 5 d→tissue [Ti] 24 h after last dose (spleen liver, MLN, ICP-MS)	[Ti] above LOD in the liver and spleen only in a few rats. Low [Ti] in MLN
[68]	80 : 20 anatase/rutile, 26.4 nm; 37.8 nm in water, 50 m^2/g, LPS-free, no dissolution in AGF or neutral solution	Sprague-Dawley rats, 6-wk.o., gavaged w/NPs in water 14 d at 0, 520.8, 1041.5, 2083 mg/kg/d or for 13 wk at 0, 260.4, 520.8, 1041.5 mg/kg/d→ADME 24 h (5/gp) and tissue [Ti]	↑[Ti] feces; no ↑[Ti] in urine, the liver, kidney, blood
[83]	30–50 nm rutile NP	Wild-type C57Bl/6J mice vs. $Nlrp^{3-/-}$ mice, normal vs. 7-d treatment w/1.5% DSS (colitis)→gavage w/ 0, 50, 500 mg/kg NP (WT) or 0 or 500 mg/kg NP ($Nlrp^{-/-}$) in drinking water once on day 8→EM of colon, spleen; colon length on day 8 Inflammatory bowel disease patients (13 w/active disease; 15 in remission) vs. 28 healthy controls→blood [Ti]	No effects of NP on colon histology in normal mice. ↑Severity colitis in WT but not $Nlrp^{-/-}$ mice. DR-accumulation of "crystals" in the spleen. ↑Blood [Ti] in patients w/active IBD vs. healthy controls and patients in remission

Table 12.1 (Continued)

References	Test article[a)]	Test system[b)]	Effects seen
[69]	NP1: 10 nm; anatase; 12.14 PDI (CPS), agglomerates up to 1 μm *NP2: 70* nm; *rutile; 95* number-% 48–154 nm and 3.50 PDI (CPS) MP: 1.8 μm, rutile, 1.28 PDI (CPS)	Nine healthy volunteers, single dose 5 mg/kg NPs or MP in water→timed urine samples over 4-d period and blood samples over 48-h period. [Ti] in urine and in the blood assessed pre- vs. post-dose	No effects dose in the blood or urine [Ti]
[79]	Anatase: 5–6 nm; 294 nm and 9.28 mV in 0.5% HPMC, 174.8 m^2/g	5-wk.o. ♀ CD-1 mice gavaged 90-d w/0, 2.5, 5, or 10 mg/kg/d NP in 0.5% HPMC→spleen and thymus [Ti] and histopathology	DR ↑spleen and thymus [Ti], up to 1.3 ppm at 10 mg/kg NP. DR ↑inflammatory macrophage infiltration, fatty degeneration, apoptosis/necrosis and inflammatory cytokine production, all doses, in spleen. Presence of large aggregates Ti in splenic macrophages (Raman spectroscopy) at ≥ 5 mg/kg
Cell-based studies			
[89]	NPs from six chewing gum samples; 40–300 nm, #fraction < 100 nm: 27.7–43.7%; by NTA: 80–160 nm, #fraction < 100 nm: 18.7–25.9%; majority anatase, −21.4 mV to −42.6 mV in water	*In vitro* digestibility (saliva, intestinal, gastric); Caco-2 or GES-1 cells (human gastric epithelial cell) viability (WST-8), ROS (DCFH-DA assay), membrane integrity (LDH) at doses of up to 200 μg/mL. Differentiation state of Caco-2 cells unstated	No effect NPs on viability or membrane integrity. ↑ROS in GES-1 and Caco-2 at 200 μg/mL. Control pure TiO_2 ↓GES viability. Incubation in digestion conditions changed #size distribution of NPs

[72]	220 nm (water), >500 nm DMEM; 20–40 mg^2/g; −20 to −10 mV	Differentiated Caco-2 cells in transwell plates, exposed to NPs w/o serum in dark 1–1000 μg/mL, acute vs. 10-d incubation→TEER, viability, cytopathology, intracellular Ca	No acute effects on TEER; ↓TEER at 1 mg/mL after 6 d w/disruption γ-catenin localization to adherens junction. No change in viability. 91% of dose did not enter or touch cells. Accumulation of NP on apical surface (6.8% dose) at 1 mg/mL in uniform layer, some transcytosis (2.2%). Change in microvilli organization ≥ 10 μg/mL (↓height), loss at 1 mg/mL. ↑ Intracellular Ca acute treatment only
[73]	T-Lite SF (rutile core + AlOH layer + PDMS coat): 347 nm in water; 391 nm in media, without serum T-Lite + AGF (DGNP): 237 nm in water, 1353 nm in media without serum, −25 mV. No effects of pH on crystal structure	Differentiated Caco-2 cells exposed to NP or DGNP in media w/o serum at concentrations up to 120 μg/mL for 4, 24, 72 h→viability (XTT, [ATP], cell counting), gene expression (microarray), cytopathology	No effects on viability or gene expression. Large aggregates of DGNP on membrane surface, embedded in microvilli. No NPs seen inside cells or on basolateral surface
[90]	305–311 nm in Caco-2 media, 300–306 nm in SW480 media	Undifferentiated Caco-2 or SW480 cells exposed to NPs in media w/10% serum or synthetic freshwater (SFW) ± natural organic matter (NOM) at 0, 0.1, 1, 10, or 100 mg/L, 24 or 48 h→ (MTT; cytokines); exposed for 4 and 24 h→ROS. NP-interference controls for all assays	No effects NPs on cell viability in media or SFW ± NOM, nor cell line, except for slight (< 10%) ↓viability in SW480 cells w/SFW ± NOM. No ROS generation either cell type. Slight ↑IL-8, both cell types, ≥ 10 mg/L
[66]	NP: 20–25 nm, rutile/anatase, 115 nm and −11 mV in cDMEM Digested NPs (DGNP, AGF→AIF): 104 nm and −35 mV in cDMEM	Undifferentiated C2BBe1 cells (Caco-2 clone) exposed to NPs or DGNPs for 24 h or 11 days at 10 $μg/cm^2$→viability over 5–12 cell passages after PBS wash (sytox red, annexin-FITC, LDH) and mitochondrial activity (MTT)	NPs only found in a fraction of treated cells. No increases necrosis or apoptosis, or differences in cell growth

Table 12.1 (Continued)

References	Test article[a)]	Test system[b)]	Effects seen
[91]	NM103 and NM104: 22–26 nm, rutile, Al/silane-coat, fractal and highly aggregated, insoluble, stable in DMEM, and 0.01% BSA, 51 and 56 m^2/g, redox-active in media	Undifferentiated Caco-2 cells exposed to 100 μg/mL NP in 0.05% BSA for 72 h→culture for 11 d after washout of NPs (acute) or 11 d continuous exposure (long-term) →cell growth assessment D11 and viability (LDH, neutral red, WST-1/WST-8) Differentiated Caco-2 cells in transwell culture exposed to NP in complete media→assess TEER for 21 d	↓Viability w/NM103 at 10 days at 3 and 10 days. No effects of NM104 on viability. No effects either NP on TEER
[37]	Anatase, food-grade NPs (FGNP): 122 nm; 23% < 100 nm; 220 nm and 0.13 PDI in H_2O; 336 nm and 0.162 PDI in media + serum Gum NPs (GNP): 141 nm; 23% < 100 nm, 290 nm 0.165 PDI in H_2O; 365 nm and 0.164 PDI in media + serum	Differentiated Caco-2 cells exposed to NPs in media at 100 ng, 10 μg, and 100 $\mu g/cm^2$ for 24 h in upright or inverted culture or under simulated microgravity conditions→SEM microscopy	↓#s microvilli on cell surface and disordered appearance, all culture conditions. Not due to NP nutrient sequestration
[64]	NP1: 99 nm; 233 nm and −20.6 mV in H_2O, 625 nm in water + BSA; 719 nm and −8.1 mV in media. Digested (saliva→AGF→AIF): 2124 nm and −15.7 mV in H_2O; 2694 nm and −4.4 mV in media NP2: 26.2 nm; 497 nm and −32.2 mV in H_2O; 1785 nm and −8.4 mV in media Digested: 2762 nm and −12.8 mV in H_2O; 2915 nm and −4.4 mV in media 99.2% pure, anatase, uncoated	Differentiated or undifferentiated Caco-2 cells exposed to 50 or 200 μg/mL NPs in media for 24 h→viability (LDH, CCK-8, WST-8), cell proliferation (undifferentiated cells only), ROS, barrier integrity (TEER, FITC-dextran, differentiated only)	No significant cytotoxicity to differentiated cells (WST-8 or LDH assay). Undifferentiated cells: ↑LDH release at high-dose NPs, ↓cell proliferation w/ digested NP1 and NP2 at 200 μg/mL. No effects on barrier integrity. Translocation <1% of dose. Cell uptake: undifferentiated > differentiated, NP1 > NP1-digested

[92]	7–10 nm, 896 nm, and 0.42 PDI in media (NTA)	3D co-culture system: differentiated Caco-2 cells+ differentiated THP-1/MUTZ-3 layer ± induced inflammation by 2d-treatment w/10 ng/mL IL-1β exposed to 1.25–625 μg/cm² NPs in media for 24 h→viability (LDH, Alamar blue), barrier integrity (TEER), IL-8 mRNA and protein Undifferentiated Caco-2 monoculture exposed to NPs at same doses 24 h→viability (LDH, Alamar blue)	No effects on viability, TEER, IL-8
[85]	Anatase: 20–1600 nm in H_2O (SEM), 13% < 100 nm; 30–1600 nm in DMEM (SEM), 33% < 100 nm	Undifferentiated Caco-2 cells exposed to 0–19.38 μg/mL NP in DMEM w/o serum for 6 or 24 h→viability (LDH)	↓LDH in media at 24 h, ≥4.75 μg/mL, non-dose responsive
[74]	A12: 95% anatase, 12 nm; 82 m²/g; 132 nm, 0.188, and −20 mV in water; 320 nm, 0.29, and −10.8 mV in cDMEM, easily dispersed in water R20: 90% rutile; 22 nm; 73 m²/g; >1000 nm, 0.8, and −19.5 mV in water; >1000 nm, 0.8, and −11.7 mV in cDMEM	Differentiated Caco-2 cells on transwell inserts exposed to 50 μg/mL NP in cDMEM, 6, 24, 48 h→qPCR and Western blotting Undifferentiated Caco-2 cells exposed to 50 μg/mL NP in cDMEM, 6, 24, 48 h→cytotoxicity (MTT, w/NP ctrl), COMET, ROS (H2DCFDA), SOD and catalase activity, GSH content	No cytotoxicity or DNA strand breaks. ↑ROS both kinds, R20 > A12. ↑GSH content and ↓SOD activity, both. Early upregulation ABC-transporter genes at 6 h→↑ABC transport protein 48 h. Upregulation GLUT-1, LRP, peptide transporters, organic anion transporters, caveolins 48 h. Intracellular accumulation NP at apical pole and adsorption NPs to plasma membrane, R20 > A12. Located in large endocytic vesicles as agglomerates
[93]	Fine MP: 10 m²/g, 400–300 nm, pure anatase Ultrafine NP: 80 : 20 anatase:rutile, 50 m²/g, 20–80 nm	Undifferentiated Caco-2 cells exposed 20 or 80 μg/cm² NP in serum-free media 4 or 24 h→viability (LDH or WST-1), COMET ± FpG (20 μg/cm² and 4 h only, light vs. dark prep), GSH content	↑LDH release: all concentrations. ↓Metabolism at 80 μg/cm² for 4 h, and at both doses at 24 h. ↑DNA break ± FpG

Table 12.1 (Continued)

References	Test article[a)]	Test system[b)]	Effects seen
[94]	TUFA/RI: pyrogenic, 77% anatase/23% rutile; 25 nm; 214.5 nm, 0.208 PDI, −21.5 mV in H_2O, and 11 800 nm in DMEM, 52.5 m^2/g TUFA/RII: pyrogenic, 77% anatase/23% rutile; 21.9 nm; 327.5 nm, 0.32 PDI, and −23.5 mV in H_2O; 12 000 nm in DMEM, 52.8 m^2/g TUFAI: anatase, 6.7 nm; 455.2 nm, 0.385 PDI, −36.5 mV in H_2O; 6500 nm in DMEM; 282.3 m^2/g TUFAII: anatase; 3.94 nm; 291.1 nm, 0.313 PDI, −38.3 mV in H_2O; 6100 nm in DMEM, 342.4 m^2/g TFA (ultrafine ctrl): anatase, 215 nm; 374 nm, 0.152 PDI, −48.9 mV in H_2O, 6200 nm in DMEM, 10 m^2/g	Undifferentiated Caco-2 cells exposed to 20 or 80 μg/cm^2 NPs in media, 4 or 24 h→cytotoxicity (LDH, WST-1) or expose to 20 μg/cm^2 NPs in media 4 h→oxidative stress (cell GSH)	↑LDH at 4 and 24 h, TUFA/R I and II only; ↓WST-1 activity w/ high-dose TUFA/RII only at 4 h ↓WST-1 activity w/all high-dose NPs but TFA at 24 h. No effects on GSH content
[95]	NP1: anatase, 11.2 nm, 201 m^2/g, 1539 nm in H_2O NP2: 14.4 nm, 120 m^2/g, 1141 nm in H_2O, anatase NP3: 144.3 nm; 200, 450, and 600 nm in H_2O, anatase. Highly agglomerated	Undifferentiated Caco-2 cells ± NF-kappa-B reporter transfection exposed to 40 μg/cm^2 NP in media up to 24 h (acute exposure) vs. chronic exposure for up to 25 d→differentiation marker RNAs and inflammatory marker genes (RT-qPCR) and NF-kB-pathway activation	No change differentiation marker expression. Transient ↑inflammatory gene expression w/NPs 1 and 2 (ICAM-1, CCL-20, COX2, IL-8) via NF-kB-pathway activation

[65]	95% Anatase, 12 nm; 132 nm, 20 mV, 0.188 PDI in water; 218.4 nm, 30.6 mV, 0.33 PDI in AGF; >1000 nm, −21.3 mV, >0.5 PDI in AIF; 320.5 nm, −10.8 mV, 0.29 PDI in DMEM + FCS	Differentiated Caco-2 cells on transwells ± co-culture HT29-MTX cells (regular epithelium) or 14-d Caco-2 cells co-cultured w/Raji B Lymphocytes (FAE) in transwells exposed to 0–500 μg/mL NPs in DMEM w/FCS up to 48 h→cytotoxicity (MTT [NP-interference ctrl], trypan blue), apoptosis induction (acridine orange/ethidium bromide), transport (dextran-FITC or radiolabeled-sucrose), gene expression (RT-PCR)	No effects NPs on TEER or paracellular permeability or viability in any culture system. Extremely low NP uptake in monoculture Significant NP uptake, both co-culture models, Caco-2/RajiB > Caco-2/HT29. NPs stable in cells up to 24 h.NP translocation in Caco-2/RajiB co-culture only
[96]	Bulk NP: 103.2 nm (TEM); 179.3 nm in water (NTA) P25: 22.8 nm (TEM); 75% anatase/25% rutile: 7.1 nm in water (NTA) Anatase NP: 16.4 nm (TEM), 143.3 nm in water (NTA) Rutile NP: 30.8 nm (TEM), 88.3 nm in water (NTA)	Undifferentiated Caco-2 cells exposed to 1 mg/L NPs in media→Ti accumulation in cells at 0, 2, 4, 6, 8, or 24 h (ICP-OES) ± nystatin or orthovanadate, LDH release	No ↑ LDH w/NPs. Nonlinear, hyperbolic ↑[Ti] in cells, steady-state at 24 h. Bulk and P25 > anatase or rutile by 60%. Endocytosed as particles
[71]	A12: 95% anatase, 12 nm, 82 m^2/g; R20: 100% rutile, 21 nm, 73 m^2/g	Caco-2 cells, undifferentiated in 96-well plates or differentiated in transwell plates, exposed to 0–500 μg/mL NPs in media, 24 h (96-well) →MTT (NP-interference ctrl) or exposed 50 μg/mL NPs in media, 6, 12, 24 h (transwell) →TEER and transport	No toxicity or translocation to basolateral pole. Some internalization at low amounts
[70]	Spherical, 18.8 nm, 89.8 m^2/g, 12.9 nm pores; 117–202 nm in H_2O; 270–376 nm in DMEM	Differentiated Caco-2 cells in transwell plates exposed to 100 μg/mL NP in cDMEM for 24–48 h→transport to basolateral compartment ([Ti] by ICP-MS after dialysis and acid digestion), “no cells” ctrl	0.4% of dose transported to basolateral side. Free transport of NPs across filter alone

Table 12.1 (Continued)

References	Test article[a)]	Test system[b)]	Effects seen
[83]	30–50 nm rutile NP vs. 360 nm/1.35 PDI anatase MP	Differentiated (?) Caco-2 and HT-29 cells in transwells or coverslips stimulated w/10 μg/mL LPS exposed to 0, 5, 20, or 100 μg/mL NP or 5, 20, or 100 μg/mL MP→ROS (DH-DCFDA), monolayer permeability (FITC-dextran beads), lysosomal integrity (acridine orange/cytochalasain B), cytokines, NLRP3-protein expression (Western blot)	DR ↑IL-1β, NLRP3-protein expression, IL-18, NP 3x > MP. Intracellular accumulation of TiO2 aggregates and ↑permeability
[97]	NPs: anatase, 20–60 nm (TEM); 220 nm (SEM); 16% < 100 nm, 3.3 m^2/g MP: 80–2000 nm (TEM); 335 nm (SEM); 10% < 100 nm; 6.7 m^2/g	Undifferentiated Caco-2 cells, exposed to 1) 1–20 μg/cm^2 NP or MP in media ± serum, 6 or 24 h→viability (neutral red uptake); 2) 2.5 μg/cm^2 NP or MP in media→8-OHDG, 2, 4, 6, 24 h; 3) 1–20 μg/cm^2 NP vs. 20 μg/cm^2 MP in media w/o serum 24 h→micronucleus assay	No effects NP or MP on viability or micronucleus formation. Slight ↑ 8-OHDG, NP and MP, 6 h time point only
[69]	NP1: 10 nm; anatase; 12.14 PDI (CPS), agglomerates up to 1 μm NP2: 70 nm; rutile; 95 number-% 48–154 nm and 3.50 PDI (CPS) MP: 1.8 μm, rutile, 1.28 PDI (CPS)	4–5 day confluent Caco-2 cultures in transwell plates exposed to 50 or 100 μg/mL NPs or MPs alone or 100 μg/mL NPs or MPs embedded in food matrix→TEER and Ti translocation to basolateral side	< 0.05% of dose translocated to basolateral side, neat NPs and MPs. < 0.02% of dose translocated to basolateral side, MP and NPs in food matrix. No effects on TEER

[98]	Anatase, 200 nm alone or conjugates w/Ca and LPS	Colonic explants from patients with ulcerative colitis (UC, $n = 28$), Crohn's disease (CD, $n = 21$), or healthy people (H, $n = 36$) stimulated *ex vivo* w/5 μg/mL NP, 1–2000 ng/mL LPS, or 2 μg/mL NP-Ca-LPS conjugate in media for 24 h vs. unstimulated control, each explant→tissue and supernatant Il-1β (ELISA)	↑IL-1 release w/NP-Ca-LPS conjugate only, 2.16× (H), 1.4× (UC). No effects CD explants. No effects NP alone
[99]	200 nm NP alone vs. NP-Ca-LPS conjugate	Isolated lamina propria macrophages from healthy ascending colon vs. inflammation sites in colon IBD patients exposed to 1 ng/mL LPS, 5 μg/mL NP, 4 mM $CaCl_2$, or conjugate in media 24, 48, 72 h→supernatant IL-1β and IL-1ra (ELISA)	No response, NP or LPS alone, IBD or healthy. ↑↑Il-1β and ↓IL-1ra after Ca or conjugate, conjugate more prolonged that Ca alone, IBD > healthy
[100]	NP: 240 m^2/g, 5 nm, 749 nm and −45 mV in water, anatase	Undifferentiated Caco-2 or RKO cells ±1-h pretreatment w/TNF (BSA ctrl) exposed to 1–100 μg/cm^2 NP in DMEM +2% FCS (cMEM) 24 h→viability (CCK-8); exposed to 5 μg/cm^2 NP in CMEM 4 h→microarray (validated by qPCR) and immunoblot	No cytotoxicity noted any test condition or effect on response to TNF stimulation. NP aggregates located in cells

a) All test articles TiO_2. Primary particle diameter in nm by TEM and/or SEM, unless otherwise noted. Hydrodynamic diameter and zeta potential (if measured) in nm and mV in stated medium via dynamic light scattering and electrophoretic light scattering, respectively, unless noted otherwise. Specific surface area provided in cm^2/g or m^2/g via Brunauer–Emmett–Teller (BET) method. NTA: nanoparticle tracking analysis; XRD: X-ray diffraction; PDI: polydispersity index; ROS: reactive oxygen species; AGF: artificial gastric fluid; AIF: artificial intestinal fluid; DR: dose-responsive; CPS: centrifugal particle sedimentation; XSDC: X-ray scanning disk centrifugation; ICP-OES: inductively coupled plasma optical emission spectroscopy; ICP-AES: inductively coupled plasma atomic emission spectroscopy; ICP-MS: inductively coupled plasma mass spectrometry.

b) Abbreviations: Ctrl: control; HPMC: hydroxypropyl methylcellulose; NP: nanoparticles; MP: microparticles; wk.o.: weeks old; Gp: group; cDMEM: DMEM + FCS; MLN: mesenteric lymph nodes; BSA: bovine serum albumin.

and pro-inflammatory cytokine production [64, 91, 93, 94], but were nontoxic in others [66, 69, 71, 74, 85, 87, 89, 90, 92, 95–97, 100]. In contrast, TiO_2-NP did not produce cytotoxicity in differentiated Caco-2 cells in monoculture [64, 65, 71–74, 91] or in double- [65] or triple co-culture [84] models. Effects of TiO_2-NP on markers of tight junction integrity and/or changes in villus height and organization were reported in some studies [37, 72], while others reported no effects on paracellular permeability or cytological structure [64, 65, 91, 92]. Increased pro-inflammatory cytokine secretion after exposure of differentiated Caco-2 cells was also noted in some studies [83, 95], although not all [92]. Pure anatase TiO_2-NP reportedly also decreased viability and increased ROS generation in GES-1 cells, a human gastric epithelial cell line, while TiO_2-NPs isolated from chewing gum had no effects [89]. Although the findings of increased ileal dextran-FITC permeability in mice after gavage with 12.5 mg/kg bw of a 12 nm anatase TiO_2-NP test material [65] would seem to corroborate the results of the studies reporting increased Caco-2 cell paracellular permeability, it is impossible to make a firm conclusion regarding this end point, due to all of the confounding factors in the design of the cell-based studies discussed earlier and the fact that these studies reported different physicochemical parameters for their test articles, making direct comparison between studies difficult.

Regardless of the species, it is clear that there is significant exposure of immune cells in regional lymph nodes to TiO_2-NPs. Accumulation of TiO_2-NPs in gut lymphoid tissue may lead to translocation of particle containing macrophages to lymphoid organs such as the spleen and thymus, as implied by a recent study conducted in mice [79], resulting in inflammation and degenerative changes in the spleens of TiO_2-exposed animals. Additional studies conducted in colonic tissue explants and isolated lamina propria macrophages from normal or inflamed human colons have reported significant induction of IL-1β release after exposure in culture to conjugates formed from LPS, Ca-ions, and TiO_2-NPs, but not after exposure to nanoparticles alone [98, 99]. Taken together, the results of these three studies imply that dietary exposure to TiO_2 particulates, in concert with the normal dietary exposure to Ca and LPS, may induce local or even systemic inflammatory activation of tissue macrophages. This, in turn, implicates dietary particulates in the etiology of IBD. However, a recent multi-center 16-week trial of low TiO_2 and silica diets in the treatment of Crohn's disease patients found no effects on markers of disease severity or rates of remission [101]. Moreover, the conjugates used in the colonic tissue studies discussed earlier were formed by ultrasonication of LPS and TiO_2-NPs in the presence of LPS; the likelihood of equivalent conjugates forming under ambient conditions in the GIT was not explored in these studies [98, 99]. Therefore, the biological relevance of these findings is uncertain at this time.

In summary, the available data from *in vivo* studies indicates that TiO_2-NPs demonstrate a species-specific toxicity to the GIT of mice but not of rats, inducing changes in the stomach and small intestine indicative of inflammation. Cell-based studies indicate that TiO_2-NPs are cytotoxic to undifferentiated, but not differentiated, intestinal cell lines. TiO_2-NPs may induce changes in paracellular permeability and/or cytokine secretion in differentiated Caco-2 cells. When coupled to LPS and Ca, TiO_2-NPs have a pro-inflammatory effect in colonic tissue *ex vivo*. The relevance of these findings to the GIT effects of consumer dietary exposure to TiO_2-NPs at the current levels is uncertain, particularly given the long history of safe use of TiO_2 as a direct color additive to food and previous conclusions of no safety concern for color additive use of TiO_2 in food by Australia's National Industrial Chemicals Notification and Assessment Scheme (NICNAS) Agency and the European Food Safety Authority (EFSA) [76]. Further mechanistic studies in animal models investigating the effects of food-grade TiO_2 on intestinal immune function will be needed to identify whether these studies are relevant to human safety.

12.4.2 Silica

Oral exposure to silica is extremely prevalent, due to its use as an anti-caking agent in various food products, an excipient in pharmaceuticals, and an agent in toothpaste [3]. Although it is not clear whether food additive-grade silica (E551 in Europe) contains nanoparticles, the processing of food containing this additive transforms a portion of the silica, up to 43%, to a nanoform (i.e. with all primary particles <100 nm); exposure to nanosilica from food, not accounting for conversion of the added silica to nanoform during processing, has been estimated to be 1.8 mg/kg bw/day [102]. Furthermore, there are now commercially available forms of nanosilica, Aerosil 200F® and 380F®, for use as food additives in the EU and other international markets, which have extremely high specific surface area values of 199 and 388 m^2/g, respectively [103]. Aluminosilicates, or clays, are commonly used in food packaging applications and may be present in waste as part of building and demolition waste, plastics, and tires [8].

Studies have examined the effects of food processing and/or digestion on food-grade SAS or non-food-grade nanosilica. One study compared the effects of incubation in artificial saliva, gastric fluid, and intestinal fluid on the size distributions and percentages of particles <100 nm of E551, SAS, or 32 nm nanosilica incorporated in coffee, soup, or pancake food matrices [104] compared to the undigested test articles diluted in water. This study reported that incubation in artificial saliva followed by incubation in AGF significantly decreased the percentages of silica particles <100 nm for all test articles in all matrices, except for SAS and E551 in pancake matrix, indicating significant agglomeration had occurred. In contrast, sequential incubation in artificial saliva, gastric fluid, and intestinal fluid (fully digested model) significantly increased the percentages of particulates <100 nm over the respective undigested controls for E551 and SAS in black coffee matrix or pancake matrix and soup with SAS. The other fully digested test articles had percentages of particles <100 nm comparable to undigested control values, indicating redispersion of the agglomerated test article upon contact with AGF. Similarly, a 28-day feeding study conducted in rats with either SAS or 10–25 nm nanosilica (NM202) reported that 80% of the silica in colons of NM202-treated rats was in the nanosize range vs. 50–100% of the colonic Si content in SAS-treated rats [105]. Therefore, the available evidence indicates that intake of either food additive-grade silica or nanosilica will result in significant exposure of the intestinal epithelium to nanoscale silica particles, with the extent of nanoscale particulate formation dependent upon the food matrix.

One study reported no or extremely limited nanoparticle uptake from the GIT [66]. One study reported increased silicon concentrations in the livers, kidneys, and spleens of rats administered 10–25 nm pyrogenic nanosilica for 84 days; however, the study authors were unable to determine whether the nanoform itself had been absorbed intact vs. dissolved silica [105]. Accumulation of 100–700 nm silicates and < 100–400 nm aluminosilicate flakes has been reported in macrophages of human GALT, with no association of these aggregates with diagnosis of IBD or histopathological evidence of inflammation in the surrounding tissue [82]. The biological significance of this finding is unclear.

The toxicity of silica to the GIT has been assessed in six studies conducted in rats, one study in mice, and eight studies conducted in cell model systems (Table 12.2). No adverse effects were noted in the GIT of rats or mice after oral administration of food-grade amorphous fumed silica with diameters ranging 7–24.7 nm or 10–25 nm diameter pyrogenic silica [105, 106]; uncoated SiO_2 nanoparticles 5–100 nm in diameter [107–110]; or 5–50 nm diameter coated SiO_2 nanoparticles [108] at doses of up to 1000 mg/kg bw/d for 90 days. In contrast, many of the studies in cell model systems reported production of ROS, cytotoxicity, and induction of DNA strand breaks after exposure of undifferentiated Caco-2 cells to 14 nm uncoated silica and silicon dioxide nanoparticles [91, 94, 113]. Another study reported an inverse dose-responsive (DR) cytotoxicity and genotoxicity of 90 nm uncoated silica particles to undifferentiated HT-26 cells [112]. Other studies reported no cytotoxicity of 10–15 nm silicon dioxide nanoparticles to undifferentiated Caco-2 cells after repeated exposures over an 11-day period [66] or after a single 24-hour exposure [100]. Coating the silicon dioxide nanoparticles with bovine serum albumin [113] and coating silica nanoparticles with carboxyl-terminated groups, PEG, or dextran ameliorated the cytotoxicity to undifferentiated Caco-2 cells, while cationic- or silica-coatings increased the cytotoxicity of silica nanoparticles [111].

Table 12.2 Toxicity data for nanosilica.

Reference	Test article[a)]	Test system[b)]	Effects seen
In vivo studies			
[105]	Food-grade SAS: hydrophilic, pyrogenic, 7 nm, 380 m^2/g, >99.8% pure, 78% < 100 nm, ≤ 15–20 wt% dissolution in AGF NM202: hydrophilic, pyrogenic, 10–25 nm, 200 m^2/g, >99.9% pure, 61% < 100 nm, ≤ 15–20 wt% dissolution in AGF	♂ Sprague-Dawley rats. Fed 100, 1000, or 2500 mg/kg/d SAS; 100, 500, or 1000 mg/kg/d NM202; or normal diet for 28 d (Expt 1); or ctrl vs. 1000 mg/kg/d NM202, or 2500 mg/kg/d SAS for 84 d. 40 wt% of total Si content in feed in nanosize (SAS; 50.5 ppm Si) vs. 100% (79.6 ppm Si) for NM-202 vs. 0% ctrl (0.95 ppm Si)	80% colon Si content in nanosized range (NM202), 50–100% for SAS; 50–100% total Si content in small intestine in nanosize range vs. 17% ctrl. No clinical signs. Small ↑jejunum villus height and crypt depth, both NMs, after 28 d. No change villus height: crypt depth ratio
[106]	NP1: 18.3 nm, 189 m^2/g; 575 nm and PDI = 1, saline NP2: 18 nm, 140 m^2/g; 286.6 nm and PDI = 0.467, saline NP3: 17.7 nm, 204 m^2/g, 143.3 nm, and PDI = 0.145 NP4: 24.7 nm, 204 m^2/g, 154.7 nm, and PDI = 0.185 in saline Na, S, Al, and Ca impurities	♂ Sprague-Dawley rats, 6–8 wk.o., gavaged w/0, 5, 10, or 20 mg/kg SAS NPs in saline at 0, 24, and 45 h→COMET assay and micronucleus assay on colon tissue	No DNA damage or clastogenic effects
[107]	NP1: 20 nm; 21 nm and −38.3 mV in water NP2: 100 nm; 91.6 nm and −45.2 mV in water	CD rats, 6-wk.o., gavaged w/0, 500, 1000, or 2000 mg NPs/kg/d in water for 90 d→sacrifice at 90 d or after 14 d recovery (ctrl and high-dose groups)	Salivation, 1 high-dose ♂ and ↑submandibular gland weight, high-dose recovery ♂ for 100 nm NP. No GI histopathology

Table 12.2 (Continued)

Reference	Test article[a)]	Test system[b)]	Effects seen
[108]	Bare NP: spherical, amorphous, 5–50 nm; 15 nm and −39 mV in PBS/BSA NP-PEG500: sphere, 8–45 nm; 29 000 nm and −26 mV in *PBS/BSA* Amino-capped NP: sphere, 5–50 nm, neutral NP-phosphate: 5–50 nm and −42.9 mV	Wistar rats, 42 d.o., gavaged w/1000 mg/kg NPs in PBS + 1 g/L BSA for 28 d	No clinical signs or GI histopathology
[109]	12 nm sphere, 33.5 nm and −44.37 mV in water	Sprague-Dawley rats, acute, 14-d repeated-dose; 0, 244.9, 489.8, and 979.5 mg/kg NP in H_2O	No clinical GI adverse effects or gross abnormalities. No systemic distribution NPs
[110]	NSP70: 70 nm, uncoated, carboxyl or amine-coated MSP: 300 or 1000 nm Spherical	BALB/c mice ♀, 6-wk.o., gavage 28 d w/0 or 2.5 mg/mouse NP or MP in water Wistar ♂ rats, 8-wk.o., everted small intestinal loop→transport of 12.5 mg/mL NP vs. MP in Tyrode's buffer, 45 min	↑Absorption coated NSP70 vs. uncoated NP and MP, carboxyl > amine in rats. No histopathological abnormalities in stomach or intestines of mice
Cell-based studies			
[66]	NP: 10–15 nm, amorphous; 96 nm and −11 mV in cDMEM; digested nanoparticle (DGNP): 109 nm and −29 mV in cDMEM	Undifferentiated Caco-2 cells, exposed to NPs or DGNPs 24 h or 11 d at 10 μg/cm^2→viability over 5–12 cell passages after PBS wash (cytox red, annexin-FITC, LDH) and mitochondrial activity (MTT)	NPs only in fraction of treated cells. No increases necrosis or apoptosis, or differences in cell growth

[91]	NM200: 14 nm and 189.2 m^2/g; NM203: 13 nm and 203.9 m^2/g, amorphous and highly aggregated, highly porous, partly soluble, Al impurities, redox-active in media	Undifferentiated Caco-2 cells 72-h exposure to 100 μg/mL NM200 or NM203 in 0.05% BSA→culture for 11 d (acute); 11 d continuous exposure (long-term)→cell growth and viability (LDH, neutral red, WST-1/WST-8 assays) Differentiated Caco-2 cells in transwell culture exposed to NM in complete media→TEER for 21 d	Undifferentiated: ↓viability w/NM203 at 3 and 10 d Differentiated: no effects
[94]	Fumed amorphous NP, 200 m^2/g, 14 nm	Undifferentiated Caco-2 cells exposed to 20 or 80 μg/cm^2 NP in serum-free media, 4 or 24 h→viability (LDH, WST-1), COMET ± FpG, GSH content	↑LDH release and metabolism. ↑DNA break w/FpG only. ↓GSH
[11]	Bare NP: 118 m^2/g, 34.42 nm and −39.2 mV in DMEM + 1% FBS, agglomerates in media by 24 h BSA-coated NP: 110–150 m^2/g, 42.06 nm and −31.6 mV in DMEM + 1% FBS, no agglomeration	Undifferentiated Caco-2 and colon-26 cells exposed to 0–300 μg/mL NPs in DMEM + 2.5% FBS 24 h→viability (CCK-8, w/NP-interference ctrl) and annexin/propidium iodide staining	↓Viability w/both NPs, bare NPs 2–4× > toxic than BSA-coated. Colon-26 > Caco-2
[111]	1.6–35 nm bare and coated NPs; cationic, silica, PEG, or dextran coatings. Fe- or Mn-doping	Undifferentiated Caco-2 cells exposed to 0–100 μg/mL NPs in media 24 h→viability (MTT w/NP-interference ctrl), BrdU staining, mitoprobe (mitochondrial membrane potential), [ATP], [Ca] via Fluo-4-direct, ROS (DCFH-DA staining, NP-interference ctrl), TNF-α, caspase 3 activity	Cytotoxicity determined by surface coating: cationic NPs and silica-coated NPs toxic; smaller > larger. COOH-terminated, PEG, and dextran-coated nontoxic. Some ↑toxicity w/Fe-doping, but not Mn-doping. Cytotoxicity due to: mitochondrial toxicity, ↑ROS, TNF, caspase activity, cytoplasmic Ca and ↓ATP

Table 12.2 (Continued)

Reference	Test article[a)]	Test system[b)]	Effects seen
[100]	NP: 124 m^2/g, 10 nm; 181 nm and −44 mV in water, amorphous	Undifferentiated Caco-2 or RKO cells ± 1-h pretreatment w/TNF (BSA ctrl) exposed to 1–100 μg/cm^2 NP in DMEM + 2% FCS 24 h→viability (CCK-8); exposed to 5 μg/cm^2 NP in media 4 h→microarray (validated by qPCR) and immunoblot	No cytotoxicity noted any test condition or effect on response to TNF stimulation
[112]	NP1: 20 nm, 22 nm in water, 25 nm in media, ± rhodamine doping NP2: 90 nm, 97 nm in water, 95 nm in media, ± TMPyP doping	Undifferentiated HT-29 cells exposed to 10–150 μg/mL NPs in media + 10% FCS→cell growth 0–50 h (cell impedance assay), cytotoxicity after 24 h (flow cytometric sulforhodamine B assay w/doped NPs), genotoxicity (γ-H2Ax [+] foci)	NP1: no effects cell growth, small ↑cell death at highest dose. No genotoxicity noted NP2: inverse DR ↓cell growth, ↑cytotoxicity, ↑genotoxicity, 10 μg/mL > 150 μg/mL. DR ↑percent fraction cells w/internalized NPs

a) All test articles nano-SiO_2, unless otherwise noted. Primary particle diameter in nm by TEM and/or SEM, unless otherwise noted. Hydrodynamic diameter and zeta potential (if measured) in nm and mV in stated medium via dynamic light scattering and electrophoretic light scattering, respectively, unless noted otherwise. Specific surface area provided in cm^2/g or m^2/g via Brunauer–Emmett–Teller (BET) method. NTA: nanoparticle tracking analysis; XRD: X-ray diffraction; PDI: polydispersity index; ROS: reactive oxygen species; AGF: artificial gastric fluid; AIF: artificial intestinal fluid; SAS: synthetic amorphous silica; XSDC: X-ray scanning disk centrifugation; ICP-OES: inductively coupled plasma optical emission spectroscopy; ICP-AES: inductively coupled plasma atomic emission spectroscopy; ICP-MS: inductively coupled plasma mass spectrometry.

b) Abbreviations: Ctrl: control; HPMC: hydroxypropyl methylcellulose; NP: nanoparticles; MP: microparticles; wk.o.: weeks old; Gp: group; cDMEM: DMEM + FCS; MLN: mesenteric lymph nodes; BSA: bovine serum albumin; DR: dose-responsive.

In summary, the available data from studies in rats indicates that nanosilica have limited bioavailability after oral administration and little toxic effect on the GIT. The overt cytotoxicity seen in studies conducted in undifferentiated cell lines does not appear to be biologically relevant and may be due to the lack of mucus barrier in these model systems.

12.4.3 Nanosilver

Nanosilver (AgNP) has a wide array of uses as an antimicrobial agent in cosmetics, medical and dental devices, textiles, water treatment filters, and coatings for polymeric materials, as well as its traditional use as a dietary supplement [5]. Colloidal and nanoparticulate Ag exerts its antimicrobial action via release of Ag + ions, which react with sulfhydryl, amino, carboxyl, phosphate, and imidazole groups of proteins, nucleic acids, and lipids, damaging and disabling them [114]. Nanosilver is also one of the most frequently occurring nanomaterials in waste due to its uses in metals, textiles, and electronics [8].

Several studies have assessed the effects of varying environmental conditions, including the conditions present within the digestive tract, on the physicochemical characteristics of nanosilver. One study that examined the effects of sequential incubation in artificial saliva, gastric fluid, and intestinal fluid on the physicochemical characteristics of 58 nm citrate-coated AgNP vs. $AgNO_3$ control noted that AgNP remained in particulate form throughout the series of incubations, although significant particle agglomeration was noted during the gastric fluid incubation [115]. These agglomerates redispersed into smaller particulates upon subsequent incubation in intestinal fluid; the $AgNO_3$ control also formed 20–30 nm particulates in intestinal fluid, composed of Ag, S, and Cl [115], indicating considerable overlap in the exposure of the intestinal epithelium to particulate Ag between nanosilver and silver salts. Aggregation of 40 nm citrate-stabilized AgNP in AGF was also reported in another study, which identified formation of AgCl salt bridges between AgNP particles as the mechanism of agglomeration [116]. It is noteworthy that the studies conducted in simulated digestive fluids did not report dissolution of AgNP under the tested conditions. Although Ag-ion release of up to 17.4% has been reported after incubation of AgNP in cell culture media [117], other studies report a high degree of stability of AgNP in culture media [118, 119]. Dissolution of nanosilver is an oxidative process that is known to increase with temperature, ionic strength, Cl-ion concentration, and oxygen tension, while sulfidation may protect AgNP from dissolution [120]. The relatively low oxygen tension in digestive fluids and/or culture media and relatively high concentration of protein may protect AgNP from dissolution within the lumen of the GIT.

Although uptake of Ag from the GIT into the systemic circulation was noted in the majority of rodent studies, oral bioavailability was <5% of the administered dose [121], and silver salts were over 10-fold more bioavailable than nanosilver [122], with the majority of the administered dose of nanosilver excreted in the feces [123]. Although studies have reported the presence of Ag nanoparticles in ileal cells after direct infusion of AgNP into the small intestine [124] or oral dosing for 90 days [125], the majority of the studies reporting increased systemic Ag concentrations after oral nanosilver exposure only measured total Ag content in various organs via ICP-MS, without reporting the form of Ag in those tissues. Therefore, it is unclear from the available data whether Ag nanoparticles reach the systemic circulation intact after oral administration.

The toxicity of nanosilver to the GIT has been assessed in 13 studies conducted in rats or mice and in cell model systems in 9 studies (Table 12.3). No adverse functional or histopathological effects of nanosilver administration were noted in the majority of studies in rats and mice [109, 125–127, 131–133], although cytostructural changes in enterocytes and/or goblet cells and indicators of inflammation in the intestinal lamina propria in mice [130] and increased intestinal mucus production, goblet cell hyperplasia, and altered mucin composition in rats were noted after administration of uncoated nanosilver [128, 129] for 21 and 28 days, respectively. The studies that did not report any toxicity to the GIT tract were also repeated-dose studies conducted with uncoated, citrate-stabilized, or polyvinylpyrrolidone-stabilized AgNP with diameters ranging from 7.9 to 323 nm. Many of the studies reporting no toxicity did not perform detailed histopathological analyses of the GIT, and Ref. [129] was the only study to assess the effects of AgNP on mucin composition and production. As such, it is possible that the studies reporting no effects of the test article on the GIT missed effects of that test article on mucus production and composition that may be biologically relevant.

In contrast, bare, citrate-stabilized, and polyoxyethylene glycerol trioleate/Tween 20-stabilized nanosilver were uniformly cytotoxic in differentiated and undifferentiated Caco-2 cells, SW480 cells, LoVo cells, and NCM460 cells, although AgNPs were less toxic than silver salts on a per mass basis in most studies [90, 117–119, 134, 136–138]. Two studies comparing the effect of particle diameter on toxicity reported that toxic potency increased with decreasing size [117, 136]. One study [135] did report that Ag nanoparticles were more toxic to differentiated Caco-2 cells on a per mass basis than Ag-ion; the reason for the difference in this result from the other studies is unclear. It seems clear from these results, however, that cell culture systems may overestimate the toxic potential of nanosilver to the GIT, most likely due to the absence of the protective mucus layer in these systems. However, most of the *in vivo* studies did not do detailed assessments of GIT histopathology, and only one study examined the effects of

Table 12.3 Toxicity data for nanosilver.

Ref.	Test article[a)]	Test system[b)]	Effects seen
In vivo studies			
[125]	0.2 wt% PVP-coated NP, spherical, 25.24 nm, 75% < 30 nm	Adult ♂ Sprague-Dawley rats, gavaged with 0, 50, 100, or 200 mg NP/kg bw/d in 0.9% saline at 4 mL/kg bw for 90 d→tissue [Ag], organ weights, ileum histopathology	Accumulation of NP in the liver and ileal cells and ↑[Ag] in all tissues, all groups. Highest [Ag] in ileum (10 ppb) > thymus (4 ppb) > all other tissues (1 ppb). Dose-responsive ↑[Ag] in feces
[126]	4 types of NPs: 22, 42, 71, or 323 nm in water	ICR mice (6-wk.o.), gavaged w/1 mg/kg all sizes NPs in water, 14 d; 28 d gavage w/42 nm NP at 0, 0.25, 0.5, 1 mg/kg	No histopathology in small intestine at 1 mg/kg in 28-d study
[127]	99.98% pure, 56 nm in vehicle (TEM)	SPF F344 rats: gavage w/0, 30, 125, or 500 mg/kg/d AgNP in 0.5% aqueous CMC, 13 wk	Pigmentation intestinal villi in 8/10 ♂ at ≥125 mg/kg and 5/10 high-dose ♀. No changes in stomach or esophageal histopathology
[128, 129]	52.7–70.9 nm in 0.5% aqueous CMC (vehicle)	SPF Sprague-Dawley rats, 5-wk.o., gavaged w/0, 30, 300, or 1000 mg/kg/d NP in vehicle for 28 d	No clinical toxicity or esophageal histopathology. NP found in lamina propria of the small and large intestine. In treated rats: disordered goblet cells in ileal mucosa w/↑#s goblet cells in ileal and rectal mucosa that had released mucus granules. Large amount mucus in crypt and ileal lumen. No change in goblet cells in proximal colon. ↓Amounts neutral and acidic mucins in goblet cells, all regions and ↓ileal sulfated mucins. Presence of sialated mucins in lamina propria in colon

Table 12.3 (Continued)

Ref.	Test article[a)]	Test system[b)]	Effects seen
[130]	3–20 nm, sphere, hydrazine and $AgNO_3$ impurities	Swiss male mice, gavage w/0, 5, 10, 15, 20 mg/kg NP in water 21 d→histology small intestine	Damaged/broken microvilli on enterocytes, ↑# inflammatory cells in lamina propria and ↑width of lamina propria, ↑mitotic figures intestinal glands
[131]	PVP-coated NP: 14 nm in 11.5 mg/mL aqueous PVP vehicle, 11% ionic Ag + (microcentrifugation→ICP-MS)	Wistar male rats, 6-wk.o., 0, 4.5, 9 mg Ag/kg/d in vehicle for 14 d Wistar rats, 6-wk.o., gavage w/0 (♀♂), 2.25 (♀), 4.5 (♀), 9 mg Ag/kg/d (♀♂) as NPs in vehicle or 9 mg Ag/kg/d Ag acetate (♀) 28 d→ileum histopathology (TUNEL staining, inflammatory cells, controls, and 9 mg/kg only)	No clinical toxicity observed. No ileal morphology changes
[123]	PVP-coated AgNP, 14 nm; 50 nm and −2 mV in water, spherical w/no aggregates, 11% as ionic Ag	Wistar rats, 5-wk.o., gavaged w/11.5 mg/mL PVP (vehicle ctrl), 12.6 mg Ag/kg/d NP, or 9 mg Ag/kg/d Ag acetate in 11.5 mg/mL PVP for 28 d→24 h urine and feces collection week 3→tissue Ag day 29	Majority of dose excreted in feces. Urine excretion 0.005% (NP) and 0.057% (acetate) of 24-h intake. [Ag] in tissues similar between groups: small intestine > stomach > kidney > liver. Ag salt > NP (40–50%)
[109]	NP-citrate: 11 nm sphere, 19 nm and −21.13 mV in water	Sprague-Dawley rats, acute, 14-d repeated-dose studies, gavaged w/0, 515.3, 1030.5, and 2061 mg/kg NP in water 13-wk study: gavaged w/0, 257.6, 515.3, 1030.5 mg/kg AgNP in water	No clinical GI toxicity in acute, 14-d, 13-wk studies or gross abnormalities in GIT. DR ↑[Ag] in the blood, liver, kidney, spleen, lung, brain
[124]	Round/ellipsoid, 34.9 nm, 8.4–80.9 nm	Male Wistar rats exposed to NPs via direct infusion into small intestine loop in saline	NP present in tunica muscularis and Peyer's patches and cross into systemic circulation

[121]	Citrate-coated NP: 7.9 nm and −17.55 mV in water	Male Sprague-Dawley rats, 7-wk.o., gavaged w/1 or 10 mg/kg NPs in water→blood Ag at 10 min and 1, 2, 4, 8, 24, 48, 96 h; 24 h feces and urine Ag; liver, lung, and kidney Ag at 24 and 96 h	Bioavailability: 1.2% (1 mg/kg) and 4.2% (10 mg/kg). No ↑[Ag] in tissue, except slight ↑ in liver at 14 h at 10 mg/kg dose. Most Ag excreted in feces
[132]	Citrate-coated NP, 7.9 nm; 3.8 nm (0 h) or 3.2 nm (24 h) in water, −17.55 mV	Male Sprague-Dawley rats, 8-wk.o., 0, 2, or 20 mg Ag/kg as NP or Ag salt in water, single gavage	Soft stools, soiled perianal region at high dose of Ag salt only, no gross pathology findings
[133]	Spherical NP Small: 14 nm and −41.4 mV in water Large: 36.6 nm and −35.5 mV in water	21-d.o. Wistar rat females, exposed to 535 μg/mL NP in drinking water 55 d→clinical signs, tissue Ag, histopathology in the stomach and intestines	No clinical signs. Tissue Ag: small intestine > kidney and liver > brain. No difference in the brain or small intestine [Ag] between NPs. Focal inflammatory cell influx in small intestine submucosa in one rat of large NP group
[122]	Uncoated NPs in 4% each, polyoxyethylene glycerol trioleate and Tween 20 w/matrix ctrl: 20 nm; #-based distribution in H_2O: 4, 13, and 59 nm PVP-coated NP: <15 nm (TEM); 2 and 49 nm in H_2O 7% soluble Ag content, stable	Sprague-Dawley male rats, 6-wk.o. gavaged 28 d w/90 mg/kg NPs, 9 mg/kg $AgNO_3$, matrix controls for each NP→weekly blood Ag→tissue Ag at necropsy D29, D36, D84	Blood [Ag] $AgNO_3$ > 10× NP, all timepoints, even when normalized to soluble Ag dose. No significant difference blood Ag between NPs. Tissue content Ag not significantly different from ctrl, except for stomach wall. Retention of Ag in the brain and testis out to D84
Cell-based studies			
[134]	< 20 nm, no characterization data presented	Caco-2/RajiB (M-cell line) co-culture vs. differentiated Caco-2 monoculture in transwell exposed to10–90 μg/mL, exposure duration unstated→viability (MTT), IL-8, NO, paracellular permeability (Lucifer yellow (LY), TEER; immunostaining tight junction proteins)	↓Viability ≥ 30 μg/mL; ↓apical IL-8, mono- and co-culture; no change basolateral IL-8. ↑NO, apical and basolateral; ↑Lucifer yellow passage and ↓TEER. ↓Occludin and ZO-1 staining in tight junctions

Table 12.3 (Continued)

Ref.	Test article[a)]	Test system[b)]	Effects seen
[135]	< 20 nm in 4% each polyoxyethylene glycerol trioleate + Tween 20, no characterization data presented	Undifferentiated Caco-2 cells exposed to 15–90 μg/mL NPs in HBSS vs. 1.5–9 μg/mL $AgNO_3$ ctrl, 3 h→MTT Caco-2/RajiB (M-cell line) co-culture vs. differentiated Caco-2 monoculture in transwell exposed to15–90 μg/mL vs. 1.5–9 μg/mL $AgNO_3$ ctrl, 24 h→ROS (DCFH-DA), paracellular permeability (Lucifer yellow, TEER; immunostaining tight junction proteins)	DR ↓viability ≥ 30 μg/mL NP and ≥ 4.5 μg/mL Ag+. ↑ROS, ≥ 45 μg/mL NP and 6 μg/mL Ag+, NP > Ag+. ↓TEER, NP and Ag+, NP > Ag+. ↑Lucifer yellow permeability and ↓tight junction protein staining, ≥ 30 μg/mL NP, mono- and co-cultures., NP > Ag+
[90]	202–227 nm in Caco-2 cell media, 201–221 nm in SW480 cell media	Undifferentiated Caco-2 or SW480 cells exposed to NPs in media w/10% serum or synthetic freshwater ± natural organic matter (SFW ± NOM) 0, 0.1, 1, 10, or 100 mg/L for 24 or 48 h (MTT; cytokines) or 4 and 24 h (ROS). NP-interference controls for all assays	No effects on viability of NP in media, either cell type. No effects on viability of NPs in SFW ± NOM in Caco-2 cells. ↓SW480 viability all concentrations NPs in SFW ± NOM. ↑IL-8 in Caco-2 cells, ≥ 10 mg/L all NPs. No change in IL-8 in SW480 cells
[136]	Two types of spherical, citrate-coated NPs, stable 24 h: 20 nm NP: 19.2 nm; 20.6 and −37.3 in 2 mM citrate buffer 100 nm NP: 99.4 nm; 93.1 nm and −57.5 mV in 2 mM citrate buffer Ion ctrl: 24-h incubation media w/AgNP→centrifuged to isolate supernatant. 100% purity	LoVo cells (human colon carcinoma line), 10 μg/mL NPs or controls for 24 h→proteomics analysis and ROS content (H2DCF-DA), protein carbonyls	↑ROS, protein carbonyls and cell stress proteins, SUMOlation proteins (20 nm NP); entry of 20 nm into cells. No entry 100 nm NP into cells

[119]	Citrate-coated NP: 20.4 nm, no aggregation/agglomeration, stable in cell media	Undifferentiated Caco-2 cells, attached vs. suspended culture, exposed to 5–40 μg/mL (Alamar blue assay) or 5–15 μg/mL (flow micronucleus assay) 24 h	Toxicity noted at ≥ 10 μg/mL. No MN formation, any condition
[118]	25.8 nm; 311.9 nm and −21.1 mV in water, 190.2 nm and −10.5 mV in cDMEM	NCM460 human colon mucosa epithelia cells exposed to in DMEM w/serum at 1 pg/L-1 g/L 24 h→viability (propidium iodide staining)	NCM460: ↓viability 1 g/L, ~80% of control
[137]	NP (10% Ag, 4% each polyoxyethylene glycol trioleate and Tween-20): 15 nm (SAXS), 28 nm in water, 50.8 nm in cDMEM, 5% Ag + release after 24-h incubation in media	Caco-2, proliferating vs. differentiated, exposed to 0.1–100 μg Ag /mL NP in media for 4, 24, or 48 h→viability (Alamar blue, DAPI), apoptosis (annexinV7, caspase activity), mitochondrial activity (JC-10, differentiated cell only), oxidative stress (dichlorofluorescein, GSH content, antioxidant rescue assay). $AgNO_3$ and surfactant-alone controls. Gene expression change (PCR) after 24-h exposure to 2.5 or 25 μg Ag/mL NP in media	Toxicity: $AgNO_3$ 10-fold > AgNP on mass basis, proliferating cells > differentiated, no effects surfactant alone. No apoptosis induction or mitochondrial depolarization. Death by necrosis. ↑ROS in serum-free media only both AgNP and $AgNO_3$. ↓GSH w/Ag in differentiated cells; antioxidants rescued proliferating cells from Ag effects. Changes in multiple genes in high-dose NP cells only
[117]	AgNP (4 types): 31, 30, 69, 112 nm (TEM); 93, 104, 115, 173 nm in media, 0 h, NTA; 117, 133, 182, 186 nm in water, NTA; 118, 140, 148, 189 nm in media, 24 h, NTA; % Ag + in media after 24 h : 17.4%, 15.8%, 5.8%, 7.6%	Undifferentiated Caco-2 cells exposed to 5–50 μg/mL in media w/o serum, 24 h→viability (WST-1) Caco-2/Raji co-culture exposed to 5 or 25 μg/mL NP for 4 h in media, no serum→translocation, barrier integrity (TEER, Lucifer yellow, FITC-dextran). $AgNO_3$ ctrl	No effects on viability, except for smallest NP at 37.5 μg/mL; clear cytotoxicity of 2.5 and 5 μg/mL $AgNO_3$. No effects on monolayer barrier integrity or difference in Ag translocation between AgNP and $AgNO_3$. NPs induced gene expression changes associated w/oxidative stress, endoplasmatic stress response, apoptosis; effects similar between AgNP and $AgNO_3$

Table 12.3 (Continued)

Ref.	Test article[a)]	Test system[b)]	Effects seen
[138]	NP (10% Ag, 4% each polyoxyethylene glycol trioleate and Tween-20): 7.02 nm, 14 nm in water Digested NP (DGNP; field flow fractionation→SAXS and DLS): saliva only: 24.5 nm; saliva + AGF: 3.4 nm; saliva + AGF + AIF: 9.39 nm and 76.1 nm, bimodal population	Undifferentiated vs. differentiated Caco-2 cells exposed to 0–25 μg/mL (UDiff) or 0–100 μg/mL (Diff) pristine vs. DGNP, digestion fluid and solvent ctrl→toxicity (DAPI and CTB), cell impedance	DR toxicity, both NP and DGNP, in undifferentiated cells; no toxicity digestive fluid ctrl. Differentiated cells: DR cytotoxicity of NP and DGNP, DGNP > NP; digestive fluid toxic at ≥ 2.5%. Adverse effects on cell impedance of DGNP, diff and udiff cells. No adverse effects NP or digestive fluid on cell impedance

a) All test articles silver nanoparticles. Primary particle diameter in nm by TEM and/or SEM, unless otherwise noted. Hydrodynamic diameter and zeta potential (if measured) in nm and mV in stated medium via dynamic light scattering and electrophoretic light scattering, respectively, unless noted otherwise. Specific surface area provided in cm^2/g or m^2/g via Brunauer–Emmett–Teller (BET) method. NP: nanoparticle; PVP: polyvinylpyrrolidine; NTA: nanoparticle tracking analysis; XRD: X-ray diffraction; PDI: polydispersity index; ROS: reactive oxygen species; AGF: artificial gastric fluid; AIF: artificial intestinal fluid; XSDC: X-ray scanning disk centrifugation; ICP-OES: inductively coupled plasma optical emission spectroscopy; ICP-AES: inductively coupled plasma atomic emission spectroscopy; ICP-MS: inductively coupled plasma mass spectrometry.

b) Abbreviations: Ctrl: control; HPMC: hydroxypropyl methylcellulose; NP: nanoparticles; MP: microparticles; wk.o.: weeks old; Gp: group; cDMEM: DMEM + FCS; MLN: mesenteric lymph nodes; BSA: bovine serum albumin; DR: dose-responsive; CMC: carboxymethyl cellulose; PVP: polyvinyl pyrrolidine.

AgNP on the mucus component of the GIT barrier. As such, the findings of the cell culture systems cannot be dismissed as completely irrelevant.

12.4.4 ZnO Nanoparticles

ZnONPs are widely used as an ingredient in cosmetics and sunscreens and are also used in animal feed and as a dietary supplement [6, 139, 140]. Studies indicate that ZnONP dissolves in AGF [66, 141], and significant Zn-ion release in cell culture media after overnight incubation has also been reported [100].

Conflicting data on the oral bioavailability of ZnONP exist, with some studies reporting increased Zn concentrations in the blood and non-GI tissues after oral ZnONP administration [68, 142–144], while others report no systemic absorption [145]. The disparate results reported in the literature may reflect the much higher doses (hundreds of milligrams) used in the studies demonstrating increased tissue Zn concentrations vs. the studies reporting no systemic absorption. However, all of the previously cited studies consistently report that the majority of the administered dose is excreted in the feces. This finding is consistent with the low oral bioavailability of nanosilver compared with its silver salt analog discussed in the previous section.

The toxicity of ZnONP to the GIT has been assessed in 8 studies using single- or repeated-dose paradigms conducted in rats or mice and in GI cell model systems in 13 studies (Table 12.4). The majority of studies noted severe adverse effects of the test article on the GIT, including clinical signs of toxicity (diarrhea, anorexia, excessive salivation) and inflammatory changes in the stomach (forestomach epithelial hyperplasia, mucosal and submucosal inflammation, ulcers of the glandular stomach) after single or repeated doses of ZnONP [141–144, 146, 147, 152]. In contrast, one study in rats reported no adverse GI-related clinical signs associated with oral administration of ZnONP [145]. No intestinal histopathology was noted in the studies.

ZnONPs and zinc salts were cytotoxic to gastric epithelial cells [141, 149] and to undifferentiated Caco-2 cells, RKO cells, and LoVo cells, producing DNA damage, inducing stress response genes, and increasing ROS production [90, 94, 97, 100, 118, 148, 150, 151, 153]. Other studies in undifferentiated Caco-2 cells [66, 91] or differentiated Caco-2 cells [91] reported no cytotoxic effects. Cytotoxic effects were independent of the measured specific surface area of the test article [94], although nanoparticles were more cytotoxic than microparticle (MP) controls [153] in one study, while another study found no difference in cytotoxic potency between ZnONP and a zinc salt control [150], possibly due to the rapid dissolution of the ZnONP in media. Interestingly, although zinc salts have been reported to produce GI distress [154] and even bleeding gastric ulcers in humans given high oral doses [155], a number of studies have reported that oral administration of zinc sulfate helps heal gastric ulcers in rodent models of stress- or chemical-induced ulcers [156, 157]. As such, the adverse gastric effects

Table 12.4 Toxicity data for zinc oxide nanoparticles.

Reference	Test article[a)]	Test system[b)]	Effects seen
In vivo studies			
[144]	10 and 120 nm, spherical; 44.8 and 187.5 nm in vehicle; 440 000 and 910 000 cm^2/g, > 95% pure, 3.6% and 2.4% dissolution in AGF after 2 h at 37 °C	Adult CD-ICR mice single gavage w/0, 1, 2, 3, 4, or 5 g NP/kg in 1% Na CMC, sacrifice after 14 d	Diarrhea at 5 g/kg 120 nm ZnONP. Inflammation in stomach lamina propria, submucosa, or serosal layer: all doses 20 nm NP, ≥ 3 g/kg 120 nm NP. Severe stomach pathology at 5 g/kg for 20 nm NP and ≥ 3 g/kg 120 nm NP
[146]	NP (58 nm) vs. MP (1080 nm), > 99.99% pure, spherical	CD-ICR mice, 5-wk.o., gavaged once w/1% Na CMC or 5 g/kg bw ZnNP or ZnMP→sacrifice after 14 d	Severe anorexia, vomiting, diarrhea w/NP only for first week. Severe NP aggregation in the intestine→mortality in NP group at day 2 and 6 after dosing. Slight stomach and intestine inflammation all Zn-doses
[141]	NP: 40 nm, 60 m^2/g, 201 nm in water, neutral/negative charge at pH ≥ 7.0, positive at pH < 4. LPS-free. Stable in AGF for 24 h→dissolution after 24 h. No re-formation NP after pH increased. No dissolution NP in water, RPMI	Sprague-Dawley rats, acute and 14-d study and 13-wk study. Doses: 0 or 2147 mg/kg (acute); 0, 536.75, 1073.5, or 2147 mg/kg/d (14-d); 0, 67.1, 134.2, 268.4, or 536.8 mg/kg/d (13-wk) in water by gavage	Acute: no adverse clinical signs 14-d: bloating and diarrhea at ≥ 1073.5 mg/kg 13-wk: no adverse GI-clinical signs or histopathology
[143]	Citrate-coated NP, 29 nm in H_2O prior to coating	Sprague-Dawley rats, 6-wk.o., gavage w/water, vehicle (HEPES/citrate), 125, 250, and 500 mg/kg ZnONP for 90 d (all doses) or 90-d + 14-d recovery (controls and high dose only)	Salivation and diarrhea during dosing. Epithelial vacuolation in the forestomach. Stomach limiting ridge lesions: basal cell hyperplasia, epithelial cell hyperplasia and vacuolation, hyperkeratosis, submucosal inflammation. Glandular stomach: erosion, infiltrating epithelial globule leukocytes, submucosal edema and inflammation, chief cells w/eosinophilic granules, ↓parietal cells

147	NP, citrate-modified (anionic) or serine-modified (cationic), 100 nm size uncoated NP	CD rats, 6-wk.o., gavaged w/0, 31.25, 125, or 500 mg NP/kg/d in vehicle for 90 d→sacrifice at 90-d or after 14-d recovery (15/sex/gp, controls and high-dose groups). Include both water and vehicle (HEPES/citrate or HEPES/serine)	Salivation (sporadic) in ≥ 125 mg/kg ♂ and high-dose ♀, both NPs. ↑Submandibular gland weight, recovery citrate-NP ♀. Stomach (both types NPs): ↑incidence squamous cell hyperplasia and vacuolation in forestomach; intracytoplasmic hyaline droplets submucosal edema and inflammation; #s of eosinophilic chief cells; and mucosal cell hyperplasia in glandular stomach, all doses, both sexes
[147]	MP not characterized. NP: 63 nm; 224.7 nm, 0.305 PDI, −30.9 mV in water; 50 m^2/g	Sprague-Dawley rats, 8–9 wk.o. gavage w/0, 5, 50, 300, 1000, 2000 mg NP or MP/kg/d in water 14 d→histopathology GIT	Inflammatory cell infiltration in stomach, all NP gps. ↑Incidence and severity of neutrophil infiltration. No histopathology for MP
[145]	3-Aminopropyl ethoxysilane-coated NP, < 35 nm. Suspended in 5% aqueous glucose: monomodal size distribution, all concentrations (100–15 000 ppm) w/peak between 10 and 20 nm, +20–30 mV. Precipitated after mix w/AGF	Sprague-Dawley rats, 7-wk.o., 0, 3, or 30 mg/kg in 5% glucose by single gavage	No clinical toxicity evident
[142]	Uncoated 1: 20 nm; 20 nm and 22 mV in water; citrate-capped 1: 20 nm; 21 nm and −28.1 mV in water Uncoated 2 : 70 nm; 67 nm and 19.6 mV in water; citrate-coated 2: 70 nm; 71 nm and −33.3 mV in water	Sprague-Dawley rats, 120–140 g, gavage w/0, 50, 300, or 2000 mg NPs/kg cnce in HEPES/citrate→tissue [Zn] in the brain, heart, kidneys, liver, lung, spleen, gonads at 1 h, 6 h, 1, 2, 3, or 7 d, ADME on days 1–10, 14	↑Zn absorption both sizes and sexes. Liver, lung, kidney w/ highest Zn levels. Small amount Zn excrete in urine, most in feces. No NPs seen in tissue; new Zn—S bonds observed. Diarrhea noted at high dose

Table 12.4 (Continued)

Reference	Test article[a)]	Test system[b)]	Effects seen
Cell-based systems			
[90]	212–260 nm (Caco-2 cell media w/10% serum), 288–303 nm (SW480 cell media w/10% serum)	Undifferentiated Caco-2 or SW480 cells, exposed to NP in media w/10% serum or synthetic freshwater ± natural organic matter (SFW ± NOM) at 0, 0.1, 1, 10, or 100 mg NP/L for 24 or 48 h (MTT; cytokines) or 4 and 24 h (ROS). NP-interference controls used for all assays	Caco-2 cells: ↓viability at 100 mg/mL in media or ≥ 10 mg/L SFW ± NOM. No ROS generation. ↑IL-8, ≥ 10 mg/L SW480 cells: ↓viability at ≥10 mg/L in media, all doses SFW + NOM, ≥ 10 mg/L SFW only. ↑ROS in SW480, ≥ 10 mg/L at 24 h. ↓IL-8 at ≥ 10 mg/L
[148]	NP: 20–60 nm, irregular-shaped, 24 m^2/g MP: 330 nm (calculated), irregular; high polydispersity, 3.2 m^2/g. Formed agglomerates in DMEM, 500–2000 nm	RKO cells in transwell culture, exposed 3 ways: direct contact in upper chamber (UC), indirect contact in lower chamber (LC), particulates put in dialysis cassette (DC). Exposed 24 h to doses of 1–100 $\mu g/cm^2$ NP or MP, zinc salt ctrl→viability (CCK-8 kit), mitochondrial potential (MitoSOX), apoptosis (annexin V/propidium iodide). NP-interference ctrl for viability assay (CCK-8 kit) only	↑[Zn] in cell media when exposed to NP or MP, NP > MP only at 100 $\mu g/cm^2$. Zn-particle dissolution > in presence of cells vs. no cells. Cytotoxicity in UC condition only, NP > MP, via apoptotic pathway in ≥ 90% cells. Disrupted mitochondrial potential, NP and MP, NP > MP, but not Zn salt ctrl
[118]	29.4 nm; 227.7 nm and 20.8 mV in water; 74.5 nm and −8.5 mV in DMEM/F12; 148.1 nm and −10.7 mV in cDMEM	NCM460 human colon mucosa epithelia cells exposed to NPs in cDMEM, 1 pg/L-1 g/L 24 h→viability (propidium iodide staining), apoptosis (TALI kit), ROS production	NCM460: ↓viability ≥100 mg/L NPs Apoptosis main mode death. ↑ROS production

[66]	Spherical NP: 10–20 nm Rod/wurtzite NP: 5–10 nm × 50–200 nm, 112 nm and −19 mV in cDMEM Digested NPs (DGNP): NPs incubated in AGF→AIF	C2BBe1 cells (Caco-2 clone) undifferentiated, exposed to NPs or DGNPs 24 h at 10 μg/cm^2→viability over 5–12 cell passages after PBS wash (sytox red, annexin-FITC, LDH) and mitochondrial activity (MTT) or repeated exposure same dose over 11 d→viability	NPs only found in a fraction of treated cells. No increases necrosis or apoptosis, or differences in cell growth
[91]	NM110: 151 nm, 12 m^2/g, uncoated, partly soluble NM111: 140 nm, triethoxycapryl silane-coated, 15 m^2/g, insoluble Fractal and highly aggregated.	Undifferentiated Caco-2 cells exposed to 100 μg/mL NM in 0.05% BSA for 72 h→culture for 11 d (acute) or 11 d continuous exposure w/replacement of treatment media on days 4 and 8 (long-term)→cell growth assessment D11 and viability (LDH, neutral red, WST-1/WST-8 assays). Differentiated Caco-2 cells in transwell culture exposed to NM in cDMEM→assess TEER for 21 d	No effects viability or TEER
[141]	40 nm, 60 ^{2}/g, 201 nm in water, neutral/negative charge at pH ≥ 7.0, positive at pH < 4. LPS-free. Stable in AGF for 24 h→dissolution after 24 h. No re-formation NP after pH increased. No dissolution NP in water, RPMI	Human epithelial gastric adenocarcinoma cells (AGS) exposed to NP or $ZnSO_3$ as source of Zn ions at doses up to 100 μg/mL for 24 h ± EDTA→cell proliferation assay	Both ZnNP and $ZnSO_3$ cytotoxic to AGS at 100 μg/mL and 50 μg/mL, respectively; toxicity abolished by EDTA
[149]	Six kinds of NPs, > 95% pure, hexagonal wurtzite, 112–149 nm and −17.4–21.4 mV in water; 148–196 nm and −5.5–6.7 mV in media	Gastric epithelial cell line (GES-1) exposed to 15 mg/L NP ± 300 mg/L vitamin C up to 24 h→viability (CCK-8, annexin-propidium iodide assay), intracellular Ca (Fluo-3 AM), DNA content (DAPI), ROS (DCFH-DA), mitochondrial potential (rhodamine 123), caspase-3 activity	↓Viability NP + vitamin C via apoptosis; ↑intracellular Ca, ROS and caspase-3 activity and ↓mitochondrial potential. No effects of NP alone. ↑[Zn] in cells in NP + vitamin C vs. NP alone. Uptake of NP in lysosome→all NP dissolved by 24 h. Vitamin C ↑endocytosis rate and dissolution rate

Table 12.4 (Continued)

Reference	Test article[a)]	Test system[b)]	Effects seen
[94]	NP1: 10 nm, 70 m^2/g NP2: 50 m^2/g, 20 nm Fine: <10 m^2/g, unknown primary particle size	Undifferentiated Caco-2 cells exposed 20 or 80 μg/cm^2 NP in serum-free media 4 or 24 h→viability (LDH or WST-1), COMET ± FpG, GSH content	↑LDH release and ↓metabolism. ↑DNA break ± FpG. ↓GSH. No relation of surface area to amount of DNA in tail
[150]	NM110: 70–100 nm (XRD), 20–350 nm (TEM), 306 nm in media, 14 m^2/g NTA: 155 nm mean and 152 nm mode in water; 210 nm mean and 169 nm mode in 1% BSA in water; 185 nm mean and 146 nm mode in 1% BSA in media	Undifferentiated Caco-2 cells, exposed to 2–128 μg/mL NP in 1% BSA in media for 24 h: (i) directly on cells, (ii) separated by 3 or 0.4 μm transwell filter, or (iii) filtered through 0.45 μm filter prior to direct exposure to cells→cell proliferation (BrdU) or cytotoxicity (WST-1, MTT), ROS (mitoSOXred, cellROX), lysosome destabilization (lysotracker green flow cytometry). $ZnCl_2$ ctrl for WST-1 assay at equal [Zn] to NP	Direct: ↑cytotoxicity ≥16 μg/mL. Filtered: ↑cytotoxicity, ≥32 μg/mL. Toxicity direct = toxicity 3 μm condition. No cytotoxicity noted under 0.3 μm condition. No difference in potency, $ZnCl_2$ vs. NP. ↑Mitochondrial ROS ≥ 4 μg/mL NP + palmitic acid, not NP alone. ↑Lysosome destabilization, ZNP ± fatty acids
[97]	45–170 nm (TEM); 128 nm (SEM); 3 m^2/g; 34% < 100 nm	Undifferentiated Caco-2 cells, exposed to 1) 1–20 μg/cm^2 NP in media ± serum 6 or 24 h→viability (neutral red uptake); 2) 2.5 μg/cm^2 NP in media→8-OHDG, 2, 4, 6, 24 h; 3) 1–20 μg/cm^2 NP in media w/o serum 24 h→micronucleus assay	DR ↓viability, significant for w/o serum condition only. Slight ↑8-OHDG, 6 h time point only. Slight ↑micronucleus formation

[148]	NP1: 26.21 nm spheres; NP2: 62.42 nm rod; NP3: 90.81 nm rod	Undifferentiated Caco-2 cells exposed to 0–100 μg/mL NPs in DMEM w/o serum 12, 24, or 36 h→MTT; exposed to 12.5 or 50 μg/mL NPs in DMEM w/o serum, 24 h→acridine orange/ethidium bromide staining; flow cytometry w/propidium iodide stain; exposed to 0–100 μg/mL NPs in DMEM w/o serum 24 h→LDH, ROS (DCFDA), GSH	Cytotoxicity IC_{50}s of NPs: 15.55 μg/mL NP1, 22.84 μg/mL NP2, 18.75 μg/mL NP3. ↑ROS and ↓GSH at ≥ 12.5 μg/mL NP1, ≥ 25 μg/mL NP2. Apoptotic mode of cell death
[151]	Mixture: 45–60 nm spherules + 70 × 170 nm rods; 34% < 100 nm; 29% 100–200 nm aggregates; 20% 300–600 nm aggregates. 98% pure (EDX). LPS-free, 3 m^2/g	72-h LoVo cell cultures exposed to 1–10 μg/cm^2 (2–23 μg/mL) NPs in media 24, 48, 72 h→microscopy, cell growth curves; exposed to 1–40 μg/cm^2 (2–92 μg/mL) NPs in media 24, 48, 72 h→cytotoxicity (WST-1 assay w/NP-interference ctrl); exposed to 10, 20, 40 μg/cm^2 NPs 24 h→Annexin-V FITC staining (mode of cell death); exposed to 1–10 μg/cm^2 NPs 72 h→SEM and Hoechst-labeling for apoptotic morphology; exposed to 5 or 10 μg/cm^2 NPs 24, 48 h→ROS (flow cytometric assay), GSH mitochondrial membrane potential; exposed to 3 or 5 μg/cm^2 NPs 24, 48 h→ supernatant IL-6, TNF-α, IL-8 (ELISA); TEM after 24-h exposure to 10 μg/cm^2 NP. Zn-ion only ctrl for cytotoxicity (NPs incubated w/media overnight)	Altered cell morphology ≥ 5 μg/cm^2 NP; DR ↑cytotoxicity ≥ 10 μg/cm^2 NP, NP > ions, apoptotic mode of death. DR ↑ROS, ↓GSH and mitochondrial potential and ↑IL-8 release at 24 h, ≥ 3 μg/cm^2

Table 12.4 (Continued)

Reference	Test article[a)]	Test system[b)]	Effects seen
[100]	NP: 24 m^2/g, 8–10 nm, 161 and 615 nm and −54 mV in water (bimodal) MP: < 44 μm, 3.2 m^2/g	Undifferentiated Caco-2 or RKO cells ±1-h pretreatment w/TNF (BSA ctrl) exposed to 1–100 μg/cm^2 NP or MP in cDMEM 24 h→viability (CCK-8); exposed to 5 μg/cm^2 NP or MP or 100 μM $ZnCl_2$ in cDMEM 4 h→microarray (validated by qPCR) and immunoblot	95× ↑soluble Zn-ion in media after o/n incubation. Cytotoxicity LC_{50}s: 27 μg/cm^2 (RKO, no TNF), 16 μg/cm^2 (RKO + TNF), 28 μg/cm^2 (Caco-2, no TNF), 27 μg/cm^2 (Caco-2 + TNF). No additional effect of NP on cell response to TNF. NP alone ↑unfolded protein response, chaperone, and stress response genes and ↓transcription regulation genes. Response to MP qualitatively similar to NP

a) All test articles ZnO nano- or microparticles, unless otherwise noted. Primary particle diameter in nm by TEM and/or SEM, unless otherwise noted. Hydrodynamic diameter and zeta potential (if measured) in nm and mV in stated medium via dynamic light scattering and electrophoretic light scattering, respectively, unless noted otherwise. Specific surface area provided in cm^2/g or m^2/g via Brunauer–Emmett–Teller (BET) method. NTA: nanoparticle tracking analysis; XRD: X-ray diffraction; PDI: polydispersity index; ROS: reactive oxygen species; AGF: artificial gastric fluid; AIF: artificial intestinal fluid; XSDC: X-ray scanning disk centrifugation; ICP-OES: inductively coupled plasma optical emission spectroscopy; ICP-AES: inductively coupled plasma atomic emission spectroscopy; ICP-MS: inductively coupled plasma mass spectrometry.

b) Abbreviations: Ctrl: control; HPMC: hydroxypropyl methylcellulose; NP: nanoparticles; MP: microparticles; wk.o.: weeks old; Gp: group; cDMEM: DMEM + FCS; MLN: mesenteric lymph nodes; BSA: bovine serum albumin; DR: dose-responsive; CMC: carboxymethyl cellulose.

of ZnONP herein may be specific to the nanoparticle form of the metal. Given the lack of histopathological changes in the intestines of orally exposed rodents, the biological relevance of the cytotoxic and DNA-damaging effects of ZnONP in undifferentiated intestinal cell models systems is unclear.

12.4.5 Carbon Nanotubes and Fullerenes

CNTs and fullerenes may be incidentally present in food or drinking water as a result of contamination by smoke residue from cooking fires or other sources [10]. Most studies found very little uptake of fullerenes, nanotubes, or nanohorns into non-GIT tissues [158–160], with most of the dose being excreted in feces.

Fourteen studies conducted in rodents and three studies conducted in intestinal cell models were located (Table 12.5). The four studies conducted with fullerenes in rats did not report any adverse clinical signs or histopathology of the GIT or elevation in colonocyte-oxidized guanine content after treatment [160–162, 165]; additional toxicity studies conducted with PVP-coated fullerenes in mice and hydroxylated fullerenes in differentiated Caco-2 cells also did not identify any adverse effects on the GIT from treatment [167, 171]. SWCNT and MWCNT were also nontoxic when administered to rats and did not increase oxidized guanine content in colonocytes, and single-walled nanohorns (SWCNHs) were nontoxic to the GIT of mice [159, 161, 166]. In contrast, both SWCNT and MWCNT induced histopathological changes in the murine GIT, indicative of toxicity. MWCNT treatment increased mucin secretion and induced glandular epithelial hypertrophy and enlargement of intestinal lymphoid follicles in one study [163] and increased forestomach cell mitotic figures and enterocyte cell mitotic and apoptotic figures in another study [164]. Dose- and time-dependent increases in cellular accumulation of fiber-like structures and damage to cellular organelles were noted in small intestine enterocytes and macrophages in mice dosed by gavage with SWCNT for 10 days [168]; these effects were associated with increased ROS production and decreased mitochondrial potential in affected cells. Other studies found no adverse effects of SWCNT or MWCNT administration on the GIT of mice [169, 170]. Adverse effects of SWCNTs on viability and barrier function in differentiated Caco-2 cells were also noted [171, 173], while MWCNT did not affect undifferentiated Caco-2 cell growth [172]. These data in aggregate seem indicative of species-specific toxicity of SWCNTs and MWCNTs to the murine GIT, similar to that of TiO_2-NPs. The reason for this apparent species difference and its relevance to humans is unclear. The one study comparing the toxicity of shortened vs. micron-length and purified vs. Fe-contaminated SWCNT in mice found no adverse effect on the GIT of any of the test articles [170], even though the mice were dosed by gavage with an extremely high dose. Of note, the studies reporting toxic effects

Table 12.5 Toxicity data for SWCNT, MWCNT, fullerenes, and single-walled carbon nanohorns.

Reference	Test article[a)]	Test system[b)]	Effects seen
In vivo studies			
[161]	C60: 0.7 nm; in corn oil: 234 nm (low dose); 40, 713, and 3124 nm (high dose); in saline: 407 nm (low dose), 621 and 5117 nm (high dose), no metals or PAHs SWCNT: 0.9–1.7 nm diameter, length < 1000 nm; in saline: 195, 797, and 5457 nm (low dose only); in corn oil: 34 and 278 nm (low dose), 1015 nm (high dose), 2% Fe-content, w/other trace metals and PAHs ROS generation in aqueous solution: SWCNT ℧ C60	Female F344 rats, 9-wk.o. gavaged w/0, 0.064, 0.64 mg/kg NP in saline or corn oil→sacrifice 24 h later→8-oxodG in colonocytes	No ↑8-oxodG content for any test condition
[162]	Fullerenes, no characterization data	Sprague-Dawley rats, 5-wk.o. gavaged w/0, 1, 10, 100, 1000 mg/kg NP in corn oil 29 d→sacrifice at 29-d or after 14-d recovery (0 or 1000 mg/kg only) →histopathology stomach, intestines	Blackish feces and stomach contents at high dose. No histopathology changes in GIT
[163]	Acid-treated MWCNT, 18–20 nm diameter, 130 m^2/g, impurities: Fe, Co, Ca, Al, Cl	CBA mice given 500 mg/kg MWCNT in feed for 6 d→sacrifice after 1, 2, 3, 4, 5, or 6 d administration→stomach and intestinal histopathology	Mast cells in lamina propria and ↑secretory activity and hypertrophy of gastric glandular epithelium. Day 3: migration of MWCNT into deep layers of gastroduodenal mucosa. CNTs located in cytoplasm and nucleus of epithelial cells. Enlarged lymphoid follicles in mucosa lamina propria

Table 12.5 (Continued)

Reference	Test article[a)]	Test system[b)]	Effects seen
[164]	MWCNT: 15–40 nm (outer diameter), 3–8 nm (inner diameter), ≥ 2 μm length; in water: 7.5–43.7 nm	Male CBA × C57Bl/6 mice, exposed to MWCNT or activated C in drinking water at 0, 0.2, 0.5 mg/L for 14 d→micronucleus assay on cells from forestomach and small intestine on day 15	No ↑MN either tissue. Significant ↑mitotic index and in forestomach w/0.5 mg/L MWCNT and mitotic index and apoptotic figures in small intestine at 0.2 mg/L MWCNT and amorphous C
[165]	Hydroxylated C60	Male Wistar rats, gavaged w/NPs in water 0, 0.1, 1, 10 mg/kg, 28 d	No clinical adverse effects
[166]	SWCNT: > 95% pure, 2 nm diameter; MWCNT: > 98% pure, ~30 nm diameter	Acute study, MWCNT, SWCNT: ♀ CD rats gavage 4× w/50 or 200 mg/kg in 20 mL/kg 0.5% gum acacia SWCNT 28-d: CD rats gavage w/0, 0.125, 1.25, or 12.5 mg/kg/d NPs in 20 mL/kg 0.5% gum acacia MWCNT: 28-d study CD-1 rats gavaged w/0, 0.5, 5, or 50 mg/kg NPs in 20 mL/kg 0.5% gum acacia	No clinical signs or GI histopathology, either study. Black feces and/or cecum contents in treated groups
[167]	PVP-C60: 127 nm and −2.2 mV in distilled water	Male C57Bl/6 gavaged w/total dose of 50 mg NP or PVP ctrl in distilled H_2O (0.5 mL) for 7 d→colon length and histopathology, fecal blood	No clinical signs, fecal blood, or changes in histopathology or colon length
[160]	SWCNHs: no metals, > 1 nm in H_2O. SWCNH-COOH: 110 nm in H_2O. "Small" CNHs: 70 nm in H_2O. Oxidized SWCNH ± PEG coat: 160 nm in H_2O w/PEG, > 1000 nm in H_2O w/o/PEG	Female BALB/cA mice ± dextran sodium sulfate-induced colitis (colitis) gavage 3/group w/0 or 2 mg NPs/0.2 mL water→cecum and colon histopathology at 24 h	No NPs in GI-tissue itself, all located in lumen. Colitis: a few NPs adhere to cecum only

[168]	SWCNT: 50–300 nm long × 0.8–1.2 nm	Kunming mice on liquid diet gavaged w/0, 5, 50, 100, 300, 400, 500 mg/kg SWCNT in saline (0.3 mL/10 g BW) 10 d→TEM intestines, [SWCNT] in enterocyte cell organelles, assays for lysosomal damage and ROS in mitochondria and lysosomes in enterocytes	DR and time-dependent ↑fiber-like structures in enterocytes and macrophages of ileum, some parts jejunum. At ≤ 300 mg/kg, exclusively localize to lysosome, no damage. At ≥ 400 mg/kg, localize to lysosomes and mitochondria, damage to both. ↑ROS and ↓potential in mitochondria. Lysosomal damage secondary to mitochondrial damage
[169]	MWCNT ± N-doping (2–4 wt%; 2–2.5 wt% Fe). Bare: < 50 nm × 100–450 μm. Doped: 20–40 nm × 100–300 μm. Oxidized in sulfuric acid	Male CD1 mice gavaged w/0, 1, 2.5, 5 mg/kg NPs in 20–150 μl PBS once→histopathology at 1, 2, 3, 7, 30 d after dose	No clinical signs or adverse intestinal histopathology. Clumps of tubes in feces at 24 h
[160]	C60 stirred continuously in olive oil 2 wk at RT→filtration thru 250 nm pore mesh after centrifuge. No characterization data	♂ Wistar rats. Acute study: gavage once w/ 4 NPs mg/kg→blood, liver, spleen, brain, urine C60 content at 48 h. Subacute: Gavage w/4 mg/kg, 7 d→ C60 tissue content.24 h and 8 d after last dose. Chronic: gavage w/water, olive oil, or 0.8 mg/mL NPs in 1 mL olive oil daily for 1 wk→weekly until end of month 2→every 2 wk until end of 7th month→survival until 25 mos. No histopathology performed	PK at 24 and 48 h: 0.18% and 0.51% of dose in the spleen, 0.14% and 0.39% of dose in the liver. Brain < 0.01% of dose. No clinical signs
[170]	SWCNTs—raw: 1 nm × 1–2 μm, 25% Fe Purified: 1 nm × 1–2 μm, < 4% Fe, 574 m^2/g Ultrashort SWCNT: 1 nm × 20–80 nm, < 1.5% Fe, 980 m^2/g	Male Swiss mice gavaged w/ 0 or 1000 mg/kg NP in water once divided into 4 doses of 1 mL/20 g bw over 4 h→histopathology 14 d after dosing	No clinical signs or GI histopathology. Black stool noted at 24 h

Table 12.5 (Continued)

Reference	Test article[a)]	Test system[b)]	Effects seen
[158]	Trimethylenemethane adducted-C60, ^{14}C-labeled	♂Fisher rats, gavaged w/18 kBq NPs in EtOH/PEG400/5% BSA→radioactivity in feces and tissues	97% radioactivity excreted in feces after 48 h. 0.2% of dose in urine
Cell-based studies			
[171]	Hydroxylated C60: 0.7 nm, 0.39% Fe; SWCNT-COOH: 4.5 nm × 0.5–1.5 μm, 0.36% Ni, 0.19% Fe SWCNT-PEG: 4.5 nm × 0.5–1.5 μm, 0.26% Ni, 0.21% Fe	Differentiated Caco-2 cells in 96-well plates (viability) or transwell (transport) inserts, exposed to 0.0156–1 mg/mL (viability) or 1 mg/mL NPs (transport) in HEPES 24 h→LDH, MTT or TEER, Lucifer yellow (LY) transport, staining for tight junction protein (ZO-1)	No effects viability. ↓TEER and ZO-1 staining w/SWCNT-COOH only. ↑LY transport at 4 h (SWCNT-COOH only) and 24 h (SWCNT-COOH and C60), reversible
[172]	MWCNT: 10 nm × 0.1–10 μm, 250–300 m^2/g, LPS-free, 10% metallic impurities. Complete sedimentation in vehicle after 3 min w/bundle formation	Undifferentiated Caco-2 cells exposed to 0.0015, 0.015, or 0.15 μg/mm^2 NP in cDMEM + 1% DMSO 72 h→7 d culture→scoring of colony formation and EM	No effects of NP on colony-forming efficiency
[173]	SWCNT-COOH, 5–10% Ni, 1.4 nm diameter, bundles: 5–5 nm × 0.5–1.5 μm	Undifferentiated or differentiated Caco-2 cells exposed to NP in DMEM 24 h up to 1 mg/mL→neutral red assay, total cell protein, MTS, LDH, trypan blue (differentiated), light microscopy (undifferentiated)	DR ↓viability to 20% ctrl at 1 mg/mL and ↑LDH at ≥ 0.2 mg/mL; ↓protein at ≥0.4 mg/mL (differentiated) and 1 mg/mL (undifferentiated). Hydropic degeneration and vacuolation

a) NP: nanoparticle; C60: fullerene; SWCNT: single-walled carbon nanotube; MWCNT: multiwalled carbon nanotube; SWCNH: single-walled carbon nanohorn. Primary particle diameter in nm by TEM and/or SEM, unless otherwise noted. Hydrodynamic diameter and zeta potential (if measured) in nm and mV in stated medium via dynamic light scattering and electrophoretic light scattering, respectively, unless noted otherwise. Specific surface area provided in cm^2/g or m^2/g via Brunauer–Emmett–Teller (BET) method. NTA: nanoparticle tracking analysis; XRD: X-ray diffraction; PDI: polydispersity index; ROS: reactive oxygen species; AFM: atomic force microscopy; CS: centrifugal sedimentation; #D: number-based distribution; WD: weight-based distribution; XSDC: X-ray scanning disk centrifugation; ICP-OES: inductively coupled plasma optical emission spectroscopy; ICP-AES: inductively coupled plasma atomic emission spectroscopy; ICP-MS: inductively coupled plasma mass spectrometry.

b) Abbreviations: Ctrl: control; HPMC: hydroxypropyl methylcellulose; NP: nanoparticles; MP: microparticles; wk.o.: weeks old; Gp: group; cDMEM: DMEM + FCS; MLN: mesenteric lymph nodes; BSA: bovine serum albumin; DR: dose-responsive; CMC: carboxymethyl cellulose.

of CNTs on the GIT of mice were repeated-dose studies, whereas the studies reporting no toxicity with CNTs or CNHs were single-dose studies. Therefore, it is possible that CNTs require repeated dosing over time in order to manifest significant toxicity to the murine GIT. If so, this finding would again be a species-specific effect as repeated dosing of the rat GIT with CNTs produced no toxicity after 28 days [166]. Further studies comparing the GIT toxicity of CNTs in mice and rats in parallel under a repeated-dosing paradigm are needed in order to resolve this issue.

12.5 Miscellaneous Nanomaterials

Some additional data on the GIT toxicity of copper/cupric oxide, iron oxide, and calcium carbonate nanomaterials is also available (Table 12.6). These nanomaterials may be used as dietary supplements or be present in food or wastewater due to degradation of materials containing these products in the environment [6, 8]. Copper and copper oxide nanoparticles are toxic to the GIT in rodents and to differentiated Caco-2 cells; clinical signs of anorexia and diarrhea, combined with morphological changes in the small intestine, were noted in rodents, while cytotoxicity, disruption of barrier function, and induction of inflammatory cytokine production was noted in Caco-2 cells [174, 177–179]. High oral doses of copper are known to produce abdominal pain, cramps, nausea, and diarrhea in humans, so the reported effects of nano-CuO in rodents in the cited studies are not surprising [154]. In contrast to the overt toxicity of copper nanoparticles, calcium carbonate and calcium citrate nanoparticles did not produce any signs of GI toxicity in either rats or mice, even at doses of up to 2.3 g/kg [182, 183]. Iron oxide nanoparticles also did not produce toxic effects in the GITs of rodents after repeated oral administration for up to 15 days; however, iron oxide nanoparticles were toxic to differentiated Caco-2 cells, although less so than iron oxide salts [175, 176, 180, 181]. Another study found that a 3 nm iron oxide nanoparticle was nontoxic to undifferentiated RKO and Caco-2 cells [100]. The same iron oxide test article that produced no toxicity in rats after 14 days of dietary administration decreased viability in differentiated Caco-2 cells [176], although it should be noted that this study did not perform extensive histopathological examinations of the GIT or assess effects on barrier function. Therefore, these data indicate that copper nanoparticles are clearly toxic to the GIT, producing cytotoxicity via increased ROS production. In contrast, $CaCO_3$ nanoparticles are clearly nontoxic, as would be expected considering the fact that enterocytes are exposed to calcium phosphate nanoparticles as part of the natural physiology of the small intestine. Iron oxide nanoparticles were benign to the GITs of rodents but were cytotoxic to intestinal cell lines, further suggesting that cytotoxicity noted in intestinal cell models may not be biologically relevant.

Table 12.6 Toxicity data for miscellaneous metal oxide nanoparticles.

Ref	Test article[a)]	Test system[b)]	Effects seen
Cell-based studies			
[174]	CuO NPs, LPS-free Rods (RNP): 10 nm × 74 nm; 18.4 m^2/g; 78% ≤ 20 nm in water (#D, CS); 97% 41–10 000 nm in water (WD, CS); 87% < 20 nm cDMEM (CS, #D); 85.2% 41–10 000 nm in media (CS; WD) Spheres (SNP): 40 nm; 6.8 m^2/g; 94% ≤ 20 nm in water #D, CS); 93% 41–10 000 nm in water (WD, CS); 99% < 20 nm in media +FCS #D, CS); 93.6% 41–10 000 nm in media + FCS (WD, CS).	Caco-2 cells, grown on filters, 24–120-h exposure in cDMEM to 5–100 μg/mL NPs or 11–215 μg/mL $CuCl_2$ ctrl→toxicity (MTS), TEER, cytokines, Cu-release, paracellular permeability (Lucifer yellow transport)	Time-dependent toxicity w/both NPs, rod ℧ sphere at ≥ 25 μg/mL, associated w/↑IL-8 (protein and RNA) and other pro-inflammatory markers and HMOX1 (ROS marker), rod ℧ sphere. Both NPs release Cu-ion, $CuCl_2$ > rod > sphere, more ions released in DHG media vs. DMEM. Rods more potent that spheres or $CuCl_2$ at ↓TEER and ↑LY permeability when normalized to released Cu-ion
[175]	Fe_2O_3 NPs 17 nm (SNP): 328 nm, 0.88 PDI, −11.7 mV, in cDMEM 53 nm (MNP): 179 nm, 0.25 PDI, −11.4 mV in cDMEM 100 nm (LNP): 147 nm, 0.19 PDI, −11.7 mV in cDMEM	Differentiated Caco-2 cells in transwell plates (TEER) or coverslips (microscopy, live/dead assay); one exposure 48 h to 0, 1 ppm, 10 ppm, 100 ppm Fe^{3+} ion in cDMEM→TEER 5, 8, 12, 14, 21, and 28 d post-exposure; live/dead assay, microarray, QT-PCR, EM	Time-dependent ↓TEER w/S and L NPs ≥ 10 ppm Fe; all NPs at 100 ppm Fe, associated w/↓expression of tight junction proteins at 14 and 28 d. No effects on viability at 10 ppm Time-dependent NP uptake. ↓#s microvilli at 10 ppm S and LNP after 5 and 14 d, SNP > LNP. Structural changes noted in microvilli

Table 12.6 (Continued)

Ref	Test article[a)]	Test system[b)]	Effects seen
[176]	Fe(III) oxo-hydroxide NPs: NPa: Tartaric acid-adipic acid coated; 3–7 nm in H_2O NPb: tartaric acid-succinic acid coated, 2.9–7.6 nm in H_2O NPc: tartaric acid-succinic acid coated, 3.7–11.6 nm in H_2O NPd: gluconic acid–adipic acid coated, 1.76–4.4 nm in H_2O NPe: tartaric acid–adipic acid coated, 2.8–7.39 nm in H_2O. 25–75% soluble in saline at pH 3: NPa > NPb and NPe > NPc and NPd	Differentiated Caco-2 cells exposed to 0.5 mM Fe in HBSS for 1 h in form of NPs vs. Fe maltolate and Fe(III) MPs ctrl→cell [Fe] and [ferritin] up to 23 h. Viability (MTT). Undifferentiated Caco-2or HT-29 cells exposed to NPs vs. Fe sulfate/ascorbate 24 h at total [Fe] of 0–8 mM in cDMEM.	↓Viability only for [Fe] ≥4 mM NPa vs. ≥ 0.5 mM Fe sulfate. Acid solubility of NPs inversely correlated w/uptake. Ferritin formation closely correlated w/total Fe uptake. Uptake: NPa, b, c > NP d, e Ʊ FeMP
[100]	NP: 222 m^2/g, 3 nm; 557, >1000 nm and −62 mV in water	Undifferentiated Caco-2 or RKO cells ±1-h pretreatment w/TNF (BSA ctrl) exposed to 1–100 μg/cm^2 NP in cDMEM 24 h→viability (CCK-8); exposed to 5 μg/cm^2 NP in cDMEM 4 h→microarray (validated by qPCR) and immunoblot	No cytotoxicity noted any test condition or effect on response to TNF stimulation. NP aggregates located in cells
Animal studies			
[177]	CuNP: 23.5 nm (AFM); 295, 000 cm^2/g CuMP: 17 μm, 399 000 cm^2/g Cu+ in solution negligible	ICR mice (8-wk.o.), OECD up/down design acute toxicity test: 500–5000 mg Cu/kg (MP), 108–1080 mg/kg (NP), 24–237 mg/kg $CuCl_2$ by gavage	No clinical signs MP. NP: anorexia, diarrhea, vomiting, all mice. Ileus at 5 g/kg MP

[178]	Cu-loaded chitosan NPs: 121.9 nm (method unstated) w/PDI = 0.857	Male Sprague-Dawley rats: basal diet, basal + 80 mg/kg $CuSO_4$, basal + 80 mg/kg chitosan, basal + 80 mg/kg NPs, or basal + 160 mg/kg NPs, 21 d→small intestine histology + enzyme activity	↑Small intestine villus heights in NP groups and ↓crypt depth in duodenum and ileum, and ↓crypt depth jejunum for 160 mg/kg group. ↑Activities trypsin, amylase, lipase, and mucosal disaccharases in small intestine contents. No effects on morphology or digestive enzymes of chitosan or $CuSO_4$ controls
[179]	CuNP: 25 nm; 5–60 nm in 1% HPMC, 6.92 m^2/g, 99% pure, 0.014% dissolution in 1% HPMC after 30 min sonication. CuMP: 17 μm, 0.5–38 μm in 1% HPMC, 0.25 m^2/g, 99% pure	Male Wistar rats gavaged w/0, 100, or 200 mg/kg NP or 200 mg/kg MP in 1% HPMC 5 d. No GI histopathology performed	Anorexia and severe diarrhea w/200 mg/kg NPs
[180]	$FePO_4$NP, 3 types: 64.2 nm, 32.6 m^2/g, amorphous/irregular sphere 30.5 nm, 68.6 m^2/g, amorphous/spherical 10.7 nm, 194.7 m^2/g, amorphous/spherical 15 and 30 min solubility in HCl ($pH = 1$): 37.8% and 73.4%, 56.8% and 79.2%, 82.5% and 84.7%. Poorly soluble at $pH > 1$	Male Sprague-Dawley rats, 21 d.o. on Fe-deficient diet for 24-d prior to administer 10 or 20 mg Fe/kg diet at NPs vs. ferrous sulfate vs. no added Fe ctrl for 15 d. 3 Fe-complete controls in parallel→histology on stomach and intestines of 3 rats/group	No adverse histopathology noted. No particulate matter seen in duodenum
[181]	FeNP, 40 nm, spherical	BALB/c anemic female mice, acute toxicity test, 0, 12 mg/kg NP, or 12 mg/kg “ferri(III) hydroxydipolymaltosum complexus” in water for 10 d or OECD acute toxicity test: 0 or 5000 mg/kg dose to female (nonanemic?)mice→sacrifice after 14 d	No clinical signs of toxicity

Table 12.6 (Continued)

Ref	Test article[a)]	Test system[b)]	Effects seen
[176]	Same as cell-based study	21-d.o. Sprague-Dawley rats, Fe-deficient, fed 20 mg Fe/kg NPa or Fe(II) sulfate in diet 14 d	No Fe deposition noted in small intestine or clinical signs
[182]	$CaCO_3$ NP: 103.08 nm, 2–5 μm aggregates; 582.9 and 1559 nm in water; 16 m^2/g $CaCO_3$MPs: 2540 nm and 1–10 μm; 6.499–30.4 μm in water:, 0.85 m^2/g calcite. Most Ca released at pH 2 (~10%)	Sprague-Dawley rats, gavaged w/0, 250, 300, 1000 mg/kg NP or MP in 10 mL/kg water for 90 d	No clinical signs or GI histopathology noted
[183]	$CaCO_3$, ovoid, MP: 3773 nm in Milli-Q water; NP: 151 nm in Milli-Q water, Ca citrate: MP: 1793 nm in Milli-Q water, NP: 398 nm in Milli-Q water	ICR mice, 8–10-wk.o. Subacute study: 0, 1.3 g/kg, 2.3 g/kg MP or NPs + Vit D3 in Milli-Q water once→necropsy 7 d later. 28-d : 0, 0.013, 0.13, 1.3 g/kg MP or NP $CaCO_3$ or 0, 0.023, 0.23, 2.3 g/kg MP or NP Ca citrate in Vit D3/Milli-Q water for 28 d	No clinical signs or gross pathology changes

a) Primary particle diameter in nm by TEM and/or SEM, unless otherwise noted. Hydrodynamic diameter and zeta potential (if measured) in nm and mV in stated medium via dynamic light scattering and electrophoretic light scattering, respectively, unless noted otherwise. Specific surface area provided in cm^2/g or m^2/g via Brunauer–Emmett–Teller (BET) method. NTA: nanoparticle tracking analysis; XRD: X-ray diffraction; PDI: polydispersity index; ROS: reactive oxygen species; AFM: atomic force microscopy; CS: centrifugal sedimentation; #D: number-based distribution; WD: weight-based distribution. NP: nanoparticle; MP: microparticle; XSDC: X-ray scanning disk centrifugation; ICP-OES: inductively coupled plasma optical emission spectroscopy; ICP-AES: inductively coupled plasma atomic emission spectroscopy; ICP-MS: inductively coupled plasma mass spectrometry.

b) Abbreviations: Ctrl: control; HPMC: hydroxypropyl methylcellulose; NP: nanoparticles; MP: microparticles; wk.o.: weeks old; Gp: group; cDMEM: DMEM + FCS; MLN: mesenteric lymph nodes; BSA: bovine serum albumin; DR: dose-responsive. CMC: carboxymethyl cellulose.

12.6 Analysis and Conclusions

This chapter has reviewed the publicly available data on the toxicity of nanomaterials to the GIT, focusing on nanomaterials that are likely to be present in drinking water and the diet, either due to use as direct or indirect food additives, use as dietary supplements, or contamination of drinking water and food crops. The database for this subset of nanomaterials is sparse, and many of the reviewed studies lacked detailed histopathological evaluation of the GIT and assessment of barrier function. Only one study assessed the effect of nanomaterials on the structure and composition of intestinal mucus. Additional factors complicating interpretation of differences in nanoparticle toxicity between studies include disparate protocols for culturing and differentiating Caco-2 cells, variability in test article characterization methods and physicochemical parameters reported for test articles, no accounting for interference of the nanomaterial test article with the assay system, and variability in dosing metric used (mass-based vs. surface area-based). Moreover, only a few studies included non-nanoscale control articles tested in parallel to the nanomaterials, and fewer still included parallel cell-based and *in vivo* studies. Due to all of these factors, it is impossible to draw firm conclusions regarding the GI toxicity of a given nanomaterial, or nanomaterials in general, from the available data set in animal models. The biological relevance of the effects reported in studies that only used cellular models is even more uncertain.

However, there are some tentative conclusions that the data set as it stands may support. First, GI absorption of nanomaterials appears to be extremely limited, even for the partially soluble metal nanoparticles like ZnO and nanosilver, with the majority of an orally administered dose being excreted in feces. There does appear to be some uptake and retention of nanomaterials into PP and mesenteric lymph nodes, which may, in turn, lead to effects on GI immunity, although the data on this point are still sparse. Notably, studies in rats indicate that direct toxicity of nanomaterials to the GIT endothelium appears to be restricted to ZnO and copper nanomaterials. Mice, in contrast, appear to be much more sensitive to the toxic effects of nanomaterials, with GIT toxicity reported for all of the nanomaterials tested in this species, with the exception of nanosilica, $CaCO_3$-NPs, and iron oxide nanoparticles.

Cell-based studies identify most of the assessed nanomaterials as cytotoxic, with undifferentiated cells being generally more vulnerable to the toxic effects of the test substance than differentiated cells. This increased sensitivity may be due to lack of a protective mucus barrier in most of these systems or physicochemical differences between the tested articles in the studies. The biological relevance of toxicity seen only in undifferentiated cell-based model systems is unclear. Additional studies that directly compare toxicity of a given nanomaterial in an undifferentiated cell-based model vs. *in vivo* systems will be necessary in resolving this issue, as would additional *in vivo* studies that examined the effects of repeated oral exposure to nanomaterials on GI structure and function in detail. Furthermore, studies of ingested nanomaterials in a food matrix are necessary to determine availability for absorption in "real-world" situations. Until those data are available, the toxicological effects of repeated exposure to nanomaterials on the GI tract must remain an open question.

References

1 International Organization for Standardization (2015). *Nanotechnologies – Vocabulary – Part 2: Nano-Objects*. Geneva: ISO.

2 US EPA. Reviewing new chemicals nder the Toxic Substances Control Act (TSCA): control of nanoscale materials under the Toxic Substances Control Act. https://www.epa.gov/reviewing-new-chemicals-under-toxic-substances-control-act-tsca/control-nanoscale-materials-under (accessed 31 January 2022).

3 Powell, J.J., Faria, N., Thomas-McKay, E., and Pele, L.C. (2010). Origin and fate of dietary nanoparticles and microparticles in the gastrointestinal tract. *J. Autoimmun.* 34: J226–J233.

4 Lorenz, C., Van Goetz, N., Scheringer, M. et al. (2011). Potential exposure of German consumers to engineered nanoparticles in cosmetics and personal care products. *Nanotoxicology* 5: 12–29.

5 Priya, T.L., Sivakumar, S., Sri, R.L. et al. (2013). A review on biogenic nanosilver—an emerging biomedical product. *J. Biodivers. Environ. Sci.* 3: 26–38.

6 Yada, R.Y., Buck, N., Canady, R. et al. (2014). Engineered nanoscale food ingredients: evaluation of current knowledge on material characteristics relevant to uptake from the gastrointestinal tract. *Compr. Rev. Food Sci. Food Safety* 13: 730–744.

7 Garnett, M.C. and Kallinteri, P. (2006). Nanomedicines and nanotoxicology: some physiological principles. *Occup. Med. (Lond.)* 56: 307–311.

8 Christensen, F.M., Christensen, T.B., Vogel, U. et al. (2015). *Better Control of Nanomaterials - A Summary of the 4-Year Danish Initiative, Environmental Project No. 1797*. Copenhagen, Denmark: The Danish Environmental Protection Agency.

9 Sk, M.P., Jaiswal, A., Paul, A. et al. (2012). Presence of amorphous carbon nanoparticles in food caramels. *Sci. Rep.* 2: 383–388.

10 Weisner, M.R., Lowry, G.V., Casman, E. et al. (2011). Meditations on the ubiquity and mutability of nano-sized materials in the environment. *ACS Nano* 5: 8466–8470.

11 Glover, R.D., Miller, J.M., and Hutchinson, J.E. (2011). Generation of metal nanoparticles from silver and copper objects: nanoparticle dynamics on surfaces and potential sources of nanoparticles in the environment. *ACS Nano* 5: 8950–8957.

12 Kararli, T.T. (1995). Comparison of the gastrointestinal anatomy, physiology, and biochemistry of humans and commonly used laboratory animals. *Biopharm. Drug Dispos.* 16: 351–380.

13 Burkitt, H.G., Young, B., and Heath, J.W. (1993). *Wheater's Functional Histology: A Text and Colour Atlas*, 3rde. Edinburgh: Churchill Livingstone.

14 McConnell, E.L., Basit, A.W., and Murdan, S. (2008). Measurements of rat and mouse gastrointestinal pH, fluid and lymphoid tissue, and implications for *in-vivo* experiments. *J. Pharm. Pharmacol.* 60: 63–70.

15 Liu, C.Y., Amidon, G.L., Berardi, R.R. et al. (1986). Comparison of gastrointestinal pH in dogs and humans: implications on the use of the beagles dog as a model for oral absorption in humans. *J. Pharm. Sci.* 75: 271–274.

16 Pelasayed, T., Bergstrom, J.H., Gustafsson, J.K. et al. (2014). The mucus and mucins of the goblet cells and enterocytes provide the first defense line of the gastrointestinal tract and interact with the immune system. *Immunol. Rev.* 260: 8–20.

17 Grajek, W. and Olejnik, A. (2004). Epithelial cell cultures *in vitro* as a model to study functional properties of food. *Pol. J. Food Nutr. Sci.* 13 (54): 5–24.

18 Yao, M., Xiao, H., and McClements, D.J. (2014). Delivery of lipophilic bioactives: assembly, disassembly, and reassembly of lipid nanoparticles. *Annu. Rev. Food Sci. Technol.* 5: 53–81.

19 Hayeshi, R., Hilgendorf, C., Artursson, P. et al. (2008). Comparison of drug transporter gene expression and functionality in Caco-2 cells from 10 different laboratories. *Eur. J. Pharm. Sci.* 35: 383–396.

20 Elder, A., Vidyasagar, S., and DeLouise, L. (2014). Physicochemical factors that affect metal and metal oxide nanoparticle passage across epithelial barriers. *Wiley Interdiscp. Rev. Nanomed. Nanobiotechnol.* 1: 434–450.

21 Florence, A.T. (1997). The oral absorption of micro- and nanoparticulates: neither exceptional nor unusual. *Pharm. Res.* 14: 259–266.

22 Ensign, L.M., Cone, R., and Hanes, J. (2012). Oral drug delivery with polymeric nanoparticles: the gastrointestinal mucus barriers. *Adv. Drug Deliv. Rev.* 64: 557–570.

23 US FDA (2003). Redbook 2000: IV.B.1. General guidelines for designing and conducting toxicity studies. http://www.fda.gov/Food/GuidanceRegulation/GuidanceDocumentsRegulatoryInformation/IngredientsAdditivesGRASPackaging/ucm078315.htm.

24 Maisel, K., Ensign, L., Reddy, M. et al. (2015). Effect of surface chemistry on nanoparticle interaction with gastrointestinal mucus and distribution in the gastrointestinal tract following oral and rectal administration in the mouse. *J. J. Controlled Release* 197: 48–57.

25 Lefebvre, D.E., Venema, K., Gombau, L. et al. (2014). Utility of models of the gastrointestinal tract for assessment of the digestion and absorption of engineered nanomaterials released from food matrices. *Nanotoxicology* 9: 523–542.

26 Leonard, F., Collnot, E.-M., and Lehr, C.-M. (2010). A three-dimensional co-culture of enterocytes, monocytes, and dendritic cells to model inflamed intestinal mucosa *in vitro*. *Mol. Pharm.* 7: 2103–2119.

27 Wiwattanapatapee, W., Carreno-Gomez, B., Malik, N., and Duncan, R. (2000). Anionic PAMAM dendrimers rapidly cross adult rat intestine *in vitro*: a potential oral delivery system. *Pharm. Res.* 17: 991–998.

28 Desai, M.P., Labhasetwar, V., Amidon, G.L., and Levy, R.J. (1996). Gastrointestinal uptake of biodegradable microparticles: effect of particle size. *Pharm. Res.* 22: 1838–1844.

29 Sakuma, S., Sudo, R., Suzuki, N. et al. (2002). Behavior of mucoadhesive nanoparticles having hydrophilic polymeric chains in the intestine. *J. Controlled Release* 81: 281–290.

30 Liu, W., Pan, H., Zhang, C. et al. (2016). Developments in methods for measuring the intestinal absorption of nanoparticle-bound drugs. *Int. J. Mol. Sci.* 17: 1–20.

31 Artursson, P., Palm, K., and Luthman, K. (2012). Caco-2 monolayers in experimental and theoretical problems of drug transport. *Adv. Drug Deliv. Rev.* 64: 280–289.

32 Christensen, J., El-Gebali, S., Natoli, M. et al. (2012). Defining new criteria for selection of cell-based intestinal models using publicly available databases. *BMC Genomics* 13: 274–285.

33 Behrens, I. and Kissel, T. (2003). Do cell culture conditions influence the carrier-mediated transport of peptides in Caco-2 cell monolayers? *Eur. J. Pharm. Sci.* 19: 433–442.

34 Esch, M.B., Mahler, G.J., Stokol, T., and Shuler, M.L. (2014). Body-on-a-chip simulation with gastrointestinal tract and liver tissues suggests that ingested nanoparticles have the potential to cause liver injury. *Lab Chip* 14: 3081–3092.

35 Mahler, G.J., Esch, M.B., Tako, E. et al. (2012). Oral exposure to polystyrene nanoparticles affects iron absorption. *Nat. Nanotechnol.* 7: 264–271.

36 Navabi, N. and McGuckin, m.A., and Linden, S.K. (2013). Gastrointestinal cell lines form polarized epithelia with adherent mucus layer when cultured in semi-wet interfaces with mechanical stimulation. *PLoS One* 8: 1–15.

37 Faust, J.J., Doudrick, K., Yang, Y. et al. (2014). Food grade titanium dioxide disrupts intestinal brush border microvilli *in vitro* independent of sedimentation. *Cell Biol. Toxicol.* 30: 169–188.
38 Moore, T.L., Rodriguez-Lorenzo, L., Hirsch, V. et al. (2015). Nanoparticle colloidal stability in cell culture media and impact on cellular interactions. *Chem. Soc. Rev.* 44: 6287–6305.
39 Wittmack, K. (2011). Excessive delivery of nanostructured matter to submersed cells caused by rapid gravitational settling. *ACS Nano* 5: 3766–3778.
40 DeLoid, G., Cohen, J.M., Darrah, T. et al. (2014). Estimating the effective density of engineered nanomaterials for *in vitro* dosimetry. *Nat. Commun.* 5: 3514–3524.
41 Summers, H.D., Rees, P., Holton, M.D. et al. (2011). Statistical analysis of nanoparticle dosing in a dynamic cellular system. *Nat. Nanotechnol.* 6: 170–174.
42 Doak, S.H., Griffiths, S.M., Manshian, B. et al. (2009). Confounding experimental considerations in nanogenotoxicology. *Mutagenesis* 24: 285–293.
43 Marquis, B.J., Love, S.A., Braun, K.L., and Haynes, C.L. (2009). Analytical methods to assess nanoparticle toxicity. *Analyst* 134: 425–439.
44 Pfaller, T., Colognato, R., Nelissen, I. et al. (2010). The suitability of different cellular *in vitro* immunotoxicity and genotoxicity methods for the analysis of nanoparticle-induced events. *Nanotoxicology* 4: 52–72.
45 Ong, K.J., Macormack, T.J., Clark, R.J. et al. (2014). Widespread nanoparticle-assay interference: implications for nanotoxicity testing. *PLoS One* 9: 1–9.
46 Kroll, A., Pillukat, M.H., Hahn, D., and Schnekenburger, J. (2012). Interference of engineered nanoparticle with *in vitro* toxicity assays. *Arch. Toxicol.* 86: 1123–1136.
47 Oostingh, G.J., Casals, E., Italiani, P. et al. (2011). Problems and challenges in the development and validation of human cell-based assays to determine nanoparticle-induced immunomodulatory effect. *Part. Fibre Toxicol.* 8: 8–29.
48 Yang, W., Shen, C., Ji, Q. et al. (2009). Food storage material silver nanoparticles interfere with DNA replication fidelity and bind with DNA. *Nanotechnology* 20: 1–7.
49 Park, M.V.D.Z., Lankveld, D.P.K., van Loveren, H., and de Jong, W.H. (2009). The status of *in vitro* toxicity studies in the risk assessment of nanomaterials. *Nanomedicine* 4: 669–685.
50 Hirsch, C., Roesslein, M., Krug, H.F., and Wick, P. (2011). Nanomaterial cell interactions: are current *in vitro* tests reliable? *Nanomedicine* 6: 837–847.
51 Petersen, E.J., Henry, T.B., Zhao, J. et al. (2014). Identification and avoidance of potential artifacts and misinterpretations in nanomaterial ecotoxicity measurements. *Environ. Sci. Technol.* 48: 4226–4246.

52 Wong, C.H.A., Sofer, Z., Kubesova, M. et al. (2014). Synthetic routes contaminate graphene materials with a whole spectrum of unanticipated metallic elements. *PNAS* 111: 13774–13779.

53 Rosslein, M., Elliott, J.T., Salit, M. et al. (2015). Use of cause-and-effect analysis to design a high-quality nanocytotoxicology assay. *Chem. Res. Toxicol.* 28: 21–30.

54 Horie, M., Kato, H., and Iwahashi, H. (2013). Cellular effects of manufactured nanoparticles: effect of adsorption ability of nanoparticles. *Arch. Toxicol.* 87: 771–781.

55 Crist, R.M., Grossman, J.H., Patri, A.K. et al. (2013). Common pitfalls in nanotechnology: lessons learned from NCI's nanotechnology characterization laboratory. *Integr. Biol.* 5: 66–73.

56 Kim, J.A., Salvati, A., Aberg, C., and Dawson, K.A. (2014). Suppression of nanoparticle cytotoxicity approaching *in vivo* serum concentrations: limitations of *in vitro* testing for nanosafety. *Nanoscale* 6: 14180–14184.

57 McClements, D.J. (2013). Edible lipid nanoparticles: digestion, absorption, and potential toxicity. *Prog. Lipid Res.* 52: 409–423.

58 Simon, P. and Joner, E. (2008). Conceivable interactions of biopersistent nanoparticles with food matrix and living systems following from their physicochemical properties. *J. Food Nutr. Res.* 47: 51–59.

59 Nel, A.E., Madler, L., Velegol, D. et al. (2009). Understanding biophysicochemical interactions at the nano-bio interface. *Nat. Mater.* 8: 543–557.

60 Moertz, S.T., Kraegeloh, A., Chanana, M., and Kraus, T. (2015). Formation of stable hybrid clusters of proteins and nanoparticles. *ACS Nano* 9: 6696–6705.

61 Dominguez-Medina, S., Kisley, L., Tauzin, L.J. et al. (2016). Adsorption and unfolding of a single protein triggers nanoparticle aggregation. *ACS Nano* 10: 2103–2112.

62 Weir, A., Westerhoff, P., Fabricius, L. et al. (2012). Titanium dioxide nanaoparticles in food and personal care products. *Environ. Sci. Technol.* 46: 2242–2250.

63 Wang, Y., Chen, Z., Ba, T. et al. (2013). Susceptibility of young and adult rats to the oral toxicity of titanium dioxide nanoparticles. *Small* 9: 1742–1752.

64 Song, Z.-M., Chen, N., Liu, J.-H. et al. (2015). Biological effect of food additive titanium dioxide nanoparticles on intestine: an *in vitro* study. *J. Appl. Toxicol.* 35: 1169–1178.

65 Brun, E., Barreau, F., Veronesi, G. et al. (2014). Titanium dioxide nanoparticle impact and translocation through *ex vivo*, *in vivo*, and *in vitro* gut epithelia. *Part. Fibre Toxicol.* 11: 13–29.

66 McCracken, C., Zane, A., Knight, D.A. et al. (2013). Minimal intestinal epithelial cell toxicity in response to short- and long-term food-relevant inorganic nanoparticle exposure. *Chem. Res. Toxicol.* 26: 1514–1525.

67 Geraets, L., Oomen, A.G., Krystek, P. et al. (2014). Tissue distribution and elimination after oral and intravenous administration of different titanium dioxide nanoparticles in rats. *Part. Fibre Toxicol.* 11: 30–51.

68 Cho, W.-S., Kang, B.-C., Lee, J.K. et al. (2013). Comparative absorption, distribution, and excretion of titanium dioxide and zinc oxide nanoparticles after repeated oral administration. *Part. Fibre Toxicol.* 10: 9–18.

69 Jones, K., Morten, J., Smith, I. et al. (2015). Human *in vivo* and *in vitro* studies on gastrointestinal absorption of titanium dioxide nanoparticles. *Toxicol. Lett.* 233: 95–101.

70 Janer, G., Mas de Molino, E., Fernandez-Rosas, E. et al. (2014). Cell uptake and oral absorption of titanium dioxide nanoparticles. *Toxicol. Lett.* 228: 103–110.

71 Brun, E., Jugenm, M.-L., Herlin-Boime, N. et al. (2011). Investigation of TiO_2 nanoparticles translocation through a Caco-2 monolayer. *J. Phys.: Conf. Ser.* 304: 1–8.

72 Koeneman, B.A., Zhang, Y., Westerhoff, P. et al. (2010). Toxicity and cellular responses of intestinal cells exposed to titanium dioxide. *Cell Biol. Toxicol.* 26: 225–238.

73 Fischella, M., Berenguer, F., Steinmetz, G. et al. (2012). Intestinal toxicity evaluation of TiO_2 degraded surface-treated nanoparticles: A combined physicochemical and toxicogenomics approach in Caco-2 cells. *Part. Fibre Toxicol.* 9: 18–31.

74 Dorier, M., Brun, E., Veronesi, G. et al. (2015). Impact of anatase and rutile titanium dioxide nanoparticles on uptake carriers and efflux pumps in Caco-2 gut epithelial cells. *Nanoscale* 7: 7352–7360.

75 Jani, P.U., McCarthy, D.E., and Florence, A.T. (1994). Titanium dioxide (rutile) particle uptake from the rat GI tract and translocation to systemic organs after oral administration. *Int. J. Pharm.* 105: 157–168.

76 Warheit, D.B., Brown, S.C., and Donner, E.M. (2015). Acute and subchronic oral toxicity studies in rats with nanoscale and pigment grade titanium dioxide particles. *Food Chem. Toxicol.* 84: 208–225.

77 Wang, J., Zhou, G., Chen, C. et al. (2007). Acute toxicity and biodistribution of different sized titanium dioxide particles after oral administration. *Toxicol. Lett.* 168: 176–185.

78 Gu, N., Hu, H., Guo, Q. et al. (2015). Effects of oral administration of titanium dioxide fine-sized particles on plasma glucose in mice. *Food Chem. Toxicol.* 86: 124–131.

79 Sang, X., Fei, M., Sheng, L. et al. (2014). Immunomodulatory effects in the spleen-injured mice following exposure to titanium dioxide nanoparticles. *J. Biomed. Mater. Res. Part A* 102A: 3562–3572.

80 Bachler, G., von Goetz, N., and Hungerbuhler, K. (2015). Using physiologically based pharmacokinetic (PBPK) modeling for dietary risk assessment of titanium dioxide (TiO_2) nanoparticles. *Nanotoxicology* 9: 373–380.

81 Thoree, V., Skepper, J., Deere, H. et al. (2008). Phenotype of exogenous microparticle-containing pigment cells of the human Peyer's patch in inflamed and normal ileum. *Inflamm. Res.* 57: 374–378.

82 Powell, J.J., Ainley, C.C., Harvey, R.S.J. et al. (1996). Characterization of inorganic microparticles in pigment cells of human gut associated lymphoid tissue. *Gut* 38: 390–395.

83 Ruiz, P.A., Moron, B., Becker, H.M. et al. (2017). Titanium dioxide nanoparticles exacerbate DSS-induced colitis: role of NLRP3 inflammasome. *Gut* 66: 1216–1224.

84 Pele, L.C., Thoree, V., Bruggraber, S.F.A. et al. (2015). Pharmaceutical/food grade titanium dioxide particles are absorbed into the bloodstream of human volunteers. *Part. Fibre Toxicol.* 12: 26–32.

85 Tassinari, R., La Rocca, C., Stecca, L. et al. (2015). *In vivo* and *in vitro* toxicological effects of titanium dioxide nanoparticles on small intestine. *AIP Conf. Proc.* 1667: 020016.

86 Nogueira, C.M., Mendes de Azevedo, W., Dagli, A.L.Z. et al. (2012). Titanium dioxide induced inflammation in the small intestine. *World J. Gastroenterol.* 18: 4729–4735.

87 Onishchenko, G.E., Erokhina, M.V., Abramchuk, S.S. et al. (2012). Effects of titanium dioxide nanoparticles on small intestinal mucosa of rats. *Bull. Exp. Biol. Med.* 154: 265–270.

88 Mohamed, H.R.H. (2015). Estimation of TiO_2 nanoparticle-induced genotoxicity persistence and possible chronic gastritis-induction in mice. *Food Chem. Toxicol.* 83: 76–83.

89 Chen, X.-X., Cheng, B., Yang, Y.-X. et al. (2013). Characterization and preliminary toxicity assay of nano-titanium dioxide additive in sugar-coated chewing gum. *Small* 9: 1765–1774.

90 Chalew, T.E.A. and Schwab, K.J. (2013). Toxicity of commercially-available engineered nanoparticles to Caco-2 and SW480 human intestinal epithelial cells. *Cell Biol. Toxicol.* 29: 101–116.

91 Farcal, L., Andon, F.T., Di Cristo, L. et al. (2015). Comprehensive *in vitro* toxicity testing of a panel of representative oxide nanomaterials: first steps towards an intelligent testing strategy. *PLoS One* 10: 1–34.

92 Susewind, J., de Souza Carvalho-Wodarz, C., Repnik, U. et al. (2015). A 3D co-culture of three human cell lines to model the inflamed intestinal mucosa for safety testing of nanomaterials. *Nanotoxicology* 10: 53–62.

93 Gerloff, K., Fenoglio, I., Carella, E. et al. (2012). Distinctive toxicity of TiO_2 rutile/anatase mixed phase nanoparticles in Caco-2 cells. *Chem. Res. Toxicol.* 25: 646–655.

94 Gerloff, K., Albrecht, C., Boots, A.W. et al. (2009). Cytotoxicity and oxidative DNA damage by nanoparticles in human intestinal Caco-2 cells. *Nanotoxicology* 3: 355–364.

95 Kruger, K., Cossais, F., Neve, H., and Klempt, M. (2014). Titanium dioxide nanoparticles activate IL8-related inflammatory pathways in human colonic epithelial Caco-2 cells. *J. Nanopart. Res.* 16: 2402–2414.

96 Gitrowski, C., Al-Jubory, A.R., and Handy, R.D. (2014). Uptake of different crystal structures of TiO_2 nanoparticles by Caco-2 intestinal cells. *Toxicol. Lett.* 226: 264–276.

97 Barone, F., De Berardis, B., Bizzarri, L. et al. (2011). Physico-chemical characteristics and cyto-genotoxic potential of ZnO and TiO_2 nanoparticles on human colon carcinoma cells. *J. Phys. Conf. Ser.* 304: 1–6.

98 Powell, J.J., Harvey, R.S.J., Ashwood, P. et al. (2000). Immune potentiation of ultrafine dietary particles in normal subjects and patients with inflammatory bowel disease. *J. Autoimmun.* 14: 99–105.

99 Evans, S.M., Ashwood, P., Warley, A. et al. (2002). The role of dietary microparticles and calcium in apoptosis and interleukin-1β release of intestinal macrophages. *Gastroenterology* 123: 1543–1553.

100 Moos, P.J., Olszewski, K., Honeggar, M. et al. (2011). Responses of human cells to ZnO nanoparticles: a gene transcription study. *Metallomics* 3: 1–28.

101 Lomer, M.C.E., Grainger, S.L., Ede, R. et al. (2005). Lack of efficacy of a reduced microparticle diet in a multi-centred trial of patients with active Crohn's disease. *Eur. J. Gastroenterol. Hepatol.* 17: 377–384.

102 Jacobs, R., van der Voet, H., and ter Braak, C.J.F. (2015). Integrated probabilistic risk assessment for nanoparticles: the case of nanosilica in food. *J. Nanopart. Res.* 17: 251–265.

103 Dekkers, S., Krystek, P., Peters, R.J.B. et al. (2010). Presence and risks of nanosilica in food products. *Nanotoxicology* 5: 393–405.

104 Peters, R., Kramer, E., Oomen, A.G. et al. (2012). Presence of nano-sized silica during in vitro digestion of foods containing silica as a food additive. *ACS Nano* 6: 2441–2451.

105 Van der Zande, M., Vandebriel, R.J., Groot, M.J. et al. (2014). Sub-chronic toxicity study in rats orally exposed to nanostructured silica. *Part. Fibre Toxicol.* 11: 8–27.

106 Tarantini, A., Huet, S., Jarry, G. et al. (2015). Genotoxicity of synthetic amorphous silica nanoparticles in rats following short-term exposure. Part One: oral route. *Environ. Mol. Mutagen* 56: 218–227.

107 Kim, Y.-R., Lee, S.-Y., Lee, E.J. et al. (2014). Toxicity of colloidal silica nanoparticles administered orally for 90 days in rats. *Int. J. Nanomedicine* 9: 67–78.

108 Buesen, R., Landsiedel, R., Sauer, U.G. et al. (2014). Effects of SiO_2, ZrO_2, and $BaSO_4$ nanomaterials with or without surface functionalization upon 28-day oral exposure to rats. *Arch. Toxicol.* 88: 1881–1906.

109 Yun, J.-W., Kim, S.-H., You, J.-R. et al. (2014). Comparative toxicity of silicon dioxide, silver and iron oxide nanoparticles after repeated oral administration to rats. *J. Appl. Toxicol.* 35: 681–693.

110 Yoshida, T., Yoshioka, Y., Takahashi, H. et al. (2014). Intestinal absorption and biological effects of orally administered amorphous silica particles. *Nanoscale Res. Lett.* 9: 532–539.

111 Bhattacharjee, S., van Opstal, E.J., Alink, G.M. et al. (2013). Surface-charge-specific interactions between polymer nanoparticles and ABC transporters in Caco-2 cells. *J. Nanopart. Res.* 15: 1695–1709.

112 Sergent, J.-A., Paget, V., and Chevillard, S. (2012). Toxicity and genotoxicity of nano-SiO_2 on human epithelial intestinal HT-29 cell line. *Ann. Occup. Hyg.* 56: 622–630.

113 Foldbjerg, R., Wang, J., Beer, C. et al. (2013). Biological effects induced by BSA-stabilized silica nanoparticles in mammalian cell lines. *Chem. Biol. Interact.* 204: 28–38.

114 Mamonova, I.A., Babushkina, I.V., Norkin, I.A. et al. (2015). Biological activity of metal nanoparticles and their oxides and their effect on bacterial cells. *Nanotechnol. Russia* 10: 128–134.

115 Walczak, A.P., Fookink, R., Peters, R. et al. (2013). Behavior of silver nanoparticles and silver ions in an *in vitro* human gastrointestinal digestion model. *Nanotoxicology* 7: 1198–1210.

116 Rogers, K.R., Bradham, K., Tolaymat, T. et al. (2012). Alterations in physical state of silver nanoparticles exposed to synthetic human stomach fluid. *Sci. Total Environ.* 420: 334–339.

117 Bouwmeester, H., Poortman, J., Peters, R.J. et al. (2011). Characterization of translocation of silver nanoparticles and effects on whole-genome gene expression using an *in vitro* intestinal epithelium co-culture model. *ACS Nano* 5: 4091–4103.

118 Giovanni, M., Tay, C.Y., Setyawati, M.I. et al. (2014). Toxicity profiling of water contextual zinc oxide, silver, and titanium dioxide nanoparticles in human oral and gastrointestinal cell systems. *Environ. Toxicol.* 30: 1459–1469.

119 Sahu, S.C., Njorage, J., Bryce, S.M. et al. (2014). Comparative genotoxicity of nanosilver in human liver HepG2 and colon Caco2 cells evaluated by a flow cytometric *in vitro* micronucleus assay. *J. Appl. Toxicol.* 34: 1226–1234.

120 Environment Directorate, OECD (2015). OECD Environment, Health and Safety Publications, No. 62: Considerations for using dissolution as a function of surface chemistry to evaluate environmental behavior of nanomaterials in risk assessments. ENV/JM/Mono(2015)44.

121 Park, K., Park, E.-J., Chun, I.K. et al. (2011). Bioavailability and toxicokinetics of citrate-coated silver nanoparticles in rats. *Arch. Pharm. Res.* 34: 153–158.

122 Van der Zande, M., Vandebriel, R.J., Van Doren, E. et al. (2012). Distribution, elimination, and toxicity of silver nanoparticles and silver ions in rats after 28-day oral exposure. *ACS Nano* 6: 7427–7442.

123 Loeschner, K., Hadrup, N., Qvordrup, K. et al. (2011). Distribution of silver in rats following 28 days of repeated oral exposure to silver nanoparticles or silver acetate. *Part. Fibre Toxicol.* 8: 18–32.

124 Platonova, T.A., Pridvorova, S.M., Zherdev, A.V. et al. (2013). Identification of silver nanoparticles in the small intestinal mucosa, liver, and spleen of rats by transmission electron microscopy. *Bull. Exp. Biol. Med.* 155: 237–241.

125 Garcia, T., Lafuente, D., Blanco, J. et al. (2016). Oral subchronic exposure to silver nanoparticles in rats. *Food Chem. Toxicol.* 92: 177–187.

126 Park, E.-J., Bae, E., Yi, J. et al. (2010). Repeated-dose toxicity and inflammatory responses in mice by oral administration of silver nanoparticles. *Environ. Toxicol. Pharmacol.* 30: 162–168.

127 Kim, Y.S., Song, M.Y., Park, J.D. et al. (2010). Subchronic oral toxicity of silver nanoparticles. *Part. Fibre Toxicol.* 7: 20–31.

128 Kim, Y.S., Kim, J.S., Cho, H.S. et al. (2008). Twenty-eight-day oral toxicity, genotoxicity, and gender-related tissue distribution of silver nanoparticles in Sprague-Dawley rats. *Inhal. Toxicol.* 20: 575–583.

129 Jeong, G.N., Jo, U.B., Ryu, H.Y. et al. (2010). Histochemical study of intestinal mucins after administration of silver nanoparticles in Sprague-Dawley rats. *Arch. Toxicol.* 84: 63–69.

130 Shahare, B., Yashpal, M., and Singh, G. (2013). Toxic effects of repeated oral exposure of silver nanoparticles on small intestine mucosa of mice. *Toxicol. Mech. Methods* 23: 161–167.

131 Hadrup, N., Loeschner, K., Bergstrom, A. et al. (2012). Sub acute oral toxicity investigation of nanoparticulate and ionic silver in rats. *Arch. Toxicol.* 86: 543–551.

132 Park, K. (2013). Toxicokinetic differences and toxicities of silver nanoparticles and silver ions in rats after single oral administration. *J. Toxicol. Environ. Health Part A* 76: 1246–1260.

133 Espinosa-Cristobal, L.F., Martinez-Castanon, G.A., Loyola-Rodriguez, J.P. et al. (2013). Toxicity, distribution, and accumulation of silver nanoparticles in Wistar rats. *J. Nanopart. Res.* 15: 1702–1714.

134 Martirosyan, A., Polet, M., and Schneider, Y.J. (2013). *In vitro* safety assessment of nanosilver in improved cell culture systems: poster presentation. *BMC Proc.* 7: P10.

135 Martirosyan, A., Bazes, A., and Schneider, Y.-J. (2014). *In vitro* toxicity assessment of silver nanoparticles in the presence of phenolic compounds—preventative agents against the harmful effect? *Nanotoxicology* 8: 573–582.

136 Verano-Braga, T., Miethling-Graff, R., Wojdyla, K. et al. (2014). Insights into the cellular response triggered by silver nanoparticles using quantitative proteomics. *ACS Nano* 8: 2161–2175.

137 Bohmert, L., Niemann, B., Lichtenstein, D. et al. (2014). Molecular mechanism of silver nanoparticles in human intestinal cells. *Nanotoxicology* 9: 852–860.

138 Bohmert, L., Girod, M., Hansen, U. et al. (2014). Analytically monitored digestion of silver nanoparticles and their toxicity on human intestinal cells. *Nanotoxicology* 8: 631–642.

139 Ahmadi, F., Ebrahimnezhad, Y., Sis, N.M., and Ghalehkandi, J.G. (2013). The effects of zinc oxide nanoparticles on performance, digestive organs, and serum lipid concentrations in broiler chickens during starter period. *Int. J. Biosci.* 3: 23–29.

140 Frohlich, E. and Roblegg, E. (2012). Models for oral uptake of nanoparticles in consumer products. *Toxicology* 291: 10–17.

141 Seok, S.H., Cho, W.-S., Park, J.S. et al. (2013). Rat pancreatitis produced by 13-week administration of zinc oxide nanoparticles: biopersistence of nanoparticles and possible solutions. *J. Appl. Toxicol.* 33: 1089–1096.

142 Baek, M., Chung, H.-E., Yu, J. et al. (2012). Pharmacokinetics, tissue distribution, and excretion of zinc oxide nanoparticles. *Int. J. Nanomedicine* 7: 3081–3097.

143 Park, H.-S., Shin, S.-S., Hong, J.-S. et al. (2014). A 90-day study of subchronic oral toxicity of 20 nm, negatively charged zinc oxide nanoparticles in Sprague-Dawley rats. *Int. J. Nanomedicine* 9: 79–92.

144 Wang, B., Feng, W., Wang, M. et al. (2008). Acute toxicological impact of nano- and submicro-scaled zinc oxide powder on healthy adult mice. *J. Nanopart. Res.* 10: 263–276.

145 Choi, J., Kim, H., Kim, P. et al. (2015). Toxicity of zinc oxide nanoparticles in rats treated by two different routes: single intravenous injection and single oral administration. *J. Toxicol. Environ. Health Part A* 78: 226–243.

146 Wang, B., Feng, W.-Y., Wang, T.-C. et al. (2006). Acute toxicity of nano- and micro-scale zinc powder in healthy adult mice. *Toxicol. Lett.* 161: 115–123.

147 Pasupuleti, S., Alapati, S., Ganapathy, S. et al. (2012). Toxicity of zinc oxide nanoparticles through oral route. *Toxicol. Ind. Health* 28: 675–686.

148 Kang, T., Guan, R., Chen, X. et al. (2013). *In vitro* toxicity of different-sized ZnO nanoparticles in Caco-2 cells. *Nanoscale Res. Lett.* 8: 496–504.
149 Wang, Y., Yuan, L., Yao, C. et al. (2014). A combined toxicity study of zinc oxide nanoparticles and vitamin C in food additives. *Nanoscale* 6: 15333–15342.
150 Cao, Y., Roursgaard, M., Kermanizadeh, A. et al. (2015). Synergistic effects of zinc oxide nanoparticles and fatty acids on toxicity to Caco-2 cells. *Int. J. Toxicol.* 34: 67–76.
151 De Berardis, B., Civitelli, G., Condello, M. et al. (2010). Exposure to ZnO nanoparticles induces oxidative stress and cytotoxicity in human colon carcinoma cells. *Toxicol. Appl. Pharmacol.* 246: 116–127.
152 Kim, Y.-R., Park, J.-I., Lee, E.J. et al. (2014). Toxicity of 100 nm zinc oxide nanoparticles: a report of 90-day repeated oral administration in Sprague-Dawley rats. *Int. J. Nanomedicine* 9: 109–126.
153 Moos, P.J., Chung, K., Woessner, D. et al. (2010). ZnO particulate matter requires cell contact for toxicity to human colon cancer cells. *Chem. Res. Toxicol.* 23: 733–739.
154 Goldhaber, S.B. (2003). Trace element risk assessment: essentiality vs. toxicity. *Regul. Toxicol. Pharmacol.* 38: 232–242.
155 Morgan, A.A. (1978). Bleeding gastric erosion after oral zinc sulphate. *Br. Med. J.* 1: 1283–1284.
156 Cho, C.H., Fong, L.Y.Y., Ma, P.C.C., and Ogle, C.W. (1987). Zinc deficiency: its role in gastric secretion and stress-induced gastric ulceration in rats. *Pharmacol. Biochem. Behav.* 26: 293–297.
157 Cho, C.H. and Ogle, C.W. (1977). The effects of zinc sulphate on vagal-induced mast cell changes and ulcers in the rat stomach. *Eur. J. Pharmacol.* 43: 315–322.
158 Yamago, S., Tokuyama, H., Nakamura, E. et al. (1995). *In vivo* biological behavior of a water-miscible fullerene: 14C-labeling, absorption, distribution, excretion, and acute toxicity. *Chem. Biol.* 2: 385–389.
159 Nakamura, M., Tahara, Y., Murakami, T. et al. (2014). Gastrointestinal actions of orally-administered single-walled carbon nanohorns. *Carbon* 69: 409–416.
160 Baati, T., Bourasset, F., Gharbi, N. et al. (2012). The prolongation of the lifespan of rats by repeated oral administration of [60] fullerene. *Biomaterials* 33: 4936–4946.
161 Folkmann, J.K., Risom, L., Jacobsen, N.R. et al. (2009). Oxidatively damaged DNA in rats exposed by oral gavage to C60 fullerenes and single-walled carbon nanotubes. *Environ. Health Perspect.* 117: 705–708.
162 Takahashi, M., Kato, H., Doi, Y. et al. (2012). Sub-acute oral toxicity study with fullerene C60 in rats. *J. Toxicol. Sci.* 37: 353–361.

163 Golokhvast, K.S., Chaika, V.V., Kuznetsoz, L.V. et al. (2013). Effects of multi-walled carbon nanotubes received orally during 6 days on the gastrointestinal tract. *Bull. Exp. Biol. Med.* 155: 753–757.

164 Sycheva, L.P., Mikhailova, R.I., Belyaeva, N.N. et al. (2015). Study of mutagenic and cytotoxic effects of multiwalled carbon nanotubes and activated carbon in six organs of mice *in vivo*. *Nanotechnol. Russia* 10: 311–317.

165 Shipelin, V.A., Trushina, E.N., Avren'eva, L.I. et al. (2013). Toxicological and sanitary characteristics of fullerenol (hydroxylated fullerene C60) in 28-day *in vivo* experiment. *Nanotechnol. Russia* 8: 799–809.

166 Matsumoto, M., Serizawa, H., Sunaga, M. et al. (2012). No toxicological effects of acute and repeated oral gavage doses of single-wall or multi-wall carbon nanotube in rats. *J. Toxicol. Sci.* 37: 463–744.

167 Yamashita, K., Yoshioka, Y., Pan, H. et al. (2013). Biochemical and hematologic effects of polyvinylpyrrolidone-wrapped fullerene C60 after oral administration. *Pharmazie* 68: 54–57.

168 Yang, Z., Zhang, Y., Yang, Y. et al. (2010). Pharmacological and toxicological target organelles and safe use of single-walled carbon nanotubes as drug carriers in treating Alzheimer disease. *Nanomedicine: NBM* 6: 427–441.

169 Carrero-Sanchez, J.C., Elias, A.L., Mancilla, R. et al. (2006). Biocompatibility and toxicological studies of carbon nanotubes doped with nitrogen. *Nano Lett.* 6: 109–116.

170 Kolosnjaj-Tabi, J., Hartman, K.B., Boudjemaa, S. et al. (2010). *In vivo* behavior of large doses of ultrashort and full-length single-walled carbon nanotubes after oral and intraperitoneal administration to Swiss mice. *ACS Nano* 4: 1481–1492.

171 Coyuco, J.C., Liu, Y., Tan, B.-J., and Chiu, G.N.C. (2011). Functionalized carbon nanomaterials: exploring the interactions with Caco-2 cells for potential oral drug delivery. *Int. J. Nanomedicine* 6: 2253–2263.

172 Ponti, J., Colognato, R., Rauscher, H. et al. (2010). Colony forming efficiency and microscopy analysis of multi-wall carbon nanotubes cell interaction. *Toxicol. Lett.* 197: 29–37.

173 Jos, A., Pichardo, S., Puerto, M. et al. (2009). Cytotoxicity of carboxylic acid functionalized single wall carbon nanotubes on the human intestinal cell line Caco-2. *Toxicol. In Vitro* 23: 1491–1496.

174 Piret, J.-P., Vankoningsloo, S., Mejia, J. et al. (2012). Differential toxicity of copper (II) oxide nanoparticles of similar hydrodynamic diameter on human differentiated intestinal Caco-2 cells monolayers is correlated in part to copper release and shape. *Nanotoxicology* 6: 789–803.

175 Kalive, M., Zhang, W., Chen, Y., and Capco, D.G. (2012). Human intestinal epithelial cells exhibit a cellular response indicating a potential toxicity upon exposure to hematite nanoparticles. *Cell Biol. Toxicol.* 28: 343–368.

176 Pereira, D.I.A., Bruggraber, S.F.A., Faria, N. et al. (2014). Nanoparticulate iron(III) oxohydroxide delivers safe iron that is well absorbed and utilized in humans. *Nanomedicine* 10: 1877–1886.

177 Chen, Z., Meng, H., Xing, G. et al. (2006). Acute toxicological effects of copper nanoparticles *in vivo*. *Toxicol. Lett.* 163: 109–120.

178 Han, Z.-Y., Du, W.-L., Huang, Q.-C. et al. (2012). Changes in small intestinal morphology and digestive enzyme activity with oral administration of copper-loaded chitosan nanoparticles in rats. *Biol. Trace Elem. Res.* 145: 355–360.

179 Liao, M. and Liu, H. (2012). Gene expression profiling of nephrotoxicity from copper nanoparticles in rats after repeated oral administration. *Environ. Toxicol. Pharmacol.* 34: 67–80.

180 Rohner, F., Ernst, F.O., Arnold, M. et al. (2007). Synthesis, characterization, and bioavailability in rats of ferric phosphate nanoparticles. *J. Nutr.* 137: 614–619.

181 Rieznichenko, L., Doroshcnko, A., Gruzina, T. et al. (2013). Iron nanoparticles effectiveness for new anti-anemic preparations development. In: *NANOCON 2013-Conference Proceedings*, 556–559. Brno, Czech Republic: IEEE.

182 Sung, J.H., Park, S.J., Jeong, M.S. et al. (2015). Physicochemical analysis and repeated-dose 90-days oral toxicity study of nanocalcium carbonate in Sprague-Dawley rats. *Nanotoxicology* 9: 603–612.

183 Huang, S., Chen, J.C., Hsu, C.W., and Chang, W.H. (2009). Effects of nano calcium carbonate and nano calcium citrate on toxicity in ICR mice and on bone mineral density in an ovariectomized mice model. *Nanotechnology* 20: 1–7.

13

The Mucosal Microbiome: Impact of Nanoparticles and Nanomaterials

Katherine M. Williams, Kuppan Gokulan, and Sangeeta Khare

National Center for Toxicological Research, US Food and Drug Administration, Division of Microbiology, Jefferson, AR 72079, USA

13.1 Introduction

The major mucus membranes in the human body are present in the gastrointestinal (GI), respiratory, and genitourinary tracts. These mucosal surfaces play many important roles: protecting delicate underlying tissues, preventing external substances from entering the body, and exerting immune responses against invading exogenous materials and organisms. Many of these mucosae also have associated populations of resident microorganisms, known as microbiota. This chapter will focus, however, on the mucosal microbiome of the GI tract. This organ was selected for a number of reasons, particularly the important immune functions, the large number of inhabiting microorganisms, the high exposure potential to external substances, and the strong evidence of a link between the microbial inhabitants of this organ and human health.

The GI tract is the body's primary route for uptake of nutrients and macromolecules and is a major immune organ, producing more than 80% of the body's antibodies [1–3]. However, the optimal function of this organ also depends upon the presence of a complex community of resident microorganisms that includes bacteria, viruses, and fungi. These gut microbiota ferment complex carbohydrates, produce beneficial short-chain fatty acids (SCFA), condition the host immune system, metabolize xenobiotics, protect the barrier function of the intestinal epithelium, and synthesize beneficial compounds, including vitamins [1, 3–5]. The human gut microbiota consists primarily of obligate anaerobes from the Firmicutes and Bacteroidetes phyla [4–6]. Common dominant genera include *Bacteroides*, *Faecalibacterium*, *Eubacterium*, *Clostridium*, and *Bifidobacterium* [3, 7, 8]. The specific composition of organisms varies from individual

Toxicology of Nanoparticles and Nanomaterials in Human, Terrestrial and Aquatic Systems, First Edition. Edited by Marc A. Williams.

to individual, with other factors such as diet, age, and early life exposure also theorized to influence the microbial activity in the GI tract [4, 6, 9]. There is also considerable evidence that the microbiota composition of the intestinal lumen is distinct from that of the intestinal mucosa [10–12]. These mucosal microorganisms often inhabit biofilms associated with mucus production on the epithelial surface [13, 14]. Mucosa-associated organisms play specialized roles in the intestinal ecosystem that include production of metabolites utilized by immune and epithelial cells [11], protection of the epithelium against invasion by pathogenic organisms [15], and conditioning of the host immune system [16, 17]. Microbiota composition may even impact mucosal barrier function by affecting properties of the mucus layer [12]. In healthy humans, these intestinal inhabitants are generally considered beneficial to their hosts [4]. However, this effect seems dependent on maintaining a stable and diverse microbiota, and a major disturbance of this community can have deleterious effects on health [4]. Disturbance of healthy gut homeostasis is also known as dysbiosis [4, 5]. Changes in gut microbiota have been linked with intestinal infection, irritable bowel disease, ulcerative colitis, Crohn's disease, and colorectal cancer [3, 4, 18]. Evidence is weaker, but growing, for links with conditions such as diabetes, obesity, metabolic syndrome, asthma, heart disease, and neurological disorders [3, 4, 18–21].

In recent decades, a surge of interest in nanotechnology has resulted in the proliferation of engineered nanomaterials (NMs) and their incorporation into everyday life. These NMs have unique properties and characteristics, and some have been used in manufactured items because of their antimicrobial effect. The increase in use of such NMs, as well as their incorporation into food packaging, cosmetics, consumer products, kitchen appliances, and water purification, raises questions about potential health effects resulting from ingestion of such materials [22]. This GI exposure could have negative effects not only on the host but also potentially on the teeming community of organisms that live in the GI tract. Determining the effects of various NMs, especially those that claim to be antibacterial, on gut microbiota is vital because there is mounting evidence that dysbiosis can encourage negative health outcomes, but far less investigation of what kind of factors can prompt such disturbances. Current research suggests that ingested substances such as antibiotics can affect the microbiota [23–25]. By contrast, far less is known about the exposure of intestinal microbiota to NMs that have antibacterial properties.

13.2 Types of Nanoparticles and Human Exposure

"Nanoparticles" and "nanomaterials" are broad terms that can encompass an immense variety of particles of various materials, sizes, and shapes. Generally, such entities are defined as a material with at least one dimension below 100 nm, but no official definition has been established [2, 26, 27]. These can include nanoparticles, nanoplates, and nanowires, composed of materials ranging from metals to organic molecules [22]. In this chapter, the term NM/NMs will be used inclusively to refer to all such materials.

Not all NMs are manufactured, or even as a result of human activity. These NMs, sometimes called "ultrafine particles," have been present to some degree for as long as the Earth has existed and are produced by volcanic activity, fire, geological action, and other natural processes [2, 22, 26–30]. Naturally occurring NMs include the protein ferritin and viruses such as norovirus (27 nm), adenoviruses (70–100 nm), and bacteriophage MS2 (28 nm) [31–34]. Naturally occurring NMs are even present in some foods [35]. Humans have been exposed to NMs throughout history, but this exposure has increased over the last century as the production of anthropogenic NMs has increased [2, 36, 37]. Apart from these natural NMs, the day-to-day activities of industrial life contribute to the production of incidental nanoparticles. These may include nanosized particles produced due to cooking smoke, diesel exhaust, welding fumes, industrial effluents, and sandblasting. Recent advances in technology have led to the invention of 3D printing. Some reports suggest that 3D printers can emit nanosized particles [38]. More recently, the rise of nanotechnology has led to the intentional manufacture of NMs specifically engineered to exhibit certain properties and/or characteristics. These materials are being used industrially, and some are being used in consumer products, raising the likelihood for human exposure [9]. This exposure has the potential to occur by several mechanisms: ingestion, inhalation, and direct contact, but this chapter will focus on potential effects of GI tract exposure through ingestion.

Potential sources of GI exposure to NMs include food, cosmetics, pharmaceuticals, drinking water, dental work, and inhaled air [1, 39]. There is also an increasing interest in incorporation of some NMs into food packaging and other food sector applications [26, 34, 40, 41]. At the moment, most ingested NMs are silicates and titanium dioxide, which are extensively used in foods [1, 2, 42]. Titanium dioxide NMs are present in consumer products such as toothpastes, cosmetics, and sunscreens. Other commonly used NMs, cerium oxide and zinc oxide, are present in coatings, pigments, paints, and personal care products such as sunscreens and cosmetics [43]. The release of engineered NMs into the environment could lead to oral exposure through contamination of the food chain and/or drinking water, and NM concentrations in surface water in some areas have been estimated to reach 15.2, 0.76, and 10 ng/L for titanium dioxide, silver, and zinc oxide, respectively [26, 44]. Exposure to silver nanoparticles is most likely to occur through the use of consumer products, which include cosmetics, food packaging, clothing, appliances, medical products, and water purifiers [9, 26, 39]. In addition, there is interest in using NMs, primarily metals and metal oxides, in dentistry to combat dental caries and periodontal disease [45, 46]. The NMs seen as highest risk for oral exposure are silver, copper, titanium dioxide, and silica [42]. Apart from metal-based NMs, there is a sharp rise in consumer use products in which different forms of carbon are used as NMs. There is promising evidence that carbon-based nanomaterials (CBN) can exert growth promoter-like activities on different crops [47–50]. The use of water-soluble carbon nanotubes (wsCNT) or multiwalled carbon nanotubes (MWCNT) on crops to promote germination rate, increase flowering, or promote overall growth of a crop raises concerns regarding the accumulation of these NMs in the edible portion of the plant. The most alarming point relevant to this chapter is the unknown effects on the intestinal microbiome if these accumulated NMs reach the intestine.

13.3 Factors Influencing Nanomaterial/Microbiota Interactions in the Intestinal Mucosal Environment

NMs can enter the human GI tract in a number of different forms and from a variety of different sources, but once inside they are universally subject to the complex and dynamic environment of the GI tract and intestinal mucosa, of which the gut microbiota are only one element. Innumerable factors can influence interactions between NMs and biological environments, but these factors tend to fall into two major categories: NM specific and environment specific. The following section briefly discusses these groups, with emphasis on factors believed to impact NM–microbiota interactions.

13.3.1 Nanomaterial-Specific Factors

Characteristics of engineered NMs such as material, size, shape, surface charge, coating, zeta potential, stability, and solubility can affect NM toxicity to inhabitants of the human GI tract [51, 52]. Many metals, such as silver and copper, are antimicrobial in bulk form, and the property is retained when materials are manufactured in nanoform [52]. Behavior of NMs in the GI tract is strongly charge dependent, and surface charge has also been shown to be vitally important in the antimicrobial effect of NMs [1, 53]. Many of these factors, such as size, coating, and surface charge, affect NM aggregation and/or agglomeration, which can also drastically affect bioavailability, rate of absorption, and toxicity [34]. One study even found that NM shape influenced antibacterial effect against *Escherichia coli*, with triangular-shaped NMs being more toxic than round and rod-shaped NMs [54]. Other potential factors important in antimicrobial activity include available surface area and particle morphology, such as NMs with rough surfaces and irregular shapes generally shown to be more reactive [55]. For metal NMs, NM characteristics that govern metal ion release will also affect antimicrobial activity [52]. However, CBN have hydrophobic surfaces that display strong adsorption affinity to organic matter [56]. The antimicrobial activity of CBN may depend on the van der Waals forces along the length axis, surface chemistry, and surface functionalizing properties [56–58].

13.3.2 Gut Environment-Specific Factors

The second set of factors that could impact NM effects on the gut microbiota are environmental characteristics of the GI tract that the NMs encounter after ingestion, such as pH, temperature, protein concentration, and presence of surfactants [45]. All of these elements have the potential to influence NM activity and toxicity. For example, pH affects the surface charge of NMs and, therefore, will influence agglomeration and other behavior [44]. An acidic environment can also accelerate the release of ions from metal NMs, leading to increased toxicity [51]. GI transit times also vary greatly between individuals and could impact direct NM exposure to microbiota [44]. Other factors, including enzymes, food material contact, and the concentration of organic molecules, can potentially reduce toxicity by altering or affecting NMs properties before they come in contact with intestinal microorganisms [1].

NMs must generally avoid binding to chyme in order to reach the epithelium, which is unlikely to occur with the majority of ingested materials [59]. Binding with chyme leads to direct excretion from the GI tract via feces, thus decreasing the bioavailability and likely lessening impact on intestinal microbiota [59]. Undoubtedly, one of the primary environmental factors that will determine NM effects on gut microbiota in the mucosal environment is the presence of mucus. Mucus is a large molecule produced by cells in the intestinal epithelium made up of cross-linked glycoproteins (mucins) that, when hydrated, forms a dense, mesh-like gel layer [59]. The intestinal mucus layer harbors large numbers of commensal bacteria, many of which exist in heterogeneous biofilm communities [60]. The intestinal mucus layer is most likely protective to gut microbiota, limiting exposure to ingested NMs [61]. Both *ex vivo* and animal studies have shown that a wide range of NMs are immobilized in mucus [59]. Mucus is extremely viscous, and NMs that cannot penetrate the mucus layer are effectively "neutralized," held until new mucus is secreted and the old mucus layer is cleared [59, 61]. However, recent work has demonstrated that it may be possible to engineer NMs to effectively bypass the mucus layer, suggesting that mucus may not be protective for all NMs [1, 59, 62]. Charge and size are the biggest factors in mucoadhesion, so modification of NMs to avoid mucus usually involves modification of one or both of these characteristics [1, 61].

13.3.3 Protein Corona

Another important factor influencing NMs in the mucosal environment is not easily classified as an environmental factor or NM-based factor but rather is better characterized as a combination of both. NMs placed into a biological environment rapidly develop a protein corona made up of diverse proteins. This is likely one reason why NMs have been reported to exhibit less toxicity in a high-protein environment [1]. This corona formation has the capacity to either stabilize or destabilize the NMs and can affect NM dispersion [36, 44]. One study found that complex protein coronas form around NMs as quickly as 30 s after introduction to human plasma. In this study, the composition of proteins was specific to each type of NM, and although the protein composition did not appear to vary greatly over time, the amount of protein was shown to increase [63]. It is believed that protein corona formation *in vivo* will even vary from individual to individual [36]. There are still many questions to be answered regarding protein corona formation in NMs, but it is important to note that many consider NMs with protein coronas to be different materials with different properties than those without, even if the base material is the same [36]. Information on protein corona impact on microbiota is virtually nonexistent, but the impact on general toxicity certainly suggests that it would have some effect. Some have theorized that acidic GI environments, such as the stomach, are likely to remove accumulated proteins from the surface of particles, but that the corona will reform when the particle proceeds into the small intestine [34].

13.4 Nanomaterial Effects on Bacterial Microbiota

In the following sections, we have attempted to provide a brief summary of available knowledge on the subject of NM effects on inhabitants of the mucosal microbiome (Table 13.1). As studies specifically examining NM effects on mucosal microbiota are rare and as microbial populations of the lumen and mucosal environment can overlap considerably [11, 12, 23], we have also included studies that examine NM effects on gut microbiota in general. However, special mention is given to those studies that have targeted or included mucosal populations. The first section summarizes the information for bacteria, which have been the subject of the vast majority of studies, while the second section presents the available information for viruses and fungi. We have limited this chapter to metal, metal oxide, and CBN, as these materials have been the primary focus of most bacteria/microbiota studies to date. Microbiota studies using other types of nanoparticles have certainly been conducted [70, 71] but are scarce, and so we have chosen to focus only on these NMs. For each type of NM, we have first

Table 13.1 Summary of published literature using fermentation studies or *in vivo* models to examine effects of metallic, metal oxide, and CBN on intestinal microbiota.

Authors/year	NM description	Size/s	Dose/s	Type of study	Study length	Major results
Metallic nanomaterials						
Das et al. [9]	Silver nanoparticles (PVP coating)	11 nm	25, 100, and 200 mg/L	Batch anaerobic fermentation of synthetic stool mix	48 h	• Increase of Gram-negative facultative anaerobes • Decrease of obligate anaerobes
Han et al. [64]	Silver nanoparticles	Not given	450 μg/mL in growth media	*In vivo* (Drosophila)	Not given	• Decrease in midgut diversity • Decrease in Gram-negative bacteria, increase in Gram-positive bacteria
Merrifield et al. [65]	Copper nanoparticles, silver nanoparticles	87 nm (CuNP), 59 nm (AgNP)	500 mg/kg in feed	*In vivo* (zebra fish)	14 d	• Significant effect of CuNP – increased overall diversity, but eliminated several species • AgNP caused only minor changes to microbiota
Hadrup et al. [66]	Silver nanoparticles (PVP coating)	14nm	2.25, 4.5, and 9 mg/kg body weight/day	*In vivo* (rat)	28 d	• No significant effects on Firmicutes and Bacteroidetes phyla • No effects on prevalence of Ag resistance genes
Williams et al. [67]	Silver nanoparticles (citrate coating)	10, 75, 110 nm	9, 18, 36 mg/kg body weight/day	*In vivo* (rat)	13 wk	• Antimicrobial effect, with smaller particle size and lower dose showing greater activity • Effect on Bacteroidetes/Firmicutes, decrease in both *Lactobacillus* and overall Gram-positive bacteria

Table 13.1 (Continued)

Authors/ year	NM description	Size/s	Dose/s	Type of study	Study length	Major results
Wilding et al. [68]	Silver nanoparticles (citrate and PVP coatings)	20 and 110 nm	10 mg/kg body weight/day	*In vivo* (mouse)	28 d	• No significant effects on gut microbial community
Metal oxide nanomaterials						
Taylor et al. [43]	ZnO, CeO_2, and TiO_2 nanoparticles	10 nm (ZnO), 21 nm (TiO_2), 10 nm (CeO_2)	0.01 μg/L (ZnO), 0.01 μg/L (CeO_2), 3 mg/L TiO_2	Simulated colon biorector (human fecal)	5 d	• Phenotypic changes, including changes in short-chain fatty acid production • Cerium oxide NM had greatest impact • Characterization/ identification of bacterial species or community groups not performed
Carbon-based nanomaterials						
Xie et al. [69]	Graphene	Not given	1, 10, and 100 μg/day	*In vivo* (mouse)	4 wk	• Increased diversity, with increase in Gram-negative bacteria and decrease in Gram-positive bacteria • Lowest dose (1 μg/day) resulted in the greatest change in gut microbiota

summarized available evidence for any antimicrobial or bacteriostatic effect and what is known about the mechanisms of such antimicrobial effect. This is based on the assumption that if a NM has demonstrated antimicrobial activity, it has the potential to affect the microbiome *in vivo*. Finally, the results of studies using more complex approaches to evaluate NM–microbiota effects, such as fermentation models and *in vivo* experiments, are presented.

13.4.1 Metallic NM

13.4.1.1 Antibacterial Activity

The effects of metallic NMs on intestinal microbiota are of concern because these materials are of high interest for incorporation into consumer products and because many metals, including silver (Ag), cobalt (Co), nickel (Ni), copper (Cu), zinc (Zn), and lead (Pb), are known to have antibacterial properties [52, 65, 72]. At the moment, most experiments studying metallic NM effects on microbiota have been conducted using silver, likely because products containing silver NMs are already available in some markets and oral human exposure is considered imminent. Silver NMs have demonstrated antibacterial activity in a number of *in vitro* studies using gut bacteria such as *E. coli*, *Bacillus subtilis*, *Enterococcus faecalis*, and *Staphylococcus aureus* [73–76]. However, specific effect varied widely between strains, indicating different susceptibilities between organisms. Silver NMs have also shown a bacteriostatic effect at fairly low doses [74]. Copper NMs have also shown similar antimicrobial effects, although effects on most strains appeared to be less robust than that of silver [76]. However, it is unclear if such effects would be observed within the gut environment. For example, exopolysaccharide components of Gram-negative bacteria have been shown capable of binding and sequestering metals, reducing their antimicrobial effect in biofilms [77]. Evidence does not seem to suggest antimicrobial activity for gold NMs, with one study finding that gold nanoparticles had no effect on the growth of *E. coli in vitro* [78].

Antimicrobial mechanisms of metallic NMs are not well understood [45]. There is also debate about whether the effects are solely a result of metal ion release or if NM-specific effects exist [51, 79]. Metallic NMs have demonstrated the ability to bind and interact with the bacterial surface membrane, and there is growing evidence that the demonstrated antibacterial activity results from NM interaction and resulting damage to critical structural elements of the bacterial membrane [52]. Studies have demonstrated the incorporation of silver nanoparticles (AgNP) into the bacterial membrane, which could prompt increases in cell permeability and death [75]. It has been suggested that the ability of certain metals, particularly silver, copper, and zinc, to bind to sulfhydryl groups in essential bacterial enzymes may be another mechanism of their antibacterial activity [80]. Silver nanoparticles are most likely to bind to phosphorus- and sulfur-containing elements on the bacterial membrane or inside the cytoplasm [45]. There is some evidence that metal ions inhibit the enzyme glucosyltransferase. This would lead to reduced polysaccharide production and could explain known effects on biofilm inhibition [46]. Silver ions are also believed to interfere with DNA replication, the expression of proteins required for ATP production, and enzymes involved in bacterial respiration [45]. Finally, metal NMs may generate reactive oxygen species (ROS) that have a cytotoxic effect [52]. In view of the complexity to distinguish between effects of ion or the NM *per se*, the use of proper positive controls is very important.

13.4.1.2 Impact in Gut Models

Only very limited studies have been conducted to assess the effect of nanoparticles on the intestinal microbiota. As mentioned earlier, the interaction of nanoparticles with the mucosa-associated microbiota is even more sparse. Table 13.1 presents a summary of study design, results, and main conclusions for all the studies. One study examined AgNP effects on an artificially generated, simplified intestinal microbiota in a static bioreactor model. Overall gas production decreased as a result of AgNP treatment, and this reduction was believed to be due to lowered CO_2 production, suggesting a reduction of metabolic activity among the microbiota [9]. AgNP treatment prompted the increase of Gram-negative facultative anaerobes, such as *E. coli* and *Raoultella* sp., and the decrease of obligate anaerobes such as *Bacteroides ovatus*, *Faecalibacterium prausnitzii*, and *Roseburia faecalis*, all of which are suspected to have important functions in the human host. Exposure to the AgCl control was also disruptive to the simulated microbiota, but the differences in response suggested AgNP-specific effects [9]. Several groups have also published *in vivo* studies examining exposure to metal NMs in a number of different hosts. A group working with AgNP in *Drosophila melanogaster* larvae found that consumption of AgNP caused a dramatic decrease in midgut microbial diversity compared with Ag microparticles and control. This shift was characterized by a drop of Gram-negative bacteria and a massive rise in Gram-positive bacteria, especially *Lactobacillus* [64]. In another study, the intestinal microbiota of zebra fish was shown to be more sensitive to the ingestion of copper nanoparticles than silver nanoparticles. Copper nanoparticles caused a significant disturbance of the intestinal community, increasing the overall diversity, but also eliminating several species that were present in controls. In comparison, fish fed silver nanoparticles had only minor changes in the microbiota compared to control diets [65]. In both of these studies, DNA for analysis was extracted from the intestinal tract of the subject animals and, therefore, likely included mucosal-associated organisms. A 14-day study incorporating AgNP into feed for pigs showed shifts in intestinal coliforms and other bacterial groups by plating and FISH, while lactobacilli remained unaffected [81].

A 28-day study of rats gavaged with PVP-stabilized AgNP found no major shifts in the Firmicutes and Bacteroidetes phyla or in the prevalence of selected silver resistance genes after exposure to AgNP or silver acetate [66]. Recently, an NCTR group published mucosa-associated gut microbiota analysis for a study in which rats were exposed to different sizes and doses of citrate-coated AgNP for 13 weeks [67]. A culture-based assessment of mucosa-associated bacteria indicated an antimicrobial effect of the AgNP, with both smaller particle size and lower AgNP dose showing greater activity. However, the larger sized particles and higher doses had a limited effect on bacterial population [67]. This effect was speculated to be the result of agglomeration in higher doses. Quantitative PCR-based analysis showed that microbial response was highly size and dose dependent, but that treatment with the 10 nm AgNP at the two lower doses appeared to alter the ratio of Bacteroidetes/Firmicutes, decrease populations of *Lactobacillus*, and cause an overall reduction of Gram-positive bacteria [67]. This study focused specifically on mucosal-associated bacteria for both culture-dependent and culture-independent methods. In contrast, a study with 28-day oral exposure to PVP- and citrate-coated AgNP at different sizes showed that exposure did not significantly alter the gut microbial community in mice [68]. The differences observed between these two studies could be due to the size of the particles, the length of the study, or the vehicle fluid used to deliver the dosage. The first study administered the AgNP in 0.1% carboxymethylcellulose, used to stabilize the AgNP during dosing, while the later study used water.

13.4.2 Metal Oxide NM

13.4.2.1 Antibacterial Activity

Due to the variety of different metal oxide NMs and fewer published works in this area, the antibacterial effects of such materials are not as well studied as other NM, such as silver. Zinc oxide nanoparticles appear to have some antibacterial activity in a number of studies [43, 82, 83]. Copper oxide nanoparticles showed antibacterial effect against *S. aureus*, *B. subtilis*, *Pseudomonas aeruginosa*, and *E. coli* [84]. Iron oxide NMs have shown antibacterial activity such as growth inhibition; however, the doses used in several of these studies were very high, raising questions about relevance for real-life application [78, 85, 86]. In a comparison of zinc oxide, copper oxide, and iron oxide NM activity against *E. coli*, *P. aeruginosa*, *S. aureus*, and *B. subtilis*, zinc oxide had the most effect, while iron oxide had the least [87]. However, in another study, copper oxide NMs were found to be more antibacterial than zinc oxide NM [46]. One study found that cerium oxide NMs do not appear to be bactericidal but may exhibit a size- and dose-dependent bacteriostatic effect [55]. Interestingly, effect was observed for *E. coli* and *B. subtilis*, but not for *Shewanella oneidensis*, a bacterial strain that reduces metals [55]. However, another work found that exposure to cerium oxide NMs reduced the viability of *E. coli* [88].

Antibacterial activity of metal oxide NMs was initially attributed to the production of ROS [79, 83], but recently researchers have begun to propose new mechanisms. For example, more recent evidence has suggested that the primary mechanism for zinc oxide NMs is the release of zinc ions from the material [89]. Copper oxide activity is likely also due to the release of ions, which are hypothesized to work by causing cell membrane damage or interfering with the structure of DNA [84]. One study presented evidence that the antibacterial effect of cerium oxide NMs was due to the intracellular generation of ROS [88], but another found no signs of major oxidative stress response and no generation of superoxide after exposure [55]. Both studies observed the adherence of NMs to the bacterial cell, suggesting that the observed responses could be due to this effect [55]. Damage to the cell membrane has been suggested as an important mechanism for other metal oxide NMs, such as zinc oxide and copper oxide [46, 83].

13.4.2.2 Impact in Gut Models

A study by Taylor et al. examined the effects of zinc oxide, titanium dioxide, and cerium oxide NM exposure in a simulated colon bioreactor model seeded with human feces. Numerous phenotypic changes were observed, including changes in hydrophobicity, electrophoretic mobility, sugar content of extracellular polymeric substance, and SCFA production, with cerium oxide NMs appearing to have the most effect. However, no characterization/identification of bacterial species or community groups was performed [43]. At the time this chapter was being compiled, there were no published *in vivo* studies using metal oxide NMs and no studies focusing specifically on mucosal microbiota of the intestine.

13.4.3 Carbon-Based NM

13.4.3.1 Antibacterial Activity

Of all the CBNs, carbon nanotubes (CNT) are the material for which the most information on antibacterial activity is available. Many studies have demonstrated antibacterial activity of CNT, the degree of which appears to depend on physical structure, with single-walled carbon nanotubes (SWCNT) generally much more antibacterial than MWCNT [90–95]. There is conflicting information about the impact of length, with one study showing that for SWCNT, longer materials were more toxic, while another demonstrated that shorter MWCNT had more effect than longer MWCNT [92, 93]. However, response may also be dose dependent, with one study showing that lower doses (10 mg/L) of various CNT structures enhanced growth of several bacterial species, while higher doses (50 mg/L) decreased viability [93]. In a comparison of different CNT structures on *Lactobacillus acidophilus*, *Bifidobacterium adolescentis*, *E. coli*, *E. faecalis*, and *S. aureus*, antibacterial effect was observed on all tested species. In this study, agglomeration of some CNT at higher doses was found to reduce toxicity to certain species compared with lower doses. The shape of the bacteria was also shown to be important, with spherical bacteria, such as cocci, showing greater susceptibility to CNT treatment than rod-shaped bacteria [92]. Interestingly, one group found that electronic structure predicted antibacterial response in SWCNT, suggesting that CBN such as CNT and graphene may have to be classified even further into subgroups before any antimicrobial consensus can be established [96]. CNT have also shown interference with biofilm formation, which raises concerns about *in vivo* effects [96]. In an *in vitro* assessment of graphite, graphite oxide, graphene oxide, and reduced graphene oxide toxicity to *E. coli*, graphene oxide had the highest antibacterial activity, followed by reduced graphene oxide and then the graphites. Toxicity was found to be time and dose dependent [97]. Graphene oxide nanowalls and reduced nanowalls deposited onto stainless steel substrates showed antibacterial activity against *E. coli* and *S. aureus* [90]. Another group found that graphene oxide and reduced graphene oxide nanosheets were bactericidal to *E. coli* [98]. However, other studies have found no antibacterial and bacteriostatic effect of graphene oxide, with one publication showing that graphene oxide exposure seemed to induce bacterial growth and biofilm formation [40, 99]. An aggregated form of C_{60} fullerenes stable in water was shown to be bacteriostatic to *E. coli* and *B. subtilis* in a low-salt, low-protein media [100, 101].

Research into antibacterial activity of CBN suggests that both physical membrane damage and induction of oxidative stress may be important mechanisms. Studies have suggested that direct contact of graphene, graphene oxide, carbon nanowalls, and CNT on bacterial cells disrupts the cellular membrane and leads to cell death [90, 92, 94–98]. One study observed incorporation of CNT into bacteria membranes using TEM, showing that loss of viability was associated with destabilization of the bacterial membrane and increased membrane fluidity [93]. CNT and fullerenes have shown a strong binding capacity with bacterial cells, although another study suggested that the activity of fullerenes could not be attributed to disruption of the bacterial membrane [79, 91, 100]. Oxidative stress likely also plays a role in the antibacterial effect of CBN, including graphene, CNT, and fullerenes [79, 94, 96, 97]. For graphene oxide and fullerenes, there is evidence that ROS production is not a major factor and that oxidative stress generated by these CBN occurs by some other mechanism [79, 97]. However, production of ROS has been observed by C_{60} fullerenes and SWCNT in other studies [22]. Metagenomic analysis in an *in vivo* graphene study showed increases in genes related to cellular membrane integrity and oxidative stress response [69], providing further evidence that antibacterial activity of CBN is related to these two mechanisms.

13.4.3.2 Impact in Gut Models

One *in vivo* study using CBN has been published by Xie et al. [69]. In a 28-day experiment, graphene exposure appeared to increase diversity in the mouse gut, with demonstrated increases in overall prevalence of Gram-negative bacteria (*Prevotella*, *Anaeroplasma*, and *Paraprevotella*) and decreases in Gram-positive bacteria (*Lactobacillus*). Interestingly, the lowest dose (1 μg/day) caused the greatest change in gut microbiota, which the authors attributed to aggregation of higher doses. Microbiota exposure to graphene also caused an increase in the detection of antibiotic resistance genes [69]. However, microbiome analysis in this study was performed on fecal samples, and was, therefore, not specific to mucosal microbiota.

13.5 Nanomaterial Effects on Viral and Fungal Microbiota

Viruses and fungi are also present in the human gut microbiota and mucosal environment [4, 67]. Viruses inhabiting the intestinal tract are diverse, present at high levels, and dominated by bacteriophages [4]. It is estimated that bacteriophages significantly outnumber bacterial cells in the gut [102]. These resident viruses are involved in the development and shaping of host immunity, biofilm formation, and in protection of host against pathogenic viruses [67, 103]. They have been associated with important physiological processes and in controlling early proliferation of cancer cells [104]. Additionally, the importance of these gut viruses, particularly phages, has been recently revisited, and these organisms are now believed to be stable residents of the gut microbiome involved in maintaining bacterial diversity of the gut, as well as modulating bacterial virulence [103]. Moreover, there is evidence that viral populations in the gut may contribute to the development of dysbiosis [4]. NMs have the potential to be applied as antiviral agents in the agriculture, medical, and veterinary fields. However, while the application of NMs as antimicrobial agents is promising, the potential side effects of intentionally consumed or accidentally ingested NMs on commensal viruses have not been properly addressed. Given that emerging research suggests that resident gut viruses exist in a balance with the gut bacteria and play beneficial roles in host health and homeostasis [105], it is important to evaluate the potential for NMs to exhibit antiviral effects. In fact, a number of *in vitro* studies have shown inhibitory activity of metallic NMs, primarily silver, against a variety of viruses, including human immunodeficiency virus (HIV-1) [106–108], hepatitis B virus [109], monkeypox virus [110], respiratory syncytial virus (RSV) [111], H1N1 influenza A virus [112], and feline calicivirus, a surrogate for human norovirus [113]. More work is needed to establish whether similar effects can be observed *in vivo*. Only a couple of studies have looked at NM effects on bacteriophages, which constitute a major part of the human gut microbiota [114, 115].

Another biotic population of the gut is the diverse array of fungal species, including yeasts [102]. The most prominent genera detected in a sequencing study were *Saccharomyces*, *Candida*, and *Cladosporium* [102]. Little information is available on the effect of NM exposure on commensal fungi in the gut, but several studies have demonstrated inhibitory effects or antimicrobial activity of NMs against different fungal species. Silver nanoparticles have been shown to inhibit the growth of yeast [73], and ZnO NMs have demonstrated activity against several fungal species [83]. One study demonstrated that CNT may have activity against *Candida albicans* [91]. It is, therefore, becoming more important to include assessments of commensal viruses and fungi in attempts to monitor and address the potential impacts of NMs on the resident gut microbes.

13.6 Conclusions

13.6.1 Antibacterial Activity of NMs

It is clear that NMs in all three groups (metallic, metal oxide, and carbon-based) have antimicrobial activity against bacteria. Metallic NMs such as silver and copper can have powerful antibacterial effects against a variety of organisms. Metal oxide NMs are a broad category with diverse properties, but there is evidence that zinc oxide and copper oxide NMs have bactericidal properties, while others may have inhibitory or bacteriostatic effects. Some CBN, such as SWCNT and graphene oxide, appeared to have antibacterial activity in several studies, but for other carbon-based materials it is less clear. These classifications are further complicated by the constant development of NMs that combine elements from more than one category [116–118]. Many of the studies discussed in this chapter provide evidence of a size-dependent effect, with smaller materials demonstrating more toxicity. It is also apparent that different organisms have different susceptibilities to the antimicrobial effects of NMs, suggesting that NMs may have disproportionate effects on certain bacterial subpopulations.

13.6.2 Evidence for NM Effects on Gut Mucosal Microbiota

While the antimicrobial potential for many NMs is clear, it is less clear how this activity and the numerous NM-specific and environment-specific factors discussed earlier interact in the actual intestinal mucosal environment. There is currently not enough published research to draw any conclusions about the potential for NM exposure to alter microbial population in the human intestinal tract. However, out of all NM categories, the evidence is strongest for silver nanoparticles, as several *in vivo* studies have shown changes in gut microbiota associated with GI dysbiosis. In general, dysbiosis is characterized by an overall reduction in bacterial diversity, a reduction in obligately anaerobic Firmicutes, and an increase in facultatively anaerobic bacteria such as Enterobaceriacae [4, 5] (Figure 13.1). These trends were observed in several *in vivo* and bioreactor-based studies with different NM types, raising concerns about health impacts of NM exposure. That being said, this pattern was not observed in all *in vivo* studies, which could be a result of NM properties, experimental model, or the route and vehicle used for the NM exposure. It has been suggested that Gram-positive bacteria may have increased resistance to the antibacterial activity of NMs because of the thicker peptidoglycan layer in their cell wall [9]. However, much of the literature reviewed in this chapter seems to suggest the opposite conclusion. Copper oxide NMs showed greater activity toward Gram-positive organisms in one study [84], as did ZnO in another [82]. This is further supported by a third study that also demonstrated greater antimicrobial effect of ZnO, CuO, and Fe_2O_3 NMs on Gram-positive bacteria [87]. Rats that were orally gavaged with silver nanoparticles exhibited a decrease in intestinal Gram-positive organisms [67]. This effect has also been observed in CBN, with graphene nanowalls [90] and CNT [91, 93] both demonstrating greater toxicity to Gram-positive than Gram-negative organisms. The mechanism behind the increased susceptibility of Gram-positive organisms is unclear, with various groups theorizing cell wall structure, the increased prevalence of NM-binding groups on the outer membrane, or a generally softer cell surface [84, 91]. This general pattern, however, may help to explain the increased prevalence of Gram-negative organisms in many of the *in vivo* treatment studies and provides more evidence that NM exposure could contribute to intestinal dysbiosis.

13.6.3 Strategies for Assessing NM–Microbiota Effects

The effects of NMs on bacteria/gut microbiota have typically been assessed using one of three strategies: *in vitro* studies using pure cultures, fermentation studies seeded with fecal organisms, and *in vivo* studies in model animals. *In vitro* studies are valuable because they allow researchers to individually evaluate NM impact on bacterial species important to human health. The current literature makes it clear that certain bacterial groups are more susceptible to NM exposure than others and *in vitro* experiments help to clarify the biochemical interactions underlying these toxic effects. NM effects on bacterial subpopulations should not be overlooked, as studies have demonstrated that low abundance species can perform specialized functions that benefit the host [6]. However, a large percentage of microbiota species have not yet been cultured in a laboratory [119–121], making it impossible

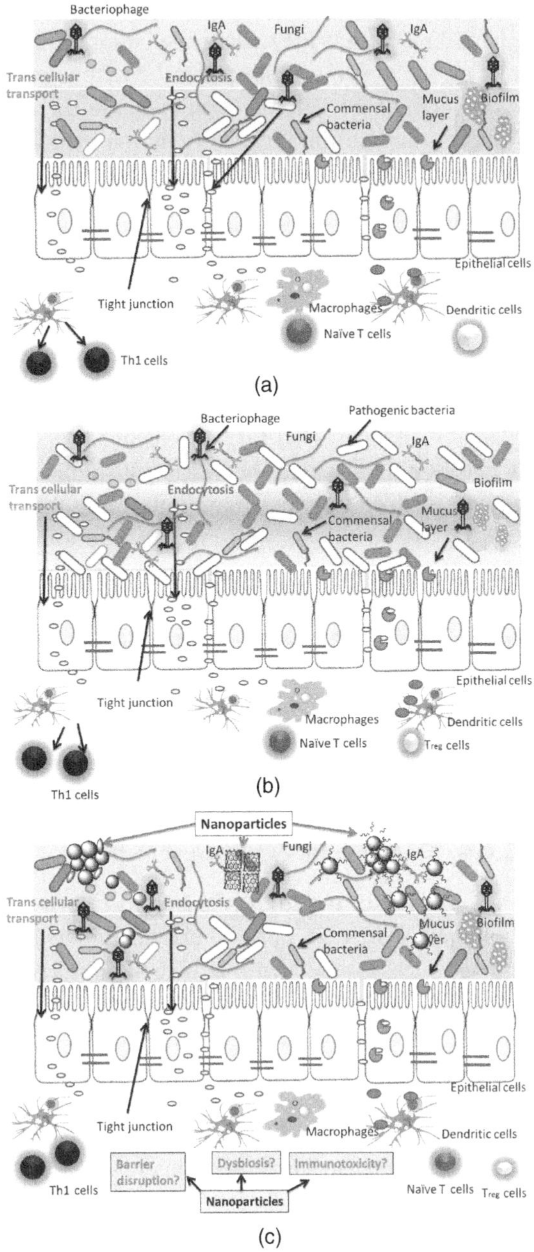

Figure 13.1 The normal, healthy intestinal mucosa (a) is characterized by diverse microflora, colonization by primarily commensal organisms, low inflammation, and intact barrier function of the intestinal epithelium. In intestinal dysbiosis (b), the mucosal environment experiences an overgrowth of certain opportunistic bacterial species. This is often accompanied by inflammation and increased gut permeability. The effects of nanomaterials (c) on the intestinal environment and microflora are still unknown, with more research required to determine whether exposure to such materials could lead to dysbiosis, disruption of the epithelial barrier, or inflammatory responses.

to use all gut organisms in these types of studies. The primary benefit of *in vivo* experiments is that they allow evaluation of both direct and indirect effects of NMs on microbiota in the mucosal environment, which is nearly impossible to replicate *in vitro*. Bacterial populations in the intestinal tract exist in a delicate homeostasis with the intestinal environment and host immune system [6]. As such, besides direct effects such as toxicity, NMs also have the potential to change the mucosal environment in a way that may affect the stability or function of the resident microbiota. Such changes could include increase/decrease of mucus production, increased inflammation, and/or damage to the epithelial barrier, leading to overgrowth of opportunistic pathogens [1, 122–124]. The ability of *in vivo* studies to examine NM effects in the host environment is also beneficial considering that most mucosa-associated organisms exist not as free-floating cells, but as mucus layer-associated biofilms. Many antimicrobials show activity against planktonic cells, but not cells in biofilms [45]. The primary disadvantage of *in vivo* studies is that they are significantly more complex, and most cannot be performed in human subjects, meaning that they generally do not evaluate human gut microbiota. Fermentation type studies performed using batch methods, or bioreactors, are growing more popular and offer an alternative to both *in vitro* and *in vitro* studies. This approach allows the evaluation of complex communities of organisms, generally obtained from fecal samples, outside the host and permits researchers to study human gut microbiota without significant risk to human subjects. However, these systems appear to be better suited for simulating NM effects on the luminal microbiome, as they lack mucus and other environment-specific factors characteristic of the intestinal mucosa.

13.6.4 Direction for Future Research

From a review of the studies presented in this chapter, it is clear that there are a number of areas in need of greater understanding regarding NM effects on human mucosa-associated microbiota. First and foremost, there is a great need for additional *in vivo* studies focused specifically on NM-induced changes in mucosa-associated populations of bacteria, viruses, and fungi. It is known that microbiota associated with the intestinal mucosa are distinct from those present in the lumen, and yet only one of the *in vivo* studies discussed in this chapter focused on evaluating NM effects on these populations. Secondly, more information is needed regarding how changes in the physical properties of the NM itself in the GI tract, specifically aggregation of NMs and corona formation, can limit antimicrobial activity, with multiple groups noting effects seen after exposure to NMs at lower doses that were not observed at higher doses. This confronts traditional dose–response relationships and has drastic implications for toxicity evaluations of gut microbiota. Another area that needs increased investigation is the impact of the anaerobic environment on NM toxicity. The vast majority of gut microbiota are obligate anaerobes that lack the ability to respire oxygen [5]. Aerobic and anaerobic bacteria respond differently to oxidative stress, which has been suggested to be a primary toxicity mechanism for many NMs [125]. Aerobic and anaerobic bacteria may also have different susceptibilities to certain metals, based upon enzymatic differences [45]. Additionally, the potential for low oxygen environments to affect NM properties has already been demonstrated, with one study showing that silver nanoparticles manufactured under anaerobic conditions remained unoxidized, did not release ions, and showed reduced toxicity to bacteria [51]. In addition to studies of NM effects on gut microbiota, more work is needed on gut microbiota metabolism and biotransformation of NMs, as such modifications have immense implications for NM use in orally administered drugs and other products [42]. Gut microbiota are known to metabolize and modify a number of compounds, including drugs, and can produce inactivated or toxic compounds [42]. Also desperately needed is a better understanding of long-term, low-dose exposure impacts on mucosal microbiota, not only via direct ingestion or oral route but also the potential for GI exposure via inhalation. These types of studies will most realistically address questions of safety resulting from exposure to NM-containing consumer products/food/food containers, which will most likely occur in small amounts over long periods of time. Such studies would elucidate chronic effects of NM exposure on microbiota composition and stability and provide insight into possibly unanticipated long-term effects, such as microbial compensation and resistance mechanisms. One study has already shown that bacteria may alter fatty acid production to counter the damaging effects of CNT on the cell membrane, suggesting that bacteria have mechanisms to compensate for NM exposure [93]. Mechanisms that can increase resistance to the antibacterial effect of metals such as silver, copper, and zinc are found in a variety of bacteria, many on transmissible elements such as plasmids [80, 126–129]. Given the prevalence of these elements, chronic exposure of the gut microbiota to metallic NMs raises concerns about the potential for proliferation of resistance genes under selective pressure. Finally, an area of consideration specific for *in vitro* studies is the choice of media in which exposure assays are performed. Several studies have noted a media-dependent effect, in which low or no NM toxicity was observed in nutrient-rich media, while toxic effect was observed in a minimal media or buffer [55, 88, 89, 99–101]. This is likely an effect of the protein corona, with the organics in the nutrient rich media coating the NMs and rendering them inactive [55].

13.6.5 Conclusion

In preparing to conclude this chapter, it should be reiterated that "nanomaterials" is a broad category encompassing an immense range of materials, sizes, coatings, sizes, and other properties, all of which can impact biological interactions. Size-dependent effects of otherwise similar materials are common, and some studies have shown that even the method of synthesis may impact NM properties [67, 84, 94, 113, 130, 131]. Given this heterogeneity and the large number of unanswered questions surrounding NM–microbiome interactions, it would be highly speculative to make any definitive statements regarding the impact of NMs on the intestinal mucosal microbiome. However, from the studies presented in this chapter, it is clear that many NMs have antibacterial activity against bacteria found in the GI tract and that there is growing evidence for disturbances in intestinal populations as a result of NM exposure. The rapid development of NM-based technologies makes human exposure to NMs inevitable. With the increased interest in NM applications for which the GI tract would be the primary target, such as drug delivery, dietary supplement formulation, and novel antimicrobials, evaluations of impact on human mucosa-associated microbiota will become an essential part of toxicology assessments of NMs.

References

1 Sutariya, V.B. and Pathak, Y. (2014). *Biointeractions of Nanomaterials*. CRC Press.

2 Chang, C. (2010). The immune effects of naturally occurring and synthetic nanoparticles. *J. Autoimmun.* 34 (3): J234–J246.

3 Alonso, V.R. and Guarner, F. (2013). Linking the gut microbiota to human health. *Br. J. Nutr.* 109: S21–S26.

4 Walker, A.W. and Lawley, T.D. (2013). Therapeutic modulation of intestinal dysbiosis. *Pharmacol. Res.* 69 (1): 75–86.

5 Winter, S.E., Lopez, C.A., and Baumler, A.J. (2013). The dynamics of gut-associated microbial communities during inflammation. *EMBO Rep.* 14 (4): 319–327.

6 Arumugam, M., Raes, J., Pelletier, E. et al. (2011). Enterotypes of the human gut microbiome. *Nature* 473 (7346): 174–180.

7 Orrhage, K. and Nord, C.E. (2000). Bifidobacteria and lactobacilli in human health. *Drugs Exp. Clin. Res.* 26 (3): 95–111.

8 Guarner, F. and Malagelada, J.-R. (2003). Gut flora in health and disease. *The Lancet* 361 (9356): 512–519.

9 Das, P., McDonald, J.A., Petrof, E.O. et al. (2014). Nanosilver-mediated change in human intestinal microbiota. *J. Nanomed. Nanotechnol.* 5 (5): 1.

10 Hollister, E.B., Gao, C., and Versalovic, J. (2014). Compositional and functional features of the gastrointestinal microbiome and their effects on human health. *Gastroenterology* 146 (6): 1449–1458.

11 Li, H., Limenitakis, J.P., Fuhrer, T. et al. (2015). The outer mucus layer hosts a distinct intestinal microbial niche. *Nat. Commun.* 6: 8292.

12 Jakobsson, H.E., Rodriguez-Pineiro, A.M., Schutte, A. et al. (2015). The composition of the gut microbiota shapes the colon mucus barrier. *EMBO Rep.* 16 (2): 164–177.

13 Williams, K., Bekele, A.Z., Gokulan, K., and Khare, S. (2015). Human intestinal microbial biofilm and its correlation with intestinal mucin secretion. In: *Biofilms in the Food Environment*, 2nde, 236–254.

14 De Vos, W.M. (2015). Microbial biofilms and the human intestinal microbiome. *NPJ Biofilms Microbiomes* 1: 15005.

15 Sassone-Corsi, M. and Raffatellu, M. (2015). No vacancy: how beneficial microbes cooperate with immunity to provide colonization resistance to pathogens. *J. Immunol.* 194 (9): 4081–4087.

16 Longman, R.S. and Littman, D.R. (2015). The functional impact of the intestinal microbiome on mucosal immunity and systemic autoimmunity. *Curr. Opin. Rheumatol.* 27 (4): 381–387.

17 Pfefferle, P.I. and Renz, H. (2014). The mucosal microbiome in shaping health and disease. *F1000 Prime Rep.* 6: 11.

18 Shreiner, A.B., Kao, J.Y., and Young, V.B. (2015). The gut microbiome in health and in disease. *Curr. Opin. Gastroenterol.* 31 (1): 69–75.

19 Momozawa, Y., Deffontaine, V., Louis, E., and Medrano, J.F. (2011). Characterization of bacteria in biopsies of colon and stools by high throughput sequencing of the V2 region of bacterial 16S rRNA gene in human. *PLoS One* 6 (2): e16952.

20 Al-Nedawi, K., Mian, M.F., Hossain, N. et al. (2015). Gut commensal microvesicles reproduce parent bacterial signals to host immune and enteric nervous systems. *FASEB J.* 29 (2): 684–695.

21 Hsiao, E.Y., McBride, S.W., Hsien, S. et al. (2013). Microbiota modulate behavioral and physiological abnormalities associated with neurodevelopmental disorders. *Cell* 155 (7): 1451–1463.

22 Oberdorster, G., Oberdorster, E., and Oberdorster, J. (2005). Nanotoxicology: an emerging discipline evolving from studies of ultrafine particles. *Environ. Health Perspect.* 113 (7): 823–839.

23 Dethlefsen, L., Huse, S., Sogin, M.L., and Relman, D.A. (2008). The pervasive effects of an antibiotic on the human gut microbiota, as revealed by deep 16S rRNA sequencing. *PLoS Biol.* 6 (11): e280.

24 Dethlefsen, L. and Relman, D.A. (2011). Incomplete recovery and individualized responses of the human distal gut microbiota to repeated antibiotic perturbation. *Proc. Natl. Acad. Sci. U.S.A.* 108 (Suppl 1): 4554–4561.

25 Perez-Cobas, A.E., Gosalbes, M.J., Friedrichs, A. et al. (2013). Gut microbiota disturbance during antibiotic therapy: a multi-omic approach. *Gut* 62 (11): 1591–1601.

26 Tiede, K., Boxall, A.B., Tear, S.P. et al. (2008). Detection and characterization of engineered nanoparticles in food and the environment. *Food Addit Contam., Part A* 25 (7): 795–821.

27 Handy, R.D., Owen, R., and Valsami-Jones, E. (2008). The ecotoxicology of nanoparticles and nanomaterials: current status, knowledge gaps, challenges, and future needs. *Ecotoxicology* 17 (5): 315–325.

28 Tepe, N. and Bau, M. (2014). Importance of nanoparticles and colloids from volcanic ash for riverine transport of trace elements to the ocean: evidence from glacial-fed rivers after the 2010 eruption of Eyjafjallajokull Volcano, Iceland. *Sci. Total Environ.* 488–489: 243–251.

29 Bisiaux, M.M., Edwards, R., Heyvaert, A.C. et al. (2011). Stormwater and fire as sources of black carbon nanoparticles to lake tahoe. *Environ. Sci. Technol.* 45 (6): 2065–2071.

30 Hough, R.M., Noble, R.R.F., Hitchen, G.J. et al. (2008). Naturally occurring gold nanoparticles and nanoplates. *Geology* 36 (7): 571–574.

31 Kapikian, A.Z. (2000). The discovery of the 27-nm Norwalk virus: an historic perspective. *J. Infect. Dis.* 181 (Suppl 2): S295–S302.

32 Golmohammadi, R., Valegård, K., Fridborg, K., and Liljas, L. (1993). The refined structure of bacteriophage MS2 at 2·8 Å resolution. *J. Mol. Biol.* 234 (3): 620–639.

33 Kennedy, M.A. and Parks, R.J. (2009). Adenovirus virion stability and the viral genome: size matters. *Mol. Ther.* 17 (10): 1664–1666.

34 Powell, J.J., Faria, N., Thomas-McKay, E., and Pele, L.C. (2010). Origin and fate of dietary nanoparticles and microparticles in the gastrointestinal tract. *J. Autoimmun.* 34 (3): J226–J233.

35 Rogers, M.A. (2016). Naturally occurring nanoparticles in food. *Curr. Opin. Food Sci.* 7: 14–19.

36 Westmeier, D., Chen, C., Stauber, R.H., and Docter, D. (2015). The bio-corona and its impact on nanomaterial toxicity. *Eur. J. Nanomed.* 7 (3): 153–168.

37 Zanker, H. and Schierz, A. (2012). Engineered nanoparticles and their identification among natural nanoparticles. *Annu. Rev. Anal. Chem.* 5: 107–132.

38 Stephens, B., Azimi, P., El Orch, Z., and Ramos, T. (2013). Ultrafine particle emissions from desktop 3D printers. *Atmos. Environ.* 79: 334–339.

39 Frohlich, E. and Roblegg, E. (2012). Models for oral uptake of nanoparticles in consumer products. *Forensic Toxicol.* 291 (1-3): 10–17.

40 Nguyen, T.H.D., Lin, M.S., and Mustapha, A. (2015). Toxicity of graphene oxide on intestinal bacteria and $CaCo_2$ cells. *J. Food Prot.* 78 (5): 996–1002.
41 Sozer, N. and Kokini, J.L. (2009). Nanotechnology and its applications in the food sector. *Trends Biotechnol.* 27 (2): 82–89.
42 Pietroiusti, A., Magrini, A., and Campagnolo, L. (2015). New frontiers in nanotoxicology: Gut microbiota/microbiome-mediated effects of engineered nanomaterials. *Toxicol. Appl. Pharmacol.*
43 Taylor, A.A., Marcus, I.M., Guysi, R.L., and Walker, S.L. (2015). Metal oxide nanoparticles induce minimal phenotypic changes in a model colon gut microbiota. *Environ. Eng. Sci.* 32 (7): 602–612.
44 Gottschalk, F., Sonderer, T., Scholz, R.W., and Nowack, B. (2009). Modeled environmental concentrations of engineered nanomaterials (TiO_2, ZnO, Ag, CNT, fullerenes) for different regions. *Environ. Sci. Technol.* 43 (24): 9216–9222.
45 Allaker, R.P. (2010). The use of nanoparticles to control oral biofilm formation. *J. Dent. Res.* 89 (11): 1175–1186.
46 Khan, S.T., Ahamed, M., Al-Khedhairy, A., and Musarrat, J. (2013). Biocidal effect of copper and zinc oxide nanoparticles on human oral microbiome and biofilm formation. *Mater. Lett.* 97: 67–70.
47 Canas, J.E., Long, M., Nations, S. et al. (2008). Effects of functionalized and nonfunctionalized single-walled carbon nanotubes on root elongation of select crop species. *Environ. Toxicol. Chem.* 27 (9): 1922–1931.
48 Tripathi, S., Sonkar, S.K., and Sarkar, S. (2011). Growth stimulation of gram (*Cicer arietinum*) plant by water soluble carbon nanotubes. *Nanoscale* 3 (3): 1176–1181.
49 Khodakovskaya, M.V., Kim, B.S., Kim, J.N. et al. (2013). Carbon nanotubes as plant growth regulators: effects on tomato growth, reproductive system, and soil microbial community. *Small* 9 (1): 115–123.
50 Lahiani, M.H., Dervishi, E., Chen, J. et al. (2013). Impact of carbon nanotube exposure to seeds of valuable crops. *ACS Appl. Mater. Interfaces* 5 (16): 7965–7973.
51 Z-m, X., Q-b, Z., Puppala, H.L. et al. (2012). Negligible particle-specific antibacterial activity of silver nanoparticles. *Nano Lett.* 12 (8): 4271–4275.
52 Lemire, J.A., Harrison, J.J., and Turner, R.J. (2013). Antimicrobial activity of metals: mechanisms, molecular targets and applications. *Nat. Rev. Microbiol.* 11 (6): 371–384.
53 McCarron, P.A., Donnelly, R.F., Marouf, W., and Calvert, D.E. (2007). Anti-adherent and antifungal activities of surfactant-coated poly (ethylcyanoacrylate) nanoparticles. *Int. J. Pharm.* 340 (1): 182–190.
54 Pal, S., Tak, Y.K., and Song, J.M. (2007). Does the antibacterial activity of silver nanoparticles depend on the shape of the nanoparticle? A study of the

Gram-negative bacterium *Escherichia coli*. *Appl. Environ. Microbiol.* 73 (6): 1712–1720.

55 Pelletier, D.A., Suresh, A.K., Holton, G.A. et al. (2010). Effects of engineered cerium oxide nanoparticles on bacterial growth and viability. *Appl. Environ. Microbiol.* 76 (24): 7981–7989.

56 Upadhyayula, V.K., Deng, S., Mitchell, M.C., and Smith, G.B. (2009). Application of carbon nanotube technology for removal of contaminants in drinking water: a review. *Sci. Total Environ.* 408 (1): 1–13.

57 Long, R.Q. and Yang, R.T. (2001). Carbon nanotubes as superior sorbent for dioxin removal. *J. Am. Chem. Soc.* 123 (9): 2058–2059.

58 Zhang, S., Shao, T., Kose, H.S., and Karanfil, T. (2012). Adsorption kinetics of aromatic compounds on carbon nanotubes and activated carbons. *Environ. Toxicol. Chem.* 31 (1): 79–85.

59 Lai, S.K., Wang, Y.Y., and Hanes, J. (2009). Mucus-penetrating nanoparticles for drug and gene delivery to mucosal tissues. *Adv. Drug Delivery Rev.* 61 (2): 158–171.

60 Pometto, A.L. III and Demirci, A. (2015). *Biofilms in the Food Environment*. John Wiley & Sons.

61 Schuster, B.S., Suk, J.S., Woodworth, G.F., and Hanes, J. (2013). Nanoparticle diffusion in respiratory mucus from humans without lung disease. *Biomaterials* 34 (13): 3439–3446.

62 Olmsted, S.S., Padgett, J.L., Yudin, A.I. et al. (2001). Diffusion of macromolecules and virus-like particles in human cervical mucus. *Biophys. J.* 81 (4): 1930–1937.

63 Tenzer, S., Docter, D., Kuharev, J. et al. (2013). Rapid formation of plasma protein corona critically affects nanoparticle pathophysiology. *Nat. Nanotechnol.* 8 (10): 772–781.

64 Han, X., Geller, B., Moniz, K. et al. (2014). Monitoring the developmental impact of copper and silver nanoparticle exposure in Drosophila and their microbiomes. *Sci. Total Environ.* 487: 822–829.

65 Merrifield, D.L., Shaw, B.J., Harper, G.M. et al. (2013). Ingestion of metal-nanoparticle contaminated food disrupts endogenous microbiota in zebrafish (*Danio rerio*). *Environ. Pollut.* 174: 157–163.

66 Hadrup, N., Loeschner, K., Bergstrom, A. et al. (2012). Subacute oral toxicity investigation of nanoparticulate and ionic silver in rats. *Arch. Toxicol.* 86 (4): 543–551.

67 Williams, K., Milner, J., Boudreau, M.D. et al. (2015). Effects of subchronic exposure of silver nanoparticles on intestinal microbiota and gut-associated immune responses in the ileum of Sprague-Dawley rats. *Nanotoxicology* 9 (3): 279–289.

68 Wilding, L.A., Bassis, C.M., Walacavage, K. et al. (2015). Repeated dose (28-day) administration of silver nanoparticles of varied size and coating does not significantly alter the indigenous murine gut microbiome. *Nanotoxicology* 1–8.
69 Xie, Y., Wu, B., Zhang, X.X. et al. (2016). Influences of graphene on microbial community and antibiotic resistance genes in mouse gut as determined by high-throughput sequencing. *Chem. Tech.* 144: 1306–1312.
70 Han, X.Y., Du, W.L., Fan, C.L., and Xu, Z.R. (2010). Changes in composition a metabolism of caecal microbiota in rats fed diets supplemented with copper-loaded chitosan nanoparticles. *J. Anim. Physiol. Anim. Nutr.* 94 (5): e138–e144.
71 Vong, L.B., Yoshitomi, T., Morikawa, K. et al. (2014). Oral nanotherapeutics: effect of redox nanoparticle on microflora in mice with dextran sodium sulfate-induced colitis. *J. Gastroenterol.* 49 (5): 806–813.
72 Yasuyuki, M., Kunihiro, K., Kurissery, S. et al. (2010). Antibacterial properties of nine pure metals: a laboratory study using *Staphylococcus aureus* and *Escherichia coli*. *Biofouling* 26 (7): 851–858.
73 Kim, J.S., Kuk, E., Yu, K.N. et al. (2007). Antimicrobial effects of silver nanoparticles. *Nanomedicine* 3 (1): 95–101.
74 Martinez-Gutierrez, F., Thi, E.P., Silverman, J.M. et al. (2012). Antibacterial activity, inflammatory response, coagulation and cytotoxicity effects of silver nanoparticles. *Nanomedicine* 8 (3): 328–336.
75 Sondi, I. and Salopek-Sondi, B. (2004). Silver nanoparticles as antimicrobial agent: a case study on *E. coli* as a model for Gram-negative bacteria. *J. Colloid Interface Sci.* 275 (1): 177–182.
76 Ruparelia, J.P., Chatterjee, A.K., Duttagupta, S.P., and Mukherji, S. (2008). Strain specificity in antimicrobial activity of silver and copper nanoparticles. *Acta Biomater.* 4 (3): 707–716.
77 Silvestry-Rodriguez, N., Bright, K.R., Slack, D.C. et al. (2008). Silver as a residual disinfectant to prevent biofilm formation in water distribution systems. *Appl. Environ. Microbiol.* 74 (5): 1639–1641.
78 Chatterjee, S., Bandyopadhyay, A., and Sarkar, K. (2011). Effect of iron oxide and gold nanoparticles on bacterial growth leading towards biological application. *J. Nanobiotechnol.* 9: 34.
79 Lyon, D.Y. and Alvarez, P.J. (2008). Fullerene water suspension (nC60) exerts antibacterial effects via ROS-independent protein oxidation. *Environ. Sci. Technol.* 42 (21): 8127–8132.
80 Spacciapoli, P., Buxton, D., Rothstein, D., and Friden, P. (2001). Antimicrobial activity of silver nitrate against periodontal pathogens. *J. Periodont. Res.* 36 (2): 108–113.

81 Fondevila, M., Herrer, R., Casallas, M. et al. (2009). Silver nanoparticles as a potential antimicrobial additive for weaned pigs. *Anim. Feed Sci. Technol.* 150 (3): 259–269.

82 Siddique, S., Shah, Z.H., Shahid, S., and Yasmin, F. (2013). Preparation, characterization and antibacterial activity of ZnO nanoparticles on broad spectrum of microorganisms. *Acta Chim. Slov.* 60 (3): 660–665.

83 Espitia, P.J.P., Soares, N.D.F., Coimbra, J.S.D. et al. (2012). Zinc oxide nanoparticles: synthesis, antimicrobial activity and food packaging applications. *Food Bioprocess Technol.* 5 (5): 1447–1464.

84 Azam, A., Ahmed, A.S., Oves, M. et al. (2012). Size-dependent antimicrobial properties of CuO nanoparticles against Gram-positive and -negative bacterial strains. *Int. J. Nanomed.* 7: 3527–3535.

85 Behera, S.S., Patra, J.K., Pramanik, K. et al. (2012). Characterization and evaluation of antibacterial activities of chemically synthesized iron oxide nanoparticles. *World J. Nano Sci. Eng.* 2 (04): 196.

86 Tran, N., Mir, A., Mallik, D. et al. (2010). Bactericidal effect of iron oxide nanoparticles on *Staphylococcus aureus*. *Int. J. Nanomed.* 5: 277–283.

87 Azam, A., Ahmed, A.S., Oves, M. et al. (2012). Antimicrobial activity of metal oxide nanoparticles against Gram-positive and Gram-negative bacteria: a comparative study. *Int. J. Nanomed.* 7: 6003–6009.

88 Kuang, Y.S., He, X., Zhang, Z.Y. et al. (2011). Comparison study on the antibacterial activity of nano- or bulk-cerium oxide. *J. Nanosci. Nanotechnol.* 11 (5): 4103–4108.

89 Li, M., Zhu, L., and Lin, D. (2011). Toxicity of ZnO nanoparticles to *Escherichia coli*: mechanism and the influence of medium components. *Environ. Sci. Technol.* 45 (5): 1977–1983.

90 Akhavan, O. and Ghaderi, E. (2010). Toxicity of graphene and graphene oxide nanowalls against bacteria. *ACS Nano* 4 (10): 5731–5736.

91 Olivi, M., Zanni, E., De Bellis, G. et al. (2013). Inhibition of microbial growth by carbon nanotube networks. *Nanoscale* 5 (19): 9023–9029.

92 Chen, H., Wang, B., Gao, D. et al. (2013). Broad-spectrum antibacterial activity of carbon nanotubes to human gut bacteria. *Small* 9 (16): 2735–2746.

93 Zhu, B., Xia, X., Xia, N. et al. (2014). Modification of fatty acids in membranes of bacteria: implication for an adaptive mechanism to the toxicity of carbon nanotubes. *Environ. Sci. Technol.* 48 (7): 4086–4095.

94 Kang, S., Herzberg, M., Rodrigues, D.F., and Elimelech, M. (2008). Antibacterial effects of carbon nanotubes: size does matter! *Langmuir* 24 (13): 6409–6413.

95 Kang, S., Pinault, M., Pfefferle, L.D., and Elimelech, M. (2007). Single-walled carbon nanotubes exhibit strong antimicrobial activity. *Langmuir* 23 (17): 8670–8673.

96 Vecitis, C.D., Zodrow, K.R., Kang, S., and Elimelech, M. (2010). Electronic-structure-dependent bacterial cytotoxicity of single-walled carbon nanotubes. *ACS Nano* 4 (9): 5471–5479.
97 Liu, S., Zeng, T.H., Hofmann, M. et al. (2011). Antibacterial activity of graphite, graphite oxide, graphene oxide, and reduced graphene oxide: membrane and oxidative stress. *ACS Nano* 5 (9): 6971–6980.
98 Hu, W., Peng, C., Luo, W. et al. (2010). Graphene-based antibacterial paper. *ACS Nano* 4 (7): 4317–4323.
99 Ruiz, O.N., Fernando, K.A., Wang, B. et al. (2011). Graphene oxide: a nonspecific enhancer of cellular growth. *ACS Nano* 5 (10): 8100–8107.
100 Lyon, D.Y., Fortner, J.D., Sayes, C.M. et al. (2005). Bacterial cell association and antimicrobial activity of a C60 water suspension. *Environ. Toxicol. Chem.* 24 (11): 2757–2762.
101 Fortner, J., Lyon, D., Sayes, C. et al. (2005). C60 in water: nanocrystal formation and microbial response. *Environ. Sci. Technol.* 39 (11): 4307–4316.
102 Hoffmann, C., Dollive, S., Grunberg, S. et al. (2013). Archaea and fungi of the human gut microbiome: correlations with diet and bacterial residents. *PLoS One* 8 (6): e66019.
103 Duerkop, B.A. and Hooper, L.V. (2013). Resident viruses and their interactions with the immune system. *Nat. Immunol.* 14 (7): 654–659.
104 Los, M., Panigrahi, S., Rashedi, I. et al. (2009). Apoptin, a tumor-selective killer. *Biochim. Biophys. Acta* 1793 (8): 1335–1342.
105 Lecuit, M. and Eloit, M. (2013). The human virome: new tools and concepts. *Trends Microbiol.* 21 (10): 510–515.
106 Elechiguerra, J.L., Burt, J.L., Morones, J.R. et al. (2005). Interaction of silver nanoparticles with HIV-1. *J. Nanobiotechnol.* 3: 6.
107 Lara, H.H., Ayala-Nunez, N.V., Ixtepan-Turrent, L., and Rodriguez-Padilla, C. (2010). Mode of antiviral action of silver nanoparticles against HIV-1. *J. Nanobiotechnol.* 8: 1.
108 Lara, H.H., Ixtepan-Turrent, L., Garza-Trevino, E.N., and Rodriguez-Padilla, C. (2010). PVP-coated silver nanoparticles block the transmission of cell-free and cell-associated HIV-1 in human cervical culture. *J. Nanobiotechnol.* 8: 15.
109 Lu, L., Sun, R.W., Chen, R. et al. (2008). Silver nanoparticles inhibit hepatitis B virus replication. *Antivir. Ther.* 13 (2): 253–262.
110 Rogers, J.V., Parkinson, C.V., Choi, Y.W. et al. (2008). A preliminary assessment of silver nanoparticle inhibition of monkeypox virus plaque formation. *Nanoscale Res. Lett.* 3 (4): 129–133.
111 Sun, L., Singh, A.K., Vig, K. et al. (2008). Silver nanoparticles inhibit replication of respiratory syncytial virus. *J. Biomed. Nanotechnol.* 4 (2): 149–158.

112 Xiang, D.X., Chen, Q., Pang, L., and Zheng, C.L. (2011). Inhibitory effects of silver nanoparticles on H1N1 influenza A virus in vitro. *J. Virol. Methods* 178 (1-2): 137–142.

113 Bekele, A.Z., Gokulan, K., Williams, K.M., and Khare, S. (2016). Dose and size-dependent antiviral effects of silver nanoparticles on feline calicivirus, a human norovirus surrogate. *Foodborne Pathog. Dis.* 13 (5): 239–244.

114 You, J., Zhang, Y., and Hu, Z. (2011). Bacteria and bacteriophage inactivation by silver and zinc oxide nanoparticles. *Colloids Surf., B* 85 (2): 161–167.

115 Liga, M.V., Bryant, E.L., Colvin, V.L., and Li, Q. (2011). Virus inactivation by silver doped titanium dioxide nanoparticles for drinking water treatment. *Water Res.* 45 (2): 535–544.

116 Akhavan, O., Abdolahad, M., Abdi, Y., and Mohajerzadeh, S. (2011). Silver nanoparticles within vertically aligned multi-wall carbon nanotubes with open tips for antibacterial purposes. *J. Mater. Chem.* 21 (2): 387–393.

117 Ma, J., Zhang, J., Xiong, Z. et al. (2011). Preparation, characterization and antibacterial properties of silver-modified graphene oxide. *J. Mater. Chem.* 21 (10): 3350–3352.

118 Tang, J., Chen, Q., Xu, L. et al. (2013). Graphene oxide–silver nanocomposite as a highly effective antibacterial agent with species-specific mechanisms. *ACS Appl. Mater. Interfaces* 5 (9): 3867–3874.

119 Walker, A.W., Duncan, S.H., Louis, P., and Flint, H.J. (2014). Phylogeny, culturing, and metagenomics of the human gut microbiota. *Trends Microbiol.* 22 (5): 267–274.

120 Fraher, M.H., O'Toole, P.W., and Quigley, E.M. (2012). Techniques used to characterize the gut microbiota: a guide for the clinician. *Nat. Rev. Gastroenterol. Hepatol.* 9 (6): 312–322.

121 Flint, H.J., Scott, K.P., Louis, P., and Duncan, S.H. (2012). The role of the gut microbiota in nutrition and health. *Nat. Rev. Gastroenterol. Hepatol.* 9 (10): 577–589.

122 Groschwitz, K.R. and Hogan, S.P. (2009). Intestinal barrier function: molecular regulation and disease pathogenesis. *J. Allergy Clin. Immunol.* 124 (1): 3–20; quiz 1–2.

123 Hyun, J.S., Lee, B.S., Ryu, H.Y. et al. (2008). Effects of repeated silver nanoparticles exposure on the histological structure and mucins of nasal respiratory mucosa in rats. *Toxicol. Lett.* 182 (1-3): 24–28.

124 Jeong, G.N., Jo, U.B., Ryu, H.Y. et al. (2010). Histochemical study of intestinal mucins after administration of silver nanoparticles in Sprague–Dawley rats. *Arch. Toxicol.* 84 (1): 63–69.

125 Imlay, J.A. (2002). How oxygen damages microbes: oxygen tolerance and obligate anaerobiosis. *Adv. Microb. Physiol.* 46: 111–153.

126 Ji, G. and Silver, S. (1995). Bacterial resistance mechanisms for heavy metals of environmental concern. *J. Ind. Microbiol.* 14 (2): 61–75.

127 Nies, D.H. (1999). Microbial heavy-metal resistance. *Appl. Microbiol. Biotechnol.* 51 (6): 730–750.

128 Silver, S. (2003). Bacterial silver resistance: molecular biology and uses and misuses of silver compounds. *FEMS Microbiol. Rev.* 27 (2-3): 341–353.

129 Randall, C.P., Gupta, A., Jackson, N. et al. (2015). Silver resistance in Gram-negative bacteria: a dissection of endogenous and exogenous mechanisms. *J. Antimicrob. Chemother.* 70 (4): 1037–1046.

130 Hartemann, P., Hoet, P., Proykova, A. et al. (2015). Nanosilver: safety, health and environmental effects and role in antimicrobial resistance. *Mater. Today* 18 (3): 122–123.

131 Iravani, S., Korbekandi, H., Mirmohammadi, S.V., and Zolfaghari, B. (2014). Synthesis of silver nanoparticles: chemical, physical and biological methods. *Res. Pharm. Sci.* 9 (6): 385–406.

14

Human Health Impacts and Immunotoxicology of Metal Nanoparticles and Nanomaterials – An Overview

Gregory P. Nichols[1] and Jason Davis[2]

[1]*GP Nichols & Company, 7527 Jenkins Rd., Knoxville, TN 37931 USA*
[2]*Oak Ridge Associated Universities, 100 ORAU Way, Oak Ridge, TN 37831 USA*

14.1 Introduction and Background

The immune system can be a sensitive target organ for toxicity due to exposure to xenobiotics. The toxic effects of certain metals on human health, particularly regarding the immune system, have been well established for decades. However, the immunotoxic effects of engineered or anthropogenic nanoparticles (NPs) containing metals are not as certain. Most toxicological studies on nanomaterials have focused on the effects of nanoparticles entering the body through accidental intake [1]. It is understood that the physicochemical properties (size, shape, charge, solubility) and surface modification of NPs influence their biocompatibility [2]. As the use of nanoparticles in biomedical applications targeting the immune system increases, such as cancer therapy, drug delivery, imaging, or diagnostics, understanding the body's immune response to nanoscale products will continue to grow in importance [1, 3].

Physicochemical properties of metal compounds are important determinants of the material's toxicity. Small quantities of many metals, including copper, magnesium, sodium, potassium, calcium, and iron (Fe), are essential for proper functioning of biological systems. At higher doses, however, metals can have toxic effects and exposure to high levels of environmental metals causes disease in humans [4]. Nanoparticles manufactured from these metals will have health effects not necessarily easily predicted from previous studies of non-nanoparticulate quantities of the same metals. As it could easily expose workers to these toxic materials, the manufacturing of metal nanoparticles (MeNPs) should be considered a serious occupational hazard.

Urban populations are often exposed to metals as constituents of ambient particulate matter (PM). The inhalation of metallic or other dusts is known to

Toxicology of Nanoparticles and Nanomaterials in Human, Terrestrial and Aquatic Systems, First Edition. Edited by Marc A. Williams.

have negative health effects. The type of lung disease caused by dust inhalation depends on the nature of the material, exposure duration, and dose. The inhalation of some metal fumes (e.g. zinc, copper) may lead to metal fume fever, a reaction with influenza-like symptoms [5]. Several metal dusts (e.g. platinum, nickel, chromium, cobalt) can lead to asthma [5], while inhalation of other metallic dusts can cause pulmonary fibrosis, and possibly lung cancer.

A large experimental database exists on the immunosuppressive properties of metals, their inorganic salts, and organometallic compounds [6]. Inorganic substances, including heavy metals, such as lead, cadmium (Cd), mercury, and nickel (Ni), have been studied, and the effects of exposure have been reported to range from suppression to enhancement of the immune system. Some metals are also reported to induce autoimmune disease. Although toxicological effects of bulk metals on human cells are well known, the effects of nanoscale metals on human cells are not as well known due to a difference in properties, including surface area and ion interaction, according to Nel and Schrand as summarized by Petrarca [7]. However, particles similar in size and composition may provide an approximation of immunological effects that could be observed following metal NP exposure.

14.2 Welding Fumes as Surrogates for Metal Nanoparticles

Welding fumes are a particularly good surrogate for predicting the immunological effects of MeNPs, given their size and composition. Welders can be exposed to concentrations of NPs up to 300 000 particles/cm^3, with approximately 60% of particles smaller than 100 nm in diameter [8, 9]. The specific welding technique or method used plays a significant role in the size of the particles generated. For example, Brand and colleagues [10] found that at least 90% of particles generated during TIG and RSW welding were smaller than 50 nm, while only 10–16% of particles generated during MMAW, MAG, and MIG welding were smaller than 50 nm.

Welding fumes comprise a variety of metals and metal oxides, including aluminum (Al), cadmium (Cd), chromium (Cr), iron (Fe), manganese (Mn), nickel (Ni), and zinc (Zn). [9, 11]. It has been reported that 19% of welders are exposed to levels above the threshold limit value (TLV) for Fe, and 62% above the TLV for Mn (as established by the American Conference of Governmental Industrial Hygienists) [12]. In a study conducted by the National Institute for Occupational Safety and Health (NIOSH), samples of fumes demonstrated that most of the Cr particles and half of the Ni and Mn particles were actually in the nanoscale range [9]. Cena, Chisholm, Keane, and Chen [9] estimated that the concentrations of nanoparticles available for deposition in the respiratory tract were in the following ranges:

- Cr: 40–105 μg/m^3
- Mn: 2.8–199 μg/m^3
- Ni: 10–51 μg/m^3

Studies have shown that particles in lungs attributed to welding activity can be as small as 10 nm [13]. Antonini and coworkers [13] analyzed post mortem (cause of death likely related to pulmonary complications) tissue samples of a welder's lungs and found that almost all particles within cells of the tissues comprised metals typically found in metal fumes. The particles were arranged in chain-like formations, and the composition was primarily nanoscale Cr, Fe, and Ni.

14.3 Immune-Related Health Effects

Prolonged exposure to welding fumes has been linked to asthma, atherosclerosis, cardiovascular disease (CVD), chronic obstructive pulmonary disease (COPD), fibroids, lung cancer, lung inflammation, neurological disorders, and oxidative damage [14, 15]. Since the primary method of exposure to welding fumes is inhalation, most health effects tend to be respiratory in nature. It has been documented that localized lung inflammation can be induced by the presence of metal particulates [16]. In addition, data strongly suggest that NPs may play a role in the development of pulmonary inflammation [8], and inflammation can lead to a variety of other disorders including fibrotic illnesses and cancer. Exposures of Cr, Fe, Mn, Ni, and silicon (Si) induce fibrogenic response in occupational settings [17]. In fact, pulmonary fibrosis is associated with siderosis, the accumulation of Fe particles in tissue [13]. In addition, chronic exposure to Al welding fumes in an occupational setting can lead to severe pneumoconiosis, pulmonary fibrosis, and a reduction in lung function according to Antonini [13] and Hull, as summarized by Taylor [18]; however, these effects are likely due to a complex mix of particulates containing Al and not just from Al alone [18]. In general, there appears to be an association between metal-containing particulates and chronic fibrotic disease [16].

Welders are also more susceptible than the general population to upper and lower respiratory tract infections [13], particularly pneumonia [12, 19]. Inhaling metal-containing particulates is associated with an increase in lower respiratory tract infections [16], but the inhalation of metal fumes in particular [12], especially Fe, Cr, and Mn [20], renders workers susceptible to pneumonia. There are two possible explanations for this. First, welding fumes lead to cytotoxic activity of lymphokine-activated killer cells, which causes a decrease in natural killer cell activity and might subject welders to increased risk of infection [12]. Second, susceptibility to pneumonia may be caused by a redox-dependent dysregulation of macrophage innate immune functions triggered by metal fumes [20]. There is evidence to suggest that eosinophils play a role in pulmonary disease [21], and curiously, Janwadkar and Ingole [22] report a case of persistent eosinophilia in a welder who was also diagnosed with an upper respiratory tract infection.

Welders can also exhibit metal fume fever (MFF), which is an inhalation fever syndrome caused by the inhalation of newly formed metal oxides [23]. Zinc oxide (ZnO) is the most common cause of MFF, but other fumes have also been implicated, including Al, Cd, Fe, and titanium dioxide (TiO_2) [11, 21]. Freshly oxidized ZnO is of particular interest, not only because ZnO exposure appears to be the primary cause of MFF, but ZnO welding fumes exist as nanoparticles [11], so there is an interesting crossover between metal oxide exposure and nanoparticulate exposure. Neutrophil activation may play a part in causing MFF [11].

The connection here among all of these descriptions is that welders exhibit unique health issues compared with other occupations; this is most likely due to a combination of exposure to particularly toxic metals, and to extremely small particles that interfere with components of the immune system.

14.4 Oxidative Stress and Immunologic Effects

The connection between oxidative stress and inflammation has been well documented [24]. A number of immunological responses following exposure to fine metal fumes have been described in welders. In a study of 44 welders, Cena and colleagues [9] found that fine and ultrafine particles from welding generated the highest number of reactive oxygen species (ROS). Erdely and colleagues [19] found that oxidative stress was increased in the mononuclear cells of welders, and Manke, Wang, and Rojanasakul [17] reported that metal-containing welding fumes elicit ROS-dependent lipid peroxidation and inflammation [17]. Welders have also demonstrated increases in IL-8, tumor necrosis factor-α (TNF-α), myeloperoxidase, matrix metalloproteinase-9, IgA, and α2-macroglobulin [22].

Metal-rich particulate matter has inhibitory effects on various components of the immune system, including circulating leukocyte population [19]. Additionally, Mn and Cr fumes suppress IgM response, an effect observed by Anderson and reported by Keila [25]. Ferrous metal fumes may cause oxidative damage and provide an environment for microorganisms to prosper [12]. Andujar and colleagues [8] evaluated the effects of Fe_2O_3, Fe_3O_4, $MnFe_2O_4$, and CrOOH welding particles on human macrophages [8]. They found that CXCL-8, IL-1β, TNF-α, CCL-2, CCL-3, CCL-4, and IL-6, CCL-7, and CCL-22 were all increased following exposure to welding nanoparticles, and the induction of macrophage migration was present in all exposures except Fe_3O_4 [8]. Exposure to metal welding fumes can cause a variety of inflammatory, immunological, and oxidative stress reactions beyond those surveyed here. For a comprehensive review of these effects, see "Immunotoxicology of arc welding fume: Worker and experimental animal studies," by Patti C. Zeidler-Erdely, Aaron Erdely, and James M. Antonini [12].

14.5 Conclusion

The fact that welding fumes are considered ultrafine particles; that at least half of inhaled fume particles are less than 100 nm in size; that fumes contain metals/metal oxides of concern; and that welders exhibit a variety of health problems unique to their occupation, makes the study of welding fumes a natural foundation for understanding immunological effects of MeNPs in humans.

14.6 Immune-Triggered Human Health Effects of Metal and Metal Oxide Nanoparticles

The majority of mechanisms contributing to the immunotoxic and immunomodulatory effects of nanometals and nanometal oxides are not completely understood. However, several mechanisms have been postulated and investigated. Bulk metals (microscale and not nanoscale particles) and metal oxides can influence immune function through a number of mechanisms, including the generation of reactive oxygen species leading to oxidative stress, induction of apoptosis, and interference with signaling pathways (Figure 14.1). Physicochemical properties of materials, such as their size, shape, and chemistry, influence the interaction of the materials and cells. For nanoparticles, other factors include surface-to-volume ratio, altered dissolution rate, shape and structure, aggregation potential, and quantum effects [27]. These factors can either facilitate or inhibit the binding of blood, cellular, and protein components which facilitate interactions with the immune cells and result in an immunological response [28].

14.7 Immune Interaction

The immune system consists of two components: innate immunity and adaptive immunity. Innate immunity is the early response phase of response to a potential pathogen, in which pathogen-associated molecular patterns activate pathogen recognition and immune response [29]. The second defense of the body is adaptive immunity, which is a more specialized and long-term system.

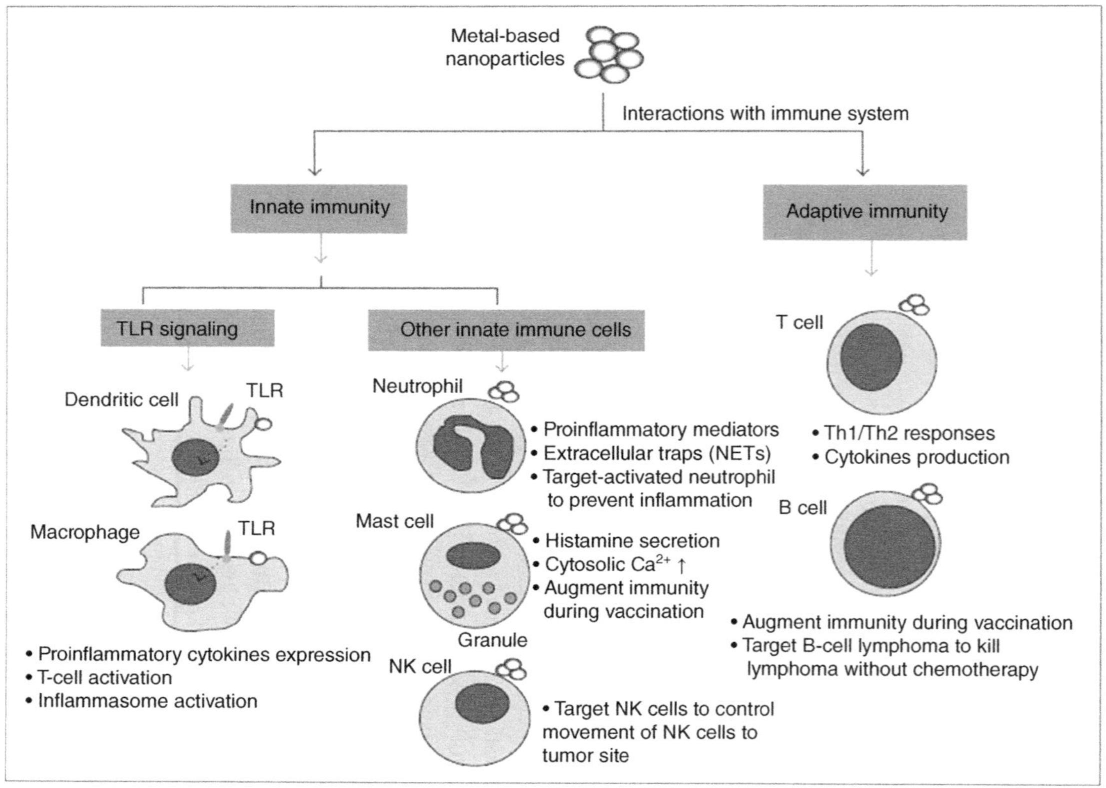

Figure 14.1 Metal NPs interaction with the immune system [26], figure 1 (p.1) / Hindawi / Licensed under CC BY 4.0.

Macrophages are typically the first immune cells to come into contact with foreign particles and identify them as potentially pathogenic, via specialized proteins known as Toll-like receptors (TLR) [30]. Once identified, the particles may be cleared by phagocytosis of the particles by macrophages, which subsequently digest the particles by generating free radicals and proteases. However, the process of phagocytosis depends on the physicochemical characteristics of the pathogen. In general, when particles of differing size share the same surface properties and composition, particles of larger size are taken up more readily by macrophages. Conversely, particles of smaller size are more likely to interact through passive mechanisms than their larger counterparts, enhancing the inflammatory response [28]. For particles of similar size, those with a surface charge are more rapidly phagocytized [31], with those that are positively charged readily interacting with the negatively charged sialic acid on the surface of macrophages [32].

14.8 Cellular Mechanisms of Injury

Internalization of metallic NPs is not limited to phagocytosis. Nanoparticles of small enough size can diffuse across cell membranes provided the particle surfaces exhibit positive ions. For other particles, membrane transit can be facilitated by ion channels and transporter proteins. Once inside the cell, the NPs can either directly or indirectly disrupt cellular function or cause damage to cellular structures (Figure 14.2).

14.9 Oxidative Stress

Mammalian cells constantly regulate the balance of antioxidants and reactive oxygen species (ROS) to achieve a steady reaction potential necessary for proper function [34]. The reduction–oxidation (redox) state of the cell is primarily a consequence of the precise balance between the levels of oxidizing ROS and endogenous thiol buffers present in the cell, such as glutathione and thioredoxin, which protect cells from oxidative damage. When proper ROS levels are maintained they are effectively used in intracellular signaling pathways for normal cell cycles including growth, differentiation, and apoptosis [35].

As the levels of ROS rise, cells may mount protective or injurious responses, such as the activation of enzymatic and non-enzymatic antioxidant defense mechanisms [36]. Overproduction of ROS activates cytokines and upregulates IL, kinases, and TNF-α as a counter reaction to oxidative stress [37]. When ROS levels exceed the levels of buffering thiols, the effects include modification of proteins

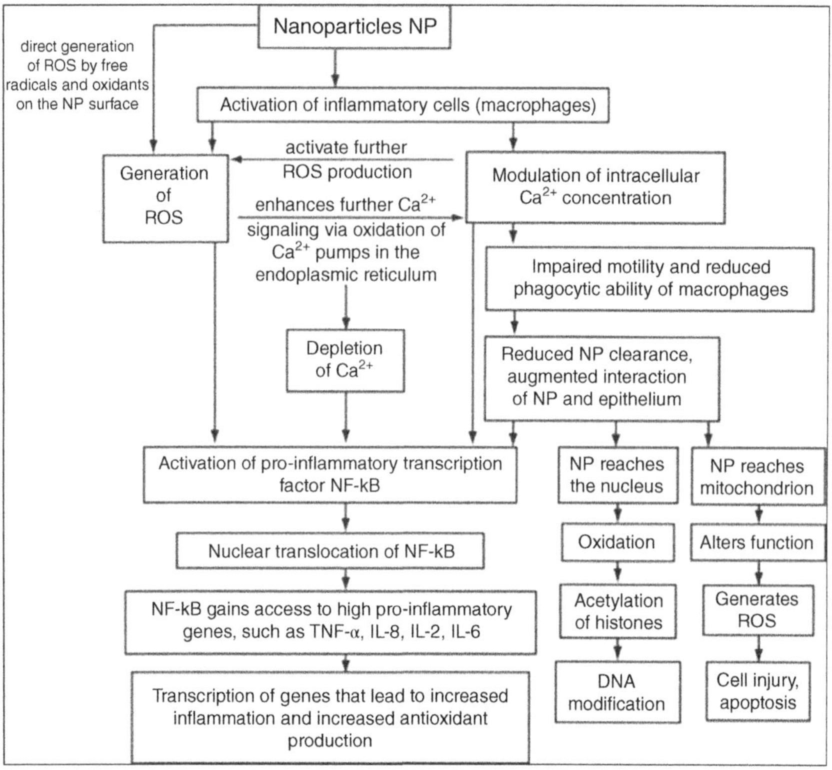

Figure 14.2 Events by which NPs cause cellular effects [33], figure 29 (p.MR49) / with permission of American Vacuum Society.

resulting in the generation of protein radicals, initiation of lipid peroxidation, DNA-strand breaks, modification of nucleic acids, modulation of gene expression through activation of redox-sensitive transcription factors, and modulation of inflammatory responses through signal transduction, leading to cell death and genotoxic effects [35].

The generation of ROS, directly or indirectly, and the subsequent production of oxidative stress is a predominant mechanism leading to nanotoxicity, including DNA damage, unregulated cell signaling, changes in cell motility, cytotoxicity, apoptosis, and cancer initiation and promotion [36]. Although the exact mechanism governing the initiation of oxidative stress by metallic and metal oxide nanoparticles is unknown, production of ROS by nanomaterials may proceed through a variety of mechanisms that depend on the physical and chemical properties of the nanomaterials [35].

Structural defects on a nanoparticle surface and formation of electronic states not present in bulk materials, resulting from the increased surface-to-volume ratio, creates reactive sites where interaction with molecular oxygen produces a superoxide anion radical (O_2^-). This superoxide can then produce other ROS such as hydroxyl ion ($HO^{\cdot}$) through the reduction of peroxide (H_2O_2). In effect, the metal or metal oxide nanoparticle acts as a catalyst for the formation of hydroxyl radicals, allowing for the radicals to be produced faster than the natural scavengers to maintain physiological equilibrium [17].

14.10 Interaction with Cellular Membranes and Proteins

Production of ROS can also be catalyzed by the release of metal cations from the surface of metal oxide nanoparticles [38]. Studies have described the ability of soluble metals and metal oxides to dissolve with or without oxidation and reduction. The mechanism for this dissolution is not fully understood, but particles sequestered in macrophage lysosomes dissolve rapidly in the acidic lysosome, allowing for the accumulation of metal ions that can destabilize the membranes of the lysosomes. Because the lower surface-to-volume ratio of a nanoparticle increases its dissolution rate [39], the ion accumulation and membrane destabilization upon exposure to metal oxide nanoparticles is enhanced when compared to bulk metal oxides [40].

Cellular uptake of nanoparticles by non-endocytotic methods may result in the disruption of the plasma membrane. Electrostatic interactions of positively charged nanoparticles with negatively charged groups on the membrane surface alter cell membranes. Membrane penetration of positively charged nanoparticles into the cytoplasm rapidly causes the formation of holes, followed by a slow closing of these perforations. Interactions between positively charged nanoparticles with negatively charged membrane also causes membrane depolarization and perturbation of membrane potential [41]. Perturbations such as these can lead to apoptosis [42].

Metallic nanoparticles interact with DNA molecules through electrostatic and hydrophobic interactions. Depending on the physicochemical properties of the NPs, damage to the DNA may be caused by the production of ROS in close proximity to the protein strands, as well as by disruption of the hydrogen bonds between the two DNA strands [41]. Without entering the cell itself, particles can interact with membrane proteins and activate signaling pathways, leading to inhibition of cell proliferation [42].

14.11 Disruption of Signaling Pathways

In addition to producing excess ROS in a cell, MeNPs can also influence immune function by direct modification of the cellular thiols. Depending on the ionic state of the metallic NP, the thiols may be either oxidized or alkylated, either of which removes the functionality of the thiols. The end result is a restructuring of the lipids in plasma membranes and proteins in both lymphocytes and antigen-presenting cells. Modification of these structures then alters communication between APCs and T cells, which in turn alters the activation of both cell types [43]. Wang, Li, and Khan [44] reported that the generation of ROS by engineered nanomaterials initiates the innate immune response through activation of the inflammasome. In particular, TiO_2, silicon dioxide (SiO_2), and silver (Ag) seem particularly active in terms of creating ROS triggering the inflammasome. In addition, certain MeNPs cause the secretion of IL-1β and caspase-1 as well as initiate complement activation.

14.12 Immune Response

When the TLRs are activated by the presence of a pathogen, the receptors recruit adapter proteins within the cytosol of the immune cell in order to present a pathogen-specific antigen on the cell surface. These antigens can then be detected by specific helper T (T_H)-cell receptors. T helper cells are subpopulations of T cells (T lymphocytes) displaying the membrane glycoprotein cluster of differentiation 4 (CD4). They are activated by the antigen presentation process when the T-cell receptor (TCR) and CD3 complex bind with the major histocompatibility complex (MHC) II molecule from the antigen presenting cell's surface. Along with co-stimulation of different cytokines, this process leads T cells to activation, proliferation and differentiation into T_H1, T_H2, T_H17, T_{reg} (regulatory T), and T_{fh} (follicular helper T) cells. This process results in the activation of humoral and cellular immunity to remove or destroy the antigen. Inappropriate activation of T helper cells can lead to autoimmune consequences that can be fatal [45, 46].

14.13 Immunosuppression

Nanoparticles can also weaken the body's immune response against infections and cancerous cells. However, these immunosuppressive properties can make nanoparticles useful in preventing transplant rejection, and in the treatment of inflammatory and autoimmune diseases. Although several mechanisms have been described for the pro-inflammatory immunostimulatory effects of nanoparticles, the immunosuppressive properties are not as well characterized [47]. Immunosuppression is not an acute toxicity that can be easily monitored in vitro, which may account for the sparsity of information on metallic NP mediated immunosuppression.

Zinc oxide nanoparticles contribute to cytotoxicity via oxidative stress from the increased production of reactive oxygen species. Pro-inflammatory cytokines, such as IL-1β, TNF-α, IL-10, and interferon-γ, are suppressed in the presence of ZnO nanoparticles. Although generally considered non-cytotoxic, TiO_2 NPs have been shown to inhibit T-cells, B-cells, macrophages, and natural killer cells in a tumor model. Iron oxide nanoparticles have been shown to block the activity of T helper cells and macrophages and reduced the expression of interferon-γ, IL-6, and TNF-α, resulting in a reduced recognition of antigens, allergens, pathogens, or foreign bodies in the blood with B cells [48]. It has also been documented that certain MeNPs are associated with suppression of immune function demonstrated by a decrease in T-cell-dependent antibody response [44].

While the generation of ROS is seen as the primary mechanism of both immune stimulation and cytotoxicity, some metallic nanoparticles act as free radical scavengers. It has been shown experimentally that cerium oxide nanoparticles catalyze the decomposition of superoxide radicals into hydrogen peroxide [49]. Depending on their physicochemical characteristics, MeNPs such as Ag have been shown to induce production of the anti-inflammatory cytokine TGF-β, which suppresses the proliferation of lymphocytes in the presence of certain cytokines (e.g. IL-6 and IL-1) [28]. However, TGF-β also induces development of T helper 17 (Th17) cells, which induce inflammation.

Some nanoparticles can suppress one immune function while stimulating another [50], or can have immunosuppressive or immunostimulatory effects that vary based on particle size and surface coatings [32]. Citrate-coated gold (Au) NPs show anti-inflammatory activity and inhibited cellular responses induced by the inflammatory cytokine interleukin-1β (IL-1β). Conversely, 200 nm Au NPs coated with poly(acrylic acid) promote inflammation in the same THP-1 cell line [28].

14.14 Inflammation and Autoimmunity

At toxic doses, NPs of cobalt (Co), ZnO, CeO_2, and TiO_2 can cause apoptosis in monocytes and macrophages, according to research from Wilhelmi, Buerki-Thurnherr, Hussain, Vandebriel, and Shavandi as reported by Petrarca [7]; however, while cell death does not occur at non-toxic doses, these metal NPs do produce inflammatory effects. Generally, the smaller the particle size, the more measurable the effect. For example, Petrarca [7] reported findings by Lim in which IL-8 was elevated and hemeoxygenase-1 and heat shock protein-70 were upregulated following the introduction of Ag NPs in macrophages. They found that these effects were more apparent when the particles were 5 nm compared with 100 nm, and this same effect was found with ZnO NPs as well. In fact, according to Chang as summarized by Luo [26], ZnO nanoparticles and quantum dots promoted inflammatory cytokines and activated a MyD88-dependent TLR signal pathway. This mechanism from Ho as reported by Luo [26] is necessary for inflammatory responses induced by NPs. For example, Petrarca (2015) reports work by Potnis showing Toll-like receptor 4 (TRL-4) is essential for activating innate immunity and cytokine production following exposure to Co NPs. Finally, Liu and colleagues, as reported by Luo [26], noted that certain types of fullerene nanocages ($Gd@C_{82}(OH)_{22}$) containing Cd increased Th1 cytokines, including IL-2, IFN-γ, and TNF-α.

Autoimmunity is an emerging area of study regarding potential immune-focused health effects of metal NPs. Evidence from *in vitro* and *in vivo* data suggest that exposure to metal NPs negatively affects the immune system and could be a leading factor associated with a population-level rise in allergies and autoimmune disorders [7]. The literature supports the notion that certain metal NPs, including Fe_2O_3, NiO, Co_3O_4, and Pd, may increase autoimmune occurrence by stimulating peripheral monocytes to produce Th1 cytokines, as described by Petrarca, Zhu, Cho, and Reale in Petrarca [7]. In addition, human peripheral blood lympho/monocytes exposed to Co NPs produce a similar pattern seen in cytokine release detected in autoimmunity and contact dermatitis, as described by Petrarca in her previous work from 2006 [7].

14.15 Sensitivity/Hypersensitivity

Metal nanoparticle species are of particular concern and interest, as some cause a very common reaction: allergic contact dermatitis (Figure 14.3), which "is one of the best-established model diseases for T cell-mediated immune responses" [52]. According to McCance [53], metal particle exposure leads to "cell-mediated tissue destruction" by T-lymphocytes, where antibodies are not involved. In addition, an antigen may not be immunogenic when less than 1000 DA, but once it becomes bound to a carrier protein, it could become immunogenic [53]. Because of this, NPs are of particular concern since they "can bind more antigen per mass unit than larger particles" [51]. A recent study highlights a case in which a chemist experienced throat irritation, flushing of the face, and skin sensitivity to other metal objects within one week after handling nanoscale Ni [54]. Unfortunately in this case, no blood or urine samples were used to make a diagnosis; the sensitivity to nickel was discovered using a skin-based test. However, this is analogous to the early theory of the immunologic reaction of beryllium documented via skin tests, and while it is known that bulk nickel does lead to sensitivity in some individuals, it is unknown if the immunological response elucidated by nanoscale nickel would follow the same mechanism [55, 56].

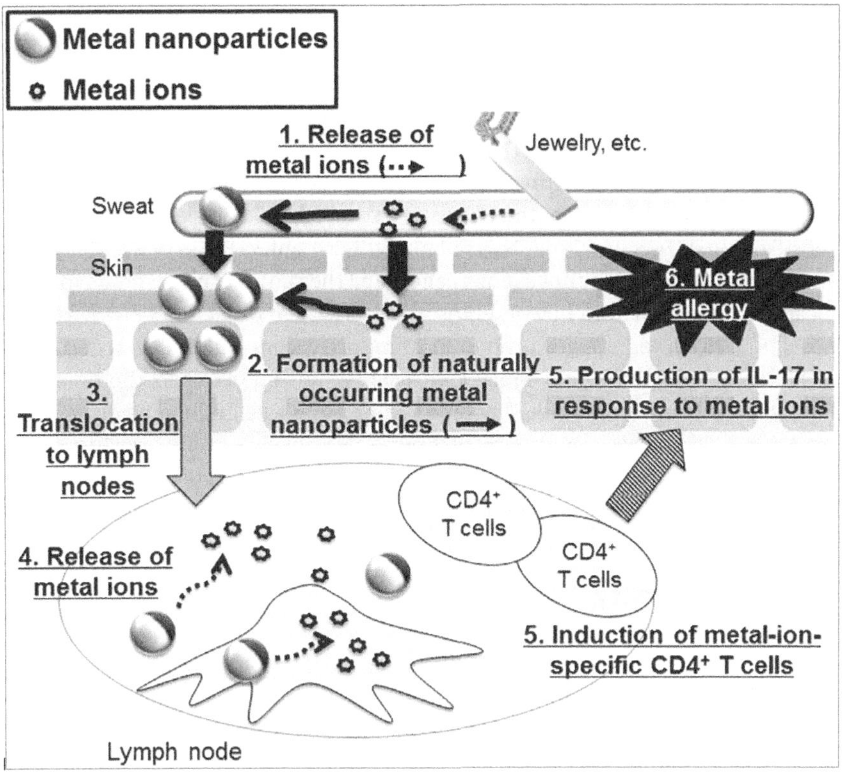

Figure 14.3 Metal NPs as a trigger for metal allergy [51], figure 1 (p.3) / Frontiers Media S.A / Licensed under CC BY 4.0.

Palladium (Pd) is a silvery-white metal used in catalytic converters, electronics, jewelry, and dental implants [57]. It has been well established that some people may have hypersensitivity reactions to Pd, prime among them is allergic contact dermatitis (ACD) [57]. The majority of data regarding this is derived from exposure to elemental Pd and Pd salts. However, a paucity of information exists for the evaluation of immunological response to Pd NPs, particularly in humans. Interestingly enough, what limited data exist show that the immunological response to Pd NPs may follow a different pathway than that of bulk Pd. Boscolo and coworkers [58] conducted an evaluation of peripheral blood mononuclear cells (PBMCs) to Pd NPs in vitro and found that IFN-γ was enhanced and TNF-α and IL-17 were inhibited. In related work, Reale and colleagues [52] also showed that IFN-γ levels were elevated after PBMCs taken from Pd-sensitized women diagnosed with ACD were exposed to Pd NPs. Increased levels of IFN-γ suggest that Pd NPs could induce a Th1 response, and there is a correlation with induction of Th1 immunity and the observation that Pd-exposed people are susceptible to developing ACD [52]. However, Iavicoli and colleagues [59] concluded that no specific Th pattern could be identified in vivo. There is strong evidence to demonstrate cross-reactivity between Pd and Ni [57], but this association as it involves Pd and Ni NPs has not been explored extensively.

Platinum (Pt) and rhodium (Rh) have been considered inert in biological systems [60], but evaluation of these metals at the nanoscale has been lacking. Mauro and colleagues [60] assessed the effects of Pt and Rh NPs on skin and found that sensitization did occur with Pt NPs but not with Rh unless the skin was disrupted. Platinum dust, especially once it has penetrated the skin, has been shown to cause local sensitivity reactions in workers as well as a systemic reaction known as platinosis [61]. However, sensitization to Pt primarily occurs with halogenated forms of the metal [60]. Rhodium is far less reactive, but on occasion rhodium salts have caused contact dermatitis as reported by Goossens and Santucci, reported in Mauro [60], and also rapid-onset asthma as described by Merget in Mauro [60]. However, apart from Mauro's evidence, these sensitivity reactions have not been studied or observed at the nanoscale.

14.16 Summary

The increasing number of consumer products containing metallic nanoparticles should serve as a call to action for researchers to continue investigating their potential adverse health effects. The wide variation in size, shape, and surface coating of nanoparticles limits broad generalizations about the interactions of nanoparticles with the immune system. As a result, risk assessment strategies for MeNPs require relevant in vitro toxicology research of new materials, with special attention given to the mechanisms of toxicity and the use of representative reference materials to correlate the toxic effect of MeNPs with their specific physico-chemical properties. Investigations performed to date have shown that the toxic potential of MeNPs is related to their physico-chemical properties, such as shape, size, charge, aggregation stage, and chemical composition.

References

1 Dobrovolskaia, M.A. and Mcneil, S.E. (2007). Immunological properties of engineered nanomaterials. *Nat. Nanotechnol.* 2 (8): 469–478.

2 Chang, C. (2010). The immune effects of naturally occurring and synthetic nanoparticles. *J. Autoimmun.* 34 (3).

3 Dobrovolskaia, M.A., Germolec, D.R., and Weaver, J.L. (2009). Evaluation of nanoparticle immunotoxicity. *Nat. Nanotechnol.* 4 (7): 411–414.

4 Shah, C.P. (2003). *Public Health and Preventative Medicine in Canada*. Toronto: Elsevier Canada.

5 Nemery, B. (1990). Metal toxicity and the respiratory tract. *J. Occup. Environ. Med.* 32 (12): 1157–1158.

6 Koller, L.D. (1980). Immunotoxicology of heavy metals. *Int. J. Immunopharmacol.* 2 (4): 269–279.

7 Petrarca, C., Clemente, E., Amato, V. et al. (2015). Engineered metal based nanoparticles and innate immunity. *Clin. Molec. Allergy.* 13 (1): 13.

8 Andujar, P., Simon-Deckers, A., Galateau-Sallé, F. et al. (2014). Role of metal oxide nanoparticles in histopathological changes observed in the lung of welders. *Part. Fibre Toxicol.* 11 (1): 23.

9 Cena, L.G., Chisholm, W.P., Keane, M.J., and Chen, B.T. (2015). A field study on the respiratory deposition of the nano-sized fraction of mild and stainless steel welding fume metals. *J. Occup. Environ. Hyg.* 12 (10): 721–728.

10 Brand, P., Lenz, K., Reisgen, U., and Kraus, T. (2013). Number size distribution of fine and ultrafine fume particles from various welding processes. *Ann. Occup. Hyg.* 57 (3): 305–313.

11 Greenberg, M.I. and Vearrier, D. (2015). Metal fume fever and polymer fume fever. *Clin. Toxicol.* 53 (4): 195–203.

12 Zeidler-Erdely, P.C., Erdely, A., and Antonini, J.M. (2012). Immunotoxicology of arc welding fume: worker and experimental animal studies. *J. Immunotoxicol.* 9 (4): 411–425.

13 Antonini, J.M., Roberts, J.R., Schwegler-Berry, D., and Mercer, R.R. (2013). Comparative microscopic study of human and rat lungs after overexposure to welding fume. *Ann. Occup. Hyg.* 57 (9): 1167–1179.

14 Liao, H.-Y., Chung, Y.-T., Lai, C.-H. et al. (2014). Sneezing and allergic dermatitis were increased in engineered nanomaterial handling workers. *Ind. Health* 52 (3): 199–215.

15 Peixe, T.S., Nascimento, E.D.S., Schofield, K.L. et al. (2015). Nanotoxicology and exposure in the occupational setting. *Occup. Dis. Environ. Med.* 03 (03): 35–48.

16 Naval Medical Research Unit Dayton, Prabhakaran K, Gunasekar PG. *In Vitro Cytotoxic Potential of Afghanistan Sand Extract*. 2013.

17 Manke, A., Wang, L., and Rojanasakul, Y. (2013). Mechanisms of nanoparticle-induced oxidative stress and toxicity. *BioMed Res. Int.* 2013: 1–15.

18 Taylor, A.J., McClure, C.D., Shipkowski, K.A. et al. (2014). Atomic layer deposition coating of carbon nanotubes with aluminum oxide alters pro-fibrogenic cytokine expression by human mononuclear phagocytes in vitro and reduces lung fibrosis in mice in vivo. *PLoS One* 9 (9): https://doi.org/10.1371/journal.pone.0106870.

19 Erdely, A., Antonini, J.M., Young, S.-H. et al. (2014). Oxidative stress and reduced responsiveness of challenged circulating leukocytes following pulmonary instillation of metal-rich particulate matter in rats. *Part. Fibre Toxicol.* 11 (1).

20 Duan, J., Kodali, V.K., Gaffrey, M.J. et al. (2015). Quantitative profiling of protein s-glutathionylation reveals redox-dependent regulation of macrophage function during nanoparticle-induced oxidative stress. *ACS Nano* 10 (1): 524–538.

21 Vallières, F., Simard, J.-C., Noël, C. et al. (2016). Activation of human AML14.3D10 eosinophils by nanoparticles: modulatory activity on apoptosis and cytokine production. *J. Immunotoxicol.* 13 (6): 817–826.

22 Janwadkar, A.S. (2017). An unusual case of Eosinophilia. *J. Clin. Diag. Res.* 11 (4): OD19–OD20.

23 Wardhana, D.E.A. (2014). Metal fume fever among galvanized welders. *Acta Medica Indonesiana* 46 (3): 256–262.

24 Casals, E., Gusta, M.F., Piella, J. et al. (2017). Intrinsic and extrinsic properties affecting innate immune responses to nanoparticles: the case of cerium oxide. *Front. Immunol.* 8 (970): 1–7.

25 Keil, D., Buck, B., Goossens, D. et al. (2016). Immunotoxicological and neurotoxicological profile of health effects following subacute exposure to geogenic dust from sand dunes at the Nellis Dunes Recreation Area, Las Vegas, NV. *Toxicol. Appl. Pharmacol.* 291: 1–12.

26 Luo, Y.H., Chang, L.W., and Lin, P. (2015). Metal-based nanoparticles and the immune system: activation, inflammation, and potential applications. *BioMed Res. Int.* 1–12. https://doi.org/10.1155/2015/143720.

27 Chang, Y.-N., Zhang, M., Xia, L. et al. (2012). The toxic effects and mechanisms of CuO and ZnO nanoparticles. *Materials* 5 (12): 2850–2871.

28 Ngobili, T.A. and Daniele, M.A. (2016). Nanoparticles and direct immunosuppression. *Exp. Biol. Med.* 241 (10): 1064–1073.

29 Look, M., Bandyopadhyay, A., Blum, J.S., and Fahmy, T.M. (2010). Application of nanotechnologies for improved immune response against infectious diseases in the developing world. *Adv. Drug Delivery Rev.* 62 (4-5): 378–393.

30 Lucarelli, M., Gatti, A.M., Savarino, G. et al. (2004). Innate defence functions of macrophages can be biased by nano-sized ceramic and metallic particles. *Europ. Cytokine Netw.* 15: 339–346.

31 Dobrovolskaia, M.A., Aggarwal, P., Hall, J.B., and Mcneil, S.E. (2008). Preclinical studies to understand nanoparticle interaction with the immune system and its potential effects on nanoparticle biodistribution. *Mol. Pharmaceutics* 5 (4): 487–495.

32 Kononenko, V., Narat, M., and Drobne, D. (2015). Nanoparticle interaction with the immune system / Interakcije nanodelcev z imunskim sistemom. *Archiv. Ind. Hyg. Toxicol.* 66 (2): 97–108.

33 Buzea, C., Pacheco, I.I., and Robbie, K. (2007). Nanomaterials and nanoparticles: sources and toxicity. *Biointerphases* 2 (4): MR17–MR71.

34 Valko, M., Leibfritz, D., Moncol, J. et al. (2007). Free radicals and antioxidants in normal physiological functions and human disease. *Int. J. Biochem. Cell Biol.* 39 (1): 44–84.

35 Fu, P.P., Xia, Q., Hwang, H.-M. et al. (2014). Mechanisms of nanotoxicity: generation of reactive oxygen species. *J. Food Drug Anal.* 22 (1): 64–75.

36 Huang, Y.-W., Wu, C.-H., and Aronstam, R.S. (2010). Toxicity of transition metal oxide nanoparticles: recent insights from in vitro studies. *Materials.* 3 (12): 4842–4859.

37 Li, J.J., Muralikrishnan, S., Ng, C.-T. et al. (2010). Nanoparticle-induced pulmonary toxicity. *Exp. Biol. Med.* 235 (9): 1025–1033.

38 Gajewicz, A., Schaeublin, N., Rasulev, B. et al. (2014). Towards understanding mechanisms governing cytotoxicity of metal oxides nanoparticles: hints from nano-QSAR studies. *Nanotoxicology* 9 (3): 313–325.

39 Misra, S.K., Dybowska, A., Berhanu, D. et al. (2012). The complexity of nanoparticle dissolution and its importance in nanotoxicological studies. *Sci. Total Environ.* 438: 225–232.

40 Teske, S. and Detweiler, C. (2015). The biomechanisms of metal and metal-oxide nanoparticles' interactions with cells. *Int. J. Environ. Res. Public Health* 12 (12): 1112–1134.

41 Mu, Q., Jiang, G., Chen, L. et al. (2014). Chemical basis of interactions between engineered nanoparticles and biological systems. *Chem. Rev.* 114 (15): 7740–7781.

42 McShan, D., Ray, P.C., and Yu, H. (2014). Molecular toxicity mechanism of nanosilver. *J. Food Drug Anal.* 22: 116–127.

43 Lynes, M.A., Pietrosimone, K., Marusov, G. et al. (2010). Metal influences on immune function. In: *Cellular and Molecular Biology of Metals* (ed. R.K. Zalups and D.J. Koropatnick), 379–414. Boca Raton: CRC Press.

44 Wang, G., Li, H., and Khan, M.F. (2012). Differential oxidative modification of proteins in MRL / and MRL/lpr mice: Increased formation of lipid peroxidation-derived aldehyde-protein adducts may contribute to accelerated onset of autoimmune response. *Free Radical Res.* 46 (12): 1472–1481.

45 Pollard, K.M., Arnush, M., Hultman, P., and Kono, D.H. (2004). Costimulation requirements of induced murine systemic autoimmune disease. *J. Immunol.* 173 (9): 5880–5887.

46 Sakaguchi, S., Sakaguchi, N., Asano, M. et al. (1995). Immunological self-tolerance maintained by activated T-cells expressing IL-2 receptor alpha-chains (CD25) - breakdown of a single mechanism of self-tolerance causes various autoimmune-diseases. *J. Immunol.* 155 (3): 1151–1164.

47 Dimitriou, N.M., Tsekenis, G., Balanikas, E.C. et al. (2017). Gold nanoparticles, radiations and the immune system: current insights into the physical mechanisms and the biological interactions of this new alliance towards cancer therapy. *Pharmacol. Therap.* 178: 1–17.

48 Jan, T.-R., Shen, L., and Wang, L. (2012). Iron oxide nanoparticles suppressed T helper 1 cell-mediated immunity in a murine model of delayed-type hypersensitivity. *Int. J. Nanomed.* 7: 2729–2737.

49 Walkey, C., Das, S., Seal, S. et al. (2015). Catalytic properties and biomedical applications of cerium oxide nanoparticles. *Environ. Sci. Nano.* 2 (1): 33–53. https://doi.org/10.1039/C4EN00138A.

50 Ilinskaya, A.N. and Dobrovolskaia, M.A. (2014). Immunosuppressive and anti-inflammatory properties of engineered nanomaterials. *Br. J. Pharmacol.* 171 (17): 3988–4000.

51 Yoshioka, Y., Kuroda, E., Hirai, T. et al. (2017). Allergic responses induced by the immunomodulatory effects of nanomaterials upon skin exposure. *Front. Immunol.* 8 (169): 1–11.

52 Reale, M., Vianale, G., Lotti, L.V. et al. (2011). Effects of palladium nanoparticles on the cytokine release from peripheral blood mononuclear cells of palladium-sensitized women. *J. Occup. Environ. Med.* 53 (9): 1054–1060.

53 McCance, K.L. and Huether, S.E. (1998). *Pathophysiology: The Biologic Basis For Disease in Adults and Children*. St. Louis, MO: Elsevier.

54 Journeay, W.S. and Goldman, R.H. (2014). Occupational handling of nickel nanoparticles: a case report. *Am. J. Ind. Med.* 57 (9): 1073–1076.

55 Kazantzis, G. (1978). The role of hypersensitivity and the immune response in influencing susceptibility to metal toxicity. *Environ. Health Perspect.* 25: 111–118.

56 Rossman, M.D. (2001). Chronic beryllium disease: a hypersensitivity disorder. *Appl. Occup. Environ. Hyg.* 16 (5): 615–618.

57 Faurschou, A., Menné, T., Johansen, J.D., and Thyssen, J.P. (2011). Metal allergen of the 21st century-a review on exposure, epidemiology and clinical manifestations of palladium allergy. *Cont. Derm.* 64 (4): 185–195.

58 Boscolo, P., Bellante, V., Leopold, K. et al. (2010). Effects of palladium nanoparticles on the cytokine release from peripheral blood mononuclear cells of non-atopic women. *J. Biol. Regul. Homeost. Agent* 24 (2): 207–2014.

59 Iavicoli, I., Fontana, L., Corbi, M. et al. (2015). Exposure to palladium nanoparticles affects serum levels of cytokines in female wistar rats. *PLoS One* 10 (11): e0143801.

60 Mauro, M., Crosera, M., Bianco, C. et al. (2015). Permeation of platinum and rhodium nanoparticles through intact and damaged human skin. *J. Nanopart. Res.* 17 (6).

61 Parrot, J.-L., Hébert, R., Saindelle, A., and Ruff, F. (1969). Platinum and platinosis. *Archiv. Environ. Health: An Int. J.* 19 (5): 685–691.

15

Vasculotoxicity of Metal-Based Nanoparticles

Maria S. Sepúlveda[1,2] *and Jiejun Gao*[1]

[1]*Department of Forestry and Natural Resources, Bindley Biosciences Center, Purdue University, West Lafayette, IN, USA*
[2]*Faculty of Life Sciences, University of Andres Bello, Santiago, Chile*

15.1 Nanoparticles in the Environment

Because of their small size (1–100 nm), nanoparticles (NPs) are highly reactive and catalytic and therefore have found their way in a myriad of applications including cosmetics, textiles, and food packaging. Government and industry around the world are developing nanotechnologies in various fields, and, as a result, the market for these products was estimated in over US$3 trillion in 2015 [1]. The most commonly used core material for engineered one-dimensional NPs are silica dioxide (SiO_2), titanium dioxide (TiO_2), and silver (Ag). Next-generation two- and three-dimensional NPs are also being produced in large amounts and applied to all sorts of products [2, 3]. Although the fate and transport of NPs in the environment is largely unknown, models estimate that the environmental concentrations of NPs in surface water, wastewater, and sediments range from 0.003 parts per trillion (fullerenes in surface water) up to 89 parts per billion (TiO_2NPs in sludge-treated soil) [4, 5]. Therefore, whether it is through the use of cosmetics, household products, medical applications, or drinking of contaminated water, exposure to NPs is likely.

NPs can enter organisms via dermal, nasal, ocular, respiratory, and gastrointestinal routes and distribute through the circulatory and lymphatic systems to major organs [6, 7]. Although recent reviews have addressed toxicity of NPs to major organs and systems [8–11], specific effects of metal-based NPs to the vascular system have been largely ignored. This is contrary to carbon-based NPs where one review published in 2009 summarized the literature on their effects on the cardiovascular system mostly induced via inflammatory pathways that lead to fibrosis and loss of function [12].

Toxicology of Nanoparticles and Nanomaterials in Human, Terrestrial and Aquatic Systems, First Edition. Edited by Marc A. Williams.

In the current review, we summarize the effects of metal or metal oxide-engineered NPs on the vertebrate vascular system. Naturally occurring NPs (such as those produced from photochemical reactions, volcanic eruptions, forest fires, or simple erosion), carbon air pollution particles, and organic NPs are not included in this review. We do not focus on effects on the heart *per se*, since we could not identify nanotoxicity studies with the NPs of interest that have focused specifically on cardiac effects. In addition, studies that report very general cardiovascular toxicity effects such as pericardial edema or changes in heart rates after NP exposure are not reported here. We start by briefly reviewing the overall mechanisms involved in the development of the vascular system in vertebrates. Next, we discuss some unique challenges in nanotoxicology research. We then separate vascular NP toxicity findings between *in vitro* and *in vivo* studies and discuss molecular mechanisms reported from different types of endothelial cells (ECs) as well as vascular responses in whole animals. Because toxicity of NPs is not only related to dose but also to size, shape, coating, and surface charge [13], we also present this information whenever possible. We end with some conclusions and needs for future research in this area.

15.2 Vascular Development

The heart and vascular system are the first to functionally develop in embryos. Vascular formation occurs in two phases: vasculogenesis, the *de novo* vascular formation via assembly of endothelial precursors called angioblasts, and angiogenesis, the sprouting of preexisting vessels by migration and proliferation of ECs [14, 15]. Normal vascular development depends on EC differentiation, migration, and proliferation, extracellular matrix remodeling, and cell–cell communication [16, 17]. ECs have the remarkable capacity of giving rise to vessels with diverse functional, morphological, and molecular signatures [17]. Proangiogenic signals induce fundamental changes in EC behavior including loosening of cell–cell junctions, leading to degradation of the surrounding basement membrane and initiating new blood vessel sprouting [17]. A small proportion of ECs (tip cells, TCs) results in newly sprouting vessels [18, 19]. ECs that trail TCs, or stalk cells (SCs), are less motile but support the extension of sprouting vessels, generating the trunk of new capillaries and maintaining connectivity within the parental vessel [20–22]. After TCs contact other ECs, tight junctions are generated and fuse with recipient vessels to form a lumen [17].

Decades of research have uncovered key signaling pathways involved in vascular development. The vascular endothelial growth factor (VEGF) and Notch signaling pathways are major controllers of EC behavior during blood vessel sprouting. VEGF signaling promotes ECs proliferation, filopodia extension, degradation of the extracellular matrix, and chemotaxis [17] via activation of multiple downstream intermediaries such as mitogen-activated protein kinases (MAPKs), phosphoinositide 3-kinases (PI3Ks), protein kinase B (PKB, also known as AKT), phospholipase Cγ, and small GTPases [17]. Overexpression of *vegf* leads to several abnormalities in heart development and embryonic lethality in mice (*vegfa, vegf*-mouse genome informatics) [23]; formation of abnormal endothelial precursor cells and disorganized vascular structures in *Xenopus* embryos (*vegf*) [24]; and ectopic vasculature development in zebra fish embryos (*vegfa, both isoforms, including* $vegf_{165}$ *and* $vegf_{121}$) [25]. Knockdown of *vegfa* in zebra fish results in blood vessel deficiencies with no or reduced numbers of circulating red blood cells and enlargement of the pericardium [26]. Mouse embryonic stem cells from embryos with dysfunctional VEGF receptor2 (VEGFR2) are unable to migrate to the correct location and form blood islands [27] and die without developing ECs and yolk sac blood islands [28].

15.3 Critical Issues in Assessing the Toxicity of NPs

The field of "nanotoxicology" is relatively new, with the first manuscripts published in late 2004 (a search using "Web of Science" (Information Science Institute, www.isi.edu) between 2004 and 2016 identified a total of 1518 manuscripts using "nanotox*" as the keyword). This body of literature has characterized a wide range of NP toxicities affecting all major organ systems [29–31]. Challenges of testing the toxicity of NPs are related to their unique physical–chemical properties, largely, their large surface area-to-volume ratio, which makes them highly biologically and chemically active [7, 32]. In addition, NPs elicit quantum effects, creating optical, electrical, and magnetic behaviors that are sometimes vastly different from bulk materials [7].

Studies have demonstrated that smaller size NPs are generally more toxic compared with larger ones. This is likely because smaller NPs can generate more reactive oxygen species (ROS) and inflammation compared with larger size particles [33–35]. For example, the cytotoxicity of gold nanoparticles (AuNPs) indicates that 15 nm are nontoxic, but ~10 times lower size (1.4 nm) increases the cytotoxicity 60- to 100-fold. Interestingly, the type of cellular response was also size dependent: 1.4 nm AuNPs caused cell death via necrosis, whereas 1.2 nm AuNPs induced programmed cell death or apoptosis instead [36]. AgNP cell toxicity has also been shown to be size dependent in macrophages, with 15 nm causing more oxidative stress and inflammation compared with 30 and 50 nm [37].

Another unique feature of NP toxicity is the formation of a biomolecular layer on their surface, referred to as "protein corona" (PC). Via van der Waals, electrostatic, hydrogen bonding, and hydrophilic/hydrophobic interactions, biomolecules such as lipids, sugars, and especially proteins will bind to the surface of NPs once in contact with biological fluids [38, 39]. The composition of this PC is important because research has shown that it can affect agglomeration, toxicokinetics, signaling, and, ultimately, toxicity [40]. For instance, negatively charged NPs will attract positively charged biomolecules, which in turn will increase interactions with the negatively charged cell membrane [41]. This also has implications for interpreting toxicity results across different animal models and exposure scenarios [42].

15.4 Vascular Toxicity of NPs *In Vitro*

Researchers have used different EC lines to investigate the vascular toxic mechanisms of NPs *in vitro*. In general, NPs can inhibit EC proliferation, cause cell membrane damage and induce apoptosis [43–50], change the permeability of ECs [45, 51, 52], and induce inflammatory response and oxidative stress [43, 46, 51–58]. NPs also inhibit EC migration and tube formation [48, 49, 59, 60].

In general, effects of NPs on ECs are core material and concentration dependent. For instance, zinc oxide (ZnO), magnesium oxide (MgO), silver (Ag), and copper oxide (CuO) NPs increased EC permeability and oxidative stress starting from 1 ppm [43, 51, 58], while AuNPs and iron oxide (Fe_3O_4) NPs were significantly less toxic at threshold concentrations of 25 and 400 ppm, respectively (Table 15.1) [51, 59]. Size of NPs has also been identified as a key factor influencing EC uptake and vascular toxicity with smaller size NPs showing higher toxic effects. NPs (50–80 nm) enter ECs via caveolin-mediated endocytosis, while larger NPs (100–200 nm) interact with surface receptors triggering the formation of clathrin-coated pits, which transport NPs into cells via clathrin-mediated endocytosis [41, 68]. Smaller AuNPs (15 nm) inhibited VEGF-induced EC proliferation and migration at a lower observed effect concentration (24 ppm), compared with 59 ppm for 50 nm AuNPs [59, 60]. Iron oxide (Fe_2O_3NPs) (22 nm) caused ECs apoptosis after exposure to 2 ppm compared with ~300 nm particles, which induced permeability of ECs but only after exposure to 50 ppm [46, 52]. However, there are also exceptions with several studies reporting no size-dependent effects on ECs after exposure to SiNPs, AgNPs, and AuNPs [43, 47, 50, 53–55, 58, 59, 61, 63]. Surface charge can also influence vascular toxicity. The kallikrein–kinin system (KKS) is involved in regulating vascular permeability. Only negatively charged AgNPs activate the plasma KKS pathway by triggering Hageman factor autoactivation [66].

Regardless of differences on physical–chemical properties, all NPs tested are antiangiogenic as they inhibit the VEGF signaling pathway affecting EC permeability, proliferation, and migration. They do this via induction of inflammation, oxidative stress, and apoptosis (Figure 15.1) [44, 48, 49, 64, 65, 67]. Inflammation

Table 15.1 Toxicity of nanoparticles to endothelial cells.

Nanoparticle	Size (nm)		Cell line	Response	References
Copper oxide (CuO)	40		HUVEC	1–100 mg/L ↓ cell survival	[44]
	180		HCMEC	1–100 mg/L ↑ permeability, inflammatory response, expression *vcam-1, icam-1, mcp-2, IL-8,* LDH	[51]
Gold (Au)	5		HUVEC	From 67 nM ↓ VEGFR2 phosphorylation, intracellular calcium release, RhoA activation leading to ↓ endothelial/fibroblast cell proliferation, permeability, and angiogenesis	[45]
	15			24 mg/L ↓ VEGF migration and tube formation by affecting cell surface ultrastructure and cytoskeleton and might affect angiogenesis via the Akt pathway	[59]
	45		HAEC	0–0.8 mg/L ↓ endothelin-1, apoptosis genes (*map3k5, mapk14, casp9*); ↑ adhesion molecule genes (*vcam1, icam1, sele*)	[61]
	50		BRPEC	59 mg/L ↓ VEGF-, IL-1β- induced proliferation and migration by suppression of Src kinase signaling pathway	[60]
Iron oxide (IO)	8 (uncoated and oleic acid-coated)		HCEC	> 30 μg NPs/cm^2 ↑ autophagy and ROS interfering with LH, CTSD and CTSB	[53]
	Fe_2O_3	22	HUEC	2–100 mg/L ↑ NO, oxidative stress, apoptosis, ↓ mitochondria membrane potential	[46]
		47	HAEC	50 mg/L no inflammation	[57]
		298	HMVEC	50 mg/L ↑ permeability, ROS oxidative stress-modulated microtubule remodeling, Akt/GSK-3β- signaling pathways	[52]
	Fe_3O_4	15–20	HUVEC	400 mg/L ↑ NO, eNOS activity, ↓ CAV1	[62]
		43	HUEC	2–100 mg/L ↑ oxidative stress, NO, apoptosis	[46]

(Continued)

Table 15.1 (Continued)

Nanoparticle	Size (nm)	Cell line	Response	References
Magnesium oxide (MgO)	39	HCMEC	1–100 mg/L ↑ permeability, inflammatory response, expression *vcam-1, icam-1, mcp-2, il-8,* LDH	[51]
Silica (Si)	20, 62	HUVEC	≥ 25 mg/L ↓ SOD, GSH-Px, mitochondrial membrane potential ≥50 mg/L, JNK, C-Jun, P53, NF-Kb signaling pathways, ↑oxidative stress, apoptosis	[47, 54, 55]
	25, 50	HCEC	>30 μg NPs/cm^2 ↑ autophagy and ROS interfering with LH, CTSD, and CTSB	[53]
	600 (mesoporous coated)	RBC	20–100 mg/L membrane deformation, internalization of NPs inside of RBCs, hemolysis	[63]
Silver (Ag)	35	HMVEC ECFC	To HMVEC: acute phase (72 h, 1 mg/L) and chronic phase (31 d, 0.5 mg/L) → loss of membrane integrity, ↑LDH, DNA damage, ROS → cell growth retarded; To ECFC: acute phase (72 h, 0.5 mg/L) → DNA damage → cell growth retarded	[58]
	40–50	PREC	5 mg/L ↓ VEGF-, IL-1β- induced endothelial permeability through inactivation of Src kinase pathway	[64, 65]
		BREC	54 mg/L ↓ VEGF-induced cell proliferation, migration, capillary-like tube formation, angiogenesis via inhibition of PI3K/Akt signaling pathways, ↑apoptosis via CASP3 activity	[48, 49]
	42	HREC	Triggering Hageman factor autoactivation ↑ plasma KKS, bradykinin, B2 receptors, VE-cadherin, paracellular permeability	[66]
	45	CEC	≤10 mg/L ↓ cell proliferation, ↑ vasoconstriction by inhibiting Ach-induced NO-mediated relaxation; 50 and 100 mg/L ↑ NO-dependent proliferation via activation of eNOS and vasodilation	[67]

	65 (citrate coated)	HUVEC	1 mg/L ↑damage cell membrane, inflammatory cytokines, adhesion molecules, chemokines, NF-kB pathways, ROS → ↑apoptosis, ↓ proliferation	[43]
	10, 50, 100 (citrate coated)	RBEC	From 1 mg/L ↑ LDH leakage, cellular membrane damage; ↓ proliferation	[50]
Titanium dioxide (TiO_2)	21	HCEC	>15 μg NPs/cm^2 ↑ autophagy and ROS interfering with LH, CTSD, and CTSB	[53]
Yttrium oxide (Y_2O_3)	20–60	HAEC	≥10 mg/L ↑inflammatory response, expression *icam-1, il-8, mcp-1*	[57]
Zinc oxide (ZnO)	45	HCMEC	1–100 mg/L ↑ permeability, expression *vcam-1, icam-1, mcp-2, il-8,* LDH	[51]
	100–200	HAEC	≥10 mg/L ↑ inflammatory response, expression *mcp-1*	[57]

Acetylcholine (Ach); bovine retinal endothelial cells (BRECs); bovine retinal pigment epithelial cells (BRPECs); caspase-3 (CASP3); caspase-9 (CASP9); cathepsins B (CTSB); cathepsins D (CTSD); caveolin-1 (CAV1); c-Jun N-terminal kinase (JNK); coronary endothelial cells (CECs); endothelial colony-forming cells (ECFC); human aortic endothelial cells (HAECs); human cardiac microvascular endothelial cells (HCMECs); human cerebral endothelial cells (HCECs); human microvascular endothelial cells (HMVECs); human retinal endothelial cells (HRECs); human umbilical vein endothelial cells (HUVECs); endothelial nitric oxide synthase (eNOS); glutathione peroxidase (GSH-Px); glycogen synthase kinase 3 beta (GSK-3β); intercellular adhesion molecule 1 (*icam-1*); interleukin 1 beta (IL-1β); interleukin 8 (*IL-8*); lactate dehydrogenase (LDH); lysosomal hydrolases (LH); mitogen-activated protein kinase 14 (*mapk14*); monocyte chemoattractant protein 2 (*mcp-2*); nitric oxide (NO); nuclear factor kappa-light-chain-enhancer of activated B cells (NF-Kb); phosphoprotein 53 (P53); porcine retinal endothelial cells (PRECs); protein kinase B (PKB, also known as Akt); proto-oncogene tyrosine-protein kinase Src (Src kinase); rat brain endothelial cell (RBEC); ras homolog gene family, membrane A (RhoA); reactive oxygen species (ROS); red blood cells (RBCs); selectin (*sele*) superoxide dismutase (SOD); vascular cell adhesion protein 1 (*vcam-1*); vascular endothelial growth factor (VEGF).

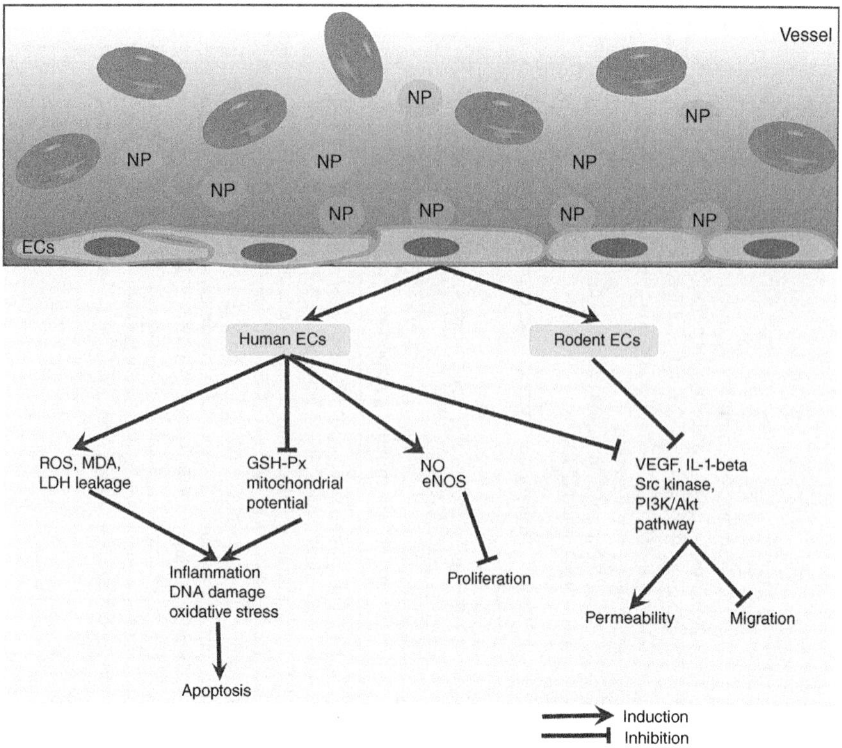

Figure 15.1 Effects of nanoparticles (NPs) on endothelial cells (ECs). Endothelial nitric oxide synthase (eNOS); glutathione peroxidase (GSH-Px); interleukin 1 beta (IL-1β); lactate dehydrogenase (LDH); malondialdehyde (MDA); nitric oxide (NO); phosphoinositide 3-kinase (PI3K); protein kinase B (PKB, also known as Akt); proto-oncogene tyrosine-protein kinase Src (Src kinase); reactive oxygen species (ROS); and vascular endothelial growth factor (VEGF).

by NPs increases endothelium permeability resulting in enhanced NP cellular uptake, and inhibition of Akt, Src kinase, Pi3k, and VEGF signaling pathways by NPs results in EC death. In Human Umbilical Vein Endothelial Cells (HUVECs), AuNPs bind to the vascular permeability factor/VEGF and basic fibroblast growth factor inhibiting endothelial and fibroblast cell proliferation due to a decrease in VEGFR-2 phosphorylation and intracellular calcium release [45]. NPs induce oxidative stress, interfering with lysosomal hydrolases (LH), cathepsins B (CTSB), and cathepsins D (CTSD), which negatively affect mitochondria membrane potential [46, 47, 52–55, 57, 62]. EC proliferation can also be inhibited by NPs via triggering production of nitric oxide (NO) and endothelial nitric oxide synthase (eNOS), which are critical for maintaining vascular homeostasis and vessel dilation, protecting from platelet aggregation, leukocyte adhesion, and controlling proliferation and migration of vascular smooth muscle cells [46, 62, 67].

15.5 Vascular Toxicity of NPs *In Vivo*

In vivo studies examining the vasculotoxicity of NPs are scarce. NPs have been identified as responsible for the formation of blood clots, leading to vessel obstruction and slowing of blood flow. Similarly to what has been reported *in vitro*, NPs have been shown to have antiangiogenic properties *in vivo* (Table 15.2).

NPs cause platelet aggregation, causing abnormal blood flow. For instance, exposure to nonporous nanospheres SiO_2 (170 nm in DI water, 450 mg/kg) of female CD-1 mice causes thrombosis of the endocardium and the lungs, leading to heart congestion and failure [71]. Besides altering platelets, NPs have also been associated with decreased red blood cells (anemia) [69, 73]. Two studies were identified where NP exposure was related with altered contractility of blood vessels. In one study, an injection of polyacrylic acid (PAA)-coated Y-Fe_2O_3 NPs (6.8 nm, 10 mg/kg) to BALB/Cj mice caused an acute decrease in mean arterial blood pressure after 12–24 h associated with a decreased contraction of small mesenteric arteries [70]. In another study, inhalation of TiO_2 NPs (21 nm) by Sprague-Dawley female rats resulted in microvascular impairments in offspring. In this study, multigenerational effects were also observed. Pups born from mothers exposed to TiO_2 NPs developed impaired endothelium-dependent dilation in coronary and uterine arterioles and had reduced maximal mitochondrial respiration in the heart and uterus as adults [74]. These results demonstrate that prenatal NP exposure can potentially lead to later-in-life adverse effects to the vascular system.

Effects on blood vessel formation is a common finding in animals exposed to NPs. Indeed, NPs change vascular permeability, inhibit vascular development, and, in some instances, lead to abnormal heart development [44, 54, 56, 66, 71, 72]. ENPs showed autophagic activity impairing the endothelium of the aortic root and negatively affected angiogenesis involving CAV-1, ICAM-1, VCAM-1, and LC3 [56, 62]. In addition, disruption of angiogenesis by NPs has been associated with both the VEGFR2-mediated autophagy and the VEGF signaling pathways in zebra fish embryos and nude mouse [44, 45, 54, 56, 72].

Our research group recently investigated the vascular effects of polyvinylpyrrolidone-coated AgNPs (PVP-AgNPs, 60 nm) on transgenic zebra fish (TG *fli1a: EGFP*) embryos. Our results show that exposure to 1 and 10 mg/L PVP-AgNPs during the period of vascular development causes a delay in the growth of intersegmental vessels (Figure 15.2). Similar results were reported in another study with TG zebra fish exposed to CuO NPs [44]. In our study, however,

Table 15.2 Toxicity of nanoparticles to the vascular system of vertebrates.

Nanoparticle		Size (nm)	Animal model/age	Response	References
Copper oxide (CuO)		40	Zebra fish TG (nacre/fli1: EGFP)/embryos	100 mg/L for 5 d ↓ number subintestinal vessels, ↓ expression *vegfAa, vegfr1*	[44]
Gold (Au)		5	Nude mice/6–8 weeks old	Two injections of 10 μL, 670 nmol/L AuNPs into mice ear on 2nd and 4th day ↓VEGF121-induced permeability and angiogenesis	[45]
		13.5	Male ICR mice	>550 μg/kg ↓RBCs, body weight, and hematocrit	[69]
Iron oxide (IO)	Fe_2O_3	6.8	BALB/Cj mice/7–10 weeks	10 mg/L for 12–24 h ↓ arterial blood pressure, contractility of small mesenteric arteries	[70]
	Fe_3O_4	15–20	BALB/c mice/15–20 g body weight	20 mg/kg for 72 h damage to endothelium in the aortic root, ↑ serum NO, eNOS, ↓ expression CAV-1	[62]
Silica (Si)		62	Zebra fish/embryos	100 mg/L for 96 h ↓ expression p-VEGFR2, p-ERK1/2 → ↓ angiogenesis, abnormal heart development	[54]
			ICR mice/8 wk and 20–22 g in body weight	178 mg/L for 14 d autophagic activity in ECS and pericytes → ↓ angiogenesis	[56]
Silica dioxide (SiO_2)		170	CD-1 mice	450 mg/kg for 10 d → thrombosis of endocardium, heart, and lung, anemia 600 mg/kg ↓ hemoglobin, ↑ alanine aminotransferase	[71]

Silver (Ag)	10 (coated with mecaptoundeonic acid)	C57BL/6 mice/10 wk old male and 22 g in body weight	500 mg/L for 1 h ↑ retinal vascular permeability, KKS signaling pathway	[66]
	50 (coated with PVP)	Zebra fish TG(fli1a:EGFP)/embryos	10 mg/L for 96 h ↓growth intersegmental vessels, ↑ VEGF signaling pathway	[72]
Titanium dioxide (TiO_2)	20	Male mature Syrian mice/2 wk old and 30 g in body weight	10, 100 mg/L for 14 d ↑ lymphocytes, ↓neutrophils	[73]
	21	Sprague-Dawley rats/250–275 g female; 300–325 g male	The parent rats were inhalation exposed to TiO_2 NPs for 7 d with daily deposition of 43.8 μg. Pups were delivered and grew into adulthood with endothelium-dependent dilation and active mechanotransduction in coronary and uterine arterioles impaired, combining with reduction in mitochondrial respiration in left ventricle and uterus	[74]

Caveolin-1 (CAV-1); endothelial cells (ECs); endothelial growth factor (EGF); endothelial nitric oxide synthase (eNOS); kallikrein–kinin system (KKS); mitogen-activated protein kinase (MAPK, also known as ERK); nitric oxide (NO); polyvinylpyrrolidone (PVP); red blood cells (RBCs); transgenic (TG); vascular endothelial growth factor (VEGF); vascular endothelial growth factor receptor 1 (VEGFR1); vascular endothelial growth factor receptor 2 (VEGFR2).

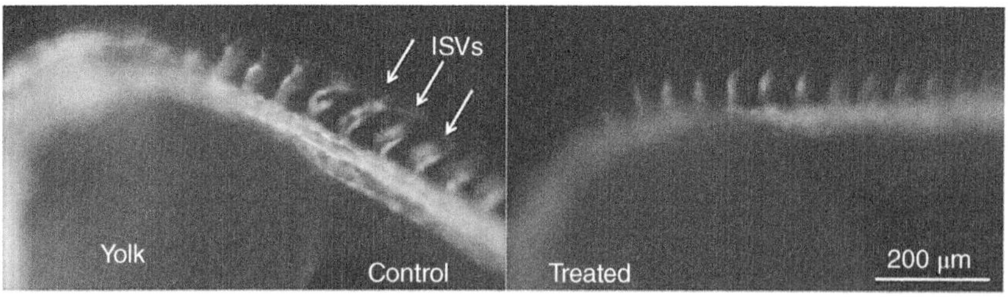

Figure 15.2 Intersegmental vessels (ISVs) of transgenic zebrafish (TG: *fli1a:EGFP*) at 24 hours post fertilization time points. Control group and 10 mg/L PVP-AgNPs treated group. The white arrows indicate ISVs. Compared with control group, ISVs of zebrafish in treated group showed slower growth and no filopodia was formed at the tips of ISVs.

expression of genes within the VEGF signaling pathway was enhanced. This apparent contradiction was explained by the induction of hypoxic conditions in the embryo via agglomeration of NPs on the surface of eggs. Hypoxia is a potent stimulant of the VEGF signaling pathway, but blood vessel development is not enhanced because no increased production in VEGF protein was observed, likely because of impaired translation due to AgNPs toxicity to the endoplasmic reticulum [72].

Similarly, the effects of NPs on vascular development also depend on their physical–chemical characteristics with smaller size NPs generally associated with more serious toxic effects [45, 62, 69, 70].

15.6 Movement of NPs Through the Blood Brain Barrier (BBB)

The blood–brain barrier (BBB) is a highly selective barrier that separates the blood from the brain extracellular fluid within the central nervous system (CNS). Brain ECs form a BBB with tight junctions, which prevent the entry of potential neurotoxins and pathogens into the brain. Importantly, NPs have been reported to be able to cross the BBB, and this movement has been associated with surface charge, with cationic NPs (97 nm, 10–20 mg/L) more easily crossing the BBB compared with neutral NPs of similar size and concentration (75 nm, 10–20 mg/L) [75]. In this same study, anionic NPs (127 nm) also crossed the BBB, and this was dose dependent with higher concentrations (20 mg/L) being more effective than lower concentrations (10 mg/L).

Size is another important factor affecting the transport of NPs across the BBB. In a study with AuNPs in rats, only the smallest size tested (10 nm) was able to cross into the brain compared with larger particles (50–250 nm) [76]. Similar results were reported in C57BL/6 mice exposed to AuNPs, where only the 10 nm size crossed the BBB reaching the retina layers compared with larger particles (10 and 100 nm) [77].

Only two studies have examined the adverse effects of NPs to the BBB. *In situ* rat brain perfusion with TiO_2NPs (25 nm, 5 mg/L, for 5 days) impair the BBB integrity and P-glycoprotein (P-gp) function, as well as modulate tight junction proteins and efflux transporters [78]. Alumina NPs can also markedly reduce the viability of human brain microvascular ECs because of altered mitochondrial potential, increased cellular oxidation, and decreased tight junction protein expression [79]. A recent study reported indirect toxic effects of NPs to the BBB [80]. CuNPs, AgNPs, and AlNPs (50–60 nm, 50 mg/kg, 7 days) induced whole-body hyperthermia, which led to BBB disruption and brain edema.

15.7 Conclusions

NPs are currently being used in a variety of commercial products; therefore, human and animal exposure is likely. In this chapter, we summarize the effects and mechanisms of metal-based NPs on the vascular system. *In vitro* studies show that NPs are antiangiogenic because they cause inflammation, oxidative stress, and apoptosis of ECs, resulting in increased permeability and decreased proliferation and migration. Mechanistic studies show that NPs induce NF-Kb, P53, IL-8, and eNOS activity while inhibiting the VEGF signaling pathway. Animal studies examining the effects of NPs on the vascular system are scarce. Although the few available studies do not disproof the antiangiogenic properties of NPs, the mechanisms and long-term effects of these changes are poorly understood. Considering the key physiological role of the vascular system, there is a need for more studies in this area, particularly assessing the impact of different NP properties (size, shape, surface coatings, etc.) on the development of the vascular system.

List of Abbreviations

Ach	acetylcholine
Ag	silver
Akt	protein kinase B (PKB, also known as Akt)
AuNPs	gold nanoparticles
BRECs	bovine retinal endothelial cells
BRPECs	bovine retinal pigment epithelial cells
CASP3	caspase-3

CASP9	caspase-9
CAV1	caveolin-1
CECs	coronary endothelial cells
CTSB	cathepsins B
CTSD	cathepsins D
CuONPs	copper oxide nanoparticles
DLS	dynamic light scattering
Ecs	endothelial cells
EGF	endothelial growth factor
eNOS	endothelial nitric oxide synthase
ENPs	engineered nanoparticles
ERK	mitogen-activated protein kinase, MAPK, also known as ERK
FCS	fluorescence correlation spectroscopy
Fe_3O_4NPs	iron oxide nanoparticles
GSH-Px	glutathione peroxidase
GSK-3β	glycogen synthase kinase 3 beta
HAECs	human aortic endothelial cells
HCECs	human cerebral endothelial cells
HCMECs	human cardiac microvascular endothelial cells
HD	hydrodynamic diameter
HMVECs	human microvascular endothelial cells
HUVECs	human umbilical vein endothelial cells
ICAM-1	intercellular adhesion molecule 1
LDH	lactate dehydrogenase
LH	lysosomal hydrolases
IL-1β	interleukin 1 beta
IL-8	interleukin 8
IONPs	iron oxide nanoparticles
ISVs	intersegmental vessels
KNK	c-Jun N-terminal kinase
KKS	kallikrein–kinin system
LDH	lactate dehydrogenase
MAPKs/ERK	mitogen-activated protein kinases
MCP-2	monocyte chemoattractant protein 2
MDA	malondialdehyde
MgONPs	magnesium oxide nanoparticles
NF-Kb	nuclear factor kappa-light-chain-enhancer of activated B cells
NO	nitric oxide
NPs	nanoparticles
NTA	nanoparticle tracking analysis
P53	phosphoprotein 53

PAA	polyacrylic acid
PC	protein corona
PI3Ks	phosphoinositide 3-kinases
PKB	protein kinase B
PRECs	porcine retinal endothelial cells
PVP	polyvinylpyrrolidone
RBCs	red blood cells
RBEC	rat brain endothelial cell
RhoA	ras homolog gene family, membrane A
ROS	reactive oxygen species
SCs	stem cells
SELE	selectin
SiO_2	silica dioxide
Src kinase	proto-oncogene tyrosine-protein kinase Src
TCs	tip cells
TEM	transmission electron microscopy
TG	transgenic
TiO_2	titanium dioxide
VCAM-1	vascular cell adhesion protein 1
VEGF	vascular endothelial growth factor
VEGFR1	vascular endothelial growth factor receptor 1
VEGFR2	vascular endothelial growth factor receptor 2
Y_2O_3NPs	yttrium oxide nanoparticles
ZnONPs	zinc oxide nanoparticles

References

1 Flynn, T. and Wei, C. (2005). The pathway to commercialization for nanomedicine. *Nanomedicine* 1 (1): 47–51.

2 Huang, X., Tan, C., Yin, Z., and Zhang, H. (2014). 25th anniversary article: hybrid nanostructures based on two-dimensional nanomaterials. *Adv. Mater.* 26 (14): 2185–2204.

3 Kerativitayanan, P., Carrow, J.K., and Gaharwar, A.K. (2015). Nanomaterials for engineering stem cell responses. *Adv. Healthc. Mater.* 4 (11): 1600–1627.

4 Gottschalk, F., Sun, T., and Nowack, B. (2013). Environmental concentrations of engineered nanomaterials: review of modeling and analytical studies. *Environ. Pollut.* 181: 287–300.

5 Mueller, N.C. and Nowack, B. (2008). Exposure modeling of engineered nanoparticles in the environment. *Environ. Sci. Technol.* 42 (12): 4447–4453.

6 Asharani, P.V., Lian Wu, Y., Gong, Z., and Valiyaveettil, S. (2008). Toxicity of silver nanoparticles in zebrafish models. *Nanotechnology* 19 (25): 255102.

7 Buzea, C., Pacheco, I.I., and Robbie, K. (2007). Nanomaterials and nanoparticles: sources and toxicity. *Biointerphases* 2 (4): MR17–MR71.

8 Simkó, M. and Mattsson, M.O. (2010). Risks from accidental exposures to engineered nanoparticles and neurological health effects: a critical review. *Part Fibre Toxicol.* 7: 42.

9 Card, J.W., Zeldin, D.C., Bonner, J.C., and Nestmann, E.R. (2008). Pulmonary applications and toxicity of engineered nanoparticles. *Am. J. Physiol. Lung Cell Mol. Physiol.* 295 (3): L400–L411.

10 Crosera, M., Bovenzi, M., Maina, G. et al. (2009). Nanoparticle dermal absorption and toxicity: a review of the literature. *Int. Arch. Occup. Environ. Health* 82 (9): 1043–1055.

11 Lettiero, B., Andersen, A.J., Hunter, A.C., and Moghimi, S.M. (2012). Complement system and the brain: selected pathologies and avenues toward engineering of neurological nanomedicines. *J. Control Release* 161 (2): 283–289.

12 Simeonova, P.P. and Erdely, A. (2009). Engineered nanoparticle respiratory exposure and potential risks for cardiovascular toxicity: predictive tests and biomarkers. *Inhal. Toxicol.* 21 (Suppl 1): 68–73.

13 Ehrhart, F., Evelo, C., and Willighagen, E. (2015). Current systems biology approaches in hazard assessment of nanoparticles. *bioRxiv* 028811. https://doi.org/10.1101/028811.

14 Ellertsdóttir, E., Lenard, A., Blum, Y. et al. (2010). Vascular morphogenesis in the zebrafish embryo. *Dev. Biol.* 341 (1): 56–65.

15 Beck, L. and D'Amore, P.A. (1997). Vascular development: cellular and molecular regulation. *FASEB J.* 11 (5): 365–373.

16 Tal, T.L., McCollum, C.W., Harris, P.S. et al. (2014). Immediate and long-term consequences of vascular toxicity during zebrafish development. *Reprod. Toxicol.* 48: 51–61.

17 Herbert, S.P. and Stainier, D.Y. (2011). Molecular control of endothelial cell behaviour during blood vessel morphogenesis. *Nat. Rev. Mol. Cell. Biol.* 12 (9): 551–564.

18 De Smet, F., Segura, I., De Bock, K. et al. (2009). Mechanisms of vessel branching: filopodia on endothelial tip cells lead the way. *Arterioscler Thromb. Vasc. Biol.* 29 (5): 639–649.

19 Gerhardt, H., Golding, M., Fruttiger, M. et al. (2003). VEGF guides angiogenic sprouting utilizing endothelial tip cell filopodia. *J. Cell. Biol.* 161 (6): 1163–1177.

20 Adams, R.H. and Alitalo, K. (2007). Molecular regulation of angiogenesis and lymphangiogenesis. *Nat. Rev. Mol. Cell. Biol.* 8 (6): 464–478.

21 Iruela-Arispe, M.L. and Davis, G.E. (2009). Cellular and molecular mechanisms of vascular lumen formation. *Dev. Cell.* 16 (2): 222–231.

22 Kamei, M., Saunders, W.B., Bayless, K.J. et al. (2006). Endothelial tubes assemble from intracellular vacuoles in vivo. *Nature* 442 (7101): 453–456.

23 Miquerol, L., Langille, B.L., and Nagy, A. (2000). Embryonic development is disrupted by modest increases in vascular endothelial growth factor gene expression. *Development* 127 (18): 3941–3946.

24 Cleaver, O., Tonissen, K.F., Saha, M.S., and Krieg, P.A. (1997). Neovascularization of the *Xenopus embryo. Dev. Dyn.* 210 (1): 66–77.

25 Liang, D., Chang, J.R., Chin, A.J. et al. (2001). The role of vascular endothelial growth factor (VEGF) in vasculogenesis, angiogenesis, and hematopoiesis in zebrafish development. *Mech. Dev.* 108 (1–2): 29–43.

26 Nasevicius, A., Larson, J., and Ekker, S.C. (2000). Distinct requirements for zebrafish angiogenesis revealed by a VEGF-A morphant. *Yeast* 17 (4): 294–301.

27 Shalaby, F., Ho, J., Stanford, W.L. et al. (1997). A requirement for Flk1 in primitive and definitive hematopoiesis and vasculogenesis. *Cell* 89 (6): 981–990.

28 Shalaby, F., Rossant, J., Yamaguchi, T.P. et al. (1995). Failure of blood-island formation and vasculogenesis in Flk-1-deficient mice. *Nature* 376 (6535): 62–66.

29 De Jong, W.H. and Borm, P.J. (2008). Drug delivery and nanoparticles: applications and hazards. *Int. J. Nanomed.* 3 (2): 133–149.

30 Borm, P.J., Robbins, D., Haubold, S. et al. (2006). The potential risks of nanomaterials: a review carried out for ECETOC. *Part Fibre Toxicol.* 3: 11.

31 Aillon, K.L., Xie, Y., El-Gendy, N. et al. (2009). Effects of nanomaterial physicochemical properties on in vivo toxicity. *Adv. Drug. Deliv. Rev.* 61 (6): 457–466.

32 Oberdörster, G., Oberdörster, E., and Oberdörster, J. (2005). Nanotoxicology: an emerging discipline evolving from studies of ultrafine particles. *Environ. Health Perspect.* 113 (7): 823–839.

33 Karlsson, H.L., Gustafsson, J., Cronholm, P., and Möller, L. (2009). Size-dependent toxicity of metal oxide particles – a comparison between nano- and micrometer size. *Toxicol. Lett.* 188 (2): 112–118.

34 Choi, O. and Hu, Z. (2008). Size dependent and reactive oxygen species related nanosilver toxicity to nitrifying bacteria. *Environ. Sci. Technol.* 42 (12): 4583–4588.

35 Scown, T.M., van Aerle, R., and Tyler, C.R. (2010). Review: do engineered nanoparticles pose a significant threat to the aquatic environment? *Crit. Rev. Toxicol.* 40 (7): 653–670.

36 Pan, Y., Neuss, S., Leifert, A. et al. (2007). Size-dependent cytotoxicity of gold nanoparticles. *Small* 3 (11): 1941–1949.

37 Carlson, C., Hussain, S.M., Schrand, A.M. et al. (2008). Unique cellular interaction of silver nanoparticles: size-dependent generation of reactive oxygen species. *J. Phys. Chem. B* 112 (43): 13608–13619.

38 Xia, X.R., Monteiro-Riviere, N.A., and Riviere, J.E. (2010). An index for characterization of nanomaterials in biological systems. *Nat. Nanotechnol.* 5 (9): 671–675.

39 Monopoli, M.P., Aberg, C., Salvati, A., and Dawson, K.A. (2012). Biomolecular coronas provide the biological identity of nanosized materials. *Nat. Nanotechnol.* 7 (12): 779–786.

40 Tenzer, S., Docter, D., Kuharev, J. et al. (2013). Rapid formation of plasma protein corona critically affects nanoparticle pathophysiology. *Nat. Nanotechnol.* 8 (10): 772–781.

41 Setyawati, M.I., Tay, C.Y., Docter, D. et al. (2015). Understanding and exploiting nanoparticles' intimacy with the blood vessel and blood. *Chem. Soc. Rev.* 44 (22): 8174–8199.

42 Shannahan, J.H., Fritz, K.S., Raghavendra, A.J. et al. (2016). Disease-induced disparities in formation of the nanoparticle-biocorona and the toxicological consequences. *Toxicol. Sci.* 152 (2): 406–416.

43 Shi, J., Sun, X., Lin, Y. et al. (2014). Endothelial cell injury and dysfunction induced by silver nanoparticles through oxidative stress via IKK/NF-κB pathways. *Biomaterials* 35 (24): 6657–6666.

44 Chang, J., Ichihara, G., Shimada, Y. et al. (2015). Copper oxide nanoparticles reduce vasculogenesis in transgenic zebrafish through down-regulation of vascular endothelial growth factor expression and induction of apoptosis. *J. Nanosci. Nanotechnol.* 15 (3): 2140–2147.

45 Mukherjee, P., Bhattacharya, R., Wang, P. et al. (2005). Antiangiogenic properties of gold nanoparticles. *Clin. Cancer Res.* 11 (9): 3530–3534.

46 Zhu, M.T., Wang, Y., Feng, W.Y. et al. (2010). Oxidative stress and apoptosis induced by iron oxide nanoparticles in cultured human umbilical endothelial cells. *J. Nanosci. Nanotechnol.* 10 (12): 8584–8590.

47 Liu, X. and Sun, J. (2010). Endothelial cells dysfunction induced by silica nanoparticles through oxidative stress via JNK/P53 and NF-kappaB pathways. *Biomaterials* 31 (32): 8198–8209.

48 Gurunathan, S., Lee, K.J., Kalishwaralal, K. et al. (2009). Antiangiogenic properties of silver nanoparticles. *Biomaterials* 30 (31): 6341–6350.

49 Kalishwaralal, K., Banumathi, E., Ram Kumar Pandian, S. et al. (2009). Silver nanoparticles inhibit VEGF induced cell proliferation and migration in bovine retinal endothelial cells. *Colloids Surf. B Biointerfaces* 73 (1): 51–57.

50 Grosse, S., Evje, L., and Syversen, T. (2013). Silver nanoparticle-induced cytotoxicity in rat brain endothelial cell culture. *Toxicol. In Vitro* 27 (1): 305–313.

51 Sun, J., Wang, S., Zhao, D. et al. (2011). Cytotoxicity, permeability, and inflammation of metal oxide nanoparticles in human cardiac microvascular endothelial cells: cytotoxicity, permeability, and inflammation of metal oxide nanoparticles. *Cell. Biol. Toxicol.* 27 (5): 333–342.

52 Apopa, P.L., Qian, Y., Shao, R. et al. (2009). Iron oxide nanoparticles induce human microvascular endothelial cell permeability through reactive oxygen species production and microtubule remodeling. *Part. Fibre Toxicol.* 6: 1.

53 Halamoda Kenzaoui, B., Chapuis Bernasconi, C., Guney-Ayra, S., and Juillerat-Jeanneret, L. (2012). Induction of oxidative stress, lysosome activation and autophagy by nanoparticles in human brain-derived endothelial cells. *Biochem. J.* 441 (3): 813–821.

54 Duan, J., Yu, Y., Li, Y., and Sun, Z. (2013). Cardiovascular toxicity evaluation of silica nanoparticles in endothelial cells and zebrafish model. *Biomaterials* 34 (23): 5853–5862.

55 Duan, J., Yu, Y., Li, Y. et al. (2013). Toxic effect of silica nanoparticles on endothelial cells through DNA damage response via Chk1-dependent G2/M checkpoint. *PLoS One* 8 (4): e62087.

56 Duan, J., Yu, Y., Li, Y. et al. (2014). Silica nanoparticles enhance autophagic activity, disturb endothelial cell homeostasis and impair angiogenesis. *Part. Fibre Toxicol.* 11: 50.

57 Gojova, A., Guo, B., Kota, R.S. et al. (2007). Induction of inflammation in vascular endothelial cells by metal oxide nanoparticles: effect of particle composition. *Environ. Health Perspect.* 115 (3): 403–409.

58 Castiglioni, S., Caspani, C., Cazzaniga, A., and Maier, J.A. (2014). Short- and long-term effects of silver nanoparticles on human microvascular endothelial cells. *World J. Biol. Chem.* 5 (4): 457–464.

59 Pan, Y., Wu, Q., Qin, L. et al. (2014). Gold nanoparticles inhibit VEGF165-induced migration and tube formation of endothelial cells via the Akt pathway. *Biomed. Res. Int.* 2014: 418624.

60 Karthikeyan, B., Kalishwaralal, K., Sheikpranbabu, S. et al. (2010). Gold nanoparticles downregulate VEGF-and IL-1β-induced cell proliferation through Src kinase in retinal pigment epithelial cells. *Exp. Eye Res.* 91 (5): 769–778.

61 Gu, C., Wu, H., Ge, G. et al. (2016). In vitro effects of hollow gold nanoshells on human aortic endothelial cells. *Nanoscale Res. Lett.* 11 (1): 397.

62 Su, L., Han, L., Ge, F. et al. (2012). The effect of novel magnetic nanoparticles on vascular endothelial cell function in vitro and in vivo. *J. Hazard. Mater.* 235–236: 316–325.

63 Zhao, Y., Sun, X., Zhang, G. et al. (2011). Interaction of mesoporous silica nanoparticles with human red blood cell membranes: size and surface effects. *ACS Nano* 5 (2): 1366–1375.

64 Sheikpranbabu, S., Kalishwaralal, K., Venkataraman, D. et al. (2009). Silver nanoparticles inhibit VEGF-and IL-1beta-induced vascular permeability via Src dependent pathway in porcine retinal endothelial cells. *J. Nanobiotechnol.* 7: 8.
65 Sheikpranbabu, S., Kalishwaralal, K., Lee, K.-J. et al. (2010). The inhibition of advanced glycation end-products-induced retinal vascular permeability by silver nanoparticles. *Biomaterials* 31 (8): 2260–2271.
66 Long, Y.M., Zhao, X.C., Clermont, A.C. et al. (2016). Negatively charged silver nanoparticles cause retinal vascular permeability by activating plasma contact system and disrupting adherens junction. *Nanotoxicology* 10 (4): 501–511.
67 Rosas-Hernández, H., Jiménez-Badillo, S., Martínez-Cuevas, P.P. et al. (2009). Effects of 45-nm silver nanoparticles on coronary endothelial cells and isolated rat aortic rings. *Toxicol. Lett.* 191 (2–3): 305–313.
68 Fröhlich, E. (2012). The role of surface charge in cellular uptake and cytotoxicity of medical nanoparticles. *Int. J. Nanomed.* 7: 5577–5591.
69 Zhang, X.D., Wu, H.Y., Wu, D. et al. (2010). Toxicologic effects of gold nanoparticles in vivo by different administration routes. *Int. J. Nanomed.* 5: 771–781.
70 Iversen, N.K., Frische, S., Thomsen, K. et al. (2013). Superparamagnetic iron oxide polyacrylic acid coated γ-Fe_2O_3 nanoparticles do not affect kidney function but cause acute effect on the cardiovascular function in healthy mice. *Toxicol. Appl. Pharmacol.* 266 (2): 276–288.
71 Yu, T., Greish, K., McGill, L.D. et al. (2012). Influence of geometry, porosity, and surface characteristics of silica nanoparticles on acute toxicity: their vasculature effect and tolerance threshold. *ACS Nano* 6 (3): 2289–2301.
72 Gao, J., Mahapatra, C.T., Mapes, C.D. et al. (2016). Vascular toxicity of silver nanoparticles to developing zebrafish (*Danio rerio*). *Nanotoxicology* 10 (9): 1363–1372.
73 Mahdieh, Y., Mahsa, S., Andishe, K. et al. (2015). The effects of TiO_2 nanoparticles on white blood cells in mice. *Der Pharm. Lett.* 7: 153–156.
74 Stapleton, P.A., Nichols, C.E., Yi, J. et al. (2015). Microvascular and mitochondrial dysfunction in the female F1 generation after gestational TiO_2 nanoparticle exposure. *Nanotoxicology* 9 (8): 941–951.
75 Lockman, P.R., Koziara, J.M., Mumper, R.J., and Allen, D.D. (2004). Nanoparticle surface charges alter blood-brain barrier integrity and permeability. *J. Drug Target.* 12 (9–10): 635–641.
76 De Jong, W.H., Hagens, W.I., Krystek, P. et al. (2008). Particle size-dependent organ distribution of gold nanoparticles after intravenous administration. *Biomaterials* 29 (12): 1912–1919.
77 Kim, J.H., Kim, K.W., Kim, M.H., and Yu, Y.S. (2009). Intravenously administered gold nanoparticles pass through the blood-retinal barrier depending

on the particle size, and induce no retinal toxicity. *Nanotechnology* 20 (50): 505101.

78 Brun, E., Carrière, M., and Mabondzo, A. (2012). *In vitro* evidence of dysregulation of blood–brain barrier function after acute and repeated/long-term exposure to TiO_2 nanoparticles. *Biomaterials* 33: 886–896.

79 Chen, L., Yokel, R.A., Hennig, B., and Toborek, M. (2008). Manufactured aluminum oxide nanoparticles decrease expression of tight junction proteins in brain vasculature. *J Neuroimmune Pharmacol.* 3 (4): 286–295.

80 Sharma, H.S. and Sharma, A. (2007). Nanoparticles aggravate heat stress induced cognitive deficits, blood-brain barrier disruption, edema formation and brain pathology. *Neurobiol. Hypertherm.* 162: 245–273.

Section IV

Future Directions and Gaps in the Knowledge

16

Knowledge Gaps, Future Directions, and the Emergence of Nanoplastics as an Environmental Threat Pollutant

Marc A. Williams and Desmond I. Bannon

Health Effects Division, Directorate of Toxicology, U.S. Army Public Health Center, Aberdeen Proving Ground, MD 21010, USA

16.1 Current Concerns and Scope of the Problem

Ultrafine particulates (UFPs) are generated in the environment by incidental anthropogenic activities and via natural environmental events. Classic examples include UFPs that are generated as by-products of fossil fuel combustion, condensation of semi-volatile substances, or industrial emissions; or as a consequence of dust and smoke plumes from volcanic activity, corrosion- and erosion-derived particles in aquatic and marine media, bush and forest fires, and other environmental sources [1–6]. When found in the ambient atmosphere, UFPs exhibit an aerodynamic diameter of less than 0.10 μm [1–6]. By contrast, nanoparticles are synthesized by processes of industrial engineering, having broad use across a range of industrial and scientific or medical applications and products to include imaging technologies, energy storage applications, personal care and cosmetic products, and therapeutic drug delivery systems [1–6]. Engineered nanoparticles typically have dimensions in the range of 1–100 nm with at least one dimension that is less than 0.1 μm. The morphology of UFPs also tends to be quite non-uniform, while nanoparticles display a regular morphology (e.g. appearing as a sheet or plate, rod, tube, sphere, or rod-like) [1–6].

While nanoparticles tend to be homogeneous in terms of their physical and organic chemistry properties, and their structural and biological characteristics, UFPs display an irregular morphology and are extraordinarily heterogenous [1–6]. For example, nanoparticles are characterized by a low organic chemical content and a highly variable content of metal impurities, whereas UFPs display a high organic chemical content and comparatively increased composition of both transition and heavy metal elements [1]. Moreover, while the major exposure pathways for UFPs are inhalation, and perhaps oral ingestion, exposure to engineered

Toxicology of Nanoparticles and Nanomaterials in Human, Terrestrial and Aquatic Systems, First Edition. Edited by Marc A. Williams.

nanoparticles includes inhalational, oral, and dermal exposure pathways, and, in the case of targeted clinical therapeutics, a direct parenteral exposure pathway [1].

Sections 2 (Terrestrial and Aquatic Systems) and 3 (Human Systems) of this book clearly showcase an emerging concern regarding the largely uncertain toxicological and environmental impacts of nanomaterials, as well as the human health impact of environmentally persistent nanoparticles and chemically complex UFPs [1–6]. At the time of this writing, the U.S. Environmental Protection Agency (EPA) maintains that UFPs cannot be regulated until additional data on their ecological impact and broader environmental toxicology are available [1–5, 7–13]. Although there are many differences in the physicochemical composition of UFPs and nanoparticles in general, one common feature of concern regarding their adverse health effects in humans is their extremely small size, which makes UFPs and nanoparticles most relevant as a danger to health through the inhalational exposure pathway.

Ambient ultrafine particles and nanoparticles share certain unique features and differences, which have helped shape our understanding of their potential adverse health effects. An appreciation of these unique differences also helps our understanding of whether or not human health and that of aquatic and terrestrial life are equally affected by UFPs and/or nanoparticles. The dynamic nature and behavior of UFPs and nanoparticles also present regulatory agencies with critical challenges when determining a risk assessment of nanoparticles and materials.

We know, for example, that UFPs and nanoparticles are environmentally persistent. This characteristic allows them to transform and interact with other chemical species in the environment and to interact with their physical and ecological environment prior to the exposure of aquatic and terrestrial organisms to those modified particles [14, 15]. Additionally, engineered/manufactured nanoparticles might transform over time when stored for prolonged periods or under variable ambient conditions [14–17]. Similarly, nanoparticles might transform in the original packaging of stored therapeutics and biological agents and throughout the life-cycle of their use, including their disposal and emergence in the environment, and their exposures to humans and aquatic and terrestrial wildlife [14–17].

A major challenge in the field of environmental nanotoxicology, and of heightened relevance to aquatic and terrestrial ecotoxicologists, is the paucity of a historically rich data set on the effects of engineered nanoparticles (ENPs), engineered nanomaterials (ENMs, defined as nanoparticles in this chapter), and UFPs. Thus, a key concern that is aligned to the "One Health Paradigm" is an appreciation that nanoparticles and nanomaterials are now used extensively across industrial, scientific, biomedical, and technological platforms, a circumstance which has served to heighten critical concerns and unanswered questions regarding their environmental fate, transport, and safety.

Using well-described and analyzed case examples, this book provides an analysis of the complex ecotoxicological and human health impact of UFPs, ENPs, and ENMs. It is clear that when compared to typical chemical toxicants, the toxicology of nanoparticles and understanding their mechanisms of action remain in their infancy and suffer from a paucity of data, particularly in terms of assessing their aquatic and terrestrial toxicology, the biological systems affected, and the observed adverse health effects following environmental exposure. Thus, the ecotoxicology of UFPs and, in particular, ENPs/ENMs, remains poorly understood, although comparatively much more is known about their potential adverse effects for human health. Accompanying these concerns is the potential for nanomaterials to adversely affect human health, aquatic and terrestrial wildlife, and ecosystem integrity (Chapters 2–15, this book; [1–3]).

While it was not our intent to provide a comprehensive book on the toxicology of nanoparticles, we have provided the reader with relevant case examples of the ecotoxicological, environmental, and human health impacts of UFPs and their engineered counterparts (see Chapter 1, Figure 1.1). In providing case examples of the observed effects of nanoparticles in aquatic, terrestrial, and human health systems, this book has revealed key knowledge and data gaps, and nano-aligned issues that are specific to the unique environments of aquatic, terrestrial, and human life.

16.2 A survey of the Identified Knowledge Gaps and Needs for Future Research

There is a considerable knowledge gap regarding the adverse effects of UFPs on human health as compared to their coarse (PM_{10}) and fine ($PM_{2.5}$) counterparts and to ENPs. References [1–6, 18–26], and comprehensively describe the adverse effects of PM and UFPs on human health. The current book has built on that literature by revealing a clear need to consider the environmental health effects and ecotoxicological knowledge gaps that challenge our ability to adequately assess the risk to health and the safety of UFPs and ENPs throughout their life-cycles.

The knowledge and data gaps identified in Chapters 2–15 of this book are sequentially summarized below and highlight the technological and experimental challenges in assessing nanoparticle and nanomaterial toxicology by means of experimental studies. The aforementioned chapters also provide insights into approaches one might consider in advancing an understanding of the chemical, toxicological, and ecological features and effects of nanomaterials that would benefit not only toxicology but risk assessment and regulatory agencies, as well.

16.3 Knowledge Gaps – Aquatic and Terrestrial Ecotoxicology

Current challenges aligned to the risk assessment framework and nanomaterial safety in aquatic ecosystems are discussed in Chapter 2. An important topic is an increased focus on environmental remediation efforts, often referred to as "nanoremediation," in an effort to help mitigate an expected increased use of nanomaterials, and thus the nanowaste that will inevitably affect aquatic systems as well as other environments or ecosystems in the broader sense (including human health). Clearly, the uncertainty regarding the ecological safety and sustainability of new nanotechnologies, including anti-fouling paints that are designed to block the formation of biofilms on marine vessels, requires urgent attention (Chapter 2).

The major challenge is that the development and industrial use of newer nanomaterials and products that display a variety of surface chemical compositions and quantum effects, coatings, and functional groups, including an increased range of sizes and shapes of those nanomaterials, will elicit altered chemical reactivity as well as increased uncertainty regarding their ecosafety in aquatic environments (Chapter 2). This chapter places these ecotoxicological considerations and the available wastewater treatment processes in the context of currently used frameworks of environmental risk assessment (ERA) of nanomaterials. Critically, this chapter identifies a knowledge gap in assessing nanomaterial ecotoxicity in sediments and translating the observations made in laboratory study to those made in the real-world environment.

The above work is complemented by critical discussions of increased industrial use of titanium dioxide (TiO_2) nanoparticles and their subsequent release to the aquatic environment (Chapter 3). This chapter discusses the emerging concern regarding the environmental fate and transport of engineered nanoparticles, particularly the sorption of natural organic matter and ionic compounds to the surfaces of metal nanoparticles. This phenomenon can alter the surface physicochemistry of TiO_2 nanoparticles, thereby promoting their agglomeration and altering their sorption capacity for organic metal substances. There is a great deal of uncertainty in assessing the bioavailability of not only the nascent TiO_2 nanoparticles but also the sorbed and altered chemical payload whose transport, bioavailability, and exposure to aquatic life is facilitated by unique physicochemical interactions (Chapter 3).

Chapter 3 also identifies a clear knowledge gap in advancing our understanding of the aquatic environmental factors that might influence sorption/desorption kinetics of organic chemicals to TiO_2 nanoparticle surfaces and the migration of TiO_2 nanoparticles in the aqueous phase. Studies aligned to these knowledge gaps should promote increased understanding of the environmental fate of

TiO_2 nanoparticles and assist in their risk assessment in aquatic environmental systems (Chapter 3).

Chapter 4 also critically analyzes the impact, environmental fate, and behavior of nanoparticles found in consumer and industrial products. This chapter recognizes that freshwater (aquatic) systems are not necessarily the final ecological sink for nanoparticles; estuarine and marine coastal environments also serve as environmental sinks for engineered nanoparticles via transport from freshwater columns or by direct input from areas of concentrated anthropogenic activities. We need to consider littoral filter-feeding bivalves in this context.

We also need to be aware of the high volumes of plastic waste that find their way to the oceans, as well as the practice of large-scale marine dumping of trash containing mostly plastics from a variety of manufacturing sources. (See below for additional information on this topic.)

Chapter 4 critically identifies knowledge gaps in our understanding of the environmental behavior, fate, transport, and ecotoxicity of engineered nanoparticles in estuarine and coastal marine systems and provides a clear refocus of research efforts in these environmental settings. There is an identified need for improved coordination of efforts that advance the understanding of how defined, engineered nanoparticles affect a variety of marine organisms.

The lack of accurately determined toxicity values, employed as points of departure for assigning relative risk of a given nanoparticle toxicant, and the need for a read-across capability or a logical grouping of nanomaterials, as defined by their physical and chemical properties and ecotoxicity, collectively impede the process of risk assessment in this setting (Chapter 4). The authors question the experimental design of studies that inadequately align to "real-world" marine environments. Such studies use inappropriate (or too-high) nanoparticle concentrations or dose-metric, or employ low ionic strength media that bear little resemblance to their natural marine environment counterparts. Of increased importance and priority to toxicologists and risk assessment professionals is the knowledge that the majority of available studies have focused solely on acute exposure settings at the expense of chronic and/or repeat exposure effects for which there is a paucity of available ecotoxicological data (Chapter 4). The authors again emphasize the critical importance of dose-metric in toxicological studies, where the inadequate characterization of real internal dose as compared to initial nominal dose fails to take into account nanoparticle agglomeration under conditions of differential salinity commonly found in aquatic, estuarine/brackish, and marine environments.

Finally, the authors identify a pressing need for better understanding of the trophic transfer and biomagnification of engineered nanoparticles across marine food-webs, laboratory modeling of prolonged ecotoxicological effects, and integrating these studies with their aquatic freshwater counterparts. Studies that integrate and harmonize approaches in modeling nanoparticle ecotoxicity in freshwater, estuarine, and marine environments will advance understanding of the risks to health from migration of nanoparticles to these compartmentalized environmental sinks (Chapter 4).

Complementing the above chapters, particularly the material discussed in Chapter 3, Chapter 5 explores the ecotoxicological impact of released engineered nanoparticles in aquatic ecosystems, using nano-TiO_2 as a key example. Whereas nanoparticles interact with both sediment and aquatic media, engineered nanoparticles interact with dissolved and particulate organic matter, hydrophobic organic compounds, and heavy metals. In the context of risk assessment, the authors emphasize the importance of not only the potential toxicity of nanoparticles, but also the modification of their inherent toxicity and changes in their bioavailability on interacting with other environmental pollutants. Consistent with the above chapters, the authors of Chapter 5 emphasize the need for additional data to strengthen understanding of the relevance of engineered nanoparticles, such as TiO_2, interacting with other organic and inorganic contaminants that might be encountered in aquatic and marine environments. The key issues are the continued uncertainty in defining differentially modified toxicities and the broader environmental effects of engineered nanoparticles found in aquatic media as compared the saline conditions of marine environments, and the paucity of data in this area. This is clearly an area warranting additional focused research.

While aquatic and marine environments are critically important sinks for environmentally persistent and reactive engineered nanoparticles and nanomaterials, Chapter 6 reminds us that soils are widely recognized as important environmental sinks for nanomaterials. While currently available standardized ecotoxicity tests appear suitable for the acute toxicological assessment of nanomaterials in soils, this chapter recognizes that careful attention is needed in the context of both dose metrics and characterization needs before, during, and after a particular test. Moreover, regarding the comprehensive assessment of the effects of soil-borne nanomaterials, the authors also recognize that currently available standardized testing approaches suffer from several limitations related to a restricted range of species that are studied, being limited to nascent manufactured or engineered nanoparticles, and not necessarily those that have interacted with environmental media and substrates; and effects measurements over short exposure periods. This chapter discusses new approaches that build on standard toxicological testing to better account for the differential effects of soil type, exposure settings, and the period of exposure. These approaches can better provide ecotoxicologically meaningful settings and outcomes, particularly given the inherently complex chemical mixture of environmentally persistent nanomaterials; subsequent testing challenges and relevancy in the context of affected soil invertebrates, plants, and microorganisms; and thus, risk assessment and regulatory frameworks (see Chapter 6).

A major concern in the risk assessment of nanomaterials is determining relative risk on exposure to the environment, the modification of the nanomaterial by its interaction with environmental media, substrates, and organisms, and, as mentioned above, the need for appropriate and meaningful regulatory policies that can account for the broad manufacture and use of nanomaterials and the evolution of nanotechnology. Like that of most other chemical toxicants, the toxicity of nanoparticles is a function of the net outcome of the exposure pathway (inhalational, oral ingestion, and dermal/ocular routes), the dose, concentration, and duration and/or frequency of exposure, and the dose metric employed to determine risk to health.

Established regulatory frameworks, including the U.S. Toxic Substances Control Act (TSCA), and the European Registration, Evaluation, Authorisation and Restriction of Chemicals (REACH), were deployed prior to the appearance of nanomaterials; consequently, neither system could distinguish between nanoparticle chemicals or their larger/bulk chemical counterparts [14, 17]. In time, the European Chemical Agency (ECHA) recognized the importance of distinguishing nanomaterial and nanoparticle forms (i.e. "nanoforms") and their inherent differences from their bulk chemicals [18, 27]. ECHA and many industrial entities increasingly recognized that "nanoforms" of any industrially engineered or synthesized material will demonstrate altered chemical and physical characteristics as compared to the same material in bulk chemical form [19]. The quantum size effects and comparatively increased surface-area-to-volume ratio also endowed their nanomaterial counterparts with unique properties, which translated to significantly altered environmental and ecotoxicity profiles of bulk chemicals as compared to their nanoform counterparts. Thus, analyzing the toxicity of nanomaterials based on their bulk form is inaccurate in predicting both the potential likelihood of adverse health effects and ecotoxicological impacts [19].

The TSCA was revised and amended in 2016 [20] to include new guidance for collecting information on the existing inventory of nanomaterials, and requirements for premanufacture notification of newly manufactured engineered nanomaterials [20]. Similarly, the ECHA later drafted and released a new set of revised and synthesized guidance documents, which became law in January 2020 [19]. These new documents make clear a number of nanomaterial-specific considerations that companies need to take into account when registering new chemicals. The ECHA requires registrants to justify the basis on which nanomaterial "nanoforms" are grouped [19]. This means that the EPA expects detailed documentation on chemical identity, manufacturing processes, and production volumes for engineered nanoforms, and any additional manufacturing processing or modifications of the nanoforms, in addition to quite extensive details on human exposure data, ecological and environmental release data, and the available health and occupational safety data [20]. Further discussion on the continued evolution of these regulatory requirements is provided later in this chapter, where data and knowledge gaps on the human health effects of nanomaterials are considered.

As discussed in Chapter 6, regulatory frameworks and policies that direct the use of standardized soil organism toxicity testing of nanoforms are aligned to those adopted for conventional chemical toxicity testing and risk assessment. The intent of standardized approaches, as set out in the frameworks of REACH, is to permit a consistent and standardized development of laboratory testing protocols employed by disparate academic, government agency, and private industry laboratories and testing centers that determine the hazard and risk to health from chemical toxicants.

Standardized testing procedures or standard operating procedures (SOPs) that are designed to test the safety of engineered nanomaterials would benefit substantially from a harmonized nanoinformatics approach. This would enable the prediction of nanomaterial properties, environmental interactions, and toxicological hazards; and the placement of toxicologically characterized nanomaterials into logical groups as defined, for example, by their primary chemistry and their known properties, including size, geometric shape, surface chemistry modifications, and so forth ([3] and Chapter 6). This approach would further permit read-across from a quality-controlled and curated dataset of well-documented laboratory findings, including the "metadata." Although this topic is beyond the scope of Chapter 6, the reader is directed elsewhere for a detailed review of nanomaterial safety, nanoinformatics, and the importance of the "metadata" [3].

Current challenges in the study of engineered nanomaterials in the context of terrestrial and aquatic life are discussed in Chapter 7, specifically the relatively poorly explored area of the environmental impacts of engineered magnetic iron oxide nanoparticles (IONPs) and the more recently developed superparamagnetic iron oxide nanoparticles (SPIONs). Given the complexity of the applications that have employed IONPs, including engineering and biotechnology, therapeutics (e.g. cancer therapy and clinical diagnostics contract agents), and environmental remediation, IONPs, like most other engineered nanomaterials, have the potential of being inadvertently or purposefully released to aquatic and terrestrial environments [7]. The authors provide clear examples of IONPs about which very little is known regarding human health effects or immediate ecotoxicological concerns. In this context, magnetite (Fe_3O_4) and its oxidized form, known as maghemite (γ-Fe_2O_3), are discussed in some detail. A major challenge, as with most other engineered nanomaterials, is the ability to quantify the concentrations of nanomaterials commonly found in the environment as a basis for modeling their toxicity (Chapter 7).

Both Chapters 6 and 7 highlight the need for novel or improved methodologies to adequately and reliably characterize the environmental fate and transport of engineered nanomaterials in complex substrates such as soil or surface and subterranean aquatic media. Notably, the authors survey the evolution of new approaches to monitor engineered nanoparticles in complex environment media, including superconducting quantum interference device magnetometry (SQUID), scanning SQUID biosusceptometry (SSB), and magnetic force microscopy (MFM), and propose that these techniques are likely to see further development and refinement ([21], and Chapter 7). Other data and knowledge gaps identified in Chapter 7 highlight the need for molecular and mechanistic studies of SPIONs, including the development of sensitive biological sensors to detect the biological toxicity of engineered nanomaterials; and studies aimed at advancing our understanding of IONP bioaccumulation and biomagnification.

In addition to the above sections that describe currently known data gaps as they relate to many of the commonly encountered engineered nanomaterials, Chapter 8 strengthens this discourse by providing a critical discussion of the health effects of carbon nanotubes (CNTs). This chapter further provides novel insights into recently discovered and novel toxicological mechanisms, pressing data gaps, the advent of novel materials (including "nano-hybrids"), and a discussion of future perspectives. This chapter highlights a need to focus future research efforts on immune modulation studies (i.e. unique toxicity testing approaches) that can complement more traditional (classic) toxicity testing, including assessment of mortality, growth, reproduction, and development and behavior (Chapter 8). This chapter highlights concerns (also discussed in Chapter 6) with regard to the design of toxicity studies and the duration of exposure. For example, in fish-feeding studies exploring the effects of CNTs, the authors of this chapter noted that in studies conducted over short periods of time (7 days or 6 weeks) to assess growth, no changes were seen. Comparative studies conducted by the authors (Chapter 8) did not see changes in growth until after week 12, which indicates that future research studies should focus on the chronic effects of CNTs (and perhaps other nanomaterials and particles) to permit sufficient time for long-term health effects such as growth to be revealed, including reproductive capacity and subsequently observable effects on the population. Additional studies have exemplified the need for careful study design and exposure duration (Chapter 8). In one study of CNT-exposed Daphnia, it was found that acute mortality was not a good predictor of chronic effects, and acute exposures did not affect mortality in this model [22]. However, chronic toxicity assays showed significant differences in mortality, reproduction, and growth [22].

Both Chapters 6 and 8, and elsewhere in the book, clearly communicate the need for careful experimental design of toxicological studies exploring the hazards and end-points of nanomaterials and nanoparticles, which behave quite differently from other chemical toxicant exposures in model systems and the environment. Studies that explored immune reactivity also inform us that future research should focus on mechanisms of action where a significant data gap is evident. Chapter 8 emphasizes that unique toxicity testing should focus on mechanisms of action, including nutrient depletion and immune modulation. In addition, this chapter analyzes the effects of CNTs on co-contaminants found in aquatic media, and the manner in which carbon nanotubes alter key toxicological end-points in fish species and perhaps other aquatic and terrestrial life.

Similarly, others have emphasized a need to carefully explore the acute aquatic toxicity of nanomaterials (Chapter 9). In this chapter, anatase and rutile forms of titanium dioxide or TiO_2 are critically discussed in the context of providing a deeper understanding of their aquatic toxicity, mechanisms, and the broader ecosystem impact of TiO_2 on aquatic communities that constitute the food chain. Chapter 9 thus provides an outstanding exemplar on traditional and more modern approaches to studying and modeling the ecotoxicology of nanoparticles and nanomaterials. These approaches are of particular relevance to those in the field of hazard identification/analysis and risk assessment.

In Chapter 8, the authors have emphasized the need for studies focusing on mechanisms of action to advance our understanding of the complex toxicity of nanoparticles and nanomaterials. In a novel and poorly understood setting, the authors of Chapter 10 focus on the cardiorespiratory toxicity of engineered nanomaterials in aquatic animals, and a growing body of evidence that indicates the importance of oxidative stress pathways in contributing to cardiorespiratory dysfunction. Cardiorespiratory function is highlighted as a key determinant of health and survival in fish. The authors of Chapter 10 remind us that, in certain settings, cardiac failure is considered a limiting factor in environmental stress tolerance pathways in fish. Additionally, one of the key conclusions from this chapter highlights the common misconception that the more commonly used engineered nanoparticles tend to display low-to-moderate toxicity as compared that of known chemical toxicants such as plasticizers, solvents, and pesticides. The authors find this line of thinking inaccurate because the acute and chronic toxicities of many engineered nanomaterials are poorly understood, meaning there is heightened uncertainty regarding the chronic impacts and environmental persistence of nanoparticulate toxicants on aquatic organisms. Chapter 10 also reveals that engineered nanoparticle exposure sensitizes fish to hypoxia and dampens their cardiorespiratory function; however, the ecotoxicological impact from chronic exposures to a diversity of nanomaterials is poorly understood (Chapter 10).

16.4 Knowledge Gaps – Adverse Health Effects in Humans

In addition to aquatic and terrestrial nanotoxicology, this book focuses on key case exemplars of direct relevance to human health and identified knowledge and data gaps. In the first chapter of this section (Chapter 11), the neurotoxicity and adverse health impact of toxic air pollutants is discussed in the context of neurodevelopmental disorders, such as autism, schizophrenia, and attention-deficit hyperactivity disorder, and their hypothesized association to exposures of ultrafine particulate matter. A clear data gap here is the pressing need for a systematic review or epidemiological meta-analysis of studies that have explored exposure to ambient UFPs from vehicular traffic sources, for example, and the putative association with development of certain neurodevelopmental disorders as described herein.

As mentioned earlier, one of the key challenges for toxicologists and risk assessors is that while the EPA has set regulatory standards for PM_{10} and $PM_{2.5}$, the regulation of UFPs requires additional numerical evidence of their toxicity before regulatory standards for these airborne pollutants are developed and adopted (Chapter 11). Given that many of the aforementioned disorders are associated with childhood development, there is a clear need to advance understanding of the underlying developmental mechanisms, the role played by chronic inflammatory pathways during development, especially the putative role played by the inflammatory cytokines, including IL-1β and IL-6; and the effects of systemic inflammation in mediating neurological pathology and associated disorders. For example, schizophrenia and autism share the same neuropathological lesion, referred to as ventriculomegaly, with evidence of white matter trauma. Although the specific relevance of this observation remains unclear, the manifestation of ventriculomegaly with evidence of white matter trauma is perhaps an effect—not a specific cause - of autism and schizophrenia per se (see Chapter 11). Although these conditions are very poorly understood, the authors suggest that the development of severe hydrocephalus might be due to disrupted cerebral spinal fluid absorption, and malformation in brain tissue development plays a major role.

Developmental and post-natal animal models have partly advanced our understanding of the underlying mechanisms of neurological disorders on *in utero* or post-natal exposure to UFPs, including inflammation and microglial activation, dampened corpus callosum myelination, ventriculomegaly, and altered glutamate and γ-aminobutyric acid (GABA) neurotransmitter levels, and inappropriate neuronal activation/transmission. A significant knowledge gap exists in determining the underlying mechanisms and the predilection of male offspring (from animal model studies) to many of the observed neurological pathologies and conditions on exposure to highly heterogeneous, air pollutant UFPs (Chapter 11). As the authors point out, air pollutant UFPs comprising multiple other chemical and biological toxicants tend to sorb to the particulates to increase their complexity, and likely introduce novel pathways of inflammatory mediator-induced neurotoxicity that require urgent study.

While the study of neurological development and the chemicals that might induce neuro-toxicological insults represents an important and emerging field of human health research, Chapters 12, 13, and 14 present complementary and emerging evidence of the role that the immune system (Chapters 12 and 14) and the gastrointestinal (GI) or mucosal microbiome (Chapter 13) play following exposure to a chemical toxicant and the subsequent adverse health outcomes. The relevance of the GI mucosal barrier and immune system (Chapters 12 and 13), and protective host immunity and subversion of host immunity on exposure to environmental toxicants (Chapters 12 and 14) are not only linked but must be considered together when determining the overall toxicological impact following oral and/or respiratory and dermal exposure to a chemical toxicant, including nanomaterials and ambient airborne ultrafine particles.

Chapter 12 carefully explores the toxicological impact of a variety of nanomaterials that display the greatest potential for an oral route of exposure via drinking water supplies and the diet, including TiO_2, silica, nanosilver, Zn oxide nanoparticles (ZnONP), fullerenes, and single- or multi-walled CNTs, and their impact on GI structural and functional integrity. This chapter clearly identifies not only the differential effects of engineered nanoparticles on GI toxicity but also the importance of selecting the most appropriate animal model (e.g. effects seen in murine models are not necessarily seen in the rat), and the disadvantages of certain *in vitro* cell culture systems in over-estimating the toxic potential of nanosilver to the GI tract.

Many studies of the GI toxicity of nanomaterials employed the use of Caco-2 cells (a human colon-derived epithelial cell-line), which are still used frequently today as relevant models of the GI system. However, Caco-2 cells lack a protective mucus layer, and there are no clear standardized test protocols on using these cells. Consequently, there is a high risk of confounding in the interpretation of the toxicological data from using Caco-2 cells as a representative *in vitro* model system (Chapter 12). In addition, comparative analysis for a particular targeted toxicant among studies and different research groups presents additional challenges that highlight the need for metadata stewardship and standardized methodological testing procedures when determining the safety and toxicity of challenging toxicants such as nanomaterials. As the authors of this chapter note, the test article is well characterized in terms of its physical chemistry properties ([1–3, 15, 23] and Chapter 12).

Observed issues with *in vitro* cell culture systems also represent a precautionary note on the use of new approach methodologies (NAMs) in estimating and/or predicting the toxicity of particular chemical or nanomaterial toxicants (Chapter 12); additional information on the advent of NAMs follows. As the authors note, the available data on the GI toxicity of a diverse set of well-known and emerging nanomaterial toxicants are relatively sparse. There is a considerable knowledge gap in understanding the manner by which ingested nanomaterials interact with and disrupt GI integrity and function, including intestinal mucus deposition and barrier function. Thus, as the authors clearly point out, there is considerable uncertainty with regard to appreciating the extent of GI toxicity on oral exposure to nanomaterials in general, at least from animal models, and interpreting the often-conflicting datasets from non-standardized *in vitro* cell culture models. Clearly, this uncertainly has marked implications for the appropriate and standardized use of NAMs in this setting.

Complementing a discussion of the available research data on nanomaterial-mediated GI toxicity (Chapter 12), the physical characteristics and biological effects of nanomaterials on host immunity and the GI microbiome are critically analyzed in Chapter 13. In particular, the authors discuss the anti-microbial effects of nanomaterials with specific emphasis on the mucosal microbiome. Clearly, strengthening our understanding of the ways by which orally ingested nanoparticle toxicants interact with, and thus disrupt, the normal functioning of the mucosal or gut microbiome is of critical interest and thus is very well surveyed and analyzed in Chapter 13. The authors recognize that there is a potentially significant variety of nascent, environmentally altered, and chemically diverse nanomaterials and nanoparticles, all of which can impact their biological interactions following oral exposure. On ingestion of such a broad variety of nanomaterials, and thus interaction with and disruption of the gut microbiome, there is clear potential for an altered gut microbiome to adversely affect immune competence and thus dampen or severely disrupt protective host immunity.

There is a clear data gap in regard to fully appreciating and determining the mechanisms of action of the gut microbiome–nanomaterial interaction and dysregulated host immunity. As the authors of Chapter 13 reveal, it is evidenced that a variety of engineered nanomaterials display antibacterial activity against the microbiota found in the GI tract. Emerging evidence also identifies clear disturbances in intestinal populations of beneficial gut bacteria on oral exposure to engineered nanomaterials. Moreover, this exposure of the mucosal system of the gut, and thus the immunotoxic potential of a growing variety and complexity of nanomaterials, is inevitable and might have far-reaching effects, such as severely disrupting protective host immune competence. Future research will clearly need to investigate the impact on the gut microbiome and host immune system of novel nanomaterial formulations found in drug delivery systems, an ever-increasing repertoire of dietary supplements, and novel anti-microbial agents. The net impact of an increasingly complex nanotoxicological challenge to the human mucosal-associated microbiome requires urgent research to enable a detailed toxicological assessment of an inevitable exposure to nanomaterials in the diet.

Continuing the theme of the immunotoxicity of engineered nanoparticles and nanomaterials, Chapter 14 provides a comprehensive overview of the interaction of engineered nanoparticles and ambient airborne ultra-fine particulates. This chapter builds on, presents, and places into context new insights into the work that we and others have previously discussed ([24–26, 28–34], Chapter 14). In particular, this chapter highlights the need to explore the relevance of occupational exposures, e.g. welding and firing fumes, and respirable metal nanoparticles present in those fumes; and the relevance of such work to occupational hazard identification and risk assessment of nanoparticulates in the occupational setting. Though this chapter critically surveys the immunotoxicological mechanisms following inhalational exposures to ultrafine particles and engineered nanoparticles, the interaction of innate and adaptive host immunity following exposures to nanomaterials is poorly understood. This knowledge gap is particularly apparent in the context of disrupted cell function and integrity and the reversibility of the immunological or cellular functional lesion on exposure to inhaled nanomaterials, where pro-oxidative and pro-inflammatory pathways are known to be critical [24–26, 28–34].

Consistent with the critical analyses provided in Chapters 12 and 13, the authors of Chapter 14 emphasize the urgent need for additional research into the toxicological effects of a growing number of consumer products and therapeutics containing metallic nanoparticles. This field of study is not only relevant to human health but is also critical in the context of aquatic and terrestrial ecotoxicology. Identified data gaps emphasize the need for mechanistic and molecular studies that align to the physico-chemical properties, including shape, size, charge, agglomeration state, and surface chemical composition. These data must be complemented by additional insights to better understand the target organs affected (e.g. damage to, and functional failure of, the lungs, liver, kidney, spleen, skin, and the mucosal microbiome system). Additionally, there is a critical need to identify and understand the mechanisms of action of new target organs of effect, e.g. the central nervous system/brain, and differential understanding of other microbiome communities that influence human health, e.g. differential analysis of the pulmonary, oral, gut, and dermal microbiomes on interaction with UFPs and ENPs – differences and similarities.

In Chapter 15 of this book, the impact of metal oxide nanomaterials on the vertebrate vascular system is critically reviewed in the context of recognizing the importance of this organ system in nutrient transport, gas exchange, and homeostatic regulation – functions that are also relevant to the immune system. Critically, this chapter reveals a hitherto-underappreciated impact of nanoparticles and nanomaterials on the vascular system, wherein *in vitro* toxicity studies showed that nanoparticles were anti-angiogenic, resulting in heightened inflammation, oxidative stress, increased vascular permeability, and decreased endothelial proliferation and migration; and inhibited the vascular endothelial-migration factor (VEGF) signaling pathway. In addition to a lack of quality and standardized *in vitro* toxicity studies, this chapter also identifies key knowledge gaps in the availability of relevant animal studies, both of which would provide key data on the mechanisms and long-term effects of the anti-angiogenic properties from oral, dermal, or inhalational exposure to nanoparticles. Finally, consistent with the findings of other chapters in this book, the authors point to a distinct knowledge gap in revealing the impact of the vast variety of nanoparticle species and their properties (e.g. size, shape, surface charge, and surface coatings/differential chemistries) on the development of the vascular system.

It is also apparent that the developmental toxicity of ultrafine particular matter and engineered nanoparticles has been inadequately addressed. Given heightened interested in the developmental immunotoxicity of environmental toxicants, and presumably nanoparticles and their ability to disrupt the adult gut microbiome and protective immunity, studies exploring these knowledge gaps are clearly warranted.

16.5 Emerging Threats and Future Directions

The emergence of an expanding array of novel nanoparticles and composites (or nanoforms) in diverse industrial and consumer settings, including personal care products, coated fabrics, military applications and materiel, energy generation, as well as in diagnostic and therapeutic settings, presents a level of complexity that requires novel approaches that can provide meaningful data to the risk assessment professional and the regulatory agency.

Concerns regarding safety and the potential for adverse human and environmental health effects on exposure to nanoparticles have long been suspected [24], and a growing body of literature supports those concerns ([1, 3, 6] and Chapters 1–15 this book). Regulatory agencies have been slow to respond, due in large part to the complexities and challenges associated with standardizing toxicological methods to characterize the hazards from exposure to nanoparticles. In the context of risk assessment and investigational toxicology of the biological effects of nanomaterials, which are increasing both in number and complexity, the challenge is clearly daunting. There are clear data gaps not only in deepening our understanding of the chemical and physical characteristics of nanoparticles but also in strengthening our subsequent understanding of the mechanisms/modes of action of both ambient UFPs and ENMs.

When the risks and safety of novel nanoparticle species and formulations are assessed, it is desirable to incorporate emerging new technologies and methodological approaches that can provide critical data in characterizing the hazard and assessing the risk of targeted or prioritized nanoparticles. This consideration includes an assessment of UFPs, ENPs, and novel ENMs, and as described briefly in the section below, the emergent concern of the environmental health impacts of micro- and nanoplastic pollutants as an emerging environmental and ecotoxicological threat and a potential threat to public health [35].

16.5.1 Nanoplastics – An Emergent Environmental Threat Pollutant

Though not a major focus of this book, plastics, particularly micro- and nanoplastics, have emerged as a major new environmental pollutant of concern. Plastics have provided significant societal benefit over the years [35–38]. Nanoplastics are derived from, or found in, a variety of food and beverage products (e.g. tubs of salt, tinned/canned foods, weathered plastic bags and nets), sources of water (e.g. tap water, bottled water, and in food preparation), personal care products (e.g. face washes and toothpastes), therapeutic and surgical or clinical products (e.g. prostheses and pharmaceuticals), and the ambient air (e.g. vehicular tire wear and tear, furniture products, and abrasion of textiles and their coatings) among many others [38]. While initially of minor concern, the increased environmental burden of these nanoplastics over the past 5–10 years has raised questions on whether they exist as innocuous physical entities or represent a rising concern as the number of particles increases exponentially.

It is estimated that global production of plastic currently exceeds 320 million tonnes (Mt) per year, over 40% of which is single-use packaging, resulting in plastic waste that evades recycling streams [35] and makes its way to the ocean. Concurrent with the increasing global consumption of plastics is the increase of their associated emissions; both are expected to continue their upward trend for the near future. The rapid accumulation of plastic waste in landfills and natural habitats, including aquatic and marine environments, signals an emerging potential hazard to wildlife.

Plastic waste is controlled by recycling or disposal in landfill sites; however, given the large volume of plastic produced, 40% of which is single-use, it is inevitable that the aquatic and marine environment receives a significant amount of plastic waste [35–37]. Thus, a significant fraction of the plastic produced each year is deposited to and persists in aquatic and marine environments, and it is thought that plastic waste in these environments has the potential to accumulate in the amount of approximately 250 Mt by the year 2025 [35]. Moreover, weathering events, such as tsunamis, hurricanes, or cyclones, contribute significantly to burdens of plastic in the aquatic environment. Once deposited in the environment, plastic debris progressively fragments into smaller pieces referred to as micro- and nanoplastics [35–38]. These are now almost homogeneously distributed throughout the terrestrial, aquatic, and atmospheric environments of the planet.

Degradation of plastic waste is a continuous process, and plastic micro- and nanoparticles are now considered ubiquitous environmental pollutants from the Arctic to the Antarctic oceans, as well as in polar ice caps [35–38]. Plastic waste exposed to solar UV irradiation is oxidized, making it brittle and easily distributed by prevailing wind currents. Mechanical action from wind and waves abrades plastic into progressively smaller particles referred to as microplastics (MPs; 0.1–1000 μm) and nanoplastics (NPs; $\leq$0.1 μm in at least one dimension), both of which are broadly distributed in aquatic and marine environments today ([39, 40] and as discussed in [35]). To put the weathering effect into perspective, a 1 cm^2 piece of polystyrene foam exposed *in situ* to visible and UV light for 14 and 56 days produced concentrations of 0.32×10^8 and 1.26×10^8 particles/mL, respectively. These particles have been shown to be ingested by plankton, severely distorting their shape and morphology. However, complementary existing fields indicate the potential for particle, chemical, and microbial hazards; current evidence shows that MPs and NPs are an emerging persistent environmental contaminant that is increasing in magnitude.

Though we now recognize that complex and exclusively anthropogenic material is a ubiquitous environmental pollutant and threat to all forms of life, just as other nanomaterials appear to be in nature [35–38], there is also a significant uncertainty regarding their real-world environmental health impact. The impact of plastic microfibers remains poorly understood. However, it is clear from the above analysis that the oceans represent a sink for micro- and nanofibers that are denser than seawater, and the sediment eventually accumulates those fibers as non-biodegradable debris [35]. An estimated 5.2 trillion plastic particles contaminate the global sea surface, whereas approximately 4 billion fibers/km^2 contaminate the deep Indian Ocean floor [35]. In the U.S. alone, it is estimated that about 8 trillion microbeads enter aquatic habitats each day via wastewater treatment plants, presenting another notable source of environmentally persistent NPs and fibers [35].

Given the lack of both a detailed research history and literature on the effects of NPs, we can hypothesize a potential adverse health impact on oral or inhalational exposure to NPs. Though additional data are needed, it is evident that MPs and NPs have the potential to affect aquatic, terrestrial, and human ecosystems, and thus disrupt biological pathways with as yet unknown consequences given the lack of data and the marked uncertainty regarding their toxicity [35–38]. However, given their persistent biofouling and capacity to accumulate continuously as an environmental pollutant, we should be concerned that NPs will most likely serve as a depot of complex environmental pollutants that will impact human health and aquatic and terrestrial life well into the future [35–38].

An emerging concern is that NPs will serve as migrating "vehicles" enabling the exposure of aquatic, terrestrial, and human life to adsorbed chemical contaminants and perhaps to attached bacterial or viral particles [35, 38]. By contrast, one could also envision this characteristic of MPs and NPs as a potential environmental remediation strategy to irreversibly sorb chemical pollutants, thus impeding their environmental bio-accessibility.

Associated with the concern that plastics could serve as migrating depots of adsorbed chemical contaminants is the realization that MPs and NPs do not biodegrade, and their reduced size due to inadvertent environmental weathering and deliberate manufacturing processes could modulate their toxicity to exposed humans and wildlife [35]. Of greater concern is the realization that MPs (and presumably NPs) adsorb and thus concentrate a variety of environmentally prevalent chemical contaminants and known toxicants, including polycyclic aromatic hydrocarbons, pesticides, and other environmentally persistent chemicals collectively referred to as per- and polyfluoroalkyl substances (PFAS); and an accumulation of heavy metals, including cadmium, lead, nickel, and others [35, 38, 41, 42]. On exposure, this means that NPs serve as a depot and carrier of a complex chemical payload capable of migrating to major organs, tissues, and individual cells in exposed fish and other aquatic and marine life, terrestrial life, mammals – and humans – with largely uncertain toxicological consequences [35, 38].

Despite the uncertainty referred to above, emerging evidence provides cause for concern that NPs provoke cytotoxic and genotoxic effects in fish and mammalian model systems [43–45], thereby provoking pro-inflammatory and pro-oxidative stress with the subsequent release of reactive oxygen species (ROS) in aquatic, terrestrial, and human health systems [46]. In murine models exploring the health effects of MPs, it was found that local innate immune cells were activated, with release of ROS and induction of cellular autophagy/cell death [46]. Moreover, at least in the human system, it is thought that approximately 90% of ingested plastic (excluding inhalational exposures) is excreted [47]. However, as we have learned in Chapters 12–14, the 10% that remains in the GI system is endowed with the capacity to cause adverse health effects. These might include disrupted GI integrity, altered homeostatic influence of the mucosal microbiome, causing dysbiosis and abnormal gut microbiotic activity in aquatic species [48] and presumably in humans, also, though the latter needs to be formally demonstrated [38, 48, 49]. Other consequences of a GI–tract–microbiome–NPs interaction might include disrupted host immune surveillance, immune competence, and immune suppression; inappropriate cytotoxicity; and perhaps allergic immune conditions and auto-immunity as seen on exposure to UFPs [1, 26, 28–34]; dysregulated cell cycling; heightened inflammatory reactions; and perhaps disease manifestations [38, 48].

In the context of human exposure, exploring the plausible impact of NPs on human health must be considered a priority. Considerable knowledge gaps also exist in terms of understanding the extent by which aquatic and terrestrial ecosystems are impacted by NPs. A growing body of evidence shows that MPs, and presumably NPs, are present in seafood and a wide variety of processed foods and liquid beverages that are consumed in the human diet [50–55]. As briefly mentioned above, a major concern regarding inhalational and oral exposure to NPs is their capacity to adsorb environmental chemical pollutants and known toxicants, e.g. PFAS, and the inherent endogenous chemical additives that are incorporated into plastics during their manufacture [43]. This complex mixture of inherent chemicals and an environmentally sorbed toxicant payload is likely to leach from the plastic polymer matrix to the surrounding environmental matrix or biological system and to the surrounding tissues and cells [56, 57]. Sustained or long-term oral and inhalational exposure to these potentially contaminant-rich "guided missiles" is likely to disrupt the gut and pulmonary microbiome in exposed humans. As mentioned above, there are similar concerns that MPs and NPs have the capacity to impact a broad geographical distribution or aquatic, marine, and terrestrial ecosystems with potentially grave consequences [35–38].

Though there is a high degree of uncertainty regarding the potential human health impact and toxicological pathways from oral or inhalational exposure to NPs, the chapters in this book certainly provide and strengthen biological plausibility to the argument (see Chapters 11–16 in regard to the human health effects). An emerging concern is the manner in which the host immune system will respond to such "foreign" and potentially "dangerous" particles, since it is evolutionarily designed to respond to non-self and foreign particulate challenges [45, 58, 59]. Studies have already shown that on exposure to NPs, relevant animal model studies and *in vitro* cell culture models display a pro-inflammatory response with heightened expression and secretion of the cytokines IL-1β, IL-6, TNFα, and IL-10, all of which are key cytokines with important immune-modulatory activities in innate and adaptive T-cell-mediated immunity [58, 59].

In the context of marine food sources and human health, a detailed assessment of the potential health risk of MPs and NPs from a variety of sources and exposure routes and a range of foods across the total diet would assist an evaluation of the causative risks of contaminated marine food on human health. Moreover, the greatest concern is the capacity of MPs and NPs to facilitate the transport of hydrophobic organic chemicals into organisms or to distant organs in fish, birds, and mammals (including humans) due in part to the high sorption capacity of plastic particles for persistent organic chemicals.

Similarly, as we discussed above, chemical additives that might leach from MPs and NPs could pose a significant environmental issue, and, depending on their unique chemical properties, could be released in the environment throughout the lifetime of the particle. Little information is known about the leaching rate of chemical additives from MPs and NPs. In high plastic debris accumulation zones, e.g. semi-enclosed and enclosed marine embayments, leaching of chemicals requires urgent hazard characterization and determination. Such leaching is exemplified by the presence of hexabromocyclodecane in not only the Styrofoam buoys used in oyster culture farms but in the sediment beds surrounding those farms, as well [60]. However, the general human population might be exposed to MPs from different sources that have not been systematically identified and researched.

While we understand the impact of macroplastics, there are significant data gaps in our understanding of the ecotoxicological and human health impact of NPs. There is also a pressing need for interim wildlife and human health risk assessments of the toxicology of NPs and the novel methods by which their risk to complex aquatic, marine, and terrestrial ecosystems might be mitigated, such as active removal programs of plastic nanowaste from environmental matrices and aquatic/marine media. Risk assessment procedures and toxicological programs should also strive to prioritize the NP species that are the most dangerous to human health and wildlife. Responsiveness to this goal will require well-designed, high-quality, standardized approaches to environmentally relevant basic toxicology research. Toxicity studies should focus in particular on the behavior, migration, and distribution of NPs in biological tissues and organs and should strive to provide accurate assessments of the chronic health effects of NPs in appropriate model systems.

16.6 Conclusions and Other Considerations

Chapter 16 has provided a top-level overview of the more pressing data and knowledge gaps in furthering our appreciation of the environmental and human health impacts of engineered nanoparticles and ambient ultrafine particulate matter. In addition, this chapter surveyed the emergent need to research and provide new knowledge on the impact of micro- and nanoplastics to the environment. Micro- and nanoplastics constitute not only an emerging and largely uncertain risk to human health but an emerging threat to aquatic and terrestrial ecosystems as well. A comprehensive toxicological and environmental risk assessment of the hazards presented by environmental plastic pollutants would be a useful start in understanding this threat.

The field of toxicology, particularly the methods employed and the means by which risk assessment documentation can be improved and strengthened, is evolving at a rapid pace [3]. In part, this evolution is driven by continued public scrutiny of the methods employed and the paths by which toxicologists characterize the hazards and safety or toxicity of a given drug, chemical, or environmental contaminant. Intensified public scrutiny and legislative policies are in many ways directing fundamental and applied toxicological studies, including risk assessment, and in the context of this book, are adapting nanotoxicity to the ethos of "Replacement, Reduction and Refinement," known as the "3Rs" principles. There is growing public pressure to reduce the reliance on *in vivo* animal models where sound documented scientific evidence fails to justify the need for such animal experimentation and *in vivo* toxicological testing.

Concordantly, as with most other chemicals and drugs requiring toxicological testing, there is a growing need for nanomaterials to be extensively characterized and measured by approaches that are more aligned to analytical and physical chemistry than to traditional toxicology. Given the complexity and minute size of nanomaterials, such approaches appear logical, since their implementation reveals important information, including the core chemistry of the analyzed nanoparticles, coatings, and surface chemistry; details on their size, structure, and dimensions; and their behavior in aqueous or other media. A re-evaluation of nanoparticle testing and analysis is required, as is a constant awareness of the need for highly standardized approaches, testing method applicability, assay and method validations, and highly standardized approaches for data analysis and interpretation [61]. There is also a growing need for well-documented, standardized quality management systems (QMSs) and data management plans (DMPs) of not only the analytical data but the metadata as well. The need for both QMSs and DMPs to systematically strengthen the quality of scientific data that is often used in decision-making processes, has simultaneously led to the globalization of novel approaches aimed at harmonization and validation of testing approaches [3]. These approaches, largely coordinated via the Organisation for Economic Co-operation and Development, include an ever-growing list of test guidance documents aimed at providing rigorously standardized toxicological testing methods, and an infrastructure that enables consistent and standardized knowledge acquisition and data aggregation across a globally diverse number of testing centers and laboratories [3, 62–65].

The above frameworks and attempts at policy-driven reform of toxicological testing (see also Chapter 1, this book), has led to the advent of NAMs, which are alternative testing strategies (ATSs) that have firm roots in *in vitro* models of biological and toxicological analysis; and *in silico* approaches, including computational and mathematical modeling, such as physiologically based pharmacokinetic modeling, quantitative structure–activity relationship models, and *in vitro*-to-*in vivo* extrapolation [3, 64, 65].

The aforementioned ATS approach and the relatively recent interest in including nanoinformatics in toxicological quality management systems and knowledge-based infrastructure are intended to strengthen existing or more traditional frameworks in the risk assessment enterprise with advancing scientific knowledge, data quality, and defensible regulation [64, 65]. In the context of assessing the risk of engineered nanoparticles to aquatic and terrestrial systems and human health, it is anticipated that ATSs will be transformative. According to the literature, each ATS, on its own, will not suffice in fully determining the toxicity of a targeted nanomaterial or predicting a particular mechanistic toxicological endpoint. Rather, ATSs or NAMs will be matrixed into test batteries based on integrated approaches to testing and assessment, next-generation risk assessment approaches, and the better-known phased approaches to toxicological testing [3, 66, 67].

Thus, we can anticipate that approaches to the synthesizing, "safer-by-design" nanomaterials that impart reduced or significantly lower risk to the environmental ecosystems and human health will depend on strengthening and increasingly using a toxicological testing infrastructure of NAMs with decreased reliance on *in vivo* animal model systems; and an accelerated, more efficient, and cost-effective path to environmental and human health risk assessment and analysis that is high-quality, defensible, ethically acceptable, and accountable in a rapidly evolving, highly sophisticated, and data-rich field of scientific study.

References

1 Li, N., Georas, S.N., Alexis, N. et al. (2016). A work group report on ultrafine particles (American Academy of Allergy, Asthma and Immunology): why ambient ultrafine and engineered nanoparticles should receive special attention for possible adverse health outcomes in human subjects (Review). *J. Allergy Clin. Immunol.* 138 (2): 386–396.

2 European Commission: Commission recommendation on a code of conduct for responsible nanosciences and nanotechnologies research and council conclusions on responsible nanosciences and nanotechnologies research. Available online: http://ec.europa.eu/research/science-society/document_library/pdf_06/nanocode-apr09_en.pdf (accessed on 09 January 2020).

3 Papadiamantis, A.G., Klaessig, C., Exner, T. et al. (2020). Metadata stewardship in nanosafety research: community-driven organisation of metadata schemas to support fair nanoscience data. *Nanomaterials (Basel)* 10 (10): 2033.

4 Oberdörster, G., Oberdörster, E., and Oberdörster, J. (2007a). Concepts of nanoparticle dose metric and response metric. *Environ. Health Perspect.* 115 (6): A290.

5 Oberdörster, G., Stone, V., and Donaldson, K. (2007b). Toxicology of nanoparticles: a historical perspective. *Nanotoxicology* 1 (1): 2–25.

6 Nel, A., Xia, T., Madler, L., and Li, N. (2006). Toxic potential of materials at the nano-level. *Science* 311: 622–627.

7 Jain, P.K., Huang, X., El-Sayed, I.H., and El-Sayed, M.A. (2008). Noble metals on the nanoscale: optical and photothermal properties and some applications in imaging, sensing, biology, and medicine. *Acc. Chem. Res.* 41: 1578–1586.

8 Liu, J., Sonshine, D.A., Shervani, S., and Hurt, R.H. (2010). Controlled release of biologically active silver from nanosilver surfaces. *ACS Nano* 4: 6903–6913.

9 Glover, R.D., Miller, J.M., and Hutchison, J.E. (2011). Generation of metal nanoparticles from silver and copper objects: nanoparticle dynamics on surfaces and potential sources of nanoparticles in the environment. *ACS Nano* 5: 8950–8957.

10 De Volder, M.F., Tawfick, S.H., Baughman, R.H., and Hart, A.J. (2013). Carbon nanotubes: present and future commercial applications. *Science* 339: 535–539.

11 Shi, H., Magaye, R., Castranova, V., and Zhao, J. (2013). Titanium dioxide nanoparticles: a review of current toxicological data. *Part. Fibre Toxicol.* 10: 15.

12 Wang, X., Mansukhani, N.D., Guiney, L.M. et al. (2015). Differences in the toxicological potential of 2D versus aggregated molybdenum disulfide in the lung. *Small* 11: 5079–5087.

13 Rudramurthy, G.R. and Swamy, M.K. (2018). Potential applications of engineered nanoparticles in medicine and biology: an update. *J. Biol. Inorg. Chem.* 23 (8): 1185–1204.

14 Lowry, G.V., Gregory, K.B., Apte, S.C., and Lead, J.R. (2012). Transformations of nanomaterials in the environment. *Environ. Sci. Technol.* 46 (13): 6893–6899.

15 Lynch, I. and Lee, R.G. (2016). In support of the inclusion of data on nanomaterials transformations and environmental interactions into existing regulatory frameworks. In: *Managing Risk in Nanotechnology: Topics in Governance, Assurance and Transfer* (ed. F. Murphy, E.M. McAlea and M. Mullins), 145–169. Cham: Springer International Publishing.

16 Mitrano, D.M., Motellier, S., Clavaguera, S., and Nowack, B. (2015). Review of nanomaterial aging and transformations through the life cycle of nano-enhanced products. *Environ. Int.* 77 (Supplement C): 132–147.

17 Monopoli, M.P., Walczyk, D., Campbell, A. et al. (2011). Physical–chemical aspects of protein corona: relevance to in vitro and in vivo biological impacts of nanoparticles. *J. Am. Chem. Soc.* 133 (8): 2525–2534.

18 Hansen, S.F., Sørensen, S.N., Skjolding, L.M. et al. (2017). Revising REACH guidance on information requirements and chemical safety assessment for engineered nanomaterials for aquatic ecotoxicity endpoints: recommendations from the EnvNano project. *Environ. Sci. Eur.* 29 (1): 14.

19 European Chemicals Agency: Get ready for new REACH requirements for nanomaterials. Available online: https://echa.europa.eu/-/get-ready-for-new-reach-requirements-for-nanomaterials (accessed on 04 February 2021).

20 Lu, X., Miousse, I.R., Pirela, S.V. et al. (2016). In vivo epigenetic effects induced by engineered nanomaterials: a case study of copper oxide and laser printer-emitted engineered nanoparticles. *Nanotoxicology* 10 (5): 629–639.

21 Hurley, K.R., Ring, H.L., Kang, H. et al. (2015). Characterization of magnetic nanoparticles in biological matrices. *Anal. Chem.* 87: 11611–11619.

22 Arndt, D.A., Moua, M., Chen, J., and Klaper, R.D. (2013). Core structure and surface functionalization of carbon nanomaterials alter impacts to daphnid mortality, reproduction, and growth: acute assays do not predict chronic exposure impacts. *Environ. Sci. Technol.* 47 (16): 9444–9452.

23 Andrea Haase and Fred Klaessig. (2018). EU-US Roadmap Nanoinformatics 2030. https://www.nanosafetycluster.eu/eu-us-nanoinformatics-2030-roadmap/ (17 February 2022).

24 Colvin, V.L. (2003). The potential environmental impact of engineered nanomaterials. *Nat. Biotechnol.* 21: 1166–1170.

25 Devlin, R.B., Smith, C.B., Schmitt, M.T. et al. (2014). Controlled exposure of humans with metabolic syndrome to concentrated ultrafine ambient particulate matter causes cardiovascular effects. *Toxicol. Sci.* 140: 61–72.

26 Bezemer, G.F., Bauer, S., Oberdörster, G. et al. (2011). Activation of pulmonary dendritic cells and Th2-type inflammatory responses on instillation of engineered, environmental diesel emission source or ambient air pollutant particles in vivo. *J. Innate Immun.* 3: 150–166.

27 Avramescu, M.L., Rasmussen, P.E., Chénier, M., and Gardner, H.D. (2017). Influence of pH, particle size and crystal form on dissolution behaviour of engineered nanomaterials. *Environ. Sci. Pollut. Res.* 24 (2): 1553–1564.

28 Karottki, D.G., Beko, G., Clausen, G. et al. (2014). Cardiovascular and lung function in relation to outdoor and indoor exposure to fine and ultrafine particulate matter in middle-aged subjects. *Environ. Int.* 73: 372–381.

29 Ostro, B., Hu, J., Goldberg, D. et al. (2015). Associations of mortality with long-term exposures to fine and ultrafine particles, species and sources: results from the California Teachers Study Cohort. *Environ. Health Perspect.* 123: 549–556.

30 Williams, M.A., Porter, M., Roman, J. et al. (2007). Ambient baltimore particulate matter directs non-classical immune activation of human myeloid dendritic cells. *J. Allergy Clin. Immunol.* 119: 488–497.

31 Porter, M., Karp, M., Killedar, S. et al. (2007). Diesel exhaust particles induce a novel pattern of pro-allergic dendritic cell activation. *Am. J. Respir. Cell Mol. Biol.* 37: 706–719.

32 Li, N., Wang, M., Barajas, B. et al. (2013). Nrf2 deficiency in dendritic cells enhances the adjuvant effect of ambient ultrafine particles on allergic sensitization. *J. Innate Immun.* 5: 543–544.

33 Williams, M.A., Rangasamy, T.E., Bauer, S.M. et al. (2008). Disruption of the Transcription Factor Nrf2 Promotes Pro- Oxidative Dendritic Cells That Stimulate Th2-Like Immuno-responsiveness on Activation by Ambient Particulate Matter. *J. Immunol.* 181: 4545–4559.

34 Williams, M.A. (ed.) (2012). *Allergens and Respiratory Pollutants: The Role of Innate Immunity*. Multiple Chapters, Series on Medical and Health Science, Technology and Policy. Oxford, UK: Biohealthcare Publishing Ltd.

35 Wright, S.L. and Kelly, F.J. (2017). Plastic and human health: a micro issue? *Environ. Sci. Technol.* 51 (12): 6634–6647.

36 Meng, Y., Kelly, F.J., and Wright, S.L. (2020). Advances and challenges of microplastic pollution in freshwater ecosystems: a UK perspective. *Environ. Pollut.* 256: 113445.

37 Barría, C., Brandts, I., Tort, L. et al. (2020). Effect of nanoplastics on fish health and performance: a review. *Mar. Pollut. Bull.* 151: 110791.

38 Teles, M., Balasch, J.C., Oliveira, M. et al. (2020). Insights into nanoplastics effects on human health. *Sci. Bull.* 65: 1966–1969.

39 Cózar, A., Echevarría, F., González-Gordillo, J.I. et al. (2014). Plastic debris in the open ocean. *Proc. Natl. Acad. Sci. USA* 111 (28): 10239–10244.

40 Lambert, S. and Wagner, M. (2016). Characterisation of nanoplastics during the degradation of polystyrene. *Chemosphere* 145: 265–268.

41 Holmes, L.A., Turner, A., and Thompson, R.C. (2012). Adsorption of trace metals to plastic resin pellets in the marine environment. *Environ. Pollut.* 160 (1): 42–48.

42 Rochman, C.M., Hentschel, B.T., and Teh, S.J. (2014). Long-term sorption of metals is similar among plastic types: implications for plastic debris in aquatic environments. *PLoS One* 9 (1): e85433.

43 de Sa, L.C., Oliveira, M., Ribeiro, F. et al. (2018). Studies of the effects of microplastics on aquatic organisms: what do we know and where should we focus our efforts in the future? *Sci. Total Environ.* 645: 1029–1039.

44 Ng, E.L., Huerta, L.E., Eldridge, S.M. et al. (2018). An overview of microplastic and nanoplastic pollution in agroecosystems. *Sci. Total Environ.* 627: 1377–1388.

45 Prata, J.C., da Costa, J.P., Lopes, I. et al. (2020). Environmental exposures to microparticles: an overview on possible human health effects. *Sci. Total Environ.* 702: 134455.

46 Suman, K.H., Haque, M.N., Uddin, M.J. et al. (2021). Toxicity and biomarkers of micro-plastic in aquatic environment: a review. *Biomarkers* 26 (1): 13–25.

47 Campanale, C., Massarelli, C., Savino, I. et al. (2020). A detailed review study on potential effects of microplastics and additives of concern on human health. *Int. J. Environ. Res. Public Health* 17 (4): 1212.

48 Facklemann, G. and Sommer, S. (2019). Microplastics and the gut microbiome: how chronically exposed species may suffer from gut dysbiosis. *Mar. Pollut. Bull.* 143: 193–203.

49 Hoffman, B.U. and Lumpkin, E.A. (2018). A gut feeling. *Science* 361 (6408): 1203–1204.

50 Li, J., Yang, D., Li, L. et al. (2015). Microplastics in commercial bivalves from China. *Environ. Pollut.* 207: 190–195.

51 Mathalon, A. and Hill, P. (2014). Microplastic fibers in the intertidal ecosystem surrounding Halifax Harbor, Nova Scotia. *Mar. Pollut. Bull.* 81 (1): 69–79.

52 Van Cauwenberghe, L. and Janssen, C.R. (2014). Microplastics in bivalves cultured for human consumption. *Environ. Pollut.* 193: 65–70.

53 Liebezeit, G. and Liebezeit, E. (2013). Non-pollen particulates in honey and sugar. *Food Addit. Contam. Part A* 30 (12): 2136–2140.

54 Liebezeit, G. and Liebezeit, E. (2014). Synthetic particles as contaminants in German beers. *Food Addit. Contam. Part A* 31 (9): 1574–1578.

55 Yang, D., Shi, H., Li, L. et al. (2015). Microplastic pollution in table salts from China. *Environ. Sci. Technol.* 49 (22): 13622–13627.

56 Browne, M.A., Niven, S.J., Galloway, T.S. et al. (2013). Microplastic moves pollutants and additives to worms, reducing functions linked to health and biodiversity. *Curr. Biol.* 23 (23): 2388–2392.

57 Rochman, C.M., Hoh, E., Kurobe, T., and Teh, S.J. (2013). Ingested plastic transfers hazardous chemicals to fish and induces hepatic stress. *Sci. Rep.* 3: 3263.

58 Lehner, R., Weber, C., Petri-Fink, A. et al. (2019). Emergence of nanoplastic in the environment and possible impact on human health. *Environ. Sci. Technol.* 53: 1748–1765.

59 Lim, S.L., Ng, C.T., Zou, L. et al. (2019). Targeted metabolomics reveals differential biological effects of nanoplastics and nanoZnO in human lung cells. *Nanotoxicology* 13 (8): 1117–1132.

60 Mukai, Y., Goto, A., Tashiro, Y. et al. (2020). Coastal biomonitoring survey on persistent organic pollutants using oysters (*Saccostrea mordax*) from Okinawa, Japan: geographical distribution and polystyrene foam as a potential source of hexabromocyclododecanes. *Sci. Total Environ.* 739: 140049.

61 Steinhäuser, K.G., Sayre, P.G., and Nowack, B. (2018). Reliability of methods and data for regulatory assessment of nanomaterial risks. *NanoImpact* 10: 68–69.

62 Organisation for Economic Co-operation and Development: Series on the Safety of Manufactured Nanomaterials: Alternative testing strategies in risk assessment of manufactured nanomaterials: Current state of knowledge and research needs to advance their use (ENV/JM/MONO(2016)63). Available online: http://www.oecd.org/officialdocuments/publicdisplaydocumentpdf/?cote=ENV/JM/MONO(2016)63&doclanguage=en (accessed on 10 February 2021).

63 Organisation for Economic Co-operation and Development: Series on the Safety of Manufactured Nanomaterials: Strategies, techniques and sampling protocols for determining the concentrations of manufactured nanomaterials in air at the workplace (ENV/JM/MONO(2017)30). http://www.oecd.org/officialdocuments/publicdisplaydocumentpdf/?cote=env/jm/mono(2017)30&doclanguage=en (accessed 10 February 2021).

64 Stone, V., Johnston, H.J., Balharry, D. et al. (2016). Approaches to develop alternative testing strategies to inform human health risk assessment of nanomaterials. *Risk Anal.* 36 (8): 1538–1550.

65 Burden, N., Aschberger, K., Chaudhry, Q. et al. (2017). The 3Rs as a framework to support a 21st century approach for nanosafety assessment. *Nano Today* 12: 10–13.

66 Berggren, E., White, A., Ouedraogo, G. et al. (2017). Ab initio chemical safety assessment: a workflow based on exposure considerations and non-animal methods. *Comput. Toxicol.* 4: 31–44.

67 Rogiers, V., Benfenati, E., Bernauer, U. et al. (2020). The way forward for assessing the human health safety of cosmetics in the EU – Workshop proceedings. *Forensic Toxicol.* 436: 152421.

Index

a

Absence of gradients and Nernstian equilibrium stripping (AGNES) technique 29
Adaptive immunity 216, 387
ADHD. *See* Attention deficit hyperactivity disorder (ADHD)
Agglomeration 26, 61–62, 82
Aggregates 61
Air pollution 239–240. *See also* Developmental CAPS exposures of mice
  and brain 240–241
  and neurodevelopmental disorders 435–436
    ADHD 242
    altered glutamate and dopamine 245
    ASD 241
    co-morbidities 242–243
    interhemispheric dysconnectivity 244
    microglial activation and inflammation 243–244
    reduction in size and myelination of CC 245
    schizophrenia 241
    ventriculomegaly 244
    white matter damage 244
Algae 85
Allergic contact dermatitis (ACD) 394, 395
Alternative testing strategies (ATSs) 445–446
Alumina NPs 413
Aluminosilicates 305
Al welding fumes 385
Alzheimer's disease 240
Ambient particulate matter (APM) 4
American Society for Testing and Materials (ASTM) International 94
Anatase and rutile samples, surface reactivity of 206–208, 434
  agglomerate size analysis 198–199
    in $NaNO_3$ 199, 200
    in SFW 199, 200
  characterization of 190–194
  ecotoxicity tests 201
    diatoms 204–205
    microcrustacean 203–204
    rotifer 201–203
  electrophoresis measurements
    in $NaNO_3$ 196–197

Toxicology of Nanoparticles and Nanomaterials in Human, Terrestrial and Aquatic Systems, First Edition. Edited by Marc A. Williams.

Anatase and rutile samples, surface reactivity of (*contd.*)
in SFW 196, 198
potentiometric titrations 194–195
surface and bulk properties 191
TEM images 193–194
XRD patterns 191–193
Animal models
of CAPS-induced neuropathology and behavioral deficits 253–256
altered ratio of glutamate and GABA 250, 252
HUCAPS system 246
impulsive-like behaviors 251–253
microglial activation 246–247
reduction in size and myelination of CC 247–251
sex-dependent differences 246
ventriculomegaly 247, 248
of IBD 284
of intestinal absorption 284–285
Anthracene 72
Anthropogenic-derived nanoparticles 3
Antibacterial activity of NMs 367
carbon-based 364–365
metallic 361–362
metal oxide 363–364
Apoptosis 393
Aromatic hydrocarbons 72
Artificial gastric fluid (AGF) 289–290
Artificial intestinal fluid (AIF) 289–290
Artificial sea water (ASW) 114–115
ASD. *See* Autism spectrum disorder (ASD) Asthma and allergic asthma 5, 7, 13, 14, 243, 354, 384, 385, 396
ATSs. *See* Alternative testing strategies (ATSs)
Attention deficit hyperactivity disorder (ADHD)
air pollution and 242
ASD and 242–243
dopamine dysfunction 243, 245
inflammation in 244
reduction in CC size 245
schizophrenia and 243
Autism spectrum disorder (ASD)
ADHD and 242–243
air pollution and 241
dopamine dysfunction 245
microglial activation and inflammation 243
reduction in size and myelination of CC 245
schizophrenia and 242
Autoimmunity 394

b

Bioaccumulation 135–136
Bioavailability 66–72, 133–135
Bisphenol A (BPA) 105
Bivalves 86–87
Blood–brain barrier (BBB), NPs effect to 412–413
Brachionus manjavacas, 202–203
Brachionus plicatilis, 201–202
Brain, air pollution and 240–241

c

$CaCO_3$ nanoparticles 332
Caenorhabditis elegans, 125

CAPS-induced neuropathology and behavioral deficits 253–256
altered ratio of glutamate and GABA 250, 252
HUCAPS system 246
impulsive-like behaviors 251–253
microglial activation 246–247
reduction in size and myelination of CC 247–251
sex-dependent differences 246
ventriculomegaly 247, 248
Carbonaceous nanomaterials 287
Carbon-based nanomaterials (CBNs) 356, 357
antibacterial activity 364–365
impact in gut models 366
Carbon–carbon NH (CCNH) 174–175
Carbon–metal NH (CMNH) 175
Carbon nanotubes (CNTs) 433
antibacterial activity 364–365
antimicrobial effects 165–166
in aquatic environments 165–166
GIT toxicity 327–332
NHs 174–175
sorptive nature of 172–173
toxicity 167–168
immune modulation 170, 172
influence on co-contaminants 172–174
nutrient depletion 168–171
Cardiorespiratory toxicity of ENMs 213–214, 434
complement 225–226
ecological perspectives 227–228
mechanisms of 215
phagocytosis 226–227
uptake-dependent mechanisms 221–223
activation of systemic immunity 225
NKA activity 223–225
uptake-independent mechanisms of 215
accumulation on cell surfaces 216
activation of cell surface receptors 216–219
ROS generation 219–221
Cd^{2+} and $nTiO_2$, interactive effects of 103–105
in ASW 114
in *M. galloprovincialis,* 106–109
representation of 104
Cell death 393
Cell surface receptors, activation of 216–219
Central nervous system (CNS) 240
CeO_2NPs
effect on freshwater organisms 32
removal of 26
Cerebral spinal fluid (CSF) 244
Charge reversal point (CRP) 188
Chemical bioaccumulation 135
Chemical safety assessment 42
Chemical stability 82
Chemodynamics 27
Chlamydomonas reinhardtii 205
Ciprofloxacin (CIP), sorption on TiO_2 NPs 64
Cladocerans 188
CNTs. *See* Carbon nanotubes (CNTs)
Complement 225–226
Conjugation 175

Contaminants, NMs interaction with 38–42
Copper and copper oxide nanoparticles (CuNPs/CuONPs) 86
  antibacterial activity 361, 363
  effect on freshwater organisms 32–33
  effect on intestinal microbiota 362, 368
  GIT toxicity 332
Corpus callosum (CC), reduction in size and myelination of 245, 247–251
CuS nanoparticles 133

d

Damage-associated molecular pattern (DAMP) receptors 218
Daphnia magna, 188, 203–204
Data management plans (DMPs) 445
Dendrites 219
Dendritic cells (DCs) 218–219, 222
Developmental CAPS exposures of mice 253–256
  altered glutamate/GABA ratios 250, 252
  excitatory–inhibitory imbalance 250, 252
  HUCAPS system 246
  impulsive-like behaviors 251–253
  microglial activation 246–247
  reduction in size and myelination of CC 247–251
  sex-dependent differences 246
  ventriculomegaly 247, 248
Diatoms 189
  toxicity test 204–205
Dicentrarchus labrax 103–104, 111–114
Disruption of signaling pathways 391–392
Dissolution 132
Dissolved organic matter (DOM) 64, 82, 83
Donnan membrane technique (DMT) 28–29
Dopamine dysfunction 243, 245
Dust inhalation 384
Dynamic energy budget (DEB) theory 30, 137
Dysbiosis 354, 368

e

Earthworms 135
Echinoderms 85–86
Ecosystems 10–11, 33, 88, 96, 102–103, 124, 127, 128–138, 147, 165, 189, 354, 427, 428, 430, 434, 441, 443–444, 446
Ecotoxicological effects of NMs
  in freshwater organisms 30–34
  in marine organisms 34–38
Electrical double layer (EDL) 187–188
Endocytosis 222, 226
Endothelial cells (ECs) 402, 404–409
Endotoxin contamination 287
Engineered nanomaterials (ENMs), cardiorespiratory toxicity of 213–214
  complement 225–226
  ecological perspectives 227–228
  mechanisms of 215
  phagocytosis 226–227

uptake-dependent mechanisms 221–225
uptake-independent mechanisms of 215–221
Engineered nanoparticles (ENPs) 4–8, 80, 425
Enterocytes 222, 281, 282
Environmental risk assessment (ERA) 42–43, 428
Eosinophils 386
Escherichia coli 136
Estuarine and coastal environments, ENPs in 79, 429
agglomeration behavior 82, 83
chemical stability 82–84
dissolved organic matter content in 82, 83
impact on living organisms
abnormal development 94
algae 85
bivalves 86–87
echinoderms 85–86
embryonal development 84–85
fish 87
gastropods 87
literature on 88–93
mechanisms of toxicity 84–85
trophic transfer 88
light intensity 83
modeling 94–95
regulatory environment 94
release routes 80–81
European Chemical Agency (ECHA) 431
European Union, regulatory environment in 94
Excitatory–inhibitory imbalance 250, 252
Exposure assessment 42–43

f

Fathead minnows (FHM) 168–170
Ferromagnetic IONPs (FM IONPs) 148, 149
Ferumoxytol 149
Fish
nanoparticle impact in 87
oxidative stress in 168
Follicle-associated epithelium (FAE) 281, 282
Food chain transfer 136
Food web 88
Freshwater aquatic organisms, NMs effect on 30–34
Fullerenes
antibacterial effect of 365
GIT toxicity 327–332

g

Gabba aminobutyric acid (GABA) 250, 252
Gastrointestinal tract (GIT) 278–279, 436–437. *See also* Mucosal microbiome
layers of 280
mucus 282–283
physiology 279–283
toxicity of nanomaterials 337–338
AgNPs 311–319
animal models 284–285
assay systems 287
assessment 286–289
cell-based test systems 285–286, 337

Gastrointestinal tract (GIT) (*contd.*)
CNTs and fullerenes 327–332
co-culture model systems 286
divergence in corona 288
IBD 284
miscellaneous metal oxide nanoparticles 332–336
nanosilica 305–311
oral toxicity testing 283–284
TiO_2-NPs 289–305
ZnONPs 319–327
Gastropods 87
General unified threshold model for survival (GUTS) 137
Gill endocytosis 222
GIT. *See* Gastrointestinal tract (GIT)
Glutamate imbalance 245, 250, 252
Glutathione-S-transferase (GST) activity 108
Glycocalyx 282
Gold nanoparticles (AuNPs)
antibacterial activity 361
anti-inflammatory activity 393
cytotoxicity of 403
effect on BBB 412–413
Graphene, antibacterial effect of 365
Gut microbiota, NM effects on. *See* Microbiota–NM/NP interactions

h
Hazard identification 42–43
Human umbilical vein endothelial cells (HUVECs) 408
Hypersensitivity 394–396
Hypoxia 412

i
IBD. *See* Inflammatory bowel disease (IBD)
ICP-MS. *See* Inductively coupled plasma-mass spectrometry (ICP-MS)
Immune cell receptors 218–219
Immune modulation 170, 172
Immune response 392
Immunosuppression 392–393
Immunotoxicology 383–384
metal NP interaction with immune system
autoimmunity 394
cell membranes and proteins 391
cellular effects 389, 390
disruption of signaling pathways 391–392
electrostatic interactions 391
immune response 392
immunosuppression 392–393
inflammatory effects 393–394
mechanism of 387–389
ROS generation 389–391
sensitivity/hypersensitivity 394–396
welding fumes
health issues 385–386
metal nanoparticles in 384–385
oxidative stress 386–387
Impulsive-like behaviors 251–253
Inductively coupled plasma-mass spectrometry (ICP-MS) 28, 290
Inflammation
metal NP interaction with immune system 393–394

neurodevelopmental disorders and 243–244
oxidative stress and 386
pulmonary 385
Inflammatory bowel disease (IBD) 284
Innate immunity 216, 387
Interhemispheric dysconnectivity 244
International Organization for Standardization (ISO) standards 94
Intersegmental vessels (ISVs), of transgenic zebrafish 412
Intestinal absorption 284–285
Intestinal villi 281
Iron oxide nanoparticles (IONPs) 147–148, 432
advantages of 148
applications of 149–151
ferromagnetic 148, 149
gaps and obstacles in 155, 158
GIT toxicity 332
immunosuppression 393
magnetic behaviors of 149
preparation of 148
superparamagnetic 148–149, 152, 155, 158
toxic effects of 151–155
aquatic organisms 152, 154, 155
terrestrial organisms 155–157
Iron oxide NMs, antibacterial effect 363
Isoelectric point 188, 196

k

Kallikrein–kinin system (KKS) 404

l

Lamina propria 280
Light intensity 83
Littorina littorea, 87
Lysosomal membrane stability (LMS) 108

m

Macrophages 389
Macroplastics 444
Maghemite 148
Magnetic iron oxide nanoparticles (IONPs). *See* Iron oxide nanoparticles (IONPs)
Magnetic nanoparticles (MNPs) 147–149
Magnetite 148
M-cells 281, 282
Medicago truncatula, 130
Metal allergy 394, 395
Metal fume fever (MFF) 386
Metallic NMs
antibacterial activity 361–362
impact in gut models 362–363
Metal nanoparticles (MeNPs)
immunotoxic effects of 383–384
autoimmunity 394
cellular effects 389, 390
disruption of signaling pathways 391–392
electrostatic interactions 391
immune response 392
immunosuppression 392–393
inflammatory effects 393–394
interaction with cellular membranes and proteins 391
mechanism of 387–389
ROS generation 389–391

Metal nanoparticles (MeNPs) (*contd.*)
sensitivity/hypersensitivity 394–396
metal allergy 394, 395
vasculotoxicity of 401–413
welding fumes 384–385
Microbiota–NM/NP interactions 369, 437–438
aerobic and anaerobic bacteria 371
antibacterial activity 376
carbon-based NM 364–366
evidence for effects on microbiota 367–368
human exposure and 354–356
influencing factors
gut environment-specific factors 357
NM-specific factors 356–357
protein corona 358
literature on 359–360
metallic NM 361–363
metal oxide NM 363–364
research on 370–371
strategies for assessing the effects of
fermentation studies 370
in vitro studies 368, 370
in vivo studies 370
Microcrustacean toxicity test 203–204
Microglial activation 243, 246–247
Microplastics (MPs) 440–444
MNPs. *See* Magnetic nanoparticles (MNPs)
Monogonont rotifers 173
Mucosal microbiome 353–354, 437–438
aerobic and anaerobic bacteria 371
factors influencing NM–microbiota interactions
gut environment-specific factors 357
NM-specific factors 356–357
protein corona 358
impact of NP and NM
antibacterial activity 376
carbon-based 364–366
evidence for 367–368
fermentation studies 370
human exposure and 354–356
literature on 359–360
metallic 361–363
metal oxide 363–364
research on 370–371
in vitro studies 368, 370
in vivo studies 370
in intestinal ecosystem 354
Mucus 357
Multi-walled carbon nanotubes (MWCNTs) 31, 278, 356
antibacterial activity 364
GIT toxicity 327–332
Myelination patterns of corpus callosum 245, 247–251
Mytilus galloprovincialis, 86–87
$nTiO_2$, interactive effects of 106
and Cd^{2+} 106–109
and TCDD 109–111

n

Nanohybrids (NHs) 174–175
Nanomaterial bioaccumulation 135–136
Nanomaterials, in aquatic environment 23
ecotoxicological effects

in freshwater organisms 30–34
in marine organisms 34–38
ERA 42–43
interaction with other contaminants 38–42
speciation of 27–30
WWTPs
processes of 24–27
removal of NMs 23–24
Nanoparticles (NPs). *See also* Engineered nanoparticles (ENPs)
in environment 401–402
natural 79–80
size of 404, 412
sources 80
Nanoparticulates 278
Nanoplastics (NPs) 440–444
Nanoremediation 22, 428
Nanoscale zerovalent iron (nZVI) 150–151
Nanosilica 305–311
Nanowastes 22, 101
Natural nanoparticles 79–80
Natural organic matter (NOM) 62, 64, 82
Neurodevelopmental disorders 435–436
air pollution and
ADHD 242
ASD 241
co-morbidities 242–243
glutamate and dopamine dysfunction 245
interhemispheric dysconnectivity 244
microglial activation and inflammation 243–244
reduction in size and myelination of CC 245
schizophrenia 241
ventriculomegaly 244
white matter damage 244
animal model (*See* Developmental CAPS exposures of mice)
New approach methodologies (NAMs) 4, 437
NHs. *See* Nanohybrids (NHs)
NKA activity 223–225
Nonimmune cell receptors 219
Notch signaling pathways 402
NP–contaminant interactions
in aquatic environment 102–103
$nTiO_2$, interactive effects of
in ASW 114–115
and BPA 105
and Cd^{2+} 103–109
in *D. labrax* 111–114
in *M. galloprovincialis,* 106–111
and PCP 106
and TCDD 109–114
toxic effects in zebrafish 105–106
representation of 102
NPs. *See* Nanoplastics (NPs)
Nutrient depletion 168–170

O

One Health 3–17
and knowledge gaps 426
One Health Initiative Task Force 9
Organic carbon content 134
Organic matter content 134–135

Organisation for Economic Co-operation and Development (OECD) 94, 445
  Guideline No. 317: Bioaccumulation in Terrestrial Oligochaetes 135
  Sponsorship Program for Representative Nanomaterials 125
Oxidation 60
  of TiO_2 NPs 62–63
Oxidative stress 166–168, 214, 219, 365
  and immunologic effects 386–387
  nanoparticle-induced 86
  NP interaction with immune system 389–391

p

Palladium NPs 395–396
Parkinson's disease 240
Pathogen-associated molecular pattern (PAMP) receptors 218
Pattern recognition receptors (PRRs) 218–220
Pentachlorophenol (PCP) 70, 106
Per- and polyfluoroalkyl substances (PFAS) 442, 443
Perfluorooctanesulfonic acid (PFOS) 64, 70–72
Peripheral blood mononuclear cells (PBMCs) 395
Peroxynitrite 220
Peyer's patches (PPs) 281
Phaeodactylum tricornutum 204–205
Phagocytosis 226–227, 389
pH *vs.* toxicity 134
Physicochemical transformation 131
Plastic waste 429, 440–441
Platelet aggregation 409
Platinum dust 396
Pneumonia 385–386
Predict(ed) environmental concentrations (PECs) 42
Protein corona (PC) 358, 403–404
Pulmonary fibrosis 385

q

Quality management systems (QMSs) 445

r

Reactive oxygen species (ROS) 63, 86, 152, 168, 219–221, 362, 386, 389–391, 442
Registration, Evaluation, Authorisation and Restriction of Chemicals (REACH) 42, 122, 431, 432
Regulatory environment 94
Respiratory tract infections 385
Rhodium NPs 396
Risk assessment 127–128, 445
Rotifers 189
  toxicity test 201–203
3Rs (Replacement, Reduction and Refinement) principles 445

s

"Safer-by-design" nanomaterials 446
Scanning mobility particle sizer (SMPS) 246
Schizophrenia
  ADHD and 243
  air pollution and 241
  ASD and 242

dopamine dysfunction 243, 245
inflammation in 244
Sensitivity 394–396
SFW. *See* Synthetic freshwater (SFW)
Signaling pathways, disruption of 391–392
Silica, GIT toxicity 305–311
Silver nanoparticles (AgNPs)
antibacterial activity 361
applications 278
bioaccumulation 135
cell toxicity 403
effect on freshwater organisms 33
effect on intestinal microbiota 362–363, 368
in *Escherichia coli* 136
GIT toxicity 311–319
impact on echinoderms 85
incorporation into bacterial membrane 361–362
light intensity *vs.*, 83
production volumes 80
products 80
PVP-AgNPs 409
sulfidation of 24
toxicity of 33, 132
toxicity to biota
algae 85
echinoderms 85
gastropods 87
trophic transfer 88, 136
vascular effects of 409
WWTPs processes 24
Single-walled carbon nanohorns (SWCNHs) 327–332
Single-walled carbon nanotubes (SWCNTs) 31, 216, 278
antibacterial activity 364
GIT toxicity 327–332
SiO_2NPs, effect on freshwater organisms 32
Small intestine 281, 283
Soil nano-ecotoxicology 122, 430–431
papers on 125, 127
risk assessment 127–128
standard test methods 138–139
choice of test species 129–131
development of 122–127
interaction of soil properties with NM properties 133–135
nanomaterial bioaccumulation 135–136
short-duration tests 136–138
short-term and long-term effects of particle aging 131–133
trophic transfer 136
Soluble intercellular adhesion molecule (siCAM) 244
Sorption 61
Species sensitivity distributions (SSDs) 127–128
Standard operating procedures (SOPs) 432
Standard soil ecotoxicology tests 138–139, 430–431
choice of test species 129–131
development of 122–127
interaction of soil properties with NM properties 133–135
limitations in 128–129
nanomaterial bioaccumulation 135–136
short-duration tests 136–138
short-term and long-term effects of particle aging 131–133
trophic transfer 136

Stomach 280
Superparamagnetic iron oxide nanoparticles (SPIONs) 148–149, 152, 155, 158, 432
Surface area, of TiO_2 NPs 64, 66
Surface reactivity of anatase and rutile samples 206–208
  agglomerate size analysis 198–199
    in $NaNO_3$ 199, 200
    in SFW 199, 200
  characterization of 190–194
  ecotoxicity tests 201
    diatoms 204–205
    microcrustacean 203–204
    rotifer 201–203
  electrophoresis measurements
    in $NaNO_3$ 196–197
    in SFW 196, 198
  potentiometric titrations 194–195
  surface and bulk properties 191
  TEM images 193–194
  XRD patterns 191–193
Synthetic amorphous silica (SAS) 277, 305–306
Synthetic freshwater (SFW)
  agglomerate size analysis 199, 200
  composition of 196
  electrophoresis measurements 196, 198
Systemic immunity, activation of 225

t

2,3,7,8-Tetrachlorodibenzo-p-dioxins (TCDD) 38, 70
  and $nTiO_2$, interactive effects of 103
    in ASW 115
    in *D. labrax* 111–114
    in *M. galloprovincialis,* 109–111
    representation of 104
T helper cells 392
$nTiO_2$, interactive effects of 430
  and BPA 105
  and Cd^{2+} 103
    in ASW 114
    in freshwater species 105
    in *M. galloprovincialis,* 106–109
    representation of 104
  and PCP 106
  and TCDD 103
    in ASW 115
    in *D. labrax* 111–114
    in *M. galloprovincialis,* 109–111
    representation of 104
  in zebrafish 105–106
Tip cells (TCs) 402
Titanium dioxide nanoparticles (TiO_2 NPs) 59–61, 428–429
  agglomeration 61–62
  anatase and rutile samples 206–208, 434
    agglomerate size analysis 198–200
    characterization of 190–194
    ecotoxicity tests 201–205
    electrophoresis measurements 196–198
    potentiometric titrations 194–195
    surface and bulk properties of 191
    TEM images 193–194
    XRD patterns of 191–193
  applications of 59
  aqueous-phase behavior of 61–63
  consumer products 355

crystalline structure of 62
dispersibility of 187
DOM coating of 64
effect on BBB 413
effect on freshwater organisms 31, 34
GIT toxicity 289–305
hydrodynamic diameter of 62
immunosuppression 393
oral exposure to 277, 289
oxidation 62–63
P25 289
production of 59, 80
products 80
release of 80
sorption of organic contaminants
bioavailability 66–72
evaluation of 65, 68–69
NOM 64
surface area 64, 66
surface area 187
thorium sorption on 60
toxicity of 31, 188–189
under UV light 62–63
vascular toxicity of 409
Toll-like receptor 4 (TRL-4) 394
Toll-like receptors (TLRs) 389, 392
Toxicity. *See also specific toxicity*
to biota 84–85
CNT 167–168
pH *vs.*, 134
standardized testing 122–127
testing, challenges of 403–404
Toxic Substances Control Act (TSCA) 277, 431
Transepithelial electrical resistance (TEER) 285
Trojan horse effect 102
Trophic transfer 88, 136, 429

u

Ultrafine particulates (UFPs) 4–8, 239, 425–427. *See also* Air pollution
Ultraviolet (UV) radiation 62–63
Ulva lactuca 87

v

Vascular development 402–403
Vascular endothelial growth factor (VEGF) 402
Vasculotoxicity 438–439
blood–brain barrier 412–413
NPs in environment 401–402
testing, challenges of 403–404
in vitro 404–409
in vivo 409–412
Ventriculomegaly (VM) 244, 247, 248

w

Wastewater treatment plants (WWTPs)
processes of 24–27
removal of NMs 23–24
Water-soluble carbon nanotubes (wsCNTs) 356
Welding fumes
immune-related health effects 385–386
metal nanoparticles in 384–385
oxidative stress 386–387
White matter damage 244
Whole cell-based capacitive biochip (WCB) 158

z

Zebra fish 70–72, 105
  intestinal microbiota of 362
  knockdown of *vegfa* in 403
  transgenic, intersegmental vessels of 412
Zeta potential 190
Zinc oxide nanoparticles (ZnONPs) 277–278
  antibacterial activity 363
  effect on freshwater organisms 32
  GIT toxicity 319–327
  homogeneous soil distribution 126
  immunosuppression 393
  toxicity to biota 85
  trophic transfer 88
  in WWTPs processes 24
ZnO welding fumes 386

Printed and bound by CPI Group (UK) Ltd, Croydon, CR0 4YY
16/08/2022
03141866-0001